LIFE

LIFE

A Transdisciplinary Inquiry

EDITED BY

Jeremy Swartz and Janet Wasko

intellect

Bristol, UK / Chicago, USA

First published in the UK in 2023 by
Intellect, The Mill, Parnall Road, Fishponds, Bristol, BS16 3JG, UK

First published in the USA in 2024 by
Intellect, The University of Chicago Press, 1427 E. 60th Street,
Chicago, IL 60637, USA

A catalogue record for this book is available from
the British Library.

Copy editor: MPS Limited
Cover designers: Wade Larsen and Jeremy Swartz
Cover photo: Gary Tepfer
Production manager: Sophia Munyengeterwa
Typesetter: MPS Limited

Hardback ISBN 978-1-78938-792-6
Paperback ISBN 978-1-78938-793-3
ePDF ISBN 978-1-78938-794-0
ePUB ISBN 978-1-78938-795-7

Printed and bound by CPI Group (UK) Ltd, Croydon, CR0 4YY

This is a peer-reviewed publication.

MIX
Paper | Supporting responsible forestry
FSC® C013604

To
H. Leslie Steeves

Contents

Media • Life • Universse

Preface to a Trilogy

Jeremy Swartz and Janet Wasko

This volume is part of a trilogy that seeks to investigate and understand conceptual frameworks of *Media • Life • Universe*. The series presents transdisciplinary inquiries emphasizing relationships of complex systems and dynamics. Our understandings of media, as well as of information and therefore life, have been limited in various ways. One of the key motivations for this series is to challenge scholars, generally, and communication and media scholars, in particular, to think and act critically, evolving beyond merely narrow conceptualizations, oversimplifications, linearities and reductionisms. By recognizing patterns, processes, networks and their relationships, it becomes increasingly possible to visualize and experience universes of communication that have emerged and are emerging. These volumes are explorations into opening more integrative and accessible *spaces of meaning*.[1]

The trilogy emerged from three conference-experience-exhibitions at the University of Oregon in Portland that posed questions challenging established and narrow interpretations of these concepts – specifically, what are media, life and universe? These gatherings cultivated *transdisciplinary* research and its impact at the heart of science, art and their environments, and they offered opportunities to build bridges, remix, regenerate and reimagine disciplinary boundaries. Our aim was to build new conversations and encourage philosophical inquiry that may inspire theory and practice going forward.

The ideas and exchanges that flowed from these discussions pointed more broadly to relationships across technology, information and communication. These connections will be examined in the trilogy, involving threads and driving forces that run throughout the volumes. In other words, we have sought to 'curate' (repair) conceptual frameworks to better understand these concepts and to invigorate further exploration. We need a more flexible notion of inquiry, one that is more adequate to the realities of emerging technologies that we are facing, not only in theory but also practice via intersections between analogue, digital, physical, virtual and hybrid frames.

We also suggest that it is important to consider grounded practical theories that enact immersion and inclusion for transdisciplinary research. The trilogy itself attempts to represent engaged reflection, or the reflexivity between theory and practice and how each informs the other as a systematic focus on problems, dilemmas and sites (Craig and Tracy 2014). To enact theory-practice, or better yet praxis, we also emphasize the value of cultivating networks, as in the *What is…?* series and other such events,[2] and stress a commitment to perspectives in problem-solving-based dialogue and community building. This trilogy represents a transdisciplinary experience, which brings together scholars from the humanities, natural sciences (physical and life sciences) and the arts. More specifically, it draws not only on media and communication studies, but also on art, biology, computer science, information studies, philosophy, political science, sociology and other disciplines.

The concept of disciplinarity itself can be traced back to at least the mid-twelfth century, while interdisciplinarity emerged in the early twentieth century. The concept of transdisciplinarity came to the fore in the 1970s as a 'critique of the standard configuration of knowledge in disciplines'. It re-emerged in the 1990s and continues to grow 'as an urgent issue relating to the solution of new, highly complex global concerns' (Bernstein 2015: 1).[3] Note the following distinctions between these various frameworks:

- *Disciplinary:* acquisition of learning or mental training or a *system* […] and relating to a branch of learning or knowledge, field study, etc.[4]
- *Interdisciplinary:* pertaining to, contributing to, or benefiting from two or more disciplines or branches of learning.[5]
- *Transdisciplinary:* a synthetic creation that encompasses work from different disciplines (Cat 2017); (*research*) new conceptual, theoretical, methodological and translational innovations that integrate and move beyond discipline-specific approaches (Harvard TREC Center 2022); (*frameworks*) the synthetic paradigms of general *systems* theory and sustainability, as well as the shift from a disease model to a new paradigm of health and wellness (National Research Council 2014: 32, emphasis added); (*emblematic practices*) the concept of gender in feminist theory (Sandford 2015) and action research.[6]

If disciplines can be understood as systems, then transdisciplinary research may also be conceptualized as a systems approach. As Fritjof Capra and Pier Luigi Luisi ([2014] 2020: i) observe, 'New emphasis has been given to complexity, networks, and patterns of organization, leading to a novel kind of "systemic" thinking'. In this sense, we suggest that the trilogy may contribute to a kind of systems *repair/curation* that can assist recoveries by reconnecting fragmented environments (Stokstad 2019). Although the trilogy may appear to present a variety of disparate frames, in actuality it is a saga of inextricable, systemic concepts.

Even communication and media scholars such as Marshall McLuhan and Herbert Schiller suggested including systems as a frame for inquiry. McLuhan once explained, 'My own interest in studying media is a "systems development" approach [...] a structural analysis of pressures and strains [...] It is concerned with the inner dynamics of the form' ([1967] 2014: 74). Schiller discussed this perspective in his 'integrated' analysis of a major media corporation, when he stated that the best way to understand the message it is selling is to adopt a systems analysis approach [...] to take [...] an entity, and to examine its many outputs as *elements in a totality* with some common features' (Schiller 1973: 99, original emphasis). More critically, McLuhan and Schiller evoke another issue that 'the "message" of any medium or technology is the change of scale, or pace, or pattern that it introduces into human affairs' (McLuhan [1964] 2003: 20).

Another initial framework for the trilogy was category analysis that stems from the philosophy of language (Lakoff 1995; Johnson 2018) as well as a range of transdisciplinary modes of inquiry, including but not limited to media genealogy and archaeology; music and dance; science, technology and society studies; and rethinking communication and information theories. If – as John Dewey ([1940] 2008: 101) recounts – a 'block universe' can be considered 'a closed universe in which there is no room for novelty and adventure', then what we require is a *philosophy of openness* that can offer dynamic impetus to remediate the emergencies we face collectively, and in turn repair and regenerate the world.

Volume I: MEDIA

> *Within the cacophony of divergent and heterogeneous attempts to define the obscure object [...] scholars today are caught by the impossibility of finding common ground for what they mean by media.*
>
> (Horn 2008: 7)

MEDIA: A Transdisciplinary Inquiry is the first volume of the trilogy and explores notions of *media as technologies* and *as systems*, with special attention to material, historical and ecological ramifications. The collection reconceptualizes media from environmental and integrative approaches, utilizing various techniques and platforms, to envision 'spaces, places and traces' of what media have been and can become.

Media today increasingly represents a wider and wider range of concepts, products, services and institutions. The very definitions of medium/media and their study continue to encompass more and more of human experience and are in a constant state of flux. What are media today? How is media studies evolving? How have technologies transformed communication and media theory, and informed praxis? What are some of the futures of media? In other words, one might ask what are mediation, emerging media/technology and emerging systems?

The volume's sections focus on Genealogy; Meanings of Media; Organs and Organization; Engagement and Extensions; Biomediations; and Repair and Meta-media.

Volume II: LIFE

> *[B]acteria communicate with our cells, telling them about a world 'out there', a chatter of which we are very little aware, not to mention the resonances of all matter in the world, constantly humming to every part of our bodies, the boundaries of which thus dissolve, because we are also resonance, a cloud of microbes, energy, smell and light.*
>
> (Lindman 2013: 63)

The second volume, *LIFE: A Transdisciplinary Inquiry*, considers how information and communication are instrumental in and for living systems. The contributors examine habitats we live in, how media increasingly constitute and permeate all avenues and forms of life – public, private, organic and synthetic. We review historical and contemporary explorations of the concept of life, and, following from the *MEDIA* volume, continue to address biosemiotics and biomediations.

Similarly, the book probes the juxtapositions between material and representational (symbolic) forms. To understand this distinction, we offer a conceptual metaphor for your consideration: differences between analogue and digital. For example, take a digital representation of an analogue waveform, whereby only a certain number of points of the wave are incorporated and the rest are approximated or ignored. Now imagine the ocean—the analogue wave is energy, manifesting as continuous or smooth, rather than as discrete or striated. So, in this way, could it be argued that digitization is an estimation of nature to fit an anthropocentric, built environment?

Life can also be viewed as an ecosystem of interactions, transactions and contexts, by recognizing patterns in the natural sciences, cognitive science and the social sciences. In contemplating our surroundings and our lifeworlds, the volume integrates discussions of embodiment together with ecomedia literacies, criticisms and sustainability-related frames.

This volume's clusters include Genealogy; Information and Ecologies; Enactions and Values; Ecomediations and Education; Syntheses and Biodesigning; and Artful Lives and Metaliving. Our hope is to weave and cultivate connections within the trilogy, expanding how we value and honour nature.

Volume III: UNIVERSE

> *Art is the opening up of the universe to becoming-other, just as science is the opening up of the universe to practical action, to becoming-useful and philosophy is the opening up of the universe to thought-becoming. Art is the most direct intensification of the resonance, and dissonance, between bodies and the cosmos, between one milieu or rhythm and another. It is that which impacts the body most directly, that which intensifies and affects most viscerally. Through the plane of composition it casts, art is the way that the universe most directly intensifies life, enervates organs, mobilizes forces. It is the passage from the house [...] the body of the living being to the universe itself.*
>
> (Grosz 2008: 23–24)

The third volume of the trilogy, *UNIVERSE: A Transdisciplinary Inquiry*, examines a range of conceptualizations of universe in relationship to communication. It explores universe through the lens of networks of relationships, highlighting links to complexity and category analysis, as well as how they may contribute to a 'pluralistic conception of an open universe' (Dewey [1948] 2008: 26). Contributors to the collection discuss issues related to cosmologies and meaning-making as well as applications across a transdisciplinary multiverse.

The book explores various scales, paces and patterns of universes. It addresses pressing questions of our time, including how do societies move forward, turning with shared agreement about the associative nature of our world, problems confronting it and/or solutions necessary to improve it? Can this be done by changing the scalar lens of our analyses? For example, consider the difference between the nanoscopic worlds of the electron microscope that have come into view, the periscopic landscapes and topologies looking across a viewing deck or holodeck, and the telescopic long lens of cosmological horizons, recognizing the earth as a system constantly (re)entwining.

The volume also examines a multiplicity of ontologies, identities, social worlds, spiritualities and universal rights, as well as the concepts of pluriverse and ecoverse respectively.

Concluding Thoughts

This has been merely a prelude to a trilogy that endeavors to reimagine inquiries of media, life and universe – a panorama where technology, information and communication catalyze, facilitate and transform. Thus, the trilogy moves toward metadisciplinarity networking a cornucopia of sources, motifs and melioristic praxes. These three volumes unfold through each other, and they can be encountered collectively, in clusters or through individual chapters. *Media • Life • Universe* are offered for contemplation and further curational co-inquiry.

NOTES

1. For neurobiologist and philosopher Owen Flanagan, these consist of a 'sextet {art, science, technology, ethics, politics, spirituality}' (2009: 11–12). Philosopher Christian Coseru 'find[s] that these logical connections open up new possibilities for enhancing, refining, and expanding the range of philosophical arguments and possibilities' (2014: 218).
2. Other examples of international transdisciplinary conference-experience-exhibitions and organizations include *documenta* (Germany, 1955–present), Ars Electronica (Austria, 1979–present), Leonardo, the International Society for Arts, Sciences, and Technology (Leonardo/ISAST) (United States, 1982–present), *ISEA Symposia* (international, 1988–present), *Transmediale* (Germany, 2001–present), Humanities, Arts, Science and Technology Alliance and Collaboratory (HASTAC) (international, 2002–present), in addition to the Philosophy of Communication section of the European Communication Research & Education Association's workshop on 'Communication, Art, Media: Probing Impacts and Intersections' (2017), the International Association for Media and Communication Research conference's Ecomedia Arts Festival (2018), and the International Communication Association conference's *Making & Doing* exhibition (2017), 'Blue Sky Workshop: Multi-Modal Research in Communication' (2019) and 'Preconference: Communicating Multi-Modally: Research and Expressive Culture' (2022).
3. Also see Ngram of disciplinary, interdisciplinary and transdisciplinary from 1550 to 2019 (https://bit.ly/3heps6t. Accessed 4 February 2020). It was created utilizing software by Michel et al. (2011). See also Osborne (2015), Jantsch (1970) and Judge (1991, 2006).
4. 'Disciplinary, adj. and n.' (*OED Online* 2023).
5. 'Interdisciplinary, adj.' (*OED Online* 2023).
6. A few examples of work on transdisciplinarity as sustainability, action research, conservation and art-science include: Stokols (2006); Christens and Perkins (2008); Brandt et al. (2013); Toomey et al. (2015); Leavy (2016); and Clark et al. (2020). For the Science.gov Alliance's search of 'sample records for transdisciplinary', see https://www.science.gov/topicpages/t/transdisciplinary.html. Accessed 4 February 2020.

REFERENCES

Bernstein, Jay Hillel (2015), 'View of transdisciplinarity: A review of its origins, development and current issues', *Journal of Research Practice*, 11:1, Article R1.

Brandt, Patric, Ernst, Anna, Gralla, Fabienne, Luederitz, Christopher, Lang, Daniel J., Newig, Jens, Reinert, Florian, Abson, David J. and Wehrden, Henrik von (2013), 'A review of transdisciplinary research in sustainability science', *Ecological Economics*, 92 (August), pp. 1–15.

Capra, Fritjof and Luisi, Pier Luigi ([2014] 2020), *Vita e Natura: Una Visione Sistemica*, Sansepolcro: Aboca.

Cat, Jordi (2017), 'The unity of science', in E. N. Zalta (ed.), *The Stanford Encyclopedia of Philosophy*, Fall ed., Metaphysics Research Lab, Stanford University, https://plato.stanford.edu/archives/fall2017/entries/scientific-unity/. Accessed 17 November 2019.

Christens, Brian and Perkins, Douglas D. (2008), 'Transdisciplinary, multilevel action research to enhance ecological and psychopolitical validity', *Journal of Community Psychology*, 36:2, pp. 214–31.

Clark, Sarah E., Magrane, Eric, Baumgartner, Thomas, Bennett, Scott E. K., Bogan, Michael, Edwards, Taylor, Dimmitt, Mark A., Green, Heather, Hedgcock, Charles, Johnson, Benjamin M., Johnson, Maria R., Velo, Kathleen and Wilder, Benjamin T. (2020), '6&6: A transdisciplinary approach to art–science collaboration', *BioScience*, 70:9, September, pp. 821–829.

Coseru, Christian (2014), 'The Bodhisattva's brain: Owen Flanagan meets critics', *Zygon®*, 49:1, pp. 208–19.

Craig, Robert T. and Tracy, Karen (2014), 'Building grounded practical theory in applied communication research: Introduction to the special issue', *Journal of Applied Communication Research*, 42:3, pp. 229–43.

Dewey, John ([1940] 2008), 'Time and individuality', in J. A. Boydston (ed.), *The Later Works, 1925–1953*, vol. 14, Carbondale, IL: Southern Illinois University Press. [First published in 1940 in *Time and Its Mysteries*, series 2, New York: New York University Press, pp. 85–109, from a lecture given on the James Arthur Foundation at New York University on 21 April 1938.]

——— ([1948] 2008), 'William James' morals and Julien Benda's: It is not pragmatism that is opportunist', in J. A. Boydston (ed.), *The Later Works: 1925–1953*, vol. 15, Carbondale, IL: Southern Illinois University Press. [First published in 1948 in *Commentary*, 5, pp. 46–50. For Benda's article, to which this is a reply, see Benda, Julien (1947), 'The attack on western morality', *Commentary*, 4 (November), pp. 416–22 (print ed., appendix 1).]

'disciplinary, adj. and n.' (2023), *OED Online*, Oxford: Oxford University Press, https://www.oed.com/dictionary/disciplinary_adj. Accessed 24 July 2023.

Flanagan, Owen (2009), *The Really Hard Problem: Meaning in a Material World*, Cambridge, MA: MIT Press.

Grosz, Elisabeth A. (2008), *Chaos, Territory, Art: Deleuze and the Framing of the Earth*, New York: Columbia University Press.

Harvard TREC (Transdisciplinary Research in Energetics and Cancer) Center (2022), 'Definitions', Harvard T. H. Chan School of Public Health, https://www.hsph.harvard.edu/trec/about-us/definitions/. Accessed 15 December 2022.

Horn, Eva (2007), 'Editor's introduction: "There are no media"', *Grey Room*, 29, pp. 7–13.

'interdisciplinary, adj. and n.' (2023), *OED Online*, Oxford: Oxford University Press, https://www.oed.com/dictionary/interdisciplinary_adj. Accessed 24 July 2023.

Jantsch, Erich (1970), 'Inter- and Transdisciplinary University: A systems approach to education and innovation', *Policy Sciences*, 1:4 (Winter), pp. 403–28.

Johnson, Mark (2018), 'Mind, metaphor, law', in *The Aesthetics of Meaning and Thought: The Bodily Roots of Philosophy, Science, Morality, and Art*, Chicago: University of Chicago Press. [First published in *Mercer Law Review*, 58:3 (2007), pp. 845–68.]

Judge, Anthony (1991), 'Metaphors as transdisciplinary vehicles of the future', *Science and Tradition: Transdisciplinary Perspectives on the Way to the 21st Century*, conference organized with UNESCO by the Union des Ingenieurs et des Techniciens utilisant la Langue Francaise, Paris, December, https://laetusinpraesens.org/docs/transveh.php. Accessed 14 April 2020. [Also appeared in Cazenave, Michel and Nicolescu, Basarab (eds) (1994), *L'homme, la Science et la Nature: Regards Transdisciplinaires*, Aix-en-Provence: Éditions Le Mail, pp. 168–204.]

——— (2006), 'Experimental challenge of cognitive fusion for ITER-8', in *Enactivating a Cognitive Fusion Reactor: Imaginal Transformation of Energy Resourcing (ITER-8)*, 31 May, https://kairos.laetusinpraesens.org/iter8_t_h_4. Accessed 16 October 2019.

Lakoff, George (1995), 'The neurocognitive self: Conceptual system research in the twenty-first century and the rethinking of what a person is', in D. Massaro and R. Solso (eds), *The Science of the Mind: 2001 and Beyond*, Oxford: Oxford University Press, pp. 221–43.

Leavy, Patricia (2016), *Essentials of Transdisciplinary Research: Using Problem-Centered Methodologies*, New York: Routledge.

Lindman, Pia (2014), 'Learning from Mould', in L. B. Larsen (ed.), *Networks, Documents of Contemporary Art*, London and Cambridge, MA: Whitechapel Gallery and MIT Press.

McLuhan, Marshall ([1964] 2003), *Understanding Media: The Extensions of Man, Critical Edition* (W. T. Gordon, ed.), Corte Madera, CA: Gingko Press.

——— ([1967] 2014), '4: The hot and cool interview', in M. Moos (ed.), *Media Research: Technology, Art and Communication: Critical Voices in Art, Theory and Culture*, New York: Routledge.

Michel, Jean-Baptiste, Shen, Yuan Kui, Aiden, Aviva Presser, Veres, Adrian, Gray, Matthew K., Google Books Team, Pickett, Joseph P., Hoiberg, Dale, Clancy, Dan, Norvig, Peter, Orwant, Jon, Pinker, Steven, Nowak, Martin A. and Aiden, Erez Lieberman (2011), 'Quantitative analysis of culture using millions of digitized books', *Science*, 331:6014, pp. 176–82.

National Research Council (2014), *Convergence: Facilitating Transdisciplinary Integration of Life Sciences, Physical Sciences, Engineering, and Beyond*, Washington, DC: National Academies Press.

Osborne, Peter (2015), 'Problematizing disciplinarity, transdisciplinary problematic', *Theory, Culture & Society*, 32:5&6, pp. 3–35.

Sandford, Stella (2015), 'Contradiction of terms: Feminist theory, philosophy and transdisciplinarity', *Theory, Culture & Society*, 32:5&6, pp. 159–82.

Schiller, Herbert I. (1973), *The Mind Managers*, Boston, MA: Beacon Press.

Stokols, Daniel (2006), 'Toward a science of transdisciplinary action research', *American Journal of Community Psychology*, 38:1&2, pp. 79–93.

Stokstad, Erik (2019), 'Connecting fragmented pieces of habitat can help endangered species recover', *Science*, 26 September, https://www.sciencemag.org/news/2019/09/connecting-fragmented-pieces-habitat-can-help-endangered-species-recover. Accessed 27 December 2019.

Toomey, Anne H., Markusson, Nils, Adams, Emily and Brockett, Beth (2015), 'Inter- and trans-disciplinary research: A critical perspective', UN Global Sustainable Development Report 2015 crowdsourced brief, https://sustainabledevelopment.un.org/content/documents/612558-Inter-%20and%20Trans-disciplinary%20Research%20-%20A%20Critical%20Perspective.pdf. Accessed 27 December 2019.

Introduction

Jeremy Swartz and Janet Wasko

Life is life, whether we can put a finger on the line that marks the boundary between it and the non-living, or whether a distinction here is far beyond our immediate powers.
— John Dewey and Arthur Bentley

Introduction

As actuality and metaphor, life can be understood as a connective tapestry that weaves together complex systems, dynamics and networks across various scales. This volume considers how information and communication are instrumental—that is, *intermediate*[1]—in and for living systems. An aim of this collection is a fuller and richer account of life through a spectrum of insights from the natural, cognitive, social sciences and the arts.

Consistent with one of the driving forces of the *Media • Life • Universe* trilogy, authors expand on the notion of medium as *environment,* with special attention given to material, historical and ecological ramifications.[2] This book continues to explore *biomediations*—if life is a medium, how is it mediated?—as well as *biosemiotics*—recognizing information simultaneously *as* something (physical), *for* something (instructional) and *about* something (semantic) (Floridi 2010: 74-5). How is the study of life evolving and what are emerging possibilities for living futures?

This project was incubated within a working group on 'Communication, Complexity, and Sustainability' in the Institute of Cognitive and Decision Sciences[3] at the University of Oregon (UO) in 2016 and then was realized as a conference-experience-exhibition held a year later in Portland.[4] Subsequently, the theme of the International Association for Media and Communication Research conference held at UO in 2018 addressed *reimagining sustainability,* echoing calls for equitable and artful resilient futures (see Appendix).[5] The intention of these gatherings was to bring people together to discuss how complex systems frameworks can reinform and transform how we study and understand life, ecosystems, natureculture, the role of information flows and their implications for worldmaking.

This collection of works engages a transdisciplinary notion of life, integrating a wide range of disciplinary-focused lenses and terminologies, too numerous to list here.[6] Although environmental and ecological accounts have been traditionally brought to our attention, a biological account has been neglected in the history and study of communication and media, for instance, by Robert T. Craig and his 'constitutive metamodel' of traditions in communication theory. However, at least since the early 90s biological approaches and a burgeoning tradition has been enacted and is evolving (Craig and Muller 2007: 497–99; Craig 2020: xvii–xviii; Floyd and Weber 2020: 5–8; Cappella 2020: 11, 15–16).

Communication and Life

Scholarship and research presented at conferences and meetings are exploring these patterns, relationships and contexts in emerging areas of inquiry.[7] There is a rich history of the use of biological and ecological *metaphors*, including fermentation, a metabolic process, most notably used in the title of the 'Ferment[s] in the Field' issues of the *Journal of Communication*.[8] Similarly, the utilization of the metaphor of 'natural life cycle' for new media evolution is also common.[9] Metaphors including the word 'ecosystem' are also employed rhetorically to elaborate connections amongst media platforms, relationships among entities and networks, etc. Another instance is 'going viral' in augmented social networks, which can be seen as a positive metaphor in contrast to a negative disease vector.

Additionally, linguistics and the philosophy of language have elaborated the bodily basis of metaphors by recognizing an organism's ongoing engagement with its environment (Johnson 2017). The conceptual blending of metaphors *along with* the biological and ecological holds together relationships across embodied, embedded and enacted views, with increasing attention to the impacts of extended apparati. In this way, awareness of critical infrastructures is fostered, explicating spaces, places and traces where physical server farms for social mediation materialize, rendering and ecologizing their reliance on cooling river waters. By acknowledging human and nonhuman embodiments in, of, for and by the natural world, these frameworks recognize biocultural diversity.

Correspondingly, science and technology studies interrogate metaphors such as technological and biological determinisms in hopes of moving 'beyond "determinisms" to capture the multifaceted complexity of technology as *communication* made durable' (Lievrouw 2014: 50, original emphasis). This focus can be seen in alignment with a perspective 'centered' on humans, similar to conceptually siloed determinations, especially regarding our technologies, our communications and our social-cultural frameworks.

Therefore, when the biological and/or the environmental is addressed, it is done so in the natural sciences, e.g. in the journals and research of *Nature Communications*,[10] and in the social sciences, e.g. via various lenses of 'environmental communication'.[11] Not until fairly recently has a material medium itself been given agency, as opposed to simply being a 'thing' that is re-sourced out of nature. In bringing a spectrum of conceptualizations of 'life' further into the fold of the field of communication and media studies, we continue to widen the scope towards transdisciplinary explorations, where natural, cognitive and social systems are entwined.[12]

Understanding Life

When asked 'What does life mean to you?', answers typically involve two general responses. The first pertains to 'everyday life', implying 'public or social life', 'private life' or 'ways of life', in other words, work, leisure and everything in between (e.g. relationships, politics, culture, education). The second conceptualization involves associations with the environment and the natural world or 'life space' (e.g. a cell, a bobcat, a forest, a river, or going for a walk).

These common understandings can be compared and contrasted with the *Oxford English Dictionary* (*OED*) etymology and definitions of life:

> Cognate with Old Frisian *lif*, life, body, person [...]
> I. The condition or attribute of living or being alive; animate existence. Opposed to death or inanimate existence.
> II. With reference to duration [e.g.] the period from birth to death.
> III. A course, condition, or manner of living. [...] The series of actions and occurrences constituting the history of an individual.
>
> (*OED Online* 2023)

> The condition that distinguishes animals and plants from inorganic matter.
>
> (*Oxford Dictionaries* 2022)

In Greek, βῐός (*bíos*) integrates many of the interpretations noted above, as well as 'mode of life'. A second definition in Greek is 'livelihood, means of living'; while the third specifies life as 'the world we live in'; and the fourth brings forward enactments of 'settled life, almost, = abode'[13] (Liddell and Scott 1940a).

Other definitions highlight ζωή (*zōḗ*), which 'separate[s] out natural life [...] from the reflective life that enables human experience' (Cassin 2014: 576). A comparable example can be found in *The Cambridge Dictionary of Philosophy*: 'mental activities such as perception and thought (*mental life*) or physical activities such as absorption, excretion, metabolism, synthesis, and reproduction (*physical life*)' (Hoffman and Rosenkrantz 1999: 504, original emphasis). These states can also be situated with 'the French *vie* (life), deriving from the Latin *vita*, [which] serves to designate' narratives, models, and histories (Cassin 2014: 576). Yet another explication comes from ecology[14]: 'A state of physical entities that utilize substances derived from outside themselves' (Allaby 2015: n.pag.).

Beyond these dictionary definitions, there has been a multitude of attempts to define life by philosophers, scientists and others.[15] Historically, encyclopedia entries, articles and books on life have included the *physiological*, *metabolic*, *biochemical*, *thermodynamic*, *evolutionary* and *autopoietic*, which are further discussed in this volume.[16] Others have suggested ways to discuss life beyond definitions, for instance, Tibor Gánti's (2003) real and potential *criteria*. And, in this volume, Mark A. Bedau offers the possibility of *hallmarks*, while Carol E. Cleland argues against definitions and suggests using *tentative criteria*.

Interestingly, *strong definitions* and *operational definitions* also have been contrasted. The former attempts 'to provide a complete set of necessary and sufficient conditions', while the latter proposes 'open–ended [...] necessary (and satisficing) provisional conditions' that 'can be applied to specific case studies' theoretically and experimentally (Bich and Green 2018: 3941, 3936).

Moreover, Carlos Mariscal (2021) has identified 'attempts at meta-categories for life definitions'. These include Bedau and Cleland (2010), Radu Popa (2004) and Edward N. Trifonov (2011), as well as Christophe Malaterre and Jean-François Chartier (2019: 4545), whose data-driven 'analyses across over 30,000 full-text biology articles' pointed to various dimensions of *lifeness* inferred from topic modelling, category analysis and sorting by subject matter. Their meta-analysis provided initial findings for these compelling dimensions: *matter-energy* (metabolism, catalysis and synthesis), *structure* (structural/cellular features), *environment* (micro/macro and related to plant/animal/human), *evolution* (variation, adaptation and speciation) and *information-program* (transcription, translation and modification) (Malaterre and Chartier 2019: 4562; see also Mariscal 2022).

But even with increasing amounts of dedicated research, there still seems to be no consensus among scholars and a definition of life is still an open question.

Systems Approaches to Life

As suggested by this discussion of categorizing and defining life, there is also a multiplicity of theoretical views about life and its origins. Many of the authors in this collection have engaged with a systems approach that embraces the natural, cognitive and social.

An etymology of the word 'system' is offered in the first chapter of this volume. Although there is a plethora of definitions for the term, a complex systems approach identifies examples of relationships between and across molecules and atoms, the brain and neurons, weather patterns and airflows. It 'studies how relationships between parts give rise to the collective [dynamic] behaviors of a system and how the system interacts and forms relationships with its environment' (Bar-Yam 2011: n.pag.).

Furthermore, complex systems are everywhere 'be they living entities, social systems, cities, or insect colonies'. These assemblages are made of functional parts and structures, and their properties stem from their interdependence and interaction (Luisi 2016: n.pag.; also see von Bertalanffy [1968] 2015; Krippendorff 2008). Thus, communication as a field can be understood to comprise 'a network of related hierarchical subsystems (disciplines) with their functional parts (topi)' (Gunaratne 2021: 131).

Emerging from a systems approach are areas that overlap biological systems and mediation, as well as information (Schiller 1988; Cook 2016; Godfrey-Smith and Sterelny 2016) and communication (Raymond 1950; Cappella 1991). There are many examples of communication and biology approaches, including different concepts, methods and variables (Knapp 2009), as well as integrations of media, information and communication.

The study of biology and communication may focus on neurobiological systems and physiological responses, as well as systems of the body and genetics (Boren and Veksler 2016: 184). This brings back into the fold a recognition of material biology and is in alignment with a systems approach (Beatty et al. 2009; Yegiyan 2017; Floyd and Weber 2020). This is most easily identifiable in the use of physiological categories delineating, for example, 'neurological, genetic, endocrine, immunological, muscular, hematological and cardiovascular' systems and 'dimensions of social behavior' (Floyd and Weber 2020: 5).[17]

Meanwhile, Kathleen Mullan Harris and Thomas W. McDade (2018: 3) note a biosocial approach which 'conceptualizes the biological and the social as mutually constituting forces'. It 'draws on models and methods from the biological, medical, behavioral and social sciences', blurring boundaries between the outside and inside of the body. They recognize that this is also 'a transdisciplinary approach to understanding human development, behavior and health' (see also Hickson and Stacks 2010; Veksler et al. 2020; Stumpf 2022).

In addition, biomedia are defined as 'informatic recontextualization of biological components and processes'. They can have ends with effects that may be medical or nonmedical (economic, technical), 'cultural, social, and political as they are scientific'. Therefore, biomedia are 'both a medium and a process of mediation' (Thacker 2010: 123, 2004). In parallel (as noted in the *MEDIA* volume), Luciano Floridi (2010) states, 'Genes do not contain information like envelopes or emails do, nor do they describe it, like a blueprint; they are more like performatives' (79), strengthening Sandra Braman's (2004) account of recombinant deoxyribonucleic acid (rDNA) as meta-technology. These biomediations can also be identified as bioinformatics, which has been defined as 'a branch of biology that is concerned with the acquisition, storage, display and analysis of biological information' as a codescript (NIH 2021: n.pag.; see also Floridi 2010: 74). Similarly, in biotechnology, Wilhelm Peekhaus (2016: 196) observes that 'the information metaphor and specifically "life as information" is given material reality that lends it to commodification' (Rajan 2006).

Philosopher of science Guenther Witzany problematizes biocommunication utilizing a biosemiotic account noted previously, but more specifically regarding plant signaling to explain 'life as a social event of cellular and sub-cellular agents that communicate, that is, based on interactions that are sign-mediated according to three levels of semiotic rules' (2019: 100). These include 'combinatorial (syntactic), context-sensitive (pragmatic) and content-specific (semantic)' (Witzany 2020: 3). Other biocommunicative approaches could include communication science and biology, human biology and culture (Condit 2000), as well as developmental biology related to models, metaphors and machines (Keller 2003).[18]

In summary, Mark Hickson III and Don W. Stacks (2010: 272) point out that the various research groups in communication 'all make separate and independent conclusions that biology is much more important than we have historically considered it to be'.

This volume presents a tapestry of relationships and contexts at different scales of emergence across *ecologies* (Bateson [1972] 2000), *enterprises* (Mattelart 1996), *metabolisms* (Nail 2023) and *registers* (Guattari [1989] 2000). The components of each of these conceptual frameworks include some sense of the natural, matter-energy as living-organism-environments; the cognitive or mental, human and nonhuman, subjectivities; and the social, relational, collectives and institutions. Félix Guattari argues for a relationship between these scales acting as driving forces for 'ethico-political articulation' ([1989] 2000: 28), similar to the metaphors of *islands* identified in urban metabolism research.[19]

Thus, conceptual frameworks interwoven together in this volume present steps towards transdisciplinarity. More implicitly, this collection constitutes an open-ended inquiry in the middle of the trilogy – exploring myriad ways in which humans understand, design and value life through the natural, cognitive, social sciences and arts.

Volume Overview

Drawing from chronicles of living systems, this volume is composed around a spectrum of perspectives from the natural, cognitive and social sciences, as well as art, architecture, design and philosophy.

Similar to the *MEDIA* volume, contributors to this collection offer an array of research, theories and practices related to life. This book is a network of nodes or chapters, organized and presented in six clusters, from genealogy to information and ecologies, from enactive values to ecomediations and education, and from syntheses and biodesigning onward to artful lives and metaliving.

By holding together histories, definitions, propagations of information, and communication, life is constantly unfolding and refolding across a range of biocultures. As this century enfolds embodied hybrid relations and meanings into ecological mediations and literacies, there are increasing demands for accountability that attenuate imaginative design practices with artful living trajectories.

Brief summaries of the clusters and chapters are offered below.

Genealogy

The first cluster of chapters highlights conceptualizations of life across multiple time periods. The first chapter explores how a systems approach from the Renaissance returned in the early 20th century through the study of communication and information. The following contribution charts a course from the ancient Greeks to contemporary research in the early 21st century. The section concludes by confronting us with the existential question: should we even be trying to define life?

Fritjof Capra presents a brief overview of systems science, design and their ecological dimensions. His historical approach begins with Leonardo da Vinci and his studies of nature, leading to present day philosophical, social and policy developments and implications. Capra's systems view of life offers a framework for integrating various patterns, including metabolism, networks, flow, emergence and cognition. He concludes with how ecological literacies and sustainable design can foster opportunities for systemic problem-solving.

Mark A. Bedau highlights phenomena of life including potential criteria such as evolvability, while recognizing some hallmarks, borderline cases and puzzles. He then explicates various accounts of life, from the genomic to the protocellular, from information processing to an open-ended evolutionary process of adaptation. Early in the chapter, Bedau notes early twenty-first century examples of these characteristics in hardware, software and wetware based artificial life.

Carol E. Cleland's chapter offers a variety of popular definitions of life, including a review of the thermodynamic, metabolic, evolutionary and self-organizing. She explains why the scientific project of defining life is mistaken and argues that life is not the sort of thing that can be successfully defined. A definition of life is more likely to hinder than facilitate the discovery of novel forms of life. Cleland's problematization of the complexities of meaning leads her to acknowledge tentative criteria as provisional space. She also provides insights into astrobiology, as well as programs and missions searching for life on various planets, including Mars.

Information and Ecologies

The next three chapters present systems in the context of information, communication and ecologies. These nodes provide insights into propagating organizations of life, plant-to-plant communication and biodiversity. The cluster begins with a multidimensional notion of information, proceeds to elucidate plant signalling and kin recognition, and consummates with a study of the human microbiome at different scales including assumptions, limitations and therapeutic applications. Each of these chapters exemplify different ecological scales of inquiry.

Stuart Kauffman, Robert K. Logan, Robert Este, Randy Goebel, David Hobill, and **Ilya Shmulevich** discuss propagating organization or the union of matter, energy, work, constraints and information in living physical systems. They argue that information should be understood simultaneously as 'Shannon information' which deals with, for example, the signal-to-noise ratio in telecommunications, as well as instructional or biotic information. In other words, information can be understood as both a thing or noun, as well as a process or verb.[20] This chapter provides context for information as physical manifestations, such as DNA, RNA and protein transcription and translation processes — giving rise to information *with* meaning. The authors draw on biology, physics and computer science to suggest that information provides the construction of constraints, which partially underlie the processes of evolution to dynamically determine the fitness of organisms within the context of a biotic universe.

Exploring the multiple pathways of information transfer and communication in plants, **Roza D. Bilas, Amanda Bretman** and **Tom Bennett** point to the key role of signalling in plant–plant communication. This emphasis on biocommunication, belowground and aboveground, is echoed by Rouhallah Sharifi and Choong-Min Ryu (2021: 1095), who describe *wired communications* involving 'direct signal transfers between plants mediated by mycorrhizal hyphae and parasitic plant stems'. They also identify *wireless communications* that include 'plant volatile emissions and root exudates elicited by microbes/insects', enabling signalling between plants with no physical contact. In other words, their research emphasizes the roles of light signals, chemical markers and how microbial organisms can alter underlying mechanisms. Furthermore, according to Niels P. R. Anten and Bin J. W. Chen (2021: 1059), plants react and respond to their neighbors; they are not passive. Kin recognition bridges mechanistic and ecological elements of plant-plant interactions, which are also discussed in terms of a spectrum of 'live and let live' between cooperation and competition.

Nicolae Morar and **Brendan J. M. Bohannan** move beyond information and plants to address the conceptual ecology of the human microbiome, a vast array of microorganisms on and in the human body. They present competing metaphors used to refer to our microbiomes, including as an organ containing our second genome; as a component of a host's immune system consisting of two distinctive parts, the self and nonself; as a superorganism or a collection of organisms that interact closely and from which functions emerge; as a metaphorical symbiont in the human holobiont, coevolving to increase their inclusive host and microbe fitness and shaping their hologenome; and as an ecosystem consisting of a community of organisms interacting with an environment, such as a forest. Each of these views include their own assumptions about the fundamental biology of the human-microbe system, with important implications and limitations, as well as potential theurapeutic applications. The synthesis of these various scales of microbiome systems offers a rich tapestry for investigating a scientifically pluralistic approach to guide the exploration of our complexity as chimeric beings.

Enactions and Values

The contributors in this cluster explore embodiment via natural, cognitive (neurosciences) and societal lenses in the study of drifting metabolisms across scales and the emergence of mind from life. Overall, the section offers an enactivist approach that culminates with how extensions/media/technologies/environments have found their way into our lives and livelihoods, and most importantly into our waste streams. In other words, the authors consider a variety of eco-nomic ways in which we are intermeshed within networks in the world.

In his chapter, **Mark L. Johnson** presents embodied cognition and its relationship to the emergence of language and mind. He recognizes allostasis as a proactive recalibration of expectations in anticipation of changing conditions and argues for the primacy of a more expansive and richer notion of aesthetics and ways that it underlies all the processes that give rise to meaning. He concludes by reflecting on the blending of science, art and the humanities as ways to develop the intelligence and values we need for a critical, reflective life.

Building on an inquiry into 'new materialism'[21] and a philosophy of movement, **Thomas Nail** has elaborated an ontology of motion. His 'kinetic' theory links to a long-forgotten threefold theory of metabolism, which includes the metabolism of nature, the human metabolism with nature and the interdependent process of social metabolism. Nail postulates that instead of metabolic rift (being torn from rootless grounds), we should embrace metabolic drift, where intra-active value is material and interwoven across systems.

Jussi Parikka's chapter connects to Johnson's conceptual scaffolding of embodiment and Nail's discussion of metabolic movements by acknowledging cognitive extensions (such as computers or metamediums) through their geological materiality. He notes how contemporary forms of waste and violence are intrinsically tied to issues related to the largely unseen international extraction of raw materials and the production of dead media (e-waste) revealed through tracing their lifecycles. Parikka also offers critical reflections on issues related to race, colonialism and news reporting on environmental disasters.

Ecomediations and Education

This cluster continues to develop an embodied, embedded, enacted and extended cognitive approach in society, acknowledging ways to address mounting problems in the education and translation of ecological literacies to future generations. This set of chapters views life through traditional media and information literacies, a journalistic approach to STEM (science, technology, engineering and mathematics), along with the language arts, as well as case studies of ecological mediations. Each chapter offers different approaches to teaching and learning, as well as how traditional media including environmental and ecological content are created by diverse producers for diverse audiences.

Divina Frau-Meigs offers a cyberist paradigm, redefining information opportunities and the international and transborder nature of network cultures. She continues by noting the relevance of a shift from information contracts (publicity) to sharing contracts (participation) and on to engagement. Frau-Meigs also addresses the contemporary information society's cyberist crisis and how health, work and education have been transformed by the pandemic. She concludes with a discussion of transliteracy and ecomedia literacies, especially for building core curricula in schools, contributing information resiliency to sustainability and democracy.

Ed Madison and others have developed an initiative for learning through journalism, presenting a tailored approach for rural communities. His work expands on an intentional approach by emphasizing student publications and diversifying STEAM (science, technology, engineering, art and mathematics) through digital storytelling and science communication. Madison explicates and explores near-peer mentorship, engaging identity-based motivation theory to postulate about futures of educational media/technologies.

In their chapter, **Salma Monani** and **Stephen Rust** reflect on more than a decade of ecological cinema, a subset of ecomedia. They bring attention to ecocritical lenses, which interrogate and decenter previous universalizing and Eurocentric trajectories. Bridging decolonial studies and affect studies, their work highlights the need for diverse voices that have been marginalized, offering emerging possibilities for directors, producers and production companies. Monani and Rust's research reinforces an account of the materiality of media, through deep contemplations of aesthetics, questioning habits and desires along with ways to contribute to sustainability and ecojustice for humans and nonhumans alike.

Syntheses and Biodesigning

This cluster challenges us with a variety of design approaches to life, bringing twenty-first century bioinspiration face-to-face with global innovation ecologies. Discussions include the history of system dynamics modelling as a field and how it has evolved along with developments in artificial neural networks; autopoiesis applied in laboratory settings and practices in the development of synthetic biology; and, implementations and problematics of bioengineers offloading digital data and storing it in life itself.[22]

Following on the historical chapters at the outset of this volume, **William Schoenberg** and **Jeremy Swartz** present an overview of system dynamics. From modelling to critiques, they address how to make use of machine learning and deep learning to improve the quality of models via structural validation. Schoenberg and Swartz address criticisms of modelling by integrating feedback loop dominance analysis to find 'loops that matter'. Their goal is to turn black box models into white box models by utilizing a variety of non-linear techniques, including feedforward artificial neural networks to address real-world problems, as well as offering resources for systems education.

Pier Luigi Luisi expands on a transdisciplinary frame across life and media, or cells and environments, further explicating systems, autopoiesis and cognition noted in earlier chapters, and taking us to the edge of synthetic biology. He reviews his experiments to realize a minimal cell, one with the smallest number of components to still act biologically, as well as procedures that utilize 'spontaneous overcrowding'. Luisi engages both top-down and bottom-up approaches, in addition to various speculations on why the origin of life is still a mystery for science. In doing so, he reinforces a systemic view of nature and evolution in multiplicity.

In their collaboration, **Mél Hogan** and **Tessa J. Brown** examine multivalent implications of biodesigning synthetic DNA to be utilized as hard drives. They present a revised history of the discovery of DNA by highlighting neglected contributors, as well as a brief exposition of genomics research — from its eugenic beginnings to the Human Genome Project and the subsequent development of clustered regularly interspaced short palindromic repeats (CRISPR) gene editing. Hogan and Brown bring to light dataists' archives, where codescript is more than merely bioinformatic content, but also requires significantly more context, revealing anxieties and priorities, discourses and infrastructures, or in other words, meanings.

Artful Lives and Metaliving

This last cluster serves as a transition into the next volume of the trilogy. The authors consider modes of communication and social organization, including harmonizing human, nonhuman and more-than-human habitats, places and spaces for symbioses across natural and built environments. Integrative approaches are brought forth through the water ways of architecture; in artful collective living, together with mountains, forests, rivers and seas; and by embodied aesthetics and ethics applied to transformations occurring every day. The contributors acknowledge the significance of stewardship, custodianship, kinship and myriad ways of valuing and honouring lifeways of past, present and future generations.

Brook Muller offers insights on the guidance of water, a medium for transporting beings, energy and thoughts. He directs our attention to reimagining city-nature relationships or urbanized watersheds, and reworking interactions between hydrological and urban landscape formations. Muller transports us from oceans to rivers, and from cities to buildings in different parts of the world, ultimately urging that our assemblies become mediums benefitting a diversity of species, that is to say, building-environment pairings that protect us from the elements, while at the same time supporting life forms nearby.

In the following chapter, **Diane Durston** immerses us in a world of satoyama and satoumi, where nature-culture exists, highlighting traditional Japanese styles of living in harmony with the land and the sea. She explores the history of these lifestyles and contemporary efforts to resuscitate them through rural regeneration. Durston concludes by stressing that initiatives dedicated to the revitalization of cooperation – from rural arts experiences and replanting forests to increasing transgenerational dialogue and well-being — are key to the flourishing of life on the planet.

Additionally, Durston is highlighted in the Appendix with Sadafumi Uchiyama, third generation Japanese gardener and curator, for their gracious facilitation of *Experience: Life • Lifeworlds • Lifeworks*, a collaboration with the Portland Japanese Garden. The *LIFEWORLDS* (2017) group exhibition expanded on an ecology of inquiry across nature and art.

In the last chapter, **Jeremy Swartz** offers a brief overview of the volume, followed by discussions of integrative systems, pattern recognition and the study of futures at various scales from the local and regional to the national and global. He then engages pluralistic approaches highlighting programs in critical futures studies and ways of worldmaking. Finally, Swartz reflects on metaliving, *metá* meaning in the *midst of* and *with* – not above or beyond – and how scholars, researchers, communities and societies are embodying – bringing forth or keeping close – their heritages, stories and worlds. It is in this way that artful co-curation honours kinships, responsibilities and accountabilities among all living systems.

NOTES

1. In *Knowing and the Known* ([1949] 2008: 244), John Dewey and Arthur Bentley explicated: 'What is called *environment* is that in which the conditions called physical are enmeshed in cultural conditions and thereby are more than "physical" in its technical sense. [...] it is their *medium*, or *milieu*, in the sense in which a medium is intermediate in the execution or carrying out of human activities, as well as being the channel through which they move and the vehicle by which they go on' (original emphasis). Around the same time, and in his purportedly lost last manuscript, Dewey explained that *instrumental* should be understood 'to designate the intermediate position and function of the subject matter of knowledge in the inclusive complex of the transactions constituting human living as an ongoing concern'. He cautioned against 'linguistic uses that give instruments a mechanical sense' and instead encouraged 'treating knowing as a mode of technology [...] exhibited and made evident in scientific inquiry'. This intermediate position is thus a mode between *means* and *ends-in-view*, distinct from a mere static *end* (Dewey 2012: 242; also see Hickman 2013).
2. This volume, *LIFE,* is reinforced by *MEDIA: A Transdisciplinary Inquiry*'s recognition of the synonymous relationships across media/technologies/environments (Swartz and Wasko 2021).
3. The group facilitated dialogues between members from architecture and allied arts (now design), biology, chemistry, environmental studies, interdisciplinary studies, journalism and communication, media studies, philosophy, psychology and sustainable materials chemistry.
4. This was the seventh in the *What is...?* Series and featured a unique coalescing of scholars, librarians, government and community officials, industry professionals, grassroots organizations, artists, designers, students and the public.

5. For examples of concepts and practices integrated into this theme, see *ecocriticism* (Starosielski and Walker 2016; Brevini and Murdock 2017; Slovic, Rangarajan and Sarveswaran 2019; Lupinacci et al. 2022), *ecomedia* (Maxwell and Miller 2012, 2017; Rust, Monani and Cubitt 2016; Iheka 2021; López et al. 2024) and *environmental movements* (Murdock 2020; Parks and Walker 2020; Atapattu et al. 2021; Carvalho et al. 2021; Caniglia et al. 2021; Grasso and Giugni 2022).
6. Disciplines and fields include architecture, art, astrobiology, bioengineering, biology, chemistry, cinema studies, cybernetics, communication, computer and information sciences, design, ecology, ecomedia, electrical engineering, environmental studies, forest and conservation sciences, garden curation, geography, journalism, materials science, media archaeology and ecology, media studies, metamedia, philosophy, physics, political economy, sociology, systems biology, system dynamics and machine learning, as well as synthetic biology.
7. There are several associations with committees, divisions and groups actively creating spaces for participation in these areas. These include the International Environmental Communication Association (IECA), whose mission is fostering inspirational communication which mitigates environmental issues, conflicts and problems with potential solutions. International communication and media associations' thematic areas include: Asociación Latinoamericana de Investigadores de la Comunicación's (ALAIC) Comunicación Y Ciudad Grupo; European Communication Research and Education Association's (ECREA) Science and Environment Communication Section; IAMCR's Environment, Science and Risk Communication Working Group; and ICA's Environmental Communication Division and Communication Science and Biology Interest Group. In addition, the Association for Education in Journalism and Mass Communication (AEJMC) features the Communicating Science, Health, Environment and Risk Division.
8. There have been many major overviews of the field in the *Journal of Communication* beginning with 'Ferment in the Field' (1983), which interestingly featured alchemaical imagery, and even a glass slide dipped in a gelatinous gel in a beaker. The fourth issue of 'Ferments in the Field' (2018), includes discussions of 'a new critical and materialist turn [which] is underway that lays bare the political-economic, labor, and environmental underpinnings of communication' (230–31). It includes Miyase Christensen and Annika Nilsson (2018) rethinking media 'as infrastructural and natural, as well as representational' (274). 'They criticize the dominant systems of environmental knowledge production and dissemination from conventional scientific communication' while pointing to materialist analysis of e-waste and its 'ecological, social, and labor costs' (Fuchs and Qiu 2018: 228, see also Parks 2020).
9. For example, several works touch upon these themes (Lehman-Wilzig and Cohen-Avigdor 2004; see also Scolari 2012, 2022; Deuze 2012, 2023; Starosielski 2021).
10. Other subjournals include communications in biology, chemistry, engineering, environment, physics, materials and medicine.

11. See *environmental communication* (Turner 2006; Cox 2015; Hansen and Cox 2015; Pezzullo and Cox 2021; Phillips and Milner 2021; Takahashi et al. 2022) and *communicology* (Catt 2017; Lanigan 2021).
12. See *environmental humanities* (Heise 2019; Horn and Bergthaller 2020), *global interventions* (Gunaratne and Gunaratne 2017; Miike and Yin 2022; Craig and Xiong 2022), *media archaeology* (Huhtamo and Galili 2020; Weibel 2023), *media ecologies* (Fuller 2005; Parikka 2014), *media ecology* (Granata 2017; Sharma and Singh 2022), *media theory* (Kittler 2007; Mitchell and Hansen 2010; Wickberg and Gärdebo 2022; Bauer 2022; Mitchell 2023) and *posthumanities* (Braidotti and Bignall 2018; Vint 2020; Jones 2023).
13. Interestingly, this last definition is reminiscent of ἶκος (*oîkos*), or 'house, not only of built houses, but of any dwelling-place' (Liddell and Scott 1940b: n.pag.). Also see a history of ecology, community and American social thought from the first half of the 20th century (Mitman 1992).
14. 'Ecology represented the borderland between the biological and social sciences through the study of interrelationships between and among individual organisms and their environment. This meaning has largely disappeared from today's disciplinary canon [...]' (Mitman 1992: 1).
15. A list of those defining life includes, but is not limited to, Aristotle, René Descartes, Immanuel Kant, Charles Darwin, Ernst Haeckel, Friedrich Engels, Thomas Henry Huxley, Alexander Oparin, Jakob von Uexküll, Erwin Schrödinger, Ernst Mayr, Rosalind Franklin, James Grier Miller, Edward O. Wilson and Leslie Orgel.
16. See Sagan 1970; Sagan, Sagan and Margulis, n.d.; see also an earlier version of the encyclopedia entry for 'life' (Weber 2018).
17. A biological approach could include, for example, analysing genetic markers and measuring chemical compounds (e.g. hormones, etc.) through blood, saliva and urine. Similarly, mediums also include the measuring and recording of underlying bioinformatic processes or specific system structures, e.g. blood pressure and heart rate of the nervous system, and/or electrical impulses in brain system scanning/mapping with functional magnetic resonance imaging (fMRI) (Boren and Veksler 2011: 6–23, 2016: 185–88, 192).
18. See also work by evolutionary biologist Andreas Wagner (2014, 2019).
19. These islands act as a metaphor for 'a global circulatory process of socio-natural relations that transforms and (re)creates urban ecosystems through the exchange of resources, capital, humans, and non-humans into and out of the spaces of global urbanization' (Newell and Cousins 2015: 22). This research includes: *industrial ecology*, where 'nature is the source of the urban metabolism's fuel and the destination for its wastes' (Wachsmuth 2012: 519); and 'focuses on quantifying material and energy flowing into, within, and out of cities, using accounting methodologies' (Newell and Cousins 2015: 21; see also Côté 2009). *Urban political ecology* or UPE presents urbanization as 'a constitutively socio-natural process' (Wachsmuth 2012: 519); 'emphasizes power relations that constitute uneven socio-ecological production' (Newell and Cousins 2015: 21). *Urban ecology* 'models the

complex patterns and processes of urban human–ecological systems' (Newell and Cousins 2015: 21). David Wachsmuth (2012) noted an important point stating that, 'urban metabolism is analogous to nature, but only analogous – the natural environment itself is simply a backdrop to a purely social process of urbanization' (519). For recent criticisms, see Hervé Corvellec et al. (2022).

20. As noted by Logan (2014: 41), these positions are held respectively by North American mathematicians, Claude Elwood Shannon and Norbert Weiner, in contrast to British physicist, Donald MacCrimmon MacKay. Separately, see Kauffman (2019) for further elaboration on why living systems should be defined by the concept of organization.
21. This was previously mentioned in Graham Murdock (2021: 45).
22. There are many sources for histories and discussions of the biotechnology industry (e.g. Braman 2004; Vettel 2013; Mitchell 2015; Krcmar, Ewoldsen and Koerner 2016; Magin 2017; Hogan 2018; Green 2022).

REFERENCES

Allaby, Michael (ed.) (2015), *A Dictionary of Ecology* (5th ed.), Oxford: Oxford University Press, https://www.oxfordreference.com/view/10.1093/acref/9780191793158.001.0001/acref-9780191793158-e-3220. Accessed 17 April 2022.

Anten, Niels P. R. and Chen, Bin J. W. (2021), 'Detect thy family: Mechanisms, ecology and agricultural aspects of kin recognition in plants', *Plant, Cell & Environment*, 44:4, pp. 1059–71.

Atapattu, Sumudu A., Gonzalez, Carmen G. and Seck, Sara L. (eds) (2021), *The Cambridge Handbook of Environmental Justice and Sustainable Development*, Cambridge, MA: Cambridge University Press.

Bar-Yam, Yaneer (2011), 'Concepts: Patterns', Cambridge, MA: New England Complex Systems Institute, https://necsi.edu/patterns. Accessed 18 January 2019.

Bateson, Gregory ([1972] 2000], *Steps to an Ecology of Mind*, Chicago: University of Chicago Press.

Bauer, Ute Meta (ed.) (2022), *Climates. Habitats. Environments*, Cambridge: MIT Press.

Beatty, Michael James, McCroskey, James C. and Floyd, Kory (2009), *Biological Dimensions of Communication: Perspectives, Methods, and Research*, Cresskill, NJ: Hampton Press.

Bedau, Mark A. and Cleland, Carol E. (eds) (2010), *The Nature of Life: Classical and Contemporary Perspectives from Philosophy and Science*, Cambridge, MA: Cambridge University Press.

Bich, Leonardo and Green, Sara (2018), 'Is defining life pointless? Operational definitions at the frontiers of biology', *Synthese*, 195:9, pp. 3919–46.

Boren, Justin P. and Veksler, Alice E. (2011), 'A decade of research exploring biology and communication: The brain, nervous, endocrine, cardiovascular and immune systems', *Communication Research Trends*, 30:4, pp. 4–31.

——— (2016), 'Biology', in K. B. Jensen and R. T. Craig (eds), *The International Encyclopedia of Communication Theory and Philosophy, Volume I*, Malden, MA: John Wiley & Sons, Inc.

Braidotti, Rosi and Bignall, Simone (eds) (2018), *Posthuman Ecologies: Complexity and Process After Deleuze*, Lanham, MD: Rowman & Littlefield.

Braman, Sandra (2004), 'The meta-technologies of information', in S. Braman (ed.), *Biotechnology and Communication: The Meta-Technologies of Information*, Mahwah, NJ: Lawrence Erlbaum Associates, Publishers, pp. 3–38.

Brevini, Benedetta and Murdock, Graham (eds) (2017), *Carbon Capitalism and Communication: Confronting Climate Crisis*, Cham: Palgrave Macmillan.

Caniglia, Beth Schaefer, Jorgenson, Andrew, Malin, Stephanie A., Peek, Lori, Pellow, David N. and Huang, Xiaorui (eds) (2021), *Handbook of Environmental Sociology*, Cham: Springer.

Cappella, Joseph N. (1991), 'The biological origins of automated patterns of human interaction', *Communication Theory*, 1:1, pp. 4–35.

——— (2020), 'Building communication science through blueprints of D. C. Dennett and Robert Pirsig', in K. Floyd and R. Weber (eds), *The Handbook of Communication Science and Biology*, New York: Routledge, pp. 10–18.

Capra, Fritjof and Luisi, Pier Luigi ([2014] 2020), *Vita e Natura: Una Visione Sistemica*, Sansepolcro: Aboca.

Carvalho, Anabela, Doyle, Julie and Russill, Chris (eds) (2021), 'Critical approaches to climate change and civic action,' *Frontiers in Communication, 6*, https://doi.org/10.3389/fcomm.2021.711897. Accessed 18 April 2023.

Cassin, Barbara (ed.) (2014), 'LIFE/LEBEN,' *Dictionary of Untranslatables: A Philosophical Lexicon*, Princeton, NJ: Princeton University Press.

Christensen, Miyase and Nilsson, Annika E. (2018), 'Media, communication, and the environment in precarious times,' *Journal of Communication, 68:2*, pp. 267–77.

Condit, Celeste Michelle (2000), 'Culture and biology in human communication: Toward a multi-causal model', *Communication Education, 49*, pp. 7–24.

Cook, Sarah (ed.) (2016), *Information, Documents of Contemporary Art*, London and Cambridge, MA: Whitechapel Gallery and MIT Press.

Corvellec, Hervé, Stowell, Alison F. and Johansson, Nils (2022), 'Critiques of the circular economy', *Journal of Industrial Ecology, 26:2*, pp. 421–32.

Côté, Raymond P. (2009), 'Industrial ecology', in D. J. Cuff and A. Goudie (eds), *The Oxford Companion to Global Change,* 1st ed., Oxford: Oxford University Press.

Cox, Robert (ed.) (2015), *Environmental Communication*, Los Angeles, CA: SAGE Publications.

Craig, Robert T. (2020), 'Series editor's foreword', in K. Floyd and R. Weber (eds), *The Handbook of Communication Science and Biology*, New York: Routledge, pp. xvii–xviii.

Craig, Robert T. and Muller, Heidi L. (2007), 'Concluding reflections', in *Theorizing Communication: Readings Across Traditions*, Thousand Oaks, MA: Sage, pp. 495–502.

Craig, Robert T. and Xiong, Bingjuan (2022), 'Traditions of communication theory and the potential for multicultural dialogue', *Journal of Multicultural Discourses*, 17:1, pp. 1–25.

Dewey, John (2012), *Unmodern Philosophy and Modern Philosophy*, P. Deen (ed.), Carbondale, IL: Southern Illinois University Press.

Dewey, John and Bentley, Arthur ([1949] 2008), *Knowing and the Known,* in J. A. Boydston (ed.), *The Later Works: 1925–1953*, vol. 16, Carbondale, IL: Southern Illinois University Press, p. 143. [First published by Boston, MA: Beacon Press, 1949.]

Deuze, Mark (2012), *Media Life*, Cambridge: Polity Press.

——— (2023), *Life in Media: A Global Introduction to Media Studies*, Cambridge, MA: MIT Press.

Floridi, Luciano (2010), *Information: A Very Short Introduction*, Oxford: Oxford University Press.

Floyd, Kory and Weber, René (eds) (2020), *The Handbook of Communication Science and Biology*, New York: Routledge.

Fuchs, Christian and Qiu, Jack Linchuan (2018), 'Ferments in the field: Introductory reflections on the past, present and future of communication studies', *Journal of Communication*, 68:2, pp. 219–32.

Fuller, Matthew (2005), *Media Ecologies: Materialist Energies in Art and Technoculture*, Cambridge, MA: MIT Press.

Gánti, Tibor (2003), *The Principles of Life*, New York: Oxford University Press.

Godfrey-Smith, Peter and Sterelny, Kim (2016), 'Biological information', in E. N. Zalta (ed.), *The Stanford Encyclopedia of Philosophy*, Summer ed., Metaphysics Research Lab, Stanford University, https://plato.stanford.edu/archives/sum2016/entries/information-biological/. Accessed 4 January 2023.

Granata, Paolo (2017), 'Medium as "metaform": An inquiry into the life of forms', in C. Anton, R. K. Logan and L. Strate (eds), *Taking up McLuhan's Cause: Perspectives on Media and Formal Causality*, Bristol: Intellect.

Grasso, Maria Teresa and Giugni, Marco (eds) (2022), *The Routledge Handbook of Environmental Movements*, New York: Routledge.

Green, Sara (2022), 'Philosophy of systems and synthetic biology', in E. N. Zalta (ed.), *The Stanford Encyclopedia of Philosophy*, Summer ed., Metaphysics Research Lab, Stanford University, https://plato.stanford.edu/entries/systems-synthetic-biology/. Accessed 4 January 2023.

Guattari, Félix ([1989] 2000), *The Three Ecologies* (trans. I. Pindar and P. Sutton), New Brunswick, NJ: The Athlone Press. [First published *Les trois écologies*, Editions Galilée in French in 1989].

Gunaratne, Shelton A. and Gunaratne, Yoke-Sim (2017), 'A new vision of reality for communication research: Call for a paradigm shift to systems view of life', *Journal of Multicultural Discourses*, 12:1, pp. 1–22.

——— (2021), 'Resolving East-West rifts in Craig's metamodel', in M. H. Rich and J. S. Robles (eds), *Practicing Communication Theory*, San Diego, CA: Cognella.

Hansen, Anders and Cox, Robert (eds) (2015), *Routledge Companion of Environment and Communication*, London: Routledge.

Harris, Kathleen Mullan and McDade, Thomas W. (2018), 'The biosocial approach to human development, behavior, and health across the life course', *RSF: The Russell Sage Foundation Journal of the Social Sciences*, 4:4, pp. 2–26.

Heise, Ursula K. (2019), 'Environmentalism, eco-cosmopolitanism, and premodern thought', in V. Nardizzi and T. J. Werth (eds), *Premodern Ecologies in the Modern Literary Imagination*, Toronto: University of Toronto Press, pp. 282–88.

Hickman, Larry A. (2013), 'John Dewey's critique of our "unmodern" philosophy', *European Journal of Pragmatism and American Philosophy, V-1*, http://journals.openedition.org/ejpap/599. Accessed 3 March 2022.

Hickson III, Mark and Stacks, Don W. (2010), 'Biological views of communication', *The Review of Communication*, 10:4, pp. 263–75 .

Hoffman, Joshua and Rosenkrantz, Gary (1999), 'life', in R. Audi (ed.), *The Cambridge Dictionary of Philosophy*, 2nd ed., New York: Cambridge University Press.

Hogan, Mél (2018), 'Big data ecologies', *ephemera*, 18.3, pp. 631–65.

Horn, Eva and Bergthaller, Hannes (2020), *The Anthropocene: Key Issues for the Humanities*, New York: Routledge.

Huhtamo, Erkki and Galili, Doron (2020), 'The pasts and prospects of media archaeology', *Early Popular Visual Culture*, 18:4, pp. 333–39.

Iheka, Cajetan (2021), *African Ecomedia: Network Forms, Planetary Politics*, Durham, NC: Duke University Press.

Johnson, Mark (2017), *Embodied Mind, Meaning, and Reason: How Our Bodies Give Rise to Understanding*, Chicago: University of Chicago Press.

Jones, Emily (2023), *Feminist Theory and International Law: Posthuman Perspectives*, New York: Taylor & Francis.

Kauffman, Stuart A. (2019), *A World Beyond Physics: The Emergence and Evolution of Life*, Oxford: Oxford University Press.

Keller, Evelyn Fox (2003), *Making Sense of Life: Explaining Biological Development with Models, Metaphors and Machines*, Cambridge, MA: Harvard University Press.

Kittler, Friedrich (2007), 'Ontologie der Medien', *Bochumer Kolloquiums Medienwissenschaft*, Bochum: Ruhr-Universtität Bochum, 11 November, https://doi.org/10.25969/mediarep/13855. Accessed 10 February 2023.

Knapp, Mark L. (2009), 'Epilogue: A "new" frontier for communication studies', in M. J. Beatty, J. C. McCroskey and K. Floyd (eds), *Biological Dimensions of Communication: Perspectives, Methods, and Research*, Cresskill, NJ: Hampton Press, Inc., pp. 267–80.

Krcmar, Marina, Ewoldsen, David R. and Koerner, Ascan (2016), 'The future of communication', in *Communication Science Theory and Research: An Advanced Introduction*, New York: Routledge, pp. 336–47.

Krippendorff, Klaus (2008), 'Generalizability', in W. Donsbach (ed.), *The International Encyclopedia of Communication,* 1st ed., Malden, MA: Blackwell Publishing, pp. 1951–54.

Lanigan, Richard L. (2021), 'Communicology, cybernetics, and chiasm: A synergism of logic, linguistics, and semiotics', in C. Vidales and S. Brier (eds), *Introduction to Cybersemiotics: A Transdisciplinary Perspective*, Cham: Springer, pp. 161–91.

Lehman-Wilzig, Sam and Cohen-Avigdor, Nava (2004), 'The natural life cycle of new media evolution: Inter-media struggle for survival in the Internet age', *New Media & Society*, 6:6, pp. 707–30.

Liddell, Henry George and Scott, Robert (1940a), 'βίος', in *A Greek-English Lexicon* (rev. and augmented by H. S. Jones and R. McKenzie), Oxford: Clarendon Press, http://www.perseus.tufts.edu/hopper/text?doc=Perseus:text:1999.04.0057:entry=bi/os1. Accessed 23 January 2021.

——— (1940b), 'οἶκος', in *A Greek-English Lexicon* (rev. and augmented by H. S. Jones and R. McKenzie), Oxford: Clarendon Press, http://www.perseus.tufts.edu/hopper/text?-doc=Perseus:text:1999.04.0057:entry=oi)%3Dkos1. Accessed 23 January 2021.

'life' (2022), *Oxford Dictionaries*, Oxford: Oxford University Press, http://www.oxforddictionaries.com/us/definition/american_english/life. Accessed 15 April 2022.

'life, n.' (2023), *OED Online*, Oxford: Oxford University Press, https://www.oed.com/dictionary/life_n. Accessed 24 July 2023.

Logan, Robert K. (2014), 'What is information? Why is it relativistic? & What is its relationship to materiality, meaning & organization?' in *What is Information?: Propagating Organization in the Biosphere, Symbolosphere, Technosphere and Econosphere*, Toronto: DEMO Publishing, pp. 25–51.

——— (2021), 'Biological dimensions of media ecology and its relationship to biosemiotics', in J. Swartz and J. Wasko (eds), *MEDIA: A Transdisciplinary Inquiry*, Bristol: Intellect Books.

López, Antonio, Ivakhiv, Adrian, Rust, Stephen, Tola, Miriam, Chang, Alenda Y. and Chu, Kiu-wai (eds) (2024), *The Routledge Handbook of Ecomedia Studies*, London: Routledge.

Luisi, Pier Luigi (2016), 'Systems thinking and its implications: The principle of interdependence', *Wall Street International*, 29 January, https://www.wsimag.com/en/19190-systems-thinking-and-its-implications. Accessed 28 May 2019.

Lupinacci, John, Happel-Parkins, Alison and Turner, Rita (eds) (2022), *Ecocritical Perspectives in Teacher Education*, Boston: Brill.

Magin, Richard L. (2017), 'Bioengineering and cybernetics: A modern caduceus', *IEEE* (Institute of Electrical and Electronics Engineers) *Pulse*, 8:1, pp. 44–47.

Mariscal, Carlos (2021), 'Life', in E. N. Zalta (ed.), *The Stanford Encyclopedia of Philosophy*, Winter ed., Metaphysics Research Lab, Stanford University, https://plato.stanford.edu/archives/win2021/entries/life/. Accessed 12 January 2022.

Mattelart, Armand (1996), 'Epilogue: New organic totalities', in *The Invention of Communication* (trans. S. Emanuel), Minneapolis, MN: University of Minnesota Press.

Maxwell, Richard and Miller, Toby (2012), *Greening the Media*, Cambridge: Oxford University Press.

Miike, Yoshitaka and Yin, Jing (eds) (2022), *The Handbook of Global Interventions in Communication Theory*, New York: Routledge.

Miller, Toby and Maxwell, Richard (2017), *Greenwashing Culture*, London: Routledge.

Mitchell, Robert E. (2015), *Bioart and the Vitality of Media*, Seattle, WA: University of Washington Press.

Mitchell, W. J. T. (ed.) (2023), 'Keywords', *The Chicago School of Media Theory*, Division of the Humanities, University of Chicago, https://web.archive.org/web/20230127202040/https://lucian.uchicago.edu/blogs/mediatheory/keywords/. Accessed 27 January 2023.

Mitchell, W. J. T. and Hansen, Mark B. N. (eds) (2010), *Critical Terms for Media Studies*, Chicago: University of Chicago Press.

Mitman, Gregg (1992), *The State of Nature: Ecology, Community, and American Social Thought,* 1900–1950, Chicago: University of Chicago Press.

Murdock, Esme G. (2020), 'A history of environmental justice: Foundations, narratives, and perspectives', in B. Coolsaet (ed.), *Environmental Justice: Key Issues*, New York: Routledge, pp. 6–17.

Murdock, Graham (2021), 'Dark materials: Media, machines, markets', in J. Swartz and J. Wasko (eds), *MEDIA: A Transdisciplinary Inquiry*, Bristol: Intellect Books.

Nail, Thomas (2023), 'Metabolism and metabolic drift', in J. Swartz and J. Wasko (eds), *LIFE: A Transdisciplinary Inquiry,* Bristol: Intellect, pp. 185–97.

Newell, Joshua P. and Cousins, Joshua J. (2015), 'The boundaries of urban metabolism: Towards a political–industrial ecology', *Progress in Human Geography*, 39:6, pp. 702–28.

NIH (National Institutes of Health) (2021), 'Science education: Glossary of terms', National Institute of Biomedical Imaging and Bioengineering (NIBIB), Department of Health & Human Services, National Institutes of Health, Bethesda, MD, https://www.nibib.nih.gov/science-education/glossary/b. Accessed 29 April 2022.

Parikka, Jussi (2014), *The Anthrobscene*, Minneapolis, MN: University of Minnesota Press.

Park, Chris and Allaby, Michael (2017), 'Industrial ecology', in C. Park and M. Allaby (eds), *A Dictionary of Environment and Conservation,* 3rd ed., Oxford: Oxford University Press.

Parks, Lisa (2020), 'Field mapping: What is the "media" of media studies?', *Television & New Media,* 21:6, pp. 642–49.

Parks, Lisa and Walker, Janet (2020), 'Disaster media: Bending the curve of ecological disruption and moving toward social justice', *Media + Environment,* 2:1, https://doi.org/10.1525/001c.13474. Accessed 18 October 2021.

Peekhaus, Wilhelm (2016), 'Biotechnology', in K. B. Jensen and R. T. Craig (eds), *The International Encyclopedia of Communication Theory and Philosophy, Volume I*, Malden, MA: John Wiley & Sons, Inc.

Pezzullo, Phaedra C. and Cox, Robert J. (2021), *Environmental Communication and the Public Sphere*, 6th ed., Los Angeles, CA: SAGE Publications.

Phillips, Whitney and Milner, Ryan M. (2021), *You Are Here: A Field Guide for Navigating Polarized Speech, Conspiracy Theories, and Our Polluted Media Landscape*, Cambridge: MIT Press.

Popa, Radu (2004), *Between Necessity and Probability: Searching for the Definition and Origin of Life*, Dordrecht: Springer Science & Business Media.

Rajan, Kaushik Sunder (2006), *Biocapital: The Constitution of Postgenomic Life*, Durham, NC: Duke University Press.

Raymond, Richard C. (1950), 'Communication, entropy, and life', *American Scientist,* 38:2, pp. 273–278.

Rust, Stephen, Monani, Salma and Cubitt, Sean (eds) (2016), *Ecomedia: Key Issues*, New York: Routledge.

Sagan, Carl (1970), 'Life', *Encyclopedia Britannica*, Chicago: Encyclopædia Britannica Incorporated, pp. 1083–1083A.

Sagan, Dorion, Sagan, Carl and Margulis, Lynn (n.d.), 'life', *Encyclopædia Britannica*, Chicago: Encyclopædia Britannica, Inc., https://www.britannica.com/science/life. Accessed 15 April 2022.

Schiller, Dan (1988), 'How to think about information', in V. Mosco and J. Wasko (eds), *The Political Economy of Information*, Madison, WI: University of Wisconsin Press, pp. 27–43.

Scolari, Carlos A. (2012), 'Media ecology: Exploring the metaphor to expand the theory', *Communication Theory,* 22, pp. 204–25.

——— (2022), 'Evolución de los medios: Mapa de una disciplina en construcción: Una revisión', *El Profesional de la Información*, 31:2, https://dialnet.unirioja.es/servlet/articulo?codigo=8402668. Accessed 5 April 2023.

Sharifi, Rouhallah and Ryu, Choong-Min (2021), 'Social networking in crop plants: Wired and wireless cross-plant communications', *Plant, Cell & Environment*, 44:4, pp. 1095–110.

Sharma, Sarah and Singh, Rianka (eds) (2022), *Re-understanding Media: Feminist Extensions of Marshall McLuhan*, Durham, NC: Duke University Press.

Slovic, Scott, Rangarajan, Swarnalatha and Sarveswaran, Vidya (eds) (2019), *Routledge Handbook of Ecocriticism and Environmental Communication*, London: Routledge.

Starosielski, Nicole (2021), *Media Hot and Cold*, Durham, NC: Duke University Press.

Starosielski, Nicole and Walker, Janet (eds) (2016), *Sustainable Media: Critical Approaches to Media and Environment*, New York: Routledge.

Stumpf, Michael P. H. (2022), 'More is different with a vengeance', *Cell Systems*, 13:8, pp. 594–97.

Takahashi, Bruno, Metag, Julia, Thaker, Jagadish and Comfort, Suzannah Evans (eds) (2021), *The Handbook of International Trends in Environmental Communication*, New York: Routledge.

Thacker, Eugene (2004), *Biomedia*, Minneapolis, MN: University of Minnesota Press.

——— (2010), 'Biomedia', in W. J. T. Mitchell and M. B. N. Hansen (eds), *Critical Terms for Media Studies*, Chicago: University of Chicago Press.

Trifonov, Edward N. (2011), 'Vocabulary of definitions of life suggests a definition', *Journal of Biomolecular Structure and Dynamics,* 29:2, pp. 259–66.

Turner, Fred (2006), *From Counterculture to Cyberculture: Stewart Brand, the Whole Earth Network, and the Rise of Digital Utopianism*, Chicago: University of Chicago Press.

Veksler, Alice E., Boren, Justin P. and Priem, Jennifer (2020), 'Social support and physiology: Current trends and future directions', in K. Floyd and R. Weber (eds), *The Handbook of Communication Science and Biology*, New York: Routledge, pp. 319–31.

Vettel, Eric J. (2013), *Biotech: The Countercultural Origins of an Industry*, Philadelphia, PA: University of Pennsylvania Press.

Vint, Sherryl (ed.) (2020), *After the Human: Culture, Theory and Criticism in the 21st Century*, vol. 6, Cambridge: Cambridge University Press.

von Bertalanffy, Ludwig ([1968] 2015), *General System Theory: Foundations, Development, Applications*, rev. ed., New York: George Braziller.

Wachsmuth, David (2012), 'Three ecologies: Urban metabolism and the society-nature opposition', *The Sociological Quarterly*, 53:4, pp. 506–23.

Wagner, Andeas (2014), *The Arrival of the Fittest: How Nature Innovates*, New York: Penguin Random House.

——— (2019), *Life Finds a Way: What Evolution Teaches Us About Creativity*, New York: Basic Books.

Weber, Bruce (2018), 'Life', in E. N. Zalta (ed.), *The Stanford Encyclopedia of Philosophy, Summer ed.*, Stanford, CA: Metaphysics Research Lab, Stanford University, https://plato.stanford.edu/archives/sum2018/entries/life/. Accessed 11 July 2020.

Weibel, Peter (ed.) (2023), *BioMedia: The Age of Media with Life-like Behavior*, Leipzig: ZKM | Center for Art and Media Karlsruhe and Spector Books.

Wickberg, Adam and Gärdebo, Johan (eds) (2022), *Environing Media*, New York: Taylor & Francis.

Witzany, Guenther (2019), 'Communication as the main characteristic of life', in V. M. Kolb (ed.), *Handbook of Astrobiology*, Boca Raton, FL: CRC Press, pp. 91–105.

——— (2020), 'What is Life?', *Frontiers in Astronomy and Space Science*, 7:7, pp. 1–20.

Yegiyan, Narine S. (2017), 'Communication and human biology', in M. Allen (ed.), *The SAGE Encyclopedia of Communication Research Methods*, Los Angeles, CA: Sage Publishing.

PART I

GENEALOGY

1

Life, Nature and Systems[1]

Fritjof Capra

A systemic understanding of life has returned in science, engaging age-old questions and concepts in remarkable ways. At the core of a 'systems view of life', there has been a fundamental change of metaphors: from seeing the world merely as mechanisms to understanding it as networks. This chapter explores *systems*, *systems thinking* and a *systems view of life* – a coherent conceptual framework that integrates life's biological, cognitive, social and ecological dimensions, and its profound philosophical, social and policy implications. The chapter also addresses systemic problem-solving through ecological literacy and design.

Conceptualizing Systems and Systems Thinking

In contemporary science, the universe is no longer seen as a mere machine composed of elementary building blocks. Scientists have discovered that the material world is a network of inseparable patterns of relationships and the planet as a whole is a living, self-regulating '*system*' (Capra and Luisi 2014). The meaning of the word derives from the Greek, σύστημα or *sústēma* as a '*whole compounded of several parts* or *members…*' (Liddell and Scott 1940e: n.pag., original emphasis), which comes from its root meaning σῠνῐ́στημῐ (*sunístēmi*) + -μᾰ (*-ma*) 'set together, combine', 'associate, unite' or 'organize, frame' (Liddell and Scott 1940d: n.pag.).[2] Therefore, to understand things systematically means literally putting them into a context to establish the nature of their relationships.

The view of the human body as a machine and of the mind as a separate entity is being replaced by one that sees not only the brain but also the immune system, bodily tissues and even each cell as a living, cognitive system. Evolution is no longer seen as merely a competitive struggle for existence but as a cooperative dance in which creativity and the constant emergence of order and novelty are driving forces. And with the emphasis on complexity, networks and patterns of organization, a science of qualities is emerging.

A systems (or systemic) view, also known as *systems thinking*, is conceptualized in terms of relationships, patterns and contexts. Thinking in terms of relationships is crucial, in particular, for the science of *ecology* – derived from the Greek *οἶκος*, 'house […] any dwelling-place' (Liddell and Scott 1940b: n.pag.) or 'household, family' (Liddell and Scott 1889) – the study of the relationships among various members of the Earth house and household. The term was coined by the German biologist Ernst Heinrich Philipp August Haeckel (1834–1919), who defined it as 'die gesammte Wissenschaft von den Beziehungen des Organismus zur umgebenden Aussenwelt' ('the entire science of the relationships of the organism to the surrounding outer world') (Haeckel 1866: 286). The word *Umwelt* ('environment') was used for the first time in 1909 by the Baltic German biologist, ecological innovator and early biosemiotician, Jakob Johann Freiherr von Uexküll (1864–1944). Systems science, however, is not limited to certain societies. It is important to recognize that many Indigenous peoples continue to embody this profound ecological awareness, understanding nature in terms of relationships and patterns (see Koohafkan and Altieri 2011; FAO 2020).

Leonardo da Vinci

Leonardo da Vinci (1452–1519) was an example of an early systemic thinker in western thought, including his scientific method, and his work in anatomy, botany, fluid dynamics, geology, mechanics and the science of flight (Capra 2013), as well as his synthesis of art and science (Capra 2007). Understanding a phenomenon meant connecting it with other phenomena through a similarity of patterns. For example, he would compare the flow of water in a landscape nourishing the vegetation, to the flow of blood nourishing the tissues of the body. Patterns of turbulence in water led him to observe similar patterns in the flow of air, and from there he went on to explore the nature of sound and the theory of music. Leonardo usually worked on several projects in parallel, and when his understanding advanced in one area, he would revise his ideas in related areas accordingly.

Thus, to appreciate his remarkable contributions, one needs to be aware of the evolution of his thinking in several parallel but interconnected disciplines. Leonardo sought to understand the nature of life.[3] His science was of living forms and radically different from the mechanistic science of Galileo Galilei (1564–1642), René Descartes (1596–1650) and Sir Isaac Newton (1643–1727).

In addition, his art served this persistent quest for life's inner secrets. In order to paint nature's living forms, he felt the need for a scientific understanding of

FIGURE 1.1: Leonardo da Vinci, *A star-of-Bethlehem and other plants*, *c*.1506–12. Red chalk, pen and ink. 19.8 × 16.0 cm (sheet of paper). Royal Collection Trust. © Her Majesty Queen Elizabeth II 2022.

their intrinsic and underlying principles. And, in order to analyse the results of his observations, he needed his artistic ability to depict them – this is the very essence of his synthesis of science and art (see Figure 1.1).

Leonardo thought of himself not only as a natural philosopher (as scientists were called in his time) and as an artist, but also as an inventor. In his view, an inventor was someone who created an assemblage of various elements into a new configuration that did not appear in his world. This description comes very close to our modern notion of a designer. Indeed, his synthesis of art, science and design used (living) nature as his mentor and model. He was not only a systemic thinker but also an ecologist and ecodesigner, which is extremely relevant for our time and is explored later in this chapter.

Systems Thinking in the Twentieth Century

By the 1930s, systems thinking had emerged simultaneously in three different fields: organismic biology (with the irreducible wholes in organisms), Gestalt psychology (in perception) and ecology (in the study of animal and plant communities). In all these fields, scientists explored *living systems*, including whole organisms, parts of organisms and communities of organisms, such as social systems and ecosystems (Capra and Luisi 2014: Chapters 2–6 and 16). The study of living systems integrates a very broad range of disciplines, and systems thinking is therefore 'transdisciplinary'. Systems science established that living systems share a set of common properties and principles of organization, which can lead to the discovery of similarities between different phenomena. This way of thinking in terms of connectedness, relationships and context was also supported by discoveries in quantum physics related to atoms, subatomic particles and entanglement.

Wholes, Parts and Relationships

When considering the concept of wholes and parts, philosophers look to antiquity and Aristotle's *Metaphysics*.[4] At the turn of the nineteenth century, Austrian philosopher, Christian von Ehrenfels (1859–1932), characterized the German word *Gestalt*, meaning organic form, in the sense of an irreducible perceptual pattern. He asserted: 'das Produkt von Einheitlichkeit des Ganzen und Mannigfaltigkeit der Teile ein größeres ist' ([1890] 1960: 63; also see [1890] 2007). I would translate this as: 'the combined effect of the unity of the whole and the multiplicity of the parts is greater'.[5] In other words, the sum is greater than the whole *and* the parts, which has evolved in subsequent decades into the catchphrase, 'the whole is greater than the sum of its parts', a key concept of systems thinkers.

Understanding this statement may require several shifts of perception. The first shift is from merely parts to wholes. Living systems are integrated wholes whose properties cannot be reduced to those of smaller parts, or in other words, a living system is more than a static configuration of components. The essential properties of a living system arise from the interactions and relationships among the parts. Thus, systems thinking requires an adjustment of focus from objects to relationships, whereas things need to be measured and weighed, and relationships need to be mapped. When we map relationships among components, we find certain repetitive configurations or patterns, which also involve a change from quantity to quality. 'Quantities can be measured; qualities need to be mapped' (Capra and Henderson 2009: 7). Finally, the study of relationships is also between a system as a whole and the surrounding larger system, its environment, its medium and its context; in other words, contextual thinking. Another important strand that emerged somewhat later pertained to a shift of perspective from structures to processes. Therefore, systems thinking means both contextual thinking *and* process thinking.

Theoretical Developments

The 1940s brought forth the formulation of actual systems theories and coherent theoretical frameworks describing the principles of the organization of living systems. These classical theories included *general systems theory* and *cybernetics*. During the 1970s and 1980s, systems thinking was elevated with the development of *complexity theory* (see Capra and Luisi 2014: Chapters 5 and 6; also see Krakauer 2019), technically known as 'nonlinear dynamics', or 'dynamical systems theory'. *Chaos theory* and *fractal geometry* are important branches of complexity theory (Hilborn 2000; Stewart 2002). While mathematics as a language is used to represent and attempt to understand the behavior of living (and non-living) systems, nonlinear mathematics is utilized to visually represent patterns and relationships of a system's complex dynamics (e.g. strange attractors, fractals and general climate models).

In the twenty-first century, the strong interest in nonlinear phenomena generated a whole series of new and powerful theories[6] that have dramatically increased our understanding of many key characteristics of life. The synthesis of these theories is a systems view of life (Capra and Luisi 2014; Capra Course n.d.).

A Systems View of Life

'What is life?' asks not only an age-old philosophical question but also has an interesting history in modern science (Schrödinger 1944). It was the title of a short book published in 1944 by physicist Erwin Schrödinger, in which he advanced clear and compelling hypotheses about the molecular structure of genes or aperiodic solids. This book stimulated biologists to think about genetics in a novel way, and in so doing opened a new frontier of science, molecular biology.

During the subsequent decades, this new field generated a series of spectacular advances, culminating in the discovery of the double-helix structure of deoxyribonucleic acid (DNA) and the genetic code (see Judson 1979, 2003). Biologists declared triumphantly that they had discovered the alphabet of a truly universal language of life. However, these advancements did not bring them any closer to answering the question posed in the title of Schrödinger's book. They had discovered the fundamental building blocks of life, but soon realized that they knew very little about the integrative processes of living organisms. How do wounds heal? How does an organism develop from a fertilized egg?

In other words, biologists knew the alphabet (codescript) of the language of life but had almost no idea of its syntax. In the past few decades, the syntax of life has indeed been discovered, and with it, a whole new conception of life has emerged.

Metabolism

Expanding on Schrödinger's question, we may ask further: What is the essential difference between a rock and a plant? When you ask biologists that question, they might say: a plant is made of cells, but there are no cells in a rock; or they might just say: a living organism is a biochemical system that contains DNA. It sounds simple: if it contains DNA, it's alive; if it doesn't contain DNA, it's not alive. But there is a catch: dead organisms also contain DNA, because it does not disappear when an organism dies. Therefore, to understand the essential nature of life, it is not enough to understand only its molecular structures.

The difference between a living and a dead organism lies in the process of *metabolism* – the ceaseless flow of energy and matter through a network of chemical reactions, which enables a living organism to continually generate, maintain, repair and perpetuate itself (see Margulis 1998: Chapters 4 and 5; Margulis and Sagan 2002: Chapters 2, 6 and 7; also see Feldman 2019).

Metabolism includes two basic aspects. The first is the continuous *flow* of energy and matter. All living systems need energy and food to sustain themselves, and all living systems produce waste. Life has evolved in such a way that organisms form communities and ecosystems in which the waste of one species is food for another so that matter cycles continually throughout. The second aspect of metabolism is the *network* of chemical reactions that processes food and forms the biochemical basis of all biological structures, functions and behaviors. One of the most important insights of a systems view of life is the recognition that networks are the basic pattern of organization of all living systems. Cells are networks of molecules; organisms are networks of (autonomous but closely coupled) cells; ecosystems are networks of organisms (understood in terms of food webs) and societies are networks of ecosystems (human, non-human, plant and machine). In other words, networks are patterns common to all life. Wherever there is life, there are networks.

Thus, we have two levels of description which involve different languages. Flows are described in the language of physics and chemistry, networks in the language of graph theory, topology, etc.

Living Networks

Detailed examination of living networks has shown that their key characteristic is that they are self-generating. Living networks continually create or recreate themselves by transforming or replacing their components. Technically, this is known as the theory of *autopoiesis*, formulated by Chilean scientists, Humberto Maturana and Francisco Varela ([1972] 1980; also see Luisi 2016: Chapters 6 and 7). In Greek αὐτός or *auto* means 'self' (Liddell and Scott 1940a) and ποίησις or *poiesis*, from the Greek root 'of poetry' (Liddell and Scott 1940c) means 'producing'. Therefore, autopoiesis can be defined as *self-producing* (Luisi 2016: 123).

In a cell, for example, all biological structures – proteins, enzymes, DNA, cell membrane, etc. – are continually produced, repaired and regenerated by the cellular network. Similarly, at the level of a multicellular organism, bodily cells are continually regenerated and recycled by the organism's metabolic network. This means that living organisms undergo continual structural changes while preserving their web-like pattern of organization. This coexistence of stability and change is indeed one of the key characteristics of life.

Philosopher of science and technology, Luisa Damiano, and biochemist, Pier Luigi Luisi, have proposed that an autopoietic system

> is organized as a network of production processes which produces the components which, through their interactions and transformations, permanently regenerate the network of processes constituting the system itself as a concrete topological unit, separated from its *medium* by a boundary and related to it through cognitive or adaptive coupling.
>
> (2010: 149, emphasis added)

Life in the social realm can also be understood in terms of networks, but here we are not dealing with merely chemical structures and reactions – we are dealing with models and meanings of communication. As Luisi (2003) states, 'Although this concept [autopoiesis] came from the analysis of a living cell, it can be metaphorically applied to social systems' (57; also see Luisi 2016: 151–52). As communication continues in social networks, multiple feedback loops are formed, which can produce shared systems of knowledge, values and rules of conduct – common contexts of meanings known as cultures, which are continually sustained by further communication (Capra and Luisi 2014: 305–09).

Flow and Emergence

Returning to the first aspect of metabolism, whereas energy and matter flow through networks, social networks correspond to a flow of ideas. In both cases, systems need to be open to information flows – as nourishment to survive.

Over the past few decades, the dynamics of this flow of energy and matter through living networks have been studied in great detail and have led to a very important discovery. Living systems generally remain in a stable state, even though energy and matter flow through them and their structures are continually changing. But every now and then, an open system will encounter a point of instability where there is either a breakdown or, more frequently, a spontaneous emergence of new forms of order (Prigogine 1980).

The detailed theory of *emergence* shows that the instabilities and jumps to new forms of organization are the result of fluctuations amplified by feedback loops. When a system encounters a small disturbance, which circulates around multiple feedback loops, and is amplified until the system as a whole becomes unstable, it 'may either break *down* or it may break *through* to a new form of order' (Capra and Luisi 2014: 319–20). This spontaneous emergence of *novelty* at critical points of instability is another hallmark of life. It has been recognized as the dynamic origin of development, learning, and evolution (Prigogine and Stengers 1984; also see Stengers 2011: Chapter V). In other words, emergent creativity – the generation of new forms – is a key property of all living systems (Prigogine 1989; also see Luisi 2016: Chapter 9).

Mind and Consciousness

One of the most important, and most radical, philosophical implications of a systems view of life is a conception of the nature of mind and consciousness, which finally overcomes the Cartesian division between matter and mind that has haunted philosophers and scientists for centuries.

In the seventeenth century, Descartes based his view of nature on the fundamental division between two independent and separate realms – that of mind, which he called the 'thinking thing' (*res cogitans*), and that of matter, the 'extended thing' (*res extensa*) (Descartes [1641] 1984). Following Descartes, scientists and philosophers continued to think of the mind as some intangible entity and were unable to imagine how this 'thinking thing' is related to the body. The decisive advance of a systems view of life has been to abandon the Cartesian view of mind as a 'thing', and to realize that mind and consciousness are not things but processes (Bateson 1979, 1991). Mental activity is a *self-organizing* property of living systems at all levels of life. The interactions of a living organism – plant, animal or human – with its environment are cognitive interactions. Thus, life and cognition are inseparably connected (Maturana and Varela [1972] 1980). For further discussion on this topic, see Mark Johnson (2023).

Systemic Problem-Solving

The critical importance of a systems view of life involves concrete applications for dealing with the problems of our multifaceted global crises (Capra and Luisi 2014: Chapters 16–18). It is becoming more evident that problems with energy, the environment, climate change, pandemics,[7] economic inequality, violence and war cannot be understood in isolation because they are all interconnected and interdependent. They are systemic problems with harmful consequences in several different areas and thus require corresponding systemic solutions, which can address several problems simultaneously.

Take, for example, agriculture. Transforming from chemical, large-scale industrial agriculture to organic, community-oriented, sustainable farming would contribute significantly to solving three problems: (1) It would greatly *reduce energy dependence* on fossil fuels to grow and process food. (2) Healthy, organically grown food would significantly *improve public health* because many chronic diseases are linked to diet. (3) Organic farming would contribute to *alleviating climate change* because organic soil is carbon-rich, drawing carbon dioxide (CO_2) from the atmosphere, and locking it up in organic matter.

Over the last few decades, research institutes and centers of learning in global civil society have proposed and developed hundreds of systemic solutions all over the world. A few examples include a shift from extractive to generative ownership (see Kelly 2013; also on the ecology of law, see Capra and Mattei 2015) and to learning from nature via *ecological design*; in other words, building and nurturing thriving communities that do not interfere with nature's inherent ability to sustain life.

Ecoliteracy and Ecodesign

Over billions of years of evolution, the Earth's ecosystems have evolved certain principles of ecology or organization to sustain the web of life. The survival of humanity depends on *ecological literacy* – an ability to understand the basic principles of ecology and to live accordingly (see Orr 1992, 2011, 2017; Barlow and Crabtree 2000; Stone and Barlow 2005; Goleman et al. 2012; McBride et al. 2013; Stone 2017).

Ecoliteracy is the first step on the pathway to sustainability (Worldwatch Institute 2017) and has emerged as a critical skill for everyone, including professionals in all spheres. It should be a key component of education at all levels. Learning opportunities and experiences must be available for understanding fundamental facts of life – for example, that one species' waste is another species' food; that matter cycles continually through networks of life; and that the energy driving ecological cycles flows from the sun and that biodiversity increases resilience.

Here we can find great inspiration from Leonardo da Vinci, who thought that gaining knowledge about a natural phenomenon meant connecting it with other phenomena through a similarity of patterns or, in other words, systems thinking. Leonardo felt that we would be wise to respect and learn from nature (Capra 2013: 324–25).

The second step is *ecodesign* – applying ecological knowledge to the redesign of technologies/media and social institutions so as to bridge the current gap between human design and ecologically sustainable systems of nature. Here again, Leonardo can be seen as inspirational as he understood nature's ingenuity as vastly superior to human design.

Design, in its broadest sense, consists of shaping flows of energy and matter for human purposes. Ecodesign, in particular, is a process in which human purposes are carefully meshed with the larger patterns and flows of the natural world (see Orr 2002). To practice design in such a context requires a fundamental shift in our attitude towards nature.

A branch of ecodesign is *biomimicry*, which is specifically concerned with nature-inspired designs, structures and processes. As Janine Benyus (1997) pointed out, we should value nature as model, measure and mentor. Like Leonardo more than 500 years ago, ecodesigners today study the patterns and flows in the natural world and try to incorporate the underlying principles into their design processes.

In recent years, there has been a dramatic rise in ecologically oriented design technologies/media and projects (Capra and Luisi 2014: 314–15 and Chapter 18), which integrate basic principles of ecology and therefore have some key common characteristics. They tend to be small-scale, community-oriented, diverse, energy efficient, non-polluting and job-creating. Technologies/media available today provide compelling evidence that the transition to a sustainable future is no longer merely technical nor conceptual, but also a problem of political will and leadership.

Finally, our great challenge is to restore and nurture thriving communities – communities that deeply respect, honour and cooperate with nature's complexity and inherent ability to sustain life.

NOTES

1. This chapter is adapted and expanded from a keynote address on 6 April 2017 at the University of Oregon Portland.
2. The *Oxford English Dictionary* (*OED*) builds on these frames to define system as 'An organized or connected group of things'; 'A group or set of related or associated things perceived or thought of as a unity or complex whole'; and 'A set of persons working together as parts of an interconnecting network', in addition to several other definitions (*OED Online* 2023). Also see Sillitto et al. (2017).
3. As Leonardo once noted: 'One who does not respect life does not deserve it'; see Leonardo (1986–90: Ms. I, folio 15r).
4. 'In the case of all things which have several parts and in which the totality is not, as it were, a mere heap, but the whole is something besides the parts, there is a cause; for even in bodies contact is the cause of unity in some cases, and in others viscosity or some other such quality' (Aristotle [n.d.] 1926: n.pag.; also see Johnson [2019]).
5. Translated by the author in 2021. Two other translations of the text into English are: 'the combined effect of the unity of the whole and manifoldness of the parts [...] is greater' (von Ehrenfels [1890] 1937: 523); and 'the unity of the whole and the multiciplicity [*sic*] of the parts is greater' (von Ehrenfels [1890] 1988: 123).
6. See Castellani and Gerrits (2021) for an abridged, illustrated historical timeline.
7. On the coronavirus disease 2019 (COVID-19) pandemic, see Capra ([2020] 2021).

REFERENCES

Aristotle ([n.d.] 1926), *Metaphysics, Book VIII, in The Internet Classics Archive* (trans. W. D. Ross), Oxford: Clarendon, https://web.archive.org/web/20160830212914/http://classics.mit.edu/Aristotle/metaphysics.8.viii.html. Accessed 7 February 2023.

Barlow, Zenobia and Crabtree, Margo (eds) (2000), *Ecoliteracy: Mapping the Terrain*, Berkeley, CA: Learning in the Real World and Center for Ecoliteracy.

Bateson, Gregory (1979), *Mind and Nature: A Necessary Unity*, New York: Dutton.

——— (1991), *A Sacred Unity: Further Steps to an Ecology of Mind*, New York: Cornelia & Michael Bessie Book.

Benyus, Janine M. (1997), *Biomimicry: Innovation Inspired by Nature*, New York: Morrow.

Capra, Fritjof (2007), *The Science of Leonardo: Inside the Mind of the Great Genius of the Renaissance*, New York: Doubleday.

——— (2013), *Learning from Leonardo: Decoding the Notebooks of a Genius*, San Francisco, CA: Berrett-Koehler.

——— ([2020] 2021), 'The COVID-19 pandemic: A systemic analysis', in F. Capra (ed.), *Patterns of Connection: Essential Essays from Five Decades*, Albuquerque, NM: University of New Mexico Press and High Road Books. [First published in 2020, *Agriculture and Human Values*, 37:3, pp. 665–66.]

——— (n.d.) 'Capra Course', https://www.capracourse.net. Accessed 1 June 2021.

Capra, Fritjof and Henderson, Hazel (2009), *Outside Insights: Qualitative Growth*, London: The Institute of Chartered Accountants in England and Wales (ICAEW) and Tomorrow's Company.

Capra, Fritjof and Luisi, Pier Luigi (2014), *The Systems View of Life*, Cambridge: Cambridge University Press.

Castellani, Brian and Gerrits, Lasse (2021), 'Map of the complexity sciences', *Art & Science Factory*, https://www.art-sciencefactory.com/images/MAP-WEB-May2021.jpg. Accessed 29 June 2021.

Damiano, Luisa and Luisi, Pier Luigi (2010), 'Towards an autopoietic redefinition of life', *Origins of Life and Evolution of Biospheres*, 40:2, pp. 145–49.

Descartes, René ([1641] 1984), *Meditations on First Philosophy: With Selections from the Objections and Replies* (trans. J. Cottingham, R. Stoothoff and D. Murdoch), Cambridge: Cambridge University Press.

FAO (Food and Agriculture Organization of the United Nations) (2020), *Indigenous Peoples and FAO: Allies for Sustainable Development in the Context of Climate Change*, Rome: FAO, http://www.fao.org/fileadmin/user_upload/faoweb/2020/Indigenous/FAO_Indigenous_peoples_narrative_MAY_2020_.pdf. Accessed 27 May 2021.

Feldman, John (dir.) (2019), *Symbiotic Earth: How Lynn Margulis Rocked the Boat and Started a Scientific Revolution*, Spencertown, NY: Hummingbird Films.

Goleman, Daniel, Bennett, Lisa and Barlow, Zenobia (2012), *Ecoliterate: How Educators are Cultivating Emotional, Social, and Ecological Intelligence*, Hoboken, NJ: John Wiley & Sons.

Haeckel, Ernst (1866), *Generelle Morphologie der Organismen: Allgemeine Grundzüge der Organischen Formen-Wissenschaft, Mechanisch Begründet Durch die von Charles Darwin Reformirte Descendenz-Theorie,* vol. 2, Berlin: Georg Reimer, https://www.biodiversitylibrary.org/item/52177#page/11/mode/1up. Accessed 5 April 2023.

Hilborn, Robert C. (2000), *Chaos and Nonlinear Dynamics: An Introduction for Scientists and Engineers*, 2nd ed., Oxford: Oxford University Press on Demand.

Johnson, Mark L. (2023), 'From life to mind', in J. Swartz and J. Wasko (eds), *LIFE: A Transdisciplinary Inquiry*, Bristol: Intellect Ltd., pp. 185–208.

Johnson, Steven (2019), 'Who said "The whole is greater than the sum of the parts"?', INCOSE [International Council on Systems Engineering] Chesapeake Chapter blog, https://www.incose-cc.org/blog/who-said-the-whole-is-greater-than-the-sum-of-the-parts. Accessed 27 May 2021. [First published 6 June 2019, SE Scholar blog.].

Judson, Horace Freeland (1979), *The Eighth Day of Creation: Makers of the Revolution in Biology*, New York: Simon & Schuster.

——— (2003), '"The greatest surprise for everyone": Notes on the 50th anniversary of the double helix', *The New England Journal of Medicine*, 348:17, pp. 1712–14.

Kelly, Marjorie (2013), 'The architecture of enterprise: Redesigning ownership for a great transition', *The Good Society*, 22:1, pp. 61–73.

Koohafkan, Parviz and Altieri, Miguel A. (2011), *Globally Important Agricultural Heritage Systems: A Legacy for the Future*, Rome: Food and Agriculture Organization of the United Nations, http://www.fao.org/3/i2232e/i2232e.pdf. Accessed 10 June 2021.

Krakauer, David C. (2019), *Worlds Hidden in Plain Sight: Thirty Years of Complexity Thinking at the Santa Fe Institute*, Santa Fe, NM: Santa Fe Institute Press.

Leonardo da Vinci (1986–90), *I manoscritti dell'Institut de France: Edizione in facsimile sotto gli auspici della Commissione Nazionale Vinciana e dell'Insitut de France*, diplomatic transcript and criticism by Augusto Marinoni, Florence: Giunti Barbèra.

Liddell, Henry George and Scott, Robert (1889), 'οἶκος' ('oîkos'), in *An Intermediate Greek-English Lexicon*, Oxford: Clarendon Press, http://www.perseus.tufts.edu/hopper/text?doc=Perseus:text:1999.04.0058:entry=oi)=kos. Accessed 10 June 2021.

——— (1940a), 'αὐτός' ('auto'), in *A Greek-English Lexicon* (rev. and augmented by H. S. Jones and R. McKenzie), Oxford: Clarendon Press, http://www.perseus.tufts.edu/hopper/text?doc=Perseus:text:1999.04.0057:entry=au)to/s. Accessed 12 June 2021.

——— (1940b), 'οἶκος' ('eco'), in *A Greek-English Lexicon* (rev. and augmented by H. S. Jones and R. McKenzie), Oxford: Clarendon Press, http://www.perseus.tufts.edu/hopper/text?-doc=Perseus:text:1999.04.0057:entry=oi)%3Dkos1. Accessed 10 June 2021.

——— (1940c), 'ποί-ησις' ('poiesis'), in *A Greek-English Lexicon* (rev. and augmented by H. S. Jones and R. McKenzie), Oxford: Clarendon Press, http://www.perseus.tufts.edu/hopper/text?doc=Perseus:text:1999.04.0057:entry=poi/hsis. Accessed 10 June 2021.

——— (1940d), 'σύστημ-α' ('system'), in *A Greek-English Lexicon* (rev. and augmented by H. S. Jones and R. McKenzie), Oxford: Clarendon Press, http://www.perseus.tufts.edu/hopper/text?doc=Perseus:text:1999.04.0057:entry=su/sthma. Accessed 4 June 2021.

Luisi, Pier Luigi (2003), 'Autopoiesis: A review and a reappraisal', *Naturwissenschaften*, 90:2, pp. 49–59.

——— (2016), *The Emergence of Life: From Chemical Origins to Synthetic Biology*, 2nd ed., Cambridge: Cambridge University Press.

Margulis, Lynn (1998), *Symbiotic Planet: A New Look at Evolution*, New York: Basic Books.

Margulis, Lynn and Sagan, Dorion (2002), *Acquiring Genomes: A Theory of the Origin of Species*, New York: Basic Books.

Maturana, Humberto R. and Varela, Francisco J. ([1972] 1980), *Autopoiesis and Cognition: The Realization of the Living*, Dordrecht: D. Reidel Publishing. [First published in 1972 as *De Maquinas y Seres Vivos*, Editorial Universitaria S.A.]

McBride, Brooke B., Brewer, Carol A., Berkowitz, Alan R. and Borrie, William T. (2013), 'Environmental literacy, ecological literacy, ecoliteracy: What do we mean and how did we get here?' *Ecosphere*, 4:5, pp. 1–20.

Orr, David W. (1992), *Ecological Literacy: Education and the Transition to a Postmodern World*, Albany, NY: SUNY Press.

——— (2002), *The Nature of Design: Ecology, Culture, and Human Intention*, New York: Oxford University Press.

——— (2011), *Hope Is an Imperative: The Essential David Orr*, Washington, DC: Island Press.

——— (2017), 'The life required: Political economy in the long emergency', in S. Shmelev (ed.), *Green Economy Reader: Lectures in Ecological Economics and Sustainability*, Cham: Springer, pp. 3–23.

Prigogine, Ilya (1980), *From Being to Becoming: Time and Complexity in the Physical Sciences*, San Francisco, CA: Freeman.

——— (1989), 'The philosophy of instability', *Futures*, 21:4, pp. 396–400.

Prigogine, Ilya and Stengers, Isabelle (1984), *Order Out of Chaos: Man's New Dialogue with Nature*, New York: Bantam.

Schrödinger, Erwin (1944), *What Is Life?: The Physical Aspect of the Living Cell*, Cambridge: Cambridge University Press.

Sillitto, Hillary, Dori, Dov, Griego, Regina M., Jackson, Scott, Krob, Daniel, Patrick, Arnold, Eileen, Martin, James and McKinney, Dorothy (2017), 'Defining "system": A comprehensive approach', in *INCOSE International Symposium*, 27:1, pp. 170–86.

Stengers, Isabelle (2011), *Cosmopolitics II* (trans. R. Bononno), Minneapolis, MN: University of Minnesota Press.

Stewart, Ian (2002), *Does God Play Dice?: The New Mathematics of Chaos*, 2nd ed., Malden, MA: Blackwell.

Stone, Michael K. (2017), 'Ecoliteracy and schooling for sustainability', in *EarthEd: Rethinking Education on a Changing Planet*, The Worldwatch Institute, Washington, DC: Island Press, pp. 35–47.

Stone, Michael K. and Barlow, Zenobia (eds) (2005), *Ecological Literacy: Educating Our Children for a Sustainable World (Bioneers)*, San Francisco, CA: Sierra Club Books.

'system, n.' (2023), *OED Online*, Oxford: Oxford University Press, https://www.oed.com/dictionary/system_n. Accessed 24 July 2023.

Uexküll, Jakob von (1909), *Umwelt and Innenwelt der Tiere*, Berlin: Julius Springer, https://www.biodiversitylibrary.org/item/15902. Accessed 5 April 2023.

von Ehrenfels, Christian ([1890] 1937), 'On Gestalt-qualities' (trans. M. Focht), *Psychological Review*, 44:6, pp. 521–24.

——— ([1890] 1988), 'On Gestalt qualities (1932)' (rev. B. Smith, trans. M. Focht) in B. Smith (ed.), *Foundations of Gestalt Theory,* Munich, Vienna: Philosophia, pp. 121–23, http://ontology.buffalo.edu/smith/book/FoGT/Ehrenfels_Gestalt_1932.pdf. Accessed 10 June 2021.

——— ([1890] 1960), 'Über Gestaltqualitäten (1932)', in F. Weinhandl (ed.), *Gestalthaftes Sehen: Ergebnisse und Aufgaben der Morphologie: Zum hundertjährigen Geburtstag von Christian von Ehrenfels*, Darmstadt: Wissenschaftliche Buchgesellschaft, pp. 61–63.

——— ([1890] 2007), 'Üeber Gestaltqualitäten', in *Vierteljahrsschrift für Wissenschaftliche Philosophie*, vol. 14, Leipzig: O. R. Reisland, pp. 249–92, https://gallica.bnf.fr/ark:/12148/bpt6k94141n/f254.image. Accessed 10 June 2021.

2

What is Life?[1]

Mark A. Bedau

The Fascination of Life

The surface of the Earth is teaming with life, and it is usually easy to recognize. A cat, a carrot, a germ are alive; a bridge, a soap bubble, a grain of sand are not. But it is notorious that biologists have no precise definition of what life is. Since biology is the science of life, one might expect life to figure prominently in contemporary biology and the philosophy of biology. In fact, though, few biologists or philosophers discuss life today. Many think that the definition of life has no direct bearing on current biological research (Sober 1992; Taylor 1992). When biologists do say something about life in general, they usually marginalize their discussions and produce something more thought-provoking than conclusive. But this is all changing now.

Today the nature of life has become a hot topic. The economic stakes for manipulating life are rising quickly. Biotechnologies like genetic engineering, cloning and high-throughput DNA sequencing have given us unprecedented powers to reconstruct and reshape life. One development has been the synthetic genomics reengineering of life to our arbitrary specifications (Gibbs 2004; Brent 2004). Special attention has fallen on Craig Venter's well-publicized effort to commercialize artificial cells that clean the environment or produce alternative fuels (Zimmer 2003). The wet artificial life race to synthesize a minimal artificial cell or protocell from scratch in a test tube (Szostak et al. 2001; Rasmussen et al. 2004; Rasmussen et al. 2007; Luisi [2006] 2016) also spotlights life, for the race requires an agreed-upon definition of life, and it must be one that reaches well beyond life's familiar forms. The social and ethical implications of creating protocells will also increase the need for understanding what life is. Controversies over the origin of life (Oparin 1964; Crick 1981; Shapiro 1986; Eigen [1857] 1992; Morowitz 1992; Dyson 1999; Luisi 1998) and over intelligent design (Pennock 2001) add more fuel to the fire.

Another development that highlights the nature of life is soft artificial life attempts to synthesize software systems with life's essential properties (Bedau 2003a). Soft artificial life has created remarkably life-like software systems, and they seem genuinely alive to some (Langton 1989; Ray 1992), but others ridicule the whole idea of a computer simulation being literally alive (Pattee 1989).

Hard artificial life achievements in hardware include the first widely available commercial robotic domestic vacuums (Brooks 2002) and the walking robots designed by evolution and fabricated by automated rapid prototyping (Lipson and Pollack 2000). These robots inevitably raise the question whether a device made only of plastic, silicon and steel could ever literally be alive. Such scientific developments increase uncertainty about how exactly to demarcate living things.

Biology makes generalizations about all possible forms of life. Those generalizations rest on the forms of life that actually exist. Biologists study a number of different model organisms, like *Escherichia coli* (a common bacterium), *Caenorhabditis elegans* (a nematode) and *Drosophila melanogaster* (a fruit fly). Picking model organisms that are as different as possible best illustrates the possible forms that life can take, and thus enables the widest generalizations about all life. But the evolution of all life on Earth still is only one large, interconnected, evolving biosphere. Thus, evolutionary biology generalizations about evolution now rest on a sample size of one. Maynard Smith (1992) pointed out that artificial life helps alleviate this problem. Actual life comes in an amazing diversity of forms. But they are just a tiny fraction of all possible forms of life. Anytime we can synthesize a system in software, hardware or wetware that exhibits life's core properties, we have a great opportunity to expand our empirical understanding of what life is.

There are three giants in the history of philosophy who advanced views about life, and their views still echo in contemporary discussion. In *De Anima*, Aristotle expressed the view that life is a nested hierarchy of capacities, such as metabolism, sensation and motion. This nested hierarchy of capacities corresponds to Aristotle's notion of 'soul' or mental capacities, so Aristotle essentially linked life and mind. As part of his wholesale replacement of Aristotelian philosophy and science, René Descartes supplanted Aristotle's position with the idea that life is just the operation of a complex but purely materialistic machine. Descartes thought that life fundamentally differed from mind, which he thought was a mode of consciousness. Descartes sketched the details of his mechanistic hypothesis about life in his *Treatise on Man*. Some generations later, Immanuel Kant's *Critique of Judgement* struggled to square Descartes's materialistic perspective with life's distinctive autonomy and purpose.

Understanding the nature of life is no mere armchair exercise. It involves investigating something real, natural and extremely complex, with huge potential creativity and power to change the face of the Earth (Margulis and Sagan 1995). This investigation will by necessity be interdisciplinary, and it will survey an almost astonishing variety of perspectives on life. Interesting and subtle hallmarks like holism, homeostasis, teleology and evolvability are thought to characterize life. But a precise definition of life remains elusive, partly because of borderline cases such as viruses and spores, and more recently artificial life creations. To add more complication, life figures centrally in a range of philosophical puzzles involving important philosophical issues such as emergence, computation and mind. So, a diversity of views about life can be expected. Some employ familiar philosophical theories like functionalism. Others use biochemical or genetic explanations and mechanisms. Still, others emphasize processes like metabolism and evolvability. The sheer diversity of views about life is itself interesting and deserves an explanation.

The Phenomena of Life

Life has various hallmarks and borderline cases and presents a variety of puzzles. The rest of this chapter is mainly devoted to explaining these phenomena.

A striking fact of life is the characteristic and distinctive *hallmarks* that it exhibits. These hallmarks are usually viewed as neither necessary nor sufficient conditions for life; they are nonetheless typical of life. Different people provide somewhat different lists of these hallmarks (e.g. Maynard Smith 1986; Farmer and Belin 1992; Mayr 1997; Gánti 2003). But most lists of hallmarks substantially overlap. Another notable point is that the hallmarks itemized on the lists are strikingly heterogeneous. A good illustration is Gánti's hallmarks (or 'criteria', as he calls them).

Gánti's hallmarks fall into two categories: real (or absolute) and potential. Real life criteria specify the necessary and sufficient conditions for life in an individual living organism. Gánti's (2003) proposed real life criteria are these:

1. *Holism*: An organism is an individual entity that cannot be subdivided without losing its essential properties. An organism cannot remain alive if its parts are separated and no longer interact.
2. *Metabolism*: An individual organism takes in material and energy from its local environment, and chemically transforms them. Seeds are dormant and so lack an active metabolism, but they can become alive if conditions reactivate their metabolism. For this reason, Gánti makes a four-part distinction between things that are alive, dormant, dead or not the kind of thing that ever could be alive.
3. *Inherent stability*: An organism maintains homeostatic internal processes while living in a changing environment. By changing and adapting to a dynamic

external environment, an organism preserves its overall structure and organization. This involves detecting changes in the environment and making compensating internal changes, with the effect of preserving overall internal organization.

4. *Active information-carrying systems*: A living system must store information that is used in its development and functioning. Children inherit this information through reproduction because the information can be copied. Mistakes in information transfer can 'mutate' this information, and natural selection can sift through the resulting genetic variance.
5. *Flexible control*: Processes in an organism are regulated and controlled so as to promote the organism's continued existence and flourishing. This control involves an adaptive flexibility and can often improve with experience.

In contrast to these 'real' criteria, Gánti also proposed 'potential' life criteria. A living individual organism can fail to possess life's potential criteria. The defining feature of potential life criteria is that if enough organisms exhibit them, then life can populate a planet and sustain itself. Gánti proposed three:

1. *Growth and reproduction*: Old animals and sterile animals and plants are all living, but none can reproduce. So, the capacity to reproduce is neither necessary nor sufficient for being a living organism. But due to the mortality of individual organisms, a population can survive and flourish only if some organisms in the population reproduce. In this sense, growth and reproduction are what Gánti calls a 'potential' rather than 'real' life criterion.
2. *Evolvability*: 'A living system must have the capacity for hereditary change and, furthermore, for evolution, i.e. the property of producing increasingly complex and differentiated forms over a very long series of successive generations' (Gánti 2003: 79). Since what evolves over time are not individual organisms but populations of them, we should rather say that living systems can be members of a population with the capacity to evolve. It is an open question today exactly which kinds of biological populations have the capacity to produce increasing complexity and differentiation.
3. *Mortality*: Living systems are mortal. This is true even of clonal asexual organisms because death can afflict both individual organisms as well as the whole clone. Systems that could never live cannot die, so death is a property of things that were alive.

Gánti's life criteria and other lists of life's hallmarks always reflect and express some preconceptions about life. This might seem to beg the question of what life is. Any non-arbitrary list of life's hallmarks was presumably constructed by someone using some criterion to rule examples in or out. But where did this criterion come from, and what assures us it is correct? Why should we be confident that any hallmarks that fit it reveal the true nature of life? Thus, lists of life's hallmarks cannot be the final word on what is life. As we learn more about life, our preconceptions change, evolve and mature. So, we should expect the same of our lists of life's hallmarks.

Another characteristic of living things is *borderline cases*, much more so than most other biological concepts. Familiar examples are viruses and prions, which self-replicate and spread even though they have no independent metabolism. Dormant seeds or spores are another kind of borderline case, the most extreme version of which might be bacteria or insects that are frozen. There are also cases that seem clearly not to be alive but yet possess the characteristic properties of living systems. Hardly anyone considers a candle flame to be alive, but by preserving its form while its constituent molecules are constantly changing, it has something like a metabolism (Maynard Smith 1986). Populations of microscopic clay crystallites growing and proliferating are another kind of borderline example, especially because they can in appropriate circumstances undergo natural selection (Bedau 1991). So is a forest fire that is spreading ('reproducing'?) from tree to tree at its edge, somewhat like the edge of a growing population of bacteria. A further kind of borderline case consists of superorganisms, which are groups of organisms, such as eusocial insect colonies, that function like a single organism. Although this is controversial, some biologists think that superorganisms should themselves be thought of as living organisms. Another kind of borderline case consists of soft artificial life creations like Tierra. Tierra is software that creates a spontaneously evolving population of computer programs that reproduce, mutate and evolve in computer memory. Tierra's inventor thinks that Tierra is literally alive (Ray 1992). This would radically violate the ordinary concept of life that most of us have. One final category of borderline cases consists of complex adaptive systems found in nature, such as financial markets or the World Wide Web. These exhibit many of the hallmarks of life, and some think that the simplest and most unified explanation of the entire range of phenomena of life is to consider these natural complex adaptive systems to be literally alive (Bedau 1996, 1998).

Puzzles about Life

A third characteristic of life is that it generates a number of *puzzles*. Seven puzzles are briefly reviewed below. Any account of life should explain the origin of these puzzles; more important, it should resolve the puzzles. Some puzzles might result simply from confusion, but others are open questions about a fundamental and fascinating aspect of the natural world.

Origins: How does life or biology arise from non-life or pure chemistry? What is the difference between a system that is undergoing merely chemical evolution, in which chemical reactions are continually changing the concentrations of chemical species, and a system that contains life? Where is the boundary between living and merely physico-chemical phenomena? How could a naturalistic process bridge the boundary, in principle or in practice? Daniel C. Dennet argues that Darwin's scheme of explanation solves this problem by appealing to 'a finite regress, in which the sought-for marvelous property (life, in this case) was acquired by slight, perhaps even imperceptible, amendments or increments' (1995: 200).

Emergence: B properties are said to emerge from A properties when the B properties both depend on, and are autonomous from, the A properties. Different kinds of dependence and autonomy generate different grades of emergence (Bedau 2003b). One is the 'strong' emergence involving in principle irreducible top-down causal powers. An example might be consciousness or qualia in the philosophy of mind (Kim 1999). If the A and B properties are simultaneous, the emergence of B from A is *synchronic*. It concerns what properties exist at a moment. Those properties might be changing, but the relationship between the A and B properties at an instant is a static snapshot of that dynamic process. By contrast, if the A properties precede the B properties, and the B properties arise over time from the A properties, then the emergence of B from A is *dynamic*. Life is the paradigm case of a dynamic form of 'weak' emergence, one that concerns macro properties that are unpredictable or underivable except by observing the process by which they are generated, or by observing a simulation of it (Bedau 1997, 2003b).

Hierarchy: Various kinds of structural hierarchies characterize life. Each organism has a hierarchical internal organization, and the relative complexity of organizations of different kinds of organisms forms another hierarchy. The simplest organisms are prokaryotic cells, which have relatively simple components. More complicated are eukaryotic cells containing complex organelles and a nucleus. Multicellular organisms are even more complicated; they have constituents (individual cells) that also are individual living entities (e.g. they can be kept alive by themselves). In addition, mammals have complex internal organs (such as the heart) that can be harvested and kept alive when an organism dies, and then surgically implanted into another living organism. Two questions arise here. First, why does life tend to generate and encompass such hierarchies? This question applies both to the hierarchy in complexity that spans all organisms together and also to the organizational hierarchy found within each individual living organism. With regard to the latter, a second question arises. Organisms are our paradigm case of something that is alive, but we also refer to organs and individual cells as alive.

For example, apoptosis is an important process by which living cells in an organism undergo programmed death, and hospitals strive to keep certain organs alive after someone dies so that they are available to be transplanted into someone else. This raises the question whether a mammal, its heart, and the cells therein are each alive in the same or different senses.

Continuum: Can things be more or less alive? Is life a black-or-white Boolean property, or a continuum property with many shades of gray? Common sense leans towards the Boolean view: a rabbit is alive and a rock isn't, end of story. But there are borderline cases like viruses that are unable to replicate without a host. And spores or frozen bacteria remain dormant and unchanging indefinitely but then come back to life when conditions become favorable. Are viruses and spores fully alive? Furthermore, when the original life forms emerged from a pre-biotic chemical soup, they differed very little from their non-living predecessors. Some conclude that there is a continuum of more or less alive things (e.g. Cairns-Smith 1985; Emmeche 1994; Dennett 1995). An alternative is to accept a sharp distinction between life and non-life, but allow that a small step could cross it. The four-fold distinction between things that are (i) inanimate and forever incapable of living, (ii) now living, (iii) dead but formerly living or (iv) dormant but capable of becoming alive again helps explain away some borderline cases by reclassifying them (e.g. seeds and spoors are dormant and not currently living). But it does not fully resolve the continuum puzzle, for there are borderline cases in the four-fold distinction, such as between being dead and alive.

Strong artificial life: Artificial life software and hardware raise the question whether our computer creations could ever literally be alive (Langton 1989; Pattee 1989; Sober 1992; Emmeche 1992; Olson 1997). On the one hand, certain distinctive carbon-based macromolecules play a crucial role in the vital processes of all known living entities; on the other hand, much of artificial life seems to presuppose that life can be realized in a suitably programmed computer. It is important to distinguish two questions here. The first concerns whether it is possible to implement a software system or create a hardware device (such as a robot) that is literally alive. The challenge here is scientific and technical. It concerns whether we can arrange the appropriate materials in such a way that they realize the appropriate set of processes. The philosophically controversial question is *in virtue of what* would such a computer or robot be alive. The 'strong' artificial life position about software is that an instantiation of artificial life software could literally be alive. There is an analogous strong position about 'hard' artificial life hardware constructions, and also about 'wet' artificial life laboratory constructions. These strong positions contrast with the uncontroversial 'weak' positions that computer models, hardware constructions and wet lab productions are useful for understanding living systems. Furthermore, the strong version of wet artificial life is intuitively plausible; we usually accept that something synthesized from scratch in the lab could be literally alive. So, the controversy about strong artificial life concerns only soft and hard artificial life.

Mind: Another puzzle is whether there is any intrinsic connection between life and mind. Plants, bacteria, insects and mammals, for example, have various kinds of sensitivity to the environment, various ways in which this environmental sensitivity affects their behaviour and various forms of inter-organism communication (e.g. Dennett 1997). These are all forms of intelligent behaviour, and the relative sophistication of these 'mental' capacities seems to correspond to, and explain the relative sophistication of, those forms of life. So, it is natural to ask whether life and mind have some deep connection. Evolution creates a genealogical connection between life and mind, of course, but they would be much more deeply unified if Randall D. Beer is right that 'it is adaptive behavior, the [...] ability to cope with the complex, dynamic, unpredictable world in which we live, that is, in fact, fundamental [to intelligence itself]' (Beer 1990: 11; see also Maturana and Varela 1987; Godfrey-Smith 1994; Clark 1997). Since all forms of life must cope in one way or another with a complex, dynamic and unpredictable world, perhaps this adaptive flexibility inseparably connects life and mind.

Accounts of Life

Let us now consider the main kinds of accounts of life in circulation today. By an account of life, I mean an attempt to state universal characteristics of all forms of life. I will briefly outline a number of them and indicate some motivations, strengths and limitations.

First, consider the skeptical position that the nature of life is largely irrelevant to biology (Sober 1992; Taylor 1992). The reason for this skepticism is that biologists can continue with their biological research whether or not life can be adequately defined, and no matter what view of life prevails in the end. One must admit, though, that developments such as attempting to make minimal artificial cells from scratch does require scientists to start to articulate their views about what is essential to life, even if these views fall short of a precise definition. So, the issue is no longer irrelevant, if it ever was. For one can set out to construct a minimal form of life only if one has at least a working hypothesis about life's minimally sufficient conditions. Otherwise, one would have no idea what to try to make.

A second form of skepticism is the view that life cannot be captured by necessary and sufficient conditions but instead consists of just a cluster of things sharing only a Wittgensteinian family resemblance. Different forms of life might share various properties or hallmarks, but the individual properties in the cluster each have exceptions. The properties would typically be possessed by living organisms but they would not be strictly necessary or sufficient. J. Doyne Farmer and Alletta d'A Belin (1992) list eight hallmarks: process; self-reproduction; information storage of self-representation; metabolization; functional interactions with the environment; interdependence of parts; stability under perturbations; and membership in a population with the ability to evolve. They then explain that a cluster conception of life arises from their despair at finding anything more precise than this list of hallmarks.

> There seems to be no single property that characterizes life. Any property that we assign to life is either too broad, so that it characterizes many non-living systems as well, or too specific, so that we can find counter-examples that we intuitively feel to be alive, but that do not satisfy it.
>
> (Farmer and Belin 1992: 818; see also Taylor 1992)

The cluster conception amounts to skepticism about the possibility of a unified theory of life.

An advantage of the cluster conception is that it offers a natural explanation for borderline cases. All cluster concepts inevitably have borderline cases. A characteristic of the cluster conception is that it cannot explain why forms of life are unified by one set of hallmarks rather than another. The cluster view must simply accept the hallmarks as given, and then identify the cluster with those hallmarks. Thus, this view can identify life's hallmarks only post hoc; it cannot predict or explain the hallmarks. Those who think that there should be an explanation for life's hallmarks will therefore find the cluster conception unsatisfying.

A third form of skepticism is directed against whether life is a natural kind. Evelyn Fox Keller (2002) says that life is a human kind, not a natural kind. Like the cluster concept view, the view that life is a human kind easily explains borderline cases. Since the concept of life changes with the progress of science and technology, one should expect its boundaries to change, thus creating borderline cases. The view also provides some general ammunition against life's puzzles, for a mutable human construct can be expected to spawn puzzles. Keller's argument that life is a human kind suggests that the present presupposition that life has an essence arose only two centuries ago, that the search for life's essence is driven by attempts to make life from non-life, and this tends to dissolve the boundary between life and non-life, and that the new concepts generated by scientific and technological progress violate older taxonomies like the life/non-life distinction (Keller 2002).

There are problems with all of these arguments. First, all modern scientific concepts like matter and energy arose at some point in human history and have evolved since then. So contingent, datable recent origin does not show that a kind is a human kind, unless it does so at one fell swoop for all scientific concepts. Second, bridging the gap in the laboratory from the non-living to the living need not dissolve the boundary between life and non-life, any more than making the first airplane dissolved the distinction between flying and not flying. Remember that we are seeking the nature of life, not just current conceptions of life.

Another way to answer the question 'What is life?' is to provide a taxonomy of living things. This is taking the question as a request for an exhaustive list of the kinds of things on the Earth that are alive. This is an interesting historical question, but one riddled with contingencies. The taxonomy is necessarily silent about forms of life that could have existed but did not. This illustrates the taxonomy view's chauvinism in assuming that life as we know it exhausts what life is or could be. Unrelated life forms that exist on an extra-terrestrial site like Europa are absent from all such taxonomies. In any case, we should welcome having our taxonomies violated by scientific and technological progress, for that is how we learn.

Some have given a biochemical definition of life. They attempt to specify the biochemical properties that any form of life must have, given the general constraints set by physics and chemistry (Pace 2001; Benner et al. 2004). This includes thermodynamic limits, energetic limits, material limits and even geographical limits. The features in a biochemical definition are sometimes called life's biochemical 'universals'. A biochemical definition always presupposes a prior account of life; it states the physical, chemical and biological possibilities for any biochemical system meeting that prior account of life. The biochemical definitions of Norman R. Pace (2001) and Steven A. Benner et al. (2004) presuppose a definition of life based on evolution, so Pace and Benner dwell on the biochemical universals for genetic capacities and emphasize molecules like DNA that can store and transmit information between generations. Biochemical definitions are often myopic and presume that all possible life forms are quite similar to the familiar ones. One could imagine starting with a different conception of life, such as the view based on metabolism, and ending up emphasizing different biochemical universals, such as those that enable open systems to retain their structure in the face of the second law of thermodynamics.

A genetic instance of a biochemical definition of life is J. Craig Venter's genomic definition of life as a minimal genome sufficient to support life (Hutchison et al. 1999). This view inherits the limitations of biochemical definitions. The genomic definition captures the simplest known set of genes sufficient for life. It does not capture genes found in every life form, for the same essential life functions can be achieved by different genes. Many people would question the molecular definition's limitation to genetic properties, on the grounds that life centrally involves much more than genes (Cho et al. 1999).

Everyone in the community of scientists making artificial cells from scratch or 'protocells' admits that the nature of life is controversial and contentious, but almost all share the goal of making a self-contained system that metabolizes and evolves (e.g. Rasmussen et al. 2004). That is, an artificial cell is viewed as any chemical system that chemically integrates three processes. The first is the process of assembling and living inside some kind of container, such as a lipid vesicle. The second is the metabolic processes that repair and regenerate the container and its contents, and enable the whole system to reproduce. Those chemical processes are shaped and directed by a third chemical process involving encoded information about the system stored in the system ('genes'). Errors ('mutations') can occur when this information is reproduced, so the systems can evolve by natural selection. The integrated-triad view of life requires that the chemical processes of containment, metabolism and evolution support and enable each other so that there is functional feedback among all three. This view of protocellular life as an integrated triad of functions accepts any biochemical realization of the triad as genuine life.

The past generation of the philosophy of mind has been dominated by functionalism: the view that mental beings are a certain kind of input–output device and that having a mind is simply having a set of internal states that causally interact (or 'function') with respect to each other and with respect to environmental inputs and behavioral outputs in a certain characteristic way. Functionalism with respect to life is the analogous view that being alive is simply realizing a network of processes that interact in a certain characteristic way. Some processes (such as information processing, metabolization, purposeful activity) operate within the organism's lifetime; other processes (such as self-reproduction and adaptive evolution) operate over many generations. These processes are always realized in some material substratum, but the substratum's material nature is irrelevant so long as the *forms* of the processes are preserved. For these reasons, functionalism is an attractive position with respect to life. Chris Langton's defense of artificial life is a classic statement of the case for functionalism with respect to life: 'Life is a property of *form*, not *matter*, a result of the organization of matter rather than something that inheres in the matter itself' (Langton 1989: 41).

> The big claim is that a properly organized set of artificial primitives carrying out the same functional roles as the biomolecules in natural living systems will support a process that is 'alive' in the same way that natural organisms are alive. Artificial Life will therefore be genuine life – it will simply be made of different stuff than the life that has evolved here on Earth.
>
> (Langton 1989: 33)

We might be unsure about the details of the processes that are definitive of life, and we might wish to reserve judgment about whether artificial life creations are genuinely alive. Nevertheless, it is hard to deny Langton's point that life's characteristic processes, metabolism, information processing and self-reproduction could be realized in a wide and potentially open-ended range of materials. Thus, the prospects for some form of functionalism with respect to life seem bright.

The main challenge for functionalism with respect to mind concerns consciousness and qualia. It is worth noting that functionalism about life does not face any analogous problems. Another challenge for functionalism with respect to mind is to explain how people's mental states are meaningful or have semantic content. Darwinian natural selection provides a naturalistic explanation of many biological functions of structures in evolved forms of life. This biological functionality gives the internal states of living creatures a kind of meaning or semantic content so that we can speak of a creature trying to find food for nourishment. Many philosophers are optimistic that the meaning problem in functionalism with respect to mind will be solved by some analogous Darwinian explanation of the biological function of mental states (e.g. Dennett 1995).

Another apparent threat to functionalism with respect to life is the suggestion that the processes involved in life are, in some relevant sense, unformalizable or non-computational (e.g. Emmeche 1992). Bedau (1999) thinks that the apparent non-computational quality of life can be explained. Advantageous traits that arise through mutations tend, ceteris paribus, to persist and spread through the population. Furthermore, trait frequencies in the population will tend, ceteris paribus, to change in a way that is generally apt for the population in its exogenously changing environment. These dynamical patterns in trait frequencies emerge as a statistical pattern from the micro-level contingencies of natural selection, mutation, drift, etc. Bedau argues that there is often a special kind of suppleness in these patterns. Such patterns in trait frequencies are not precise and exceptionless universal generalizations, but instead hold, for the most part, only ceteris paribus. Furthermore, those regularities have exceptions that sometimes 'prove the rule' in the sense that they are a byproduct of trying to achieve some deeper adaptive goal. For example, Bedau describes a system in which mutation rates can evolve and shows that the mutation rates tend to evolve so as to keep the population's gene pool at the 'edge of disorder'; but this regularity has exceptions, some of which are due to the operation of a deeper regularity about mutation rates evolving so as to optimally balance evolutionary 'memory' and 'creativity' (for details, see Bedau 1999). In this sort of way, supple regularities reflect an underlying capacity to respond appropriately in an open-ended variety of contexts. This explains a certain kind of unformalizability of life processes, though it also allows life to be captured in appropriate computer models.

Functionalism leaves unanswered exactly which processes play what role in the functional characterization of life. Persisting in the face of the second law of thermodynamics by means of metabolism is the defining process of life according to Erwin Schrödinger's influential account:

> When is a piece of matter said to be alive? When it goes on 'doing something', moving, exchanging material with its environment, and so forth, and that for a much longer period than we would expect an inanimate piece of matter to 'keep going' under similar circumstances [...] It is by avoiding the rapid decay into the inert state of 'equilibrium' that an organism appears so enigmatic; [...] How does the living organism avoid decay? The obvious answer is: By eating, drinking, breathing and (in the case of plants) assimilating. The technical term is metabolism.
>
> (Schrödinger [1944] 1969: 74–76)

Metabolism-centered views of life attract many (Margulis and Sagan 1995; Boden 1999). They are closely related to views that focus on autopoiesis (Varela et al. 1974; Maturana and Varela 1987).

The view that metabolism is life's central process has some clear advantages, such as explaining our intuition that a crystal is not alive (there is a metabolic flux of molecules only at the crystal's edge, not inside it). Also, the fact that metabolism is needed to combat entropy implies that metabolism is at least a necessary condition of all physical life forms. Metabolism also naturally explains the fourfold distinction between the non-living, living, dead and dormant. The non-living cannot metabolize in principle, and the living are now metabolizing. The dead were once living and metabolizing, but now they are decaying. The dormant were once living but now do not metabolize but they could resume metabolizing given the right circumstances.

The main drawback of metabolism as an all-encompassing account of life is that many metabolizing entities seem intuitively not to be alive or to involve life in any way. Standard examples include a candle flame, a vortex and a convection cell (Maynard Smith 1986; Bagley and Farmer 1992). Such examples by themselves do not prove conclusively that metabolism is insufficient for life, for pre-theoretic intuitive judgments can be wrong. The question is whether on balance metabolism adequately explains life's hallmarks and resolves life's puzzles.

Some think that the central feature underlying all life is the open-ended evolutionary process of adaptation. The idea is that what distinguishes life is its automatic and open-ended capacity (within limits) to adapt appropriately to unpredictable changes in the environment. From this perspective, what is distinctive of life is the way in which adaptive evolution automatically fashions new and intelligent strategies for surviving and flourishing as local contexts change. Maynard

Smith (1975: 96f; see also Mayr 1982; Cairns-Smith 1985) succinctly explains the justification for the view that life crucially depends on the evolutionary process of adaptation:

> We shall regard as alive any population of entities which has the properties of multiplication, heredity and variation. The justification for this definition is as follows: any population with these properties will evolve by natural selection so as to become better adapted to its environment. Given time, any degree of adaptive complexity can be generated by natural selection.

These remarks suggest how the process of adaptive evolution could explain life's hallmarks, borderline cases and puzzles (see Bedau 1998).

There are a few characteristic criticisms of such evolution-centered views. One is purported counterexamples of creatures that are alive but cannot give birth (mules, old people, etc.) and so cannot contribute to the process of evolution. The typical response is to require that organisms be produced by an evolutionary process, but not that they necessarily can affect further evolution. Another kind of purported counterexample is a clearly non-living system, such as a population of clay crystallites or a free market economy, which evolves by natural selection. Some think that we should accept these unintuitive examples because evolution-centered views provide such a compelling explanation of life's hallmarks, borderline cases and puzzles (e.g. Bedau 1998).

Not all of these positions are competing; many are consistent. For example, functionalism is consistent with the protocell integrated-triad account of minimal life. Also, accounts of the nature of life each entail a biochemical characterization of life, and many accounts of life overlap. The problem of understanding life is to identify exactly which of these accounts is true.

The Problem of Understanding Life

How should we compare and evaluate accounts of the nature of life? One straightforward answer is simply to see how well each explains the phenomena of life. This amounts to doing three things: explaining life's hallmarks, explaining the borderline cases and resolving the puzzles about life. The problem of understanding life is the problem of explaining these three things.

One initial difficulty is confusion about what question is at stake. Some investigations think the key test for any account of life is to fit it with our pre-theoretic intuitions about which things are alive and which are not (e.g. Boden 1999). But one should ask why we should emphasize such intuitions.

A good theory of life might make us reconceptualize and recategorize life. This might change our attitudes about exactly which cases are the ones in which life is present. Thus, although they have some weight, our pre-theoretic intuitions are not inviolable.

One could also ask about the *meaning of the word* 'life' in today's English. But the stereotypes associated with the term 'life' are commonplaces and reflect the lowest common denominator of our current shared picture of life. So, we are not likely to learn much about life by relying on what 'life' means.

Nor are we likely to learn much by analysis of the *concept* of life. As with the meaning of 'life', our current concept of life will reflect our current understanding of life. If we want to learn the real nature of the phenomena with life's hallmarks, borderline cases and puzzles, we should study the natural phenomena themselves, not our words or concepts. And we should expect our understanding of the phenomena of life to evolve and sometimes improve.

Explaining the phenomena of life involves at least a rough view of life's essence or nature, and perhaps even a rough definition of life. Scientific essentialism, originating from Saul Kripke (1980), is the philosophical view that the essence of natural kinds like water and gold is their underlying causal powers, which are discovered by empirical science (see Bealer 1987). The essence of substances like water and gold turns out to be their underlying chemical composition. Life, on the other hand, is a certain kind of flexible process, not a fixed chemical substance. So unlike water or gold, life's nature would presumably be captured by the characteristic network of processes (such as metabolism, reproduction and sensation) that explains its characteristic causal powers. In this regard, life is more like heat, which is a certain process in matter (high molecular kinetic energy). A specific temperature (say 23°C) is a specific kind of process that can occur in all kinds of matter. Life is also a kind of process that can occur in different kinds of material, but unlike temperature not *all* kinds of material can be alive. Mapping the biochemical constraints on the kinds of substances that could instantiate life yields a biochemical definition of life (recall above). Note that scientific essentialism about life might be true, even if contemporary science has reached no consensus about life. Scientific essentialism is a philosophical view about the method by which life's essence would be discovered; it does not include a view about the specific content of that essence. The details of the scientific essentialist definition of life might need to await further scientific progress.

It is unclear whether living things have any features that make them essentially alive. In Dennett's opinion, for example, the life/non-life distinction is a matter of degree and life is too 'interesting' to have an essence (1995: 201). In fact, contemporary biology and philosophy of biology thoroughly embrace a Darwinian anti-essentialism according to which species have no essence and their members share no necessary and sufficient properties. Instead, the similarities among the members of a species are only statistical. Species are no more than a cloud or clump in an abstract possible feature space. Although some sub-regions of possible feature space are unoccupied because they are maladaptive, it is an accident exactly which of the acceptable sub-regions are occupied. No sub-regions are any more natural than any other; none are privileged by fixed and immutable Platonic essences. The generalization of this anti-essentialism probably helps account for why so many philosophers are attracted to the cluster concept of life, for that seems like a direct consequence of anti-essentialism.

Darwinian anti-essentialism is directed against a narrow notion of essence that embraces exceptionless necessary and sufficient conditions and excludes borderline cases. Borderline cases are one of the hallmarks of life, so the nature of life must be broad and flexible enough to embrace them. One could embrace Darwinian anti-essentialism but still accept scientific essentialism about life. On this view, the 'essence' of life would be whatever process explains the phenomena of life, including life's hallmarks, borderline cases and puzzles. Life would not be defined by exceptionless conditions but empirically. It is unfortunate that contemporary philosophical terminology obscures that Darwinian anti-essentialism and scientific essentialism about life are compatible.

Carol E. Cleland and Christopher F. Chyba (2002) argue that it is too early to formulate definitions of life because our current understanding of life is too limited. They conclude that we should put off formulating definitions until scientists can tell much more about the different forms that life could take. Now might nevertheless be the right time to construct tentative and testable hypotheses about the phenomena of life. These hypotheses will likely be false, but they can aid our search for better theories (Wimsatt 1987). When we have good theories of life in hand, we can extract their implied definitions of life. So, the quest for the definition of life is better recast as the quest for the nature of life (Bedau and Cleland 2010).

Life is one of the most fundamental and complex aspects of nature. So, accounts of life are rich and interesting, with a complicated structure. They come in many forms, including skepticism, detailed biochemical and molecular descriptions, and abstract functionalism, and they emphasize fundamental biological processes like metabolism and evolution. The criteria for evaluation include their ability to explain life's hallmarks and borderline cases and their ability to resolve the puzzles about life. Many of the main accounts of life still lack substantial development and careful evaluation along a number of these dimensions. Thus, the problem of understanding life is still wide open (Varela et al. 1974).

NOTE

1. Republished with permission of John Wiley & Sons, from Mark Bedau (2008), 'What is life?', in Sajptra Sarlar amd Amua Plutynski (eds), *A Companion to the Philosophy of Biology*, permission conveyed through Copyright Clearance Center, Inc.

REFERENCES

Bagley, Richard J. and Farmer, J. Doyne (1992), 'Spontaneous emergence of a metabolism', in C. Langton, C. Taylor, J. D. Farmer and S. Rasmussen (eds), *Artificial Life II*, Redwood City, CA: Addison-Wesley, pp. 93–140.

Bealer, George (1987), 'The philosophical limits of scientific essentialism', in J. Tomberlin (ed.), *Philosophical Perspectives*, Atascadero, CA: Ridgeway, pp. 289–365.

Bedau, Mark A. (1991), 'Can biological teleology be naturalized?', *The Journal of Philosophy*, 88, pp. 647–55.

——— (1996), 'The nature of life', in M. Boden (ed.), *The Philosophy of Artificial Life*, New York: Oxford University Press, pp. 332–57.

——— (1997), 'Weak emergence', in J. Tomberlin (ed.), *Philosophical Perspectives: Mind, Causation, and World*, vol. 11, Oxford: Blackwell, pp. 375–99.

——— (1998), 'Four puzzles about life', *Artificial Life*, 4, pp. 125–40.

——— (1999), 'Supple laws in biology and psychology', in V. Hardcastle (ed.), *Where Biology Meets Psychology: Philosophical Essays*, Cambridge, MA: MIT Press, pp. 287–302.

——— (2003a), 'Artificial life: Organization, adaptation, and complexity from the bottom up', *Trends in Cognitive Science*, 7:11, November, pp. 505–12.

——— (2003b), 'Downward causation and autonomy in weak emergence', *Principia Revista Inernacional de Epistemologica*, 6:1, pp. 5–50.

Bedau, Mark A. and Cleland, Carol E. (2010), *The Nature of Life: Classical and Contemporary Perspectives from Philosophy and Science*, Cambridge: Cambridge University Press.

Beer, Randall D. (1990), *Intelligence as Adaptive Behavior: An Experiment in Computational Neuroethology*, Boston, MA: Academic Press.

Benner, Steven A, Ricardo, Alonso and Carrigan, Matthew A. (2004), 'Is there a common chemical model for life in the universe?', *Current Opinion in Chemical Biology*, 8, pp. 679–89.

Boden, Margaret A. (1999), 'Is metabolism necessary?', *British Journal for the Philosophy of Science*, 50, pp. 231–48.

Brent, Roger (2004), 'A partnership between biology and engineering', *Nature Biotechnology*, 22:10, pp. 1211–14.

Brooks, Rodney (2002), *Flesh and Machines: How Robots Will Change Us*, New York: Pantheon.

Cairns-Smith, Graham (1985), *Seven Clues to the Origin of Life*, Cambridge: Cambridge University Press.

Cho, Mildred K., Magnus, David, Caplan, Arthur L., McGee, Daniel and The Ethics of Genomics Group (1999), 'Ethical considerations in synthesizing a minimal genome', *Science*, 286:5447, pp. 2087–90.

Clark, Andy (1997), *Being There: Putting Brain, Body, and World Together Again*, Cambridge, MA: MIT Press.

Cleland, Carol E. and Chyba, Christopher F. (2002), 'Defining "life"', *Origins of Life and Evolution of the Biosphere*, 32, pp. 387–93.

Crick, Francis (1981), *Life Itself: Its Origin and Nature*, New York: Simon & Schuster.

Dennett, Daniel C. (1995), *Darwin's Dangerous Idea*, New York: Simon & Schuster.

——— (1997), *Kinds of Minds: Towards an Understanding of Consciousness*, New York: Basic Books.

Dyson, Freeman (1999), *Origins of Life*, 2nd ed., Cambridge: Cambridge University Press.

Eigen, Manfred ([1857] 1992), *Steps Toward Life: A Perspective on Evolution*, Oxford: Oxford University Press.

Emmeche, Claus (1992), 'Life as an abstract phenomenon: Is artificial life possible?', in F. Varela and P. Bourgine (eds), *Towards a Practice of Autonomous System*, Cambridge, MA: Bradford Books and MIT Press, pp. 466–74.

——— (1994), *The Garden in the Machine: The Emerging Science of Artificial Life*, Princeton, NJ: Princeton University Press.

Farmer, J. Doyne and Belin, Alletta d'A (1992), 'Artificial life: The coming evolution', in C. Langton, C. Taylor, J. D. Farmer and S. Rasmussen (eds), *Artificial Life II*, Redwood City, CA: Addison-Wesley, pp. 815–40.

Gánti, Tibor (2003), *The Principles of Life, with a commentary by James Grisemer and Eörs Szathmáry*, New York: Oxford University Press.

Gibbs, W. Wayt (2004), 'Synthetic life', *Scientific American*, 290:5 (April), pp. 75–81.

Godfrey-Smith, Peter (1994), 'Spencer and Dewey on life and mind', in R. Brooks and P. Maes (eds), *Artificial Life IV*, Cambridge, MA: MIT Press and Bradford Books, pp. 80–89.

Hutchison III, Clyde A., Peterson, Scott N., Gill, Steven R., Cline, Robin T., White, Owen, Fraser, Claire M., Smith, Hamilton O. and Venter, J. Craig (1999), 'Global transposon mutagenesis and a minimal Mycoplasma genome', *Science*, 286:5447 (10 December), pp. 2165–69.

Keller, Evelyn Fox (2002), *Making Sense of Life: Explaining Biological Development with Models, Metaphors, and Machines*, Cambridge, MA: Harvard University Press.

Kim, Jaegwon (1999), 'Making sense of emergence', *Philosophical Studies*, 95, pp. 3–36.

Kripke, Saul (1980), *Naming and Necessity*, Cambridge, MA: Harvard University Press.

Langton, Christopher G. (ed.) (1989), *Artificial Life*, Redwood City, CA: Addison-Wesley, pp. 1–47.

Lipson, Hod and Pollack, Jordan (2000), 'Automatic design and manufacture of robotic lifeforms', *Nature*, 406:6799 (31 August) pp. 974–78.

Luisi, Pier Luigi (1998), 'About various definitions of life', *Origins of Life and Evolution of the Biosphere*, 28, pp. 613–22.

——— ([2006] 2016), *The Emergence of Life: From Chemical Origins to Synthetic Biology*, 2nd ed., Cambridge: Cambridge University Press.

Margulis, Lynn and Sagan, Dorion (1995), *What is Life?*, Berkeley, CA: University of California Press.

Maturana, Humberto R. and Varela, Francisco J. ([1987] 1992), *The Tree of Knowledge: The Biological Roots of Human Understanding*, 2nd ed., Boston, MA: Shambhala.

Maynard Smith, John (1975), *The Theory of Evolution*, 3rd ed., New York: Penguin.

——— (1986), *The Problems of Biology*, New York: Oxford University Press.

——— (1992), 'Byte-sized evolution', *Nature*, 355:6363, pp. 772–73.

Mayr, Ernst (1982), *The Growth of Biological Thought*, Cambridge, MA: Harvard University Press.

——— (1997), 'What is the meaning of "life"?', in *This is Biology: The Science of the Living World*, Cambridge, MA: Harvard University Press, pp. 1–23.

Morowitz, Harold J. (1992), *Beginnings of Cellular Life: Metabolism Recapitulates Biogenesis*, New Haven, NJ: Yale University Press.

Olson, Eric T. (1997), 'The ontological basis of strong artificial life', *Artificial Life*, 3:1, pp. 29–39.

Oparin, Aleksandr Ivanovich (1964), *Life: Its Nature, Origin, and Development* (trans. A. Synge), New York: Academic Press.

Pace, Norman R. (2001), 'The universal nature of biochemistry', *Proceedings of the National Academy of Science USA*, 98:3, pp. 805–08.

Pattee, Howard H. (1989), 'Simulations, realization, and theories of life', in C. Langton (ed.), *Artificial Life*, Redwood City, CA: Addison-Wesley, pp. 63–78.

Pennock, Robert T. (2001), *Intelligent Design Creationism and Its Critics: Philosophical, Theological and Scientific Perspectives*, Cambridge, MA: MIT Press.

Rasmussen, Steen, Bedau, Mark A., Chen, Liaohai, Deamer, David, Krakauer, David C., Packard, Norman H. and Stadler, Peter F. (eds) (2007), *Protocells: Bridging Nonliving and Living Matter*, Cambridge, MA: MIT Press.

Rasmussen, Steen, Chen, Liaohai, Deamer, David, Krakauer, David C., Packard, Norman H., Stadler, Peter F. and Bedau, Mark A. (2004), 'Transitions from nonliving to living matter', *Science*, 303:5660, pp. 963–65.

Ray, Tom (1992), 'An approach to the synthesis of life', in C. Langton, C. Taylor, J. D. Farmer and S. Rasmussen (eds), *Artificial Life II*, Redwood City, CA: Addison-Wesley, pp. 371–408.

Schrödinger, Erwin ([1944] 1969), *What is Life? The Physical Aspect of the Living Cell*, Cambridge: Cambridge University Press.

Shapiro, Robert (1986), *Origins: A Skeptic's Guide to the Creation of Life on Earth*, New York: Summit Books.

Sober, Elliot (1992), 'Learning from functionalism – Prospects for strong artificial life', in C. Langton, C. Taylor, J. D. Farmer and S. Rasmussen (eds), *Artificial Life II*, Redwood City, CA: Addison-Wesley, pp. 749–65.

Szostak, Jack W., Bartel, David P. and Luisi, Pier Luigi (2001), 'Synthesizing life', *Nature*, 409 (18 January), pp. 387–90.

Taylor, Charles (1992), '"Fleshing out" artificial life II', in C. Langton, C. Taylor, J. D. Farmer and S. Rasmussen (eds), *Artificial Life II*, Redwood City, CA: Addison-Wesley, pp. 25–38.

Varela, Francisco J., Maturana, Humberto R. and Uribe, Ricardo B. (1974), 'Autopoiesis: The organization of living systems, its characterization and a model', *Biosystems*, 5, pp. 187–96.

Wimsatt, William C. (1987), 'False models as means to truer theories', in M. Niteckiand and A. Hoffman (eds), *Neutral Modes in Biology*, Oxford: Oxford University Press, pp. 23–55.

Zimmer, Carl (2003), 'Tinker, tailor: Can Venter stitch together a genome from scratch?', *Science*, 299 (14 February), pp. 1006–07.

3

Why Life Cannot Be Defined[1]

Carol E. Cleland

Faced with the question 'what is life?', many scientists, and some philosophers, advance definitions of life. Defining life is especially popular among astrobiologists, many of whom are convinced that one cannot successfully search for truly novel forms of microbial life without a definition of life: How else will one recognize it if one encounters it? The extensive discussion of definitions of life in the *Handbook of Astrobiology* (Kolb 2019) provides a salient illustration of this attitude. Along the same lines, a recent version of the *NASA Astrobiology Strategy* (Hays 2015: 145) contains a large section devoted to 'Key Research Questions for Defining Life'.[2] This chapter explains why the scientific project of defining life is mistaken. Life is not the sort of thing that can be successfully defined. In truth, a definition of life is more likely to hinder than facilitate the discovery of novel forms of life.

The most influential contemporary definitions include thermodynamic, metabolic, evolutionary and autopoietic definitions of life. As will become apparent, these definitions, which come in a variety of incarnations of varying degrees of abstractness (including purely informational), face robust counterexamples, and attempts to resolve them invariably generate new problems. It is thus hardly surprising that there is no consensus among scientists or philosophers on a definition of life. But, incessant counterexamples to definitions of life are just the tip of the iceberg. It turns out that definitions are inherently incapable of answering 'what is' questions about natural categories.

Some categories of things seem natural in the sense that they would exist even if there had been no human beings to think about them. Water, bird and gold provide good illustrations. Philosophers refer to groupings of this sort as 'natural kinds'.[3] Other groupings are non-natural in the sense that they depend upon human interests and concerns.[4] Examples of non-natural kinds include garbage, money and art. The classification of a category of things as natural or non-natural may change in light of empirical and theoretical developments. On the basis of macroscopic mineralogical characteristics, such as hardness and colour, jade was once classified as a natural kind. With the discovery that it includes molecularly distinct mineralogical compounds, jade was reclassified as a non-natural kind comprising two natural kinds (jadeite and nephrite). Similarly, in the 1970s, it was discovered that bacteria – which were viewed as a natural biological kind encompassing all prokaryotes (unicellular microorganisms lacking a membrane-bound nucleus) – consist of organisms differing biochemically and genetically from each other more than from single-celled eukaryotes (which have membrane-bound nuclei). This discovery resulted in major revisions to the highest level of biological taxonomy, with the Kingdom Monera (at the time also known as Bacteria) being replaced by two new domains of life, Archaea and Bacteria, and all eukaryotes (protists, fungi, plants and animals) being lumped together into a third domain, Eukarya. Whether life is a natural kind, like water, or a non-natural (human) kind, like jade, is an empirical question. The goal of this chapter is to show that *if* life is a natural kind (which is taken for granted by those championing definitions of life) attempts to define life are mistaken.

So, what exactly is a definition? Following earlier work (Cleland 2012; Cleland and Chyba 2002, 2007) that explores the logical character of definition, this chapter concludes that definitions are the wrong tools for answering the question 'what is life?' (also see Machery 2012). Definitions are concerned with language and concepts, as opposed to things in the world of nature. The best (most informative) definitions, which I dub 'ideal', dissect the concept associated with a word or expression into sub-concepts providing necessary and sufficient conditions for the application of the term; a necessary condition (being unmarried) is a condition that must be present for the correct application of a general term (e.g. 'bachelor') and a sufficient condition is a condition or set of conditions (being unmarried, adult, human and male) that secures the correct application of the term. Ideal definitions thus fully determine membership in the class of items designated by a term, which helps to explain their appeal. An astrobiologist could decisively settle the question of whether an extraterrestrial geochemical system is a living thing merely by consulting the 'right' ideal definition of life.

However, ideal definitions cannot live up to their promises. They are an ideal to which one can aspire but not something that can be actually achieved. The best definitions are subject to problems of vagueness, and plagued with counterexamples and borderline cases, which makes them especially poor tools for answering scientific questions about natural phenomena, such as life, that are not well understood. In this context, it is worth mentioning that it is not true, as some philosophers are fond of claiming, that scientists do not intend their definitions of life to be logically rigorous. Scientists not only seek necessary and sufficient conditions for life but also frequently get bogged down dealing with counterexamples and borderline cases. The work of astronomer Carl Sagan, synthetic biologist Pier Luigi

Luisi, theoretical biologist Stuart Kauffman and biochemist Tibor Gánti provide especially salient illustrations because they either advance or explicitly discuss the need for supplying necessary and sufficient conditions when 'defining' life.

Our current concept of life is based on a single example of life, and there are reasons for suspecting that it may be unrepresentative, which means that it is unlikely to shed much light on the general nature of life. The situation is even worse than this, however. In the second half of the twentieth century, philosopher Hilary Putnam (1973, 1975) mounted a devastating attack against definitional approaches to explicating natural kinds, showing that the meanings of natural kind terms cannot be captured by definitions.[5] This powerful but subtle argument (reviewed in Cleland 2019: Chapter 2) revolutionized philosophical thought about the capacities and limitations of definition but is not well known among scientists. To cut a long story short, natural kind terms *cannot* be defined. I have argued that what is needed to answer the question 'what is life?' is a universal *theory* of life (Cleland 2019: Chapter 3). Yet, we currently lack such a theory: To formulate one we need samples of life as we don't know it. And this presents astrobiologists with a potential dilemma: How will they recognize truly novel forms of life should they be fortunate enough to encounter them? The solution that I propose (Cleland 2019: Chapter 8) to this dilemma is to use sets of *tentative* criteria to identify potentially biological anomalies, phenomena that resemble familiar Earth life in provocative ways and yet also differ from it in perplexing ways. It is those phenomena that are most difficult to classify as living vs. nonliving that hold promise of moving us beyond our Earth-centric concepts of life.

Popular Definitions of Life

The history of attempts to define life is very long, but most scholars believe that it dates back to the writings of Aristotle. Scientific interest in defining life was rekindled in the latter part of the twentieth century by Sagan's now classic essay in *Encyclopædia Britannica* (Sagan 1970). Writing during a period of rapid advances in molecular biology, computer technology and space science, Sagan reviewed the then most widely accepted scientific definitions of life – physiological, metabolic, biochemical, thermodynamic and Darwinian (which he somewhat misleadingly called 'genetic') – and showed that each faces intractable counterexamples,[6] either including things that are clearly not alive (e.g. candle flames and quartz crystals) or excluding things (e.g. dormant seeds and spores, and mules) that are clearly biological. Sagan's essay produced a flurry of new activity on definitions of life. Some of the definitions that he discussed (e.g. physiological) were abandoned. Others were reformulated in hopes of undermining the strongest counterexamples advanced against them. New definitions of life were also introduced, for example, autopoietic. The upshot is that we currently have an impressively large and diverse collection of definitions (e.g. see Pályi et al. 2002; Popa 2004: Appendix B; Trifonov 2011).

Yet despite these efforts, there is no widely accepted definition of life (Chyba and McDonald 1995; Cleland and Chyba 2002, 2007; Tsokolov 2009). Although reworked and refined in light of contemporary scientific advances, the most popular contemporary definitions (thermodynamic, metabolic, evolutionary and autopoietic) still face serious problems, ranging from robust counterexamples to being too *ad hoc*, i.e. designed expressly for the purpose of fortifying a favoured view of the nature of life against what is *prima facie* a devastating counterexample. A brief survey of these definitions and their problems is illuminating.

Thermodynamic Definitions

Thermodynamic definitions of life are popular among biophysicists. Commonly associated with the writings of physicist Erwin Schrödinger (1944), a basic thermodynamic definition might characterize a living system as one that creates and maintains local order (organismal bodies and ecosystems) by extracting energy ('drinking order') from the environment over an extended period of time. As Sagan noted, however, many nonliving physical systems create order by extracting energy from their environment. Mineral crystals, such as quartz or diamond, and hurricanes provide good illustrations.

More nuanced thermodynamic definitions of life have been developed in the hopes of avoiding counterexamples of this sort (e.g. Hoelzer et al. 2006; Lineweaver 2006; Schneider and Kay 1994; Schulze-Makuch and Irwin 2008). But they are not free of troubling counterexamples either. In this light, Kauffman (e.g. Kauffman 2000: 14–22), who is convinced that life is a novel thermodynamic phenomenon, makes a radical suggestion: A new 'fifth law of thermodynamics', applying to complex, far from equilibrium, chemical reaction systems that are 'self-constructing' is required for understanding life. Kauffman develops a concept of 'autonomous agency' based on this idea and attempts to define life in terms of it (for example, see Kauffman 2003; Kauffman and Clayton 2006). From my perspective, one of the more interesting aspects of his work is the resurrection of the old Aristotelian notion of self-causation in a more contemporary scientific guise.

The details of Kauffman's definition are unimportant for our purposes. What is important is that his reasons for proposing a new law of thermodynamics – namely, the inability to explicate life in terms of established thermodynamic considerations – should count *against* the viability of a thermodynamic definition of life. To propose a vaguely described new general law of thermodynamics *solely* for the special purpose of salvaging a thermodynamic account of life is *ad hoc*. In this context, it is important to keep in mind that there is nothing *ad hoc* about pursuing Kauffman's conjecture as a *research project* – formulating more precisely described candidates for a fifth law of thermodynamics and exploring a diversity of physicochemical systems for empirical support. This highlights an important difference between definition-based and theory-based approaches to universal biology (discussed in much greater detail in Cleland 2019: Chapter 3).

Metabolic Definitions

Metabolic and thermodynamic definitions of life are closely related, the primary difference between them being that the former focus on the flow of material – vs. energy, interpreted statistically as local order – between a physical system and its environment. Sagan's nuts and bolts version of a metabolic definition characterizes life as 'an object with a definite boundary, continually exchanging some of its materials with its surroundings, but without altering its general properties at least over some period of time' (Sagan 1970: n.pag.). As Sagan observes, however, this definition faces compelling counterexamples. It includes nonliving phenomena such as candle flames, which have a definite shape and maintain themselves by continually exchanging material with their environment – extracting oxygen and combining it with wax, and releasing carbon dioxide and water in the process – and excludes viable seeds and spores, which may lie dormant for hundreds or even thousands of years before germinating.

More sophisticated chemical-metabolic definitions have been developed in efforts to ward off counterexamples of these sorts (e.g. Dyson 1999; Feinberg and Shapiro 1980; Hazen 2005). Some restrict the possible chemistries of metabolic processes to organic (carbon-containing) molecules. Basing a definition of life too closely on the chemistry of familiar Earth life is problematic, however, insofar as it excludes forms of life that are conceptually and perhaps even scientifically plausible. The Horta of *Star Trek: The Original Series* fame provides an especially colourful illustration. In the episode 'The Devil in the Dark' (Pevney 1967), the Horta, an amoeba-like blob of rock, confounds Captain Kirk and his crew by burrowing through solid rock and behaving in a life-like manner. The Horta turns out to be a truly alien, chemical form of life based on silicon rather than carbon. Mr. Spock discovers, via a Vulcan mind meld (of course), that it is a 'mother' defending her eggs from miners.

It is of course an open question whether life forms differing this radically from Earth life in basic chemical composition are biologically possible, but surely one does not want to rule them out *merely by definition*. Besides, there are theoretical reasons for thinking that silicon-based life forms such as the Horta may not be as scientifically implausible as sometimes supposed. Schulze-Makuch and Irwin (2008) argue that under physical and chemical conditions very different from those on Earth (viz., little free oxygen or liquid water, and very low temperatures), silicon may be able to form longer and more complex molecular structures; molecular size and complexity have long been thought to be a requirement for life (Pace 2001). They conjecture that there might even be silicon-based life forms on Titan, a moon of Saturn. As biochemist William Bains observes, however, Mr. Spock's experience with Titanians would not be very pleasant. Even a small whiff of their breath would stink to high heaven, and if a Titanian were beamed aboard the starship Enterprise, it would boil, burst into flames, and kill everyone in the transporter room (Heward 2010).

A less extreme but far more worrisome illustration of how a metabolic definition of life based too closely on the biochemistry of familiar Earth life could hinder the recognition of extraterrestrial life is provided by the Viking spacecraft missions to Mars in the 1970s. The purpose was to search for microbial life.[7]

The Viking biology package consisted of three experiments. As the Biology Team Leader, Harold Klein, recounts, these experiments were designed to test a number of 'metabolic models of possible Martian biology' (Klein 1978: 157) centred 'on the assumption that Martian biology would be based on carbon chemistry' (Klein 1991: 259). In other words, they were implicitly based on an Earth-centric chemical-metabolic definition of life. The basic idea was to search for evidence of Earth-like microbial metabolism. A miniature gas chromatograph mass spectrometer (GCMS) for detecting organic molecules was also included.

The experiment that produced the most baffling results was the Labeled Release (LR) experiment (see Cleland 2019: Chapter 8 for a detailed discussion). In the LR experiment, Martian soil was robotically introduced into a test chamber and injected with a carbohydrate solution, labelled with ^{14}C (a radioactive isotope of carbon), that is readily metabolizable by a wide range of culturable microorganisms. To the delight of the Viking biology team, the soil started evolving $^{14}CO_2$ – just what one would expect from Earth microorganisms. When a sample of Martian soil was heated to 160°C for three hours, more than enough to kill any known microorganisms, and then inoculated with the nutrient solution, no gas was released, strongly suggesting that the initial reaction was biological. When a sample of Martian soil, that had evolved gas after an initial inoculation, was given a second helping of nutrients, however, the anticipated burst of new activity from hungry Martian microbes failed to occur. Even more mysteriously, $^{14}CO_2$ left over from the initial reaction began disappearing (Levin and Straat 1977). The biology team was stunned. They could not explain these results in terms of any known chemical process, biological or nonbiological. In the end, they concluded that it must not have been biological on the basis of the anomalous (for Earth microbes) results and, most importantly, the failure of the Viking GCMS to find any organic molecules in the soil to its limits of detection with sample heating up to 500°C (Biemann et al. 1976, 1977).

A consensus was quickly reached by the majority of the astrobiology community that the mysterious results of the LR experiment were produced by some sort of inorganic oxidant in the Martian soil. Various possibilities, including hydrogen peroxide, superoxides and exotic oxidation states of iron, and, most recently, perchlorate, have been explored over the years. None are able to fully explain all the results of the LR experiment (see Cleland 2019: Chapter 8). In addition, it has been shown that the Viking GCMS, which was the linchpin in the original argument against biology, could not have detected as many as 10^6 bacterial cells per gram of soil, more than are found in some samples of Antarctic soil (Bada 2001; Glavin et al. 2001; Navarro-Gonzáles et al. 2006). Yet NASA has not placed a high priority on examining the Martian surface for the presence of a powerful inorganic oxidant. Why? Because (as a NASA scientist involved in the mission informed me) they were looking for life and what they found was not life. In other words, the definition undergirding the Viking life detection instrument package was not satisfied, and hence no life. This attitude illustrates both the potential power of an accepted definition of life to blind scientists to alternative forms of life and the dangers of basing a definition of life too closely on the biochemistry of familiar Earth life. In short, we might not recognize an alternative form of life if we search for it with an inadequate definition of life.[8]

Evolutionary Definitions

Evolutionary accounts currently dominate scientific and philosophical thought about the nature of life. Most evolutionary definitions of life are founded upon naturalist Charles Darwin's theory of evolution by natural selection. A few are more inclusive, postulating either a more general evolutionary mechanism, subsuming natural selection as a special case, for example, philosopher Mark Bedau's (1998) 'supple adaptation', or identifying natural selection as one of several evolutionary mechanisms, for example, Eva Jablonka and Marion J. Lamb (2005), who argue that epigenetic inheritance represents a neo-Lamarckian mechanism of evolution.[9] Our focus in this section, however, is on Darwinian definitions.

The most abstract versions of the Darwinian definition of life follow Richard Dawkins (1983) and Daniel Dennett (1995) in holding not only that the capacity to undergo evolution by natural selection is universal to life ('universal Darwinism') but also that it is independent of the material stuff that instantiates life. Darwinian definitions of this type are especially popular among artificial life (ALife) researchers, who frequently view evolution by natural selection as an algorithmic process, reducible to a system of equations. As an illustration, Steen Rasmussen (1992) and Tom S. Ray (2009) contend that a sufficiently complex computer simulation exhibiting Darwinian evolution would be truly alive. Indeed, Ray contends that he has already created digital organisms in the informational structure of a computer. On Rasmussen's and Ray's version of the Darwinian definition, it is not the computer that is alive but the computational processes themselves; the computer's role is analogous to that of glassware that may someday be used in the artificial synthesis of chemical life in a laboratory.

Most biologists and biochemists, however, find the idea of an immaterial computer simulation of life actually *being* alive counterintuitive. Few biochemists believe that a computer simulation of biochemistry is biochemistry itself.

A computer simulation of photosynthesis does not produce authentic carbohydrates (*chemical* compounds of carbon, hydrogen and oxygen); at best, it may be said to yield simulated carbohydrates. Analogously, why would one think that a computer simulation of life is actually alive, as opposed to merely simulating life? The point is that the only unequivocal example of life available to us is familiar biological Earth life, and it is chemical. One cannot rule out someday discovering that the essential characteristics of life are independent of the stuff that realizes them. It is important to keep in mind, however, that there currently exists no empirical support whatsoever for this claim. All that a computer simulation of life establishes is that some of the high-level, functional characteristics of life can be successfully emulated on a computer, which is hardly surprising because computers can simulate the purely structural and functional characteristics of virtually any physical system, for example, bridges, landslides, hurricanes and the human circulatory system. In this context, it is especially important not to be seduced by the starry-eyed claim that computers are on the threshold of providing us with the first truly alien creatures. To recapitulate, for the claim that a computer simulation of a living thing is itself alive rests upon the dubious assumption that the basic properties of life are independent of their physical embodiment, and this assumption is exactly what is at issue. (For more details, see Cleland 2019: Chapter 7.)

In light of these concerns, many biologists and biochemists explicitly restrict Darwinian definitions of life to chemical systems. The 'chemical Darwinian definition', which became NASA's official 'working definition' of life in the 1990s, is the prototype and remains popular (NRC 2007). According to this definition, 'life is a self-sustained chemical system capable of undergoing Darwinian evolution' (Joyce 1994). As Gerald F. Joyce explains, the notion of Darwinian evolution subsumes the processes of self-reproduction, material continuity over a historical lineage, genetic variation and natural selection. The requirement that the system be self-sustaining refers to the fact that living systems contain all the genetic information necessary for their own constant production (i.e. metabolism) (xi–xii).

Although it is often interpreted as such, the chemical Darwinian definition is not restricted to organic molecules. As a result, it is open to the possibility of silicon-based life on Titan, or perhaps even sulphur-based life in the sulphuric acid clouds of Venus (Grinspoon 2003: 283–86). As the next sub-section discusses, however, it is difficult to design instrument packages for detecting organic (let alone inorganic) forms of extraterrestrial life on the basis of the chemical Darwinian definition. This undermines its utility for searching for extraterrestrial life. But this is only a practical problem. The definition faces much more serious problems.

The chemical Darwinian definition cannot accommodate the possibility that cellular life on Earth passed through a stage in which it could reproduce in a rudimentary fashion but could not yet engage in Darwinian evolution. According to some metabolism-first models of the origin of life, the earliest forms of Earth life were autocatalytic chemical reaction systems consisting of networks of encapsulated, small organic molecules, primarily amino acids, cofactors and peptides. Protocells in these reaction networks could divide but, lacking nucleic acid-type genetic systems, were not yet capable of Darwinian evolution. Furthermore, some advocates of the RNA World, a genes-first theory of the origins of life, hold a similar view. Carl Woese (1998) and Kalin Vetsigian et al. (2006), for instance, point out that Darwinian evolution requires a very complex genetic apparatus with a universal genetic code coupled to a very precise mechanism for translating genetic information into organismal traits. It is unlikely that a hereditary apparatus of this complexity emerged in one fell swoop with the first protocells. Woese conjectures that cellular life, as we know it today, was preceded by collections of cellular entities ('progenotes') capable of primitive cell division (by pinching off) but with genetic systems (consisting of small mobile genetic elements subject to rampant mutation and haphazard transfer from one progenote to another) too inaccurate to support Darwinian evolution. On the chemical Darwinian definition of life, neither progenotes nor their communities qualify as living things. Yet if such entities existed (or perhaps still exist on some other world), it would surely be a mistake to deny them the status of living things solely because they fail to satisfy a preferred, preconceived, definition of life.[10]

Another potential difficulty, which the original version of the chemical Darwinian definition shares with many other evolutionary definitions of life, concerns the status of individual organisms, such as a bacterium, cat or tree, as living things.

Biological evolution occurs at the level of populations of organisms. Individual organisms (whether fertile and reproducing, fertile and nonreproducing, sterile members of fertile species or sterile hybrids) do not *themselves* evolve. Most champions of evolutionary definitions recognize that individual organisms provide the paradigm for a living thing and hence that an inability to classify individual organisms as living things undermines the conceptual plausibility of a definition of life. A strategy for circumventing this difficulty is to distinguish life from the entities (individual organisms) that participate in it. Bedau (1998), for example, defines life as an evolutionary process ('supple adaptation') in which individual organisms (fertile as well as sterile) participate as derivative ('secondary') forms of life. Similarly, biochemist Steven A. Benner (2010) distinguishes 'life' as an evolutionary (population-level) system from its 'parts', which include individual organisms that are said to 'be alive'. As Benner puts it, 'one rabbit may be alive even though he or she is not life' (1021).

Efforts to shore up evolutionary definitions of life in this manner are not problem-free, however. Consider the very real (but thus far hypothetical) possibility of artificially synthesizing a microorganism from scratch (bottles of nonliving chemicals) that cannot reproduce and is unable to hybridize (swap genes) with other microorganisms. Such microbes would not qualify as alive on either Bedau's or Benner's definition of life; it does not participate in an evolving (population-level) system.

A well-worn strategy for resolving difficulties of this sort is to modify the definition so that the troublesome entities are included within its scope. One might, for example, stretch the concept of evolution to include being constructed artificially by a creature that itself is the product of Darwinian evolution. This does not fully solve the problem, however, because we cannot rule out the possibility of designing robots to synthesize microorganisms of the sort in question; robots designed by humans would not be direct products of evolution by natural selection. One could, of course, resolve this difficulty by stretching the concept of evolution even further to include microorganisms synthesized by robots, constructed by robots, ..., constructed by creatures that are the products of Darwinian evolution. But now we have another problem, namely, the concept of evolution has been stretched so far beyond the original biological notion that it is not clear that it qualifies as Darwinian anymore. The central problem should be obvious. Whatever strategy one comes up with – including biting the bullet and declaring that the artificial microbes concerned are not living things – is extremely *ad hoc*, that is, motivated solely by a desire to protect a preferred definition from what would otherwise be intractable counterexamples. Any definition of life can always be salvaged in this contrived manner but at a high cost, namely, undermining its capacity to shed light on the nature of life.

Defining Life as Self-Organized Complexity

In recent years, newer definitions focusing on the self-organizing properties of organisms have been developed. The most influential are autopoietic definitions (Luisi 1993; Maturana and Varela 1973) and Gánti's 1971 chemoton [or chemical automaton] model of life (see Gánti 2003). Kauffman's definition of life in terms of autonomous agency should also be included. As Luisi observes, all three modes of definition 'share a common view about minimal life as a distributed emergent property based on an organized network of reactions and/or processes' (Luisi 2003: 53). They differ primarily on how they understand self-organized complexity.

Kauffman cashes it out in terms of physics, the product of a heretofore unrecognized, new law of thermodynamics. Autopoietic definitions and the chemoton model, in contrast, regard it as a logical property of a system (Luisi 2003: 53). Luisa Damiano and Pier Luigi Luisi (2010) define life as 'a system capable of self-production and self-maintenance through a regenerative network of processes which takes place within a boundary of its own making and regenerates itself through cognitive or adaptive interactions with the medium' (149). Gánti (2003) develops a formal model of life, as 'fluid automata', described by a system of equations.

The most obvious difficulty with all three approaches is that the vague concepts of self-organization (e.g. emergence, self-referentiality, self-sustainability and auto-maintenance) invoked to explain life are treated as primitive, raising the hoary old problem of goal-directed self-causation. The basic idea seems to be that at a certain level of organization, novel functional properties spontaneously arise, as a matter of either fundamental physics or logic, rendering the system capable of developing increasingly greater organized complexity. Advocates of these definitions seem to believe that the idea of novel properties arising spontaneously at higher levels of organization is an asset, but it is a weakness, generating, among other things, the threat of backward causation in which newly acquired novel properties causally interact with the conditions that brought them about, thereby preventing their occurrence. In truth, these approaches to explicating the nature of life have much in common with older views such as vitalism and British emergentism, with the downside that less attention is paid to the physical and metaphysical problems posed by the notion of goal-directed self-causation.

In addition to the problem of making good sense of goal-directed self-causation, autopoietic definitions, which have received more attention than the other two, are notable for their permissiveness. They classify many systems that would otherwise be considered paradigmatically nonliving (and hence, compelling counterexamples) as living things. Some provocative illustrations are whole ecosystems, human institutions (such as corporations) and even Earth itself (Lovelock 2000; Margulis and Sagan 1995);[11] computer simulations are also commonly included on the grounds that the difference between life and nonlife is merely a matter of how the components of a system are organized and integrated. That a proposed definition admits so many *prima facie* nonliving entities (seeming counterexamples) as living individuals would normally count against it. Turning the tables and insisting that they are living things *just because* a favourite definition says so is logically problematic.

As earlier, it is important to keep in mind that there is nothing problematic about attempting to develop a coherent concept of autopoiesis and exploring its potential for shedding light on unambiguous cases of life. Furthermore, research within the context of an empirically successful scientific theory sometimes changes classification schemes in significant ways. It is thus foolish to insist that our current classifications of entities as living and nonliving are the final word. But such a project is very different from formulating a logically rigorous definition that (in virtue of supplying necessary and sufficient conditions for life) fixes what is and is not a living thing in one fell swoop.

The point is the best definitions are brittle. One of the strengths of scientific theories is the open-endedness and flexibility of their classification schemes, which allows them to evolve in response to new empirical discoveries and theoretical insights.

NOTES

1. This is an adapted version of Chapter 2 from Carol Cleland (2019), *The Quest for a Universal Theory of Life: Searching for Life as We Don't Know It*, published with permission from Cambridge University Press, reproduced with permission of the Licensor through PLSclear.
2. In addition to astrobiology, origins of life and artificial life researchers devote significant efforts to the issue of defining life in the belief that an understanding of how to create life, whether via natural processes, in a laboratory, or in the information structures of a computer, requires defining life; some origins of life researchers hold the opposite view, contending that one cannot 'define' life without understanding how it originated. (For further discussion, see Cleland 2019: Chapters 5 and 7.)
3. As discussed in Cleland (2019: Chapter 4), naturalism about kinds does not presuppose that there is a unique way of carving nature at its joints. Nature is very complex. As John Dupré (1993) argues, there may be many highly plausible ways of dividing it into kinds. It does not follow, however, that these diverse systems of classification are equally good for the purpose of yielding scientifically fruitful, general theories, even supposing that they are useful for other purposes, such as cooking or horticulture. Furthermore [...] naturalism about the classifications of our most successful scientific theories does not (as some philosophers believe) imply the strong metaphysical doctrine of essentialism. Somewhat ironically, essentialism is closely bound up with the claim, rejected in this chapter, that natural kinds have defining properties (essences).

4. Many philosophers who are hostile to essentialism reject this distinction, contending that all scientific categories are ultimately human kinds. This view seems highly problematic given the remarkable empirical successes of modern science (for more discussion, see Cleland 2019: Chapter 6; Bedau and Cleland 2010).
5. Saul Kripke (1972) developed a similar line of argument but, while he applied it to natural kind terms, his central focus was proper names. Putnam's work provides a far more extensive and sustained attack on the notion that natural kind terms can be defined.
6. A counterexample to a definition is something that violates one of the stipulated conditions for the application of the term being defined.
7. It is important to distinguish searching for habitable environments from searching for life. A number of missions have explored the habitability of planets and moons within our solar system by looking for evidence of liquid water, an atmosphere with the right chemical composition, etc. But the presence of these conditions does not provide direct evidence of life (past or present). For the latter, one needs to look for biosignatures (telltale traces of life). NASA's Mars 2020 mission included the Perseverance rover, which has the Scanning Habitable Environments with Raman and Luminescence for Organics and Chemicals (SHERLOC) instrument package dedicated to looking for biosignatures. The ExoMars (Exobiology on Mars) programme's mission sponsored by the European Space Agency (ESA) in 2028 features the Rosalind Franklin rover dedicated to searching for biosignatures. It includes the Mars Organic Molecule Analyzer (MOMA), a mass spectrometer, which utilizes an ultraviolet laser to help identify the origins of soil samples.
8. I am not of course claiming that the Viking spacecrafts found Martian life; on my view, the jury is still out.
9. Epigenetic inheritance involves chemical changes that affect the way in which genes are *expressed* (as organismal traits) without changing the primary structure of DNA; they do not involve mutations of genetic material. These changes are acquired by an organism as a result of interactions with its environmental and are passed down to its offspring. To the extent that they can be said to amount to the inheritance of acquired characteristics, they have a Lamarckian character. It is important to keep in mind, however, that the changes in organismal traits in question are not, as Lamarck proposed, directly inherited as such. What is inherited epigenetically are chemical markers attached to DNA that affect (suppress or activate) the way in which genes coded for by DNA are expressed in organismal traits. The classic example is that mice born to stressed mothers are more susceptible to stress than those born to mothers that were not subject to stress.
10. In this context, it is worth noting that one cannot successfully defend the chemical Darwinian definition on the grounds that a genetic system is required for an autocatalytic chemical network to achieve the level of organized complexity required of a primitive metabolism. For as discussed in Cleland (2019: Chapter 5), a genes-first theory of the *origins* of life is fully compatible with a metabolic account of the *nature* of life.

11. The claim that the Earth itself is a living individual is known as the Gaia hypothesis; it was first proposed by James E. Lovelock (1972) and co-developed with Lynn Margulis (Lovelock and Margulis 1974; see also Margulis 2004).

REFERENCES

Bada, Jeffrey L. (2001), 'State-of-the-art instruments for detecting extraterrestrial life', *Proceedings of the National Academy of Sciences of the USA*, 98:3, pp. 797–800.

Bedau, Mark A. (1998), 'Four puzzles about life', *Artificial Life*, 4, pp. 125–40.

Bedau, Mark A. and Cleland, Carol E. (eds) (2010), *The Nature of Life: Classical and Contemporary Perspectives from Philosophy and Science*, Cambridge: Cambridge University Press.

Benner, Steven A. (2010), 'Defining life', *Astrobiology*, 10:10, pp. 1021–30.

Biemann, Klaus, Oró, Joan, Toulmin III, Prestley, Orgel, Leslie E., Nier, Alfred O., D. M. Anderson, P. G. Simmonds, D. Flory, A. V. Diaz, D. R. Rushneck and Biller, Jose A. (1976), 'Search for organic and volatile inorganic compounds in two surface samples from the Chryse Planitia region of Mars', *Science*, 194, pp. 72–76.

Biemann, Klaus, Oró, Joan, Toulmin III, Prestley, Orgel, Leslie E., Nier, Alfred O., Anderson, D. M., Simmonds, P. G., Flory, D., Diaz, A. V. Rushneck, D. R., Biller, Jose A. and Lafleur, A. L. (1977), 'The search for organic substances and inorganic volatile compounds in the surface of Mars', *Journal of Geophysical Research*, 82:28, pp. 4641–58.

Chyba, Christopher F. and McDonald, Gene D. (1995), 'The origin of life in the solar system: Current issues', *Annual Review of Earth and Planetary Sciences*, 23, pp. 215–49.

Cleland, Carol E. (2012), 'Life without definitions', *Synthese*, 185, pp. 125–44.

——— (2019), *The Quest for a Universal Theory of Life: Searching for Life as We Don't Know It*, Cambridge: Cambridge University Press.

Cleland, Carol E. and Chyba, Christopher F. (2002), 'Defining "life"', *Origins of Life and Evolution of the Biosphere*, 32, pp. 387–93.

——— (2007), 'Does "life" have a definition?', in W. T. Sullivan III and J. A. Baross (eds), *Planets and Life: The Emerging Science of Astrobiology*, Cambridge: Cambridge University Press, pp. 119–31.

Damiano, Luisa and Luisi, Pier Luigi (2010), 'Towards an autopoietic redefinition of life', *Origins of Life and Evolution of the Biosphere*, 40:2, pp. 145–49.

Dawkins, Richard (1983), 'Universal Darwinism', in D. S. Bendall (ed.), *Evolution from Molecules to Men*, Cambridge: Cambridge University Press, pp. 403–25.

Dennett, Daniel C. (1995), *Darwin's Dangerous Idea: Evolution and the Meanings of Life*, New York: Simon and Schuster.

Dupré, John (1993), *The Disorder of Things: Metaphysical Foundations of the Disunity of Science*, Cambridge, MA: Harvard University Press.

Dyson, Freeman J. (1999), *Origins of Life*, Cambridge: Cambridge University Press.

Feinberg, Gerald and Shapiro, Robert (1980), *Life Beyond Earth: Intelligent Earthlings, Guide to the Universe*, New York: William Morrow.

Gánti, Tibor (2003), *The Principle of Life*, with commentary by J. Griesemer and E. Swathmáry, New York: Oxford University Press.

Glavin, Daniel P., Schubert, Michael, Botta, Oliver, Kminek, Gerhard and Bada, Jeffrey L. (2001), 'Detecting pyrolysis products from bacteria on Mars', *Earth and Planetary Science Letters*, 185, pp. 1–5.

Grinspoon, David (2003), *Lonely Planets: The Natural Philosophy of Alien Life*, New York: Harper Collins.

Hazen, Robert M. (2005), *Genesis: The Scientific Quest for Life's Origin*, Washington, DC: Joseph Henry Press.

Hays, Lindsey, L. (editor-in-chief) (2015), *NASA Astrobiology Strategy*, National Aeronautics and Space Administration, Washington, DC: NASA.

Heward, Anita (2010), 'Life on Titan: Stand well back and hold your nose!', *PhysOrg.com*, 14 April, https://phys.org/news/2010-04-life-titan-nose.html. Accessed 5 April 2023.

Hoelzer, Guy A., Smith, Eric and Pepper, John W. (2006), 'On the logical relationship between natural selection and self organization', *Journal of Evolutionary Biology*, 19, pp. 1785–94.

Jablonka, Eva and Lamb, Marion J. (2005), *Evolution in Four Dimensions: Genetic, Epigenetic, Behavioral, and Symbolic Variation in the History of Life*, Cambridge, MA: MIT Press.

Joyce, Gerald F. (1994), 'Foreword', in D. W. Deamer and G. R. Fleischaker (eds), *Origins of Life: The Central Concepts*, Boston, MA: Jones & Bartlett, pp. xi–xii.

Kauffman, Stuart (2000), *Investigations*, Oxford: Oxford University Press.

——— (2003), 'The emergence of autonomous agents', in N. H. Gregersen (ed.), *From Complexity to Life: On the Emergence of Life and Meaning*, Oxford: Oxford University Press, pp. 47–71.

Kauffman, Stuart and Clayton, Philip (2006), 'On emergence, agency, and organization', *Biology and Philosophy*, 21, pp. 501–21.

Klein, Harold P. (1978), 'The Viking biology investigations: Review and status', *Origins of Life and Evolution of the Biosphere*, 9:2, pp. 157–60.

——— (1991), 'The Viking biology experiments: Epilogue and prologue', *Origins of Life and Evolution of the Biosphere*, 21, pp. 255–61.

Kolb, Vera (ed.) (2019), 'Section II: Definition and nature of life', in *Handbook of Astrobiology*, Boca Raton, FL: CRC Press, Taylor & Francis, pp. 55–106.

Kripke, Saul A. (1972), *Naming and Necessity*, Cambridge, MA: Harvard University Press.

Levin, Gilbert V. and Straat, Patricia Ann (1977), 'Recent results from the Viking labeled release experiment on Mars', *Journal of Geophysical Research*, 82:28, pp. 4663–67.

Lineweaver, Charles H. (2006), 'We have not detected extraterrestrial life, or have we?', in J. Seckbach (ed.), *Life as We Know It*, Dordrecht: Springer, pp. 445–57.

Lovelock, James E. (1972), 'Gaia as seen through the atmosphere', *Atmospheric Environment*, 6, pp. 579–80.
——— (2000), *The Ages of Gaia: A Biography of Our Living Earth*, 2nd ed., Oxford: Oxford University Press.
Lovelock, James E. and Margulis, Lynn (1974), 'At atmospheric homeostasis by and for the biosphere: The Gaia hypothesis', *Tellus*, 26, pp. 1–10.
Luisi, Pier Luigi (1993), 'The chemical implementation of autopoiesis', in G. R. Fleischaker, C. Stephano and P. L. Luisi (eds), *Self-Production of Supramolecular Structures: From Synthetic Structures to Models of Minimal Living Systems*, Dordrecht: Kluwer, pp. 179–97.
——— (2003), 'Autopoiesis: A review and a reappraisal', *Naturwissenschaften*, 90:2, pp. 49–59.
Machery, Edouard (2012), 'Why I stopped worrying about the definition of life . . . and why you should as well', *Synthese*, 185, pp. 145–64.
Margulis, Lynn (2004), 'Gaia by any other name', in S. H. Schneider, J. R. Miller, E. Crist and P. J. Boston (eds), *Scientists Debate Gaia: The Next Century*, Cambridge, MA: The MIT Press, pp. 7–12.
Margulis, Lynn and Sagan, Dorion (1995), *What Is Life?*, Berkeley, CA: University of California Press.
Maturana, Humberto and Varela, Francisco (1973), *Autopoiesis and Cognition: The Realization of the Living*, Dordrecht: Reidel.
Navarro-Gonzáles, Rafael, Navarro, Karina F., de la Rosa, José, Iniguez, Enrique, Molina, Paola, Miranda, Luis D., Morales, Pedro, Cienfuegos, Edith, Coll, Patrice, F. Raulin, François, Amils, Ricardo and McKay, Christopher P. (2006), 'The limitations on organic detection in Mars-like soils by thermal volatilization-gas chromatography-MS and their implications for the Viking results', *Proceedings of the National Academy of Sciences*, 103:44, pp. 16089–94.
National Research Council (NRC) (2007), *The Limits of Organic Life in Planetary Systems*, Washington, DC: National Academy of Sciences.
Pace, Norman R. (2001), 'The universal nature of biochemistry', *Proceedings of the National Academy of Sciences of the USA*, 98:3, pp. 805–80.
Pályi, Gyula, Zucchi, Claudia and Caglioti, Luciano (2002), 'Introduction: Definitions of life', in G. Pályi, C. Zucchi and L. Caglioti (eds), *Fundamentals of Life*, New York: Elsevier, pp. 2–13.
Pevney, Joseph (dir.) (1967), 'The Devil in the Dark', *Star Trek: The Original Series*, Season 1 Episode 25, 9 March, USA: National Broadcasting Company (NBC).
Popa, Radu (2004), *Between Necessity and Probability: Searching for the Definition and Origin of Life*, Berlin: Springer Verlag.
Putnam, Hilary (1973), 'Meaning and reference', *Journal of Philosophy*, 70:19, pp. 699–711.
——— (1975), 'The meaning of "meaning"', in K. Gunderson (ed.), *Language, Mind, and Knowledge: Minnesota Studies in the Philosophy of Science*, vol. VII, Minneapolis, MN: University of Minnesota Press, pp. 131–93. [Reprinted in Putnam 1979.]

Rasmussen, Steen (1992), 'Aspects of information, life, reality, and physics', *Artificial Life*, 2, pp. 767–74. [Reprinted in Langton and Taylor 2003.]

Ray, Tom S. (2009), 'Artificial life programs and evolution', in M. Ruse and J. Travis (eds), *Companion to Evolution*, Cambridge, MA: Harvard University Press, pp. 429–33.

Sagan, Carl (1970), 'Life', in *Encyclopædia Britannica*, 14th ed., Chicago: Encyclopædia Britannica Incorporated.

Schneider, Eric D. and Kay, James J. (1994), 'Life as a manifestation of the second law of thermodynamics', *Mathematics and Computer Modelling*, 19:6–8, pp. 25–48.

Schrödinger, Erwin (1944), *What is Life? The Physical Aspect of the Living Cell*, Cambridge: Cambridge University Press.

Schulze-Makuch, Dirk and Irwin, Louis N. (2008), *Life in the Universe: Expectations and Constraints*, 2nd ed., Berlin: Springer Verlag.

Trifonov, Edward N. (2011), 'Vocabulary of definitions of life suggests a definition', *Journal of Biomolecular Structure and Dynamics*, 29:2, pp. 259–66.

Tsokolov, Serhiy A. (2009), 'Why is the definition of life so elusive? Epistemological considerations', *Astrobiology*, 9:4, pp. 401–12.

Vetsigian, Kalin, Woese, Carl and Goldenfield, Nigel (2006), 'Collective evolution and the genetic code', *Proceedings of the National Academy of Sciences of the USA*, 103:28, pp. 10696–701.

Woese, Carl (1998), 'The universal ancestor', *Proceedings of the National Academy of Sciences*, 95:12, pp. 6854–59.

PART II

INFORMATION AND ECOLOGIES

4

Propagating Organization: An Enquiry[1]

Stuart Kauffman, Robert K. Logan, Robert Este,

Randy Goebel, David Hobill and Ilya Shmulevich

> *An organized being is then not a mere machine, for that has merely moving power, but it possesses in itself* formative *power of a self-propagating kind which it communicates to its materials though they have it not of themselves; it organizes them, in fact, and this cannot be explained by the mere mechanical faculty of motion.*
>
> – Immanuel Kant, *Critique of Judgement*

Introduction

Our broad aim is to understand propagating organization as exemplified by the vast organization of the co-evolving biosphere. Our effort is a rather mysterious undertaking, for we entirely lack a theory of organization of process, yet the biosphere, from the inception of life to today manifestly propagates organization of process. Indeed, we believe that the evolving universe as a whole also manifests the propagation of organization. We shall focus most of our efforts on the biotic case but undertake an initial extension of our analysis to the abiotic case as well.

The role of information in biology, what it 'is', how it accumulates and how it is 'used', has been directly addressed by mainstream biologists and philosophers of biology. By and large, the biological concept of information derives from the DNA, RNA, protein processes of 'coding', transcription and translation. Yet in the broader sense that we seek to articulate, information in terrestrial life is likely to be one of the key unifying concepts in the emerging field of systems biology. As part of the propagating organization within living cells, the cell operates as an information-processing unit, receiving information from its environment, propagating that information through complex molecular networks, and using the information stored in its DNA and cell-molecular systems to mount the appropriate response. Indeed, biology is acquiring many characteristics of an information science (Hood and Galas 2003).

It is sometimes the case that science progresses by finding the concepts and language to 'see that which is directly in front of us'. Such is the case with the present enquiry. We are persuaded that we are not wholly successful, but hope that we shall have at least started a far broader discussion.

Two predecessors to this [chapter] can be found in *Investigations* (Kauffman 2000) and 'Emergence, autonomous agents, and organization' (Kauffman and Clayton 2006). At its core, *Investigations* seeks to understand the physical nature of agency itself, and proposes that a molecular autonomous agent, able to act on its own behalf in an environment, is an autocatalytic system carrying out at least one thermodynamic work cycle. Much follows from this tentative definition, which implies that an *autonomous agent* is an open non-equilibrium chemical system, and finds general biotic importance in the fact that work cycles link spontaneous and non-spontaneous (exergonic and endergonic) processes. This linkage has built up the enormous complexity of the biosphere.

Further analysis reveals this work to be the constrained release of energy into a few degrees of freedom. But if one asks where the constraints themselves come from – as in the example of a cylinder and piston that confine the expansion of the working gas in the head of the cylinder to yield the translational motion of the piston, hence the release of energy into a few degrees of freedom – one finds that it typically takes work to construct the constraints.[2]

Thus, we arrive at the first surprise – it takes constraints on the release of energy for work to happen but work for the constraints themselves to come into existence. This circle of work and constraint shall turn out to be part of the theory of propagating organization that we shall discuss.

Most importantly, contemporary cells are both collectively autocatalytic and do work cycles, in part to construct constraints on the release of energy. When released, this energy constitutes further work that drives non-spontaneous processes, builds structures, drives processes and also builds further constraints on the release of energy, which when released can build still more such constraints. In short, cells carry out propagating work linking spontaneous processes, constraints, work and non-spontaneous processes, and more broadly as we shall see, the propagating organization of process. In doing so, the cell carries out a set of interlocked tasks that achieve a closure of tasks whereby the cell literally builds a rough copy of itself. We know this, yet we have no clear way to say what we know. This closure of work, constraints, tasks and information, as we shall see below, is a new state of matter, energy, information and organization that constitutes the living state.

The new insight that we explore in this chapter is that the constraints that allow autonomous agents to channel free energy into work are connected to information: in fact, simply put, the constraints *are* the information, are partially causal in the diversity of what occurs in cells and are part of the organization that is propagated.

In Kauffman and Clayton (2006), the tentative definition of autonomous agent is extended to include construction of boundaries enclosing the agent, discrimination of 'yuck' (meaning poison) or 'yum' (meaning food), and at least one choice of action: flee (or not), approach (or not). Our language is teleological. We believe that autonomous agents constitute the minimal physical system to which teleological language rightly applies.

It is important that our definition of a molecular autonomous agent applies to terrestrial life, but is, in principle, far broader. The concepts identify a new class of far from equilibrium chemical thermodynamic systems, and we suspect, could form the basis of life in a variety of molecular instantiations. For example, already Dane Sievers and Güntervon von Kiedrowski (1994) and David H. Lee et al. (1997a, b) have made collectively self-reproducing DNA and peptide systems. Quite directly, Lee et al. (1997a, b) have shown that self-reproduction does not depend upon the double helix structure of DNA and RNA. Thus, self-reproduction on a basis other than template replication, transcription and translation has been achieved. Further, work on the origin of life based on self-reproducing liposomes (Mavelli and Luisi 1996), the theory of the probable emergence of collectively autocatalytic sets of molecules (Kauffman 1993) and autocatalysis in organic reaction mixtures (Smith and Morowitz 2004) begins to suggest a broad physical basis for life in the cosmos. Molecular autonomous agents have yet to be created, but work cycles and molecular motors are accomplished experimental facts.

If we succeed in creating or finding life which is radically different from contemporary earth life, the way will open up for general biology, and a new union of physics, chemistry, biology and the information sciences. Core to this, we feel, will be an understanding of propagating organization of process.

We comment that it is unlikely that very early life was based on DNA, RNA and proteins via transcription and translation, given the huge complexity of the molecular apparatus to achieve these events, including encoded enzymes that charge transfer RNA with the correct amino acids to achieve translation. We emphasize this because we wish to place a discussion of information and its relation to work, constraint and propagating organization in a wider context than the contemporary debate among philosophers of biology and biologists about the information status of the DNA → RNA → protein chain.

In turn, a general biology will necessarily confront us with a discussion of evolution by heritable variation and natural selection, perhaps typically without the familiar concepts of DNA, gene frequencies, alterations of gene frequencies as the microevents of microevolution and the diverse philosophic opinions that have ranged over these issues. We will have to explore the general conditions allowing for evolution and the emergence of biospheres.

Darwinian Adaptations and Preadaptations

Were one to have asked Charles Darwin what the function of the heart is, he would presumably have responded that the function of the heart is to pump blood. But the heart has a wealth of other causal consequences, such as heart sounds. Heart sounds are not the function of the heart. That is, the causal consequence of the heart that matters, the virtue for which it was selected, was the pumping of blood. So, the function of a part (or organ) of an organism is typically, if not always, a subset of its causal consequences. This has major implications. Among these, the function of a part (or organ) of an organism cannot be analysed except in the context of the whole organism in its selective environment. But further, this fact is just one of the reasons that biology cannot be reduced to physics. In Kauffman and Clayton (2006), it is argued that, if we grant the physicist a theory of everything, say string theory to cite one example, and the capacity to deduce upwards to all that occurs in the universe – an impossibility given throws of the quantum dice – the physicist could deduce all the causal features of the heart, but would have no way to pick out the pumping of blood as the relevant causal property which is the function of the heart and which is the property that gave rise to the evolutionary emergence of this organ. To do so, the physicist would have to discuss whole organisms as causal agents in their own right, evolving under natural selection in changing environments. That is, the physicist would have to become a biologist and talk biology talk. Thus, biology cannot be reduced to physics, rather physics has to be lifted up to biology.

A second reason we feel biology is emergent with respect to physics is that Darwin's natural selection is utterly neutral with respect to the physical basis of heritable variation and hence natural selection. Life might be based on DNA, RNA and proteins, or might be based on autocatalytic organic chemical reaction systems, and/or polymer systems that create a bounding membrane. This implies that a physicist armed with a theory of everything might (actually could not) deduce that a specific molecular autonomous agent would have offspring of differential reproductive success; the physicist cannot deduce Darwin's natural selection itself which transcends any specific realization of it. Indeed, for small changes in the constants of nature, life might still be possible, hence Darwin's 'higher order' or emergent law cannot even be reduced to the physics of this universe.

In short, for these and other reasons, we wish to join forces with those who argue for a limitation of reductionism and the reality of emergence with respect to the furniture of the universe (Silberstein 2003).

Darwin had many brilliant insights. Among these is what is now called a Darwinian preadaptation. Here the central concept is that a causal property of a part of an organism that is not of selective significance in the normal environment might become useful in a different environment, and hence become subject to selection. It is critical to point out, first, that Darwinian preadaptations have occurred repeatedly in evolution, and second, that such an evolutionary step results in the emergence in the biosphere of a novel function. For example, lungs evolved from the swim bladders of certain early fish. The swim bladders, partially filled with water and partially with air, adjusted the height in the water column to establish neutral buoyancy of the fish. But the swim bladder, with air in it, was preadapted for use as a lung, and air breathing was a novel functionality with its own causal consequences that allowed life to conquer land thereby changing the universe.

We now raise a central question discussed in *Investigations*. Is it possible to say ahead of time what all possible Darwinian preadaptations are for human beings, or for the whole biota of the contemporary biosphere for that matter? The answer appears to be 'No'. We cannot finitely prestate all possible Darwinian preadaptations. Part of the difficulty, or impossibility, in doing so is that we cannot even begin the task of prestating what all possible selective environments will be. That is, there appears to be no finitely stateable procedure which would allow us to enumerate all possible selective environments.

Part of the challenge is that the concept of such environments is systematically vague. It is not even clear how to begin on the project of listing all possible environments for all actual, let alone possible, organisms. While we do not know how to prove our claim, we believe it to be true and shall assume that it is.

We point out that the property or causal consequence which becomes the subject of a Darwinian preadaptation need not be a mutant property. It might be a normal feature of the organism, but normally of no selective significance until the new environment is encountered. Therefore, an attempt to enumerate the possible preadaptations by trying to count the number of mutations possible to a genome is irrelevant. Darwinian preadaptations cannot, in general, be prestated.

Much follows from the claim that we cannot finitely prestate all possible Darwinian preadaptations of all contemporary organisms. First, it means in a radical sense that we cannot predict the future evolution of the biosphere. We literally have no idea of what such preadaptations may be. Second, it means that a frequency interpretation of probability statements does not apply to possible probability statements about the evolution of the biosphere. In the normal frequency interpretation of probability, say that a fair coin will be heads about 5000 times out of 10,000 coin flips, one can finitely prestate all possible outcomes. This is not possible for the evolving biosphere. Third, and dramatically, the incapacity to say ahead of time what the relevant preadaptations will be means that we cannot write down a stateable set of variables in equations whose dynamics captures the evolution of the biosphere. But all our mathematical techniques in physics begin with a prestatement of the full set of variables and the configuration space of the system. This is true in Newtonian dynamics, statistical mechanics, general relativity and in quantum mechanics if one does not believe in hidden variables. If one believes in hidden variables then because they are hidden they cannot be prestated, hence the caveat for quantum mechanics.

But we cannot prestate the configuration space of the biosphere. Now a classical physicist might argue that, if we take the solar system, it is just a large classical $3N$ dimensional system where N is the number of particles in the solar system and the current biosphere within the rest of the solar system is a point in that space. Let us grant the move. Then, we rejoin, the physicist has no way to pick out the collective variables, the lungs and hearts and wings, and features of the environment that are the relevant causal variables for the ongoing evolution of the biosphere. Thus, again we see that we cannot write down causal laws with a prestated set of (collective) variables for the evolution of the biosphere.

We shall not discuss it further here, but the same incapacity to prestate the evolution of the economy and its technology also arises, as does the incapacity to prestate the evolution of human culture. But all this has the deepest implications. Reductionist science is powerful but is limited. This sets us free in astonishing ways, for organisms live their lives forward, they do not deduce them. We appear to live in a universe in which our reductionistic worldview is inadequate: there is the emergence of life, and value as we discuss below. Human language and culture also represent propagating organization (Logan 2006, 2007). Moreover, we live in and partially co-create a ceaselessly 'creative' biosphere, economy and human culture. This glimmers a scientific world view beyond reductionism with broad potential societal ramifications (Kauffman 2006).

Shannon Information

Claude Elwood Shannon's (1948) information theory has been a brilliant mathematical construct. At its core, Shannon envisioned a Source with a set of messages, symbol strings, over which a well-defined probability distribution might be attributed. Then he envisioned a (perhaps noisy) channel over which information is transmitted. He then envisioned a receiver and, importantly, a decoder. Shannon's move was to calculate the entropy of the set of messages at the Source. The information that propagated down the channel and was received at the receiver removed uncertainty with respect to the entropy of the Source. This reduction of uncertainty, hence the lowering of the entropy of the Source, constitutes the amount of information transmitted. One interpretation, not given by Shannon himself who abjured to say what information 'is', is that information is just the reduction in uncertainty at the receiver. This definition leaves open exactly what the claim might mean. It might be the reduction of uncertainty in a human receiver's mind, for example.

Importantly, and widely recognized, is the fact that *Shannon information* considers the amount of information, nominally in bits [or binary digits], but is devoid of semantics. There is no sense of what information is 'about' in Shannon information.

Now we ask whether Shannon information applies to the evolution of the biosphere. We answer that it does not. In particular, Shannon information requires that a prestated probability distribution (frequency interpreted) be well stated concerning the message ensemble, from which its entropy can be computed. But if Darwinian preadaptations cannot be prestated, then the entropy calculation cannot be carried out ahead of time with respect to the distribution of features of organisms in the biosphere. This, we believe, is a sufficient condition to state that Shannon information does not describe the information content in the evolution of the biosphere.

There are further difficulties with Shannon information and the evolving biosphere. What might constitute the 'Source'? Start at the origin of life or the last common ancestor. What is the source of something like 'messages' that are being transmitted in the process of evolution from that Source? The answer is entirely unclear. Further, what is the transmission channel? Contemporary terrestrial life is based on DNA, RNA and proteins via the genetic code. It is insufficient to state that the channel is the transmission of DNA from one generation to the next. Instead, one would have to say that the actual 'channel' involves successive life cycles of whole organisms. For sexual organisms, this involves the generation of the zygote, the development of the adult from that zygote, the pairing of that adult with a mate and a further life cycle. Hence, part of one answer to what the 'channel' might be is that the fertilized egg is a channel with the Shannon information to yield the subsequent adult. But it has turned out that even if all orientations of all molecules in the zygote were utilized, there is not enough information capacity to store the information to yield the adult. This move was countered by noting that, if anything, development is rather more like an algorithm than an information channel (Apter and Wolpert 1965). In short, a channel to transmit Shannon information along life cycles does not exist, so again, Shannon information does not seem to apply to the biosphere.

It seems central to point out that the evolution of the biosphere is not the transmission of information down some channel from some source, but rather the persistent, ongoing, co-construction, via propagating organization, heritable variation and natural selection, of the collective biosphere. Propagating organization requires work. It is important to note that Shannon ignored the work requirements to transmit 'abstract' information, although it might be argued that the concept of constraints is implicit in the restrictions on the messages at the Source. While we mention this, we have no clear understanding physically of what such constraints are.

One might be tempted to argue that a Shannon-like information theory could be applied to the vast set of selective events that have led to the specific DNA sequences that are in contemporary organisms. But does this move work? Can we specify a finite ensemble of possible DNA sequences out of which the present DNA sequences have been derived? If we consider all DNA sequences longer than, say 1000 nucleotides, it would take vastly large repetitions of the history of the universe for the universe to construct one copy of each possibility. This cannot physically constitute the ensemble. Is the ensemble the set of DNA sequences that have been explored in the actual evolution of the biosphere, some accepted, most rejected? This approach initially seems promising but has the obvious difficulty that we cannot specify the ensemble explored in 3.8 billion years, hence do not and cannot know the Shannon information content of the biosphere. A further difficulty with this approach is that it measures the information content of the biosphere as a function of the number of DNA sequences 'tried' in evolution. But very different numbers of attempted mutations might have led to the same biosphere, hence quantitating the information of the biosphere by the number of attempted DNA mutations is not in direct correspondence to any specific biosphere.

We conclude that a Shannon information content analysis of the information content of the evolving biosphere is not legitimate.

Schrödinger's Aperiodic Crystal: 'Instructional' Information as Constraint or Boundary Condition

In *What is Life*?, Erwin Schrödinger ([1944] 1992) was concerned with the order in organisms and hence the physical basis of the gene. He argued, based on X-ray mutation induction frequency, that the gene must have a few hundred to a few thousand atoms, and pointed out that statistical mechanical equilibria cannot account for the stability of the organism over generations. He then posited that quantum mechanics in the form of chemical bonds was the answer. Then he brilliantly pointed out that the order of life cannot be based on a periodic crystal, for such a crystal cannot say a lot, or carry much information. He placed his bet on aperiodic crystals which can, in strong contrast, say a lot, or carry much information, even a microcode which will somehow specify the adult.

He was brilliantly right and presaged DNA and the genetic code. Now we come to the critical issue. In just what sense can an aperiodic crystal 'say a lot?' Schrödinger did not himself say more than suggest that the aperiodic crystal can contain a microcode.

We believe Schrödinger was deeply correct, and that the proper and deep understanding of his intuition is precisely that an aperiodic solid crystal can contain a wide variety of microconstraints, or micro boundary conditions, that help cause a wide variety of different specific events to happen in the cell or organism. Therefore, we starkly identify information, which we here call *instructional information* or *biotic information*, not with Shannon, but with constraints or boundary conditions. The amount of information will be related to the diversity of constraints and the diversity of processes that they can partially cause to occur. By taking this step, we embed the concept of information in the ongoing processes of the biosphere, for they are causally relevant to that which happens in the unfolding of the biosphere.

We therefore conclude that constraints are information and, as we argue below, information is constraints which we term as instructional or biotic information to distinguish it from Shannon information. We use the term instructional information because of the instructional function this information performs and we sometimes call it biotic information because this is the domain it acts in, as opposed to human telecommunication or computer information systems where Shannon information operates. This step, identifying information as a constraint or boundary condition, is perhaps the central step in our analysis. We believe it applies in the unfolding biosphere and the evolving universe, expanding and cooling and breaking symmetries, that we will discuss below.

Is this interpretation right? It certainly seems right. Precisely what the DNA molecule, an aperiodic solid, does, is to 'specify' via the heterogeneity of its structural constraints on the behaviour of RNA polymerase, the transcription of DNA into messenger RNA. Importantly, this constitutes the copying or propagating of information. Also, importantly, typically, the information contained in aperiodic solids requires complex solids, i.e. molecules, whose construction requires the linking of spontaneous and non-spontaneous, exergonic and endergonic, processes. These linkages are part of the work cycles that cells carry out as they propagate organization.

It is essential to note that the set of constraints in a contemporary cell is not merely the DNA and RNA, but lies also in the specific stereochemistry of a vast horde of specific molecular species. So, when an enzyme binds two substrates and holds them in proximity, lowering the potential energy barrier to their joining, the enzyme is acting as a constraint on the motion of the two substrates, hence as a catalyst. The working of a cell is, in part, a complex web of constraints, or boundary conditions, which partially direct or cause the events which happen. Importantly, the propagating organization in the cell is the structural union of constraints as instructional information, the constrained release of energy as work, the use of work in the construction of copies of information, the use of work in the construction of other structures and the construction of further constraints as instructional information. This instructional information further constrains the further release of energy in diverse specific ways, all of which propagates organization of process that completes a closure of tasks whereby the cell reproduces.

Our discussion here has some of the flavour of Sterelny and Griffiths (1999) in their discussion of an extended concept of information beyond DNA, RNA and protein sequences. On the other hand, none of those who have written on the concept of information in biology have taken up the struggle to relate it to constraints, work, and propagating organization of process such as that in reproducing cells.

The Relativity of Information

We have argued that the Shannon conception of information is not directly suited to describe the information of autonomous agents that propagate their organization. We have also defined a new form of information, instructional or biotic, as the constraints that direct the flow of free energy to do work.

The reader may legitimately ask the question 'isn't information just information?', i.e. an invariant like the speed of light. Our response to this question is no, and to then clarify what seems arbitrary about the definition of information. Instructional or biotic information is a useful definition for biotic systems, just as Shannon information was useful for telecommunication channel engineering, and Andrey Nikolaevich Kolmogorov (Shiryayev 1993) information was useful for the study of information compression with respect to Turing machines.

The definition of information is relative and depends on the context in which it is to be considered. There appears to be no such thing as absolute information that is an invariant that applies to all circumstances. Just as Shannon defined information in such a way as to understand the engineering of telecommunication channels, our definition of instructional or biotic information best describes the interaction and evolution of biological systems and the propagation of organization. Information is a tool, and as such, it comes in different forms. We therefore would like to suggest that information is not an invariant, but rather a quantity that is relative to the environment in which it operates. It is also the case that the information in a system or structure is not an intrinsic property of that system or structure; rather it is sensitive to history and environment. To drive home this point, we will now examine the historic context in which Shannon information emerged.

Before delving into the origin of Shannon information, we will first examine the relationship of information and materiality. Information is about material things and furthermore is instantiated in material things, but is not material itself. Information is an abstraction we use to describe the behaviour of material things and often is thought as something that controls, in the cybernetic sense, material things. So, what do we mean when we say the constraints are information and information is constraints.

'The constraints are information' is a way to describe the limits on the behaviour of an autonomous agent who acts on its own behalf, but is nevertheless constrained by the internal logic that allows it to propagate its organization. This is consistent with N. Katherine Hayles' (1999: 72) description of the way information is regarded by information science: 'It constructs information as the site of mastery and control over the material world'. She claims, and we concur, that information science treats information as separate from the material base in which it is instantiated. This suggests that there is nothing intrinsic about information, but rather it is merely a description of or a metaphor for the complex patterns of behaviour of material things. In fact, the key is to what degree information is a completely vivid description of the objects in question.

This understanding of the nature of information arises from Shannon's original formulation of information:

> The fundamental problem of communication is that of reproducing at one point either exactly or approximately a message selected at another point. Frequently the messages have meaning; that is they refer to or are correlated according to some system with certain physical or conceptual entities. These semantic aspects of communication are irrelevant to the engineering problem. The significant aspect is that the actual message is one selected from a set of possible messages. The system must be designed to operate for each possible selection, not just the one that will actually be chosen since this is unknown at the time of design.
>
> If the number of messages in the set is finite then this number or any monotonic function of this number can be regarded as a measure of the information produced when one message is chosen from the set, all choices being equally likely.
>
> ([1948] 1949: 31–32)

A number of problems for biology emerge from this view of information. The first is that the number of possible messages is not finite because we are not able to prestate all possible preadaptations from which a particular message can be selected and therefore the Shannon measure breaks down. Another problem is that for Shannon the semantics or meaning of the message does not matter, whereas in biology the opposite is true. Biotic agents have purpose and hence meaning.

The third problem is that Shannon information is defined independent of the medium of its instantiation. This independence of the medium is at the heart of a strong AI [artificial intelligence] approach in which it is claimed that human intelligence does not require a wet computer, the brain, to operate but can be instantiated onto a silicon-based computer. In the biosphere, however, one cannot separate the information from the material in which it is instantiated. DNA is not a sign for something else – it is the actual thing in itself, which regulates other genes, generates messenger RNA [or mRNA], which in turn control the production of proteins. Information on a computer or a telecommunication device can slide from one computer or device to another and then via a printer to paper and not really change, Marshall McLuhan's 'the medium is the message' aside. This is not true of living things. The same genotype does not always produce the same phenotype.

According to the Shannon definition of information, a structured set of numbers, like the set of even numbers, has less information than a set of random numbers because one can predict the sequence of even numbers. By this argument, a random soup of organic chemicals has more information than a structured biotic agent. The biotic agent has more meaning than the soup, however. The living organism with more structure and more organization has less Shannon information. This is counterintuitive to a biologist's understanding of a living organism. We therefore conclude that the use of Shannon information to describe a biotic system would not be valid. Shannon information for a biotic system is simply a category error.

A living organism has meaning because it is an autonomous agent acting on its own behalf. A random soup of organic chemicals has no meaning and no organization. We may therefore conclude that a central feature of life is organization – organization that propagates.

Semiosis: A Special Case of Constraint as Information

We next consider the minimal physical conditions for semiosis. We shall not concern ourselves with fully human linguistic symbols but with the semiosis of our minimal molecular autonomous agent. Consider an agent that is confronted by molecules in its environment, which constitute yuck or yum. To respond to these environmental features, the agent, assumed to be bounded (Kauffman and Clayton 2006), must also have yuck and yum receptors capable in the simplest case of 'recognizing' molecules of yuck or yum, and responding appropriately by avoiding yuck and eating yum. Assume such molecular machinery exists in the agent. They, of course, exist in prokaryotic and eukaryotic cells. We wish to say that the agent confronting yuck or yum receives information 'about' yuck or yum. This appears to constitute the minimal physical system to which semiotic information might apply. And it is worth noting that the 'meaning', or semiotic content of the yuck and yum molecules is built into the propagating organization of the cell. The cell, we want to say, has embodied knowledge and know-how with respect to the proper responses to yuck and yum, which was assembled for the agent and its descendants by heritable variation and natural selection.

The existence of yuck and yum as semiotic signs is a subcase of constraint as information. How does the agent detect yuck? A concrete case would be that a yuck molecule binds a yuck receptor, constraining the receptor's motions, which in turn acts as a constraint in unleashing a cell signalling cascade leading to motion away from yuck. Further, if yuck is present below a detection threshold, it will not be detected by the agent. Hence that threshold, and the receptor itself, act as a constraint partially determining the behaviour of the agent in fleeing or not fleeing.

One can construct an underlying set theory analysis for yuck and yum semantics in two equivalent ways: The first posits a set of instances and a set of properties to which each instance is assigned. The second posits a set of instances and detectors, or classifying operators, that classify 'properties' of instances. Note that in the second case, those properties need not themselves be discussed because the detectors do the job. If the second stance is taken, then detectors, 'yuck' and 'not yuck' suffice and no extension beyond instructional information is required. If the second stance suffices, we want to say not only that constraints are information but also that information is constraints. We recognize that this second step is arguable and do not analyse this issue further here.

Semiotic information cannot itself embody 'agentness' for it has no agency, but identified agents can be observed to respect the semiotic interpretation like yuck and yum. This inspectable behaviour provides the opportunity to attribute constraint-directed behaviour to the agent organism.

Another important point in this attempt to understand propagating organization is that the semiotic behaviour can identify a source of free energy, yum in this case, from which work can be extracted and propagate in the cell. This behaviour is part of a theory that unifies matter, energy, information and propagating organization.

We end this section with the description of a final interesting feature of the yum receptor. A wide variety of molecules might bind to the yum receptor with modest affinity, hence mimic true yum molecules. So, the yum receptor can be 'fooled'.

This might allow another agent to emit a poison that mimics the yum molecule, fools the receptor and leads to the death of the agent. So evolves the biosphere. Now ask, can a Shannon channel be 'fooled?' Clearly noise can be present in the channel. Due to noise, a 1 value can replace a 0 value in the constrained sense of 1 and 0 as subsets of the physical carriers of 1 and 0. But the Shannon channel cannot be fooled: 'fooling' is a semantic property of detectors, hence not present in a Shannon channel. Therefore, while one might be tempted to measure the amount of semiotic information using a Shannon-like approach, the fact that semiosis in an organism can be fooled suggests that a symbol-based Shannon move is inappropriate.

We conclude that semiotic information in molecular agents such as organisms is a special case of information as constraint. For semiotic information to be 'about' something, and to be extracted, it appears that a constraint must be present in one or more variables that are themselves causally derived from that which the information is about.

Like the threshold level of yum needed for detection, to use the information, the extracted semiotic information must do work on some system. That work might copy the information, for example, into a record, or might construct constraints on the release of energy which is further work. Here, semiotic information becomes part of propagating organization.

We comment that in standard semiotic analyses with human agents and language, there are three elements to semiotic information, namely,

1. The subject of the information or the agent being informed;
2. The object of the information or what the information is about; and
3. The possibly arbitrary, sign or symbol referring to the object.
4. With Jacques Monod (1971) in *Chance and Necessity*, we add that allosteric chemistry allows arbitrary molecules to cause events. If we wish to call such molecules 'symbols' that 'refer to' 'yum', the standard semiotic analysis just noted applies to molecular autonomous agents. Note that Monod's example is broader than DNA, RNA and proteins. It is the general arbitrariness of allosteric chemistry that allows arbitrary molecules to cause events. Information is thus broader than coding.

Heritable Variation and Natural Selection as Assembly Processes

We have now grounded biotic information as 'instructional information' or constraint, or boundary condition, that partially causes subsequent events in the unfolding of the biosphere. In this view, information is not an abstraction but is causally efficacious in the biosphere and, we argue below, in the unfolding of the abiotic universe. And we have grounded semiotic information as information detected about external (to the agent) features of the environment about which it learns. These semiotic cases are also cases of constraints, or boundary conditions, detecting and categorizing inputs and partially causing subsequent events. We note again that we remain neutral for the moment about whether information needs to be extended beyond instructional information for a set theory analysis of the categorization of objects. At the level of complex molecules, as noted above, the universe has not had time to create all possible versions. For example, the universe has not had time to create all possible proteins to length 200 [amino acids], by about 10^{67} repetitions of the history of the universe.

Consider a simple set of organic molecules and all the reactions they can collectively undergo. Call the initial set of molecules the *Actual*. Now among the reactions that might happen, some may lead to molecular species that are not present in the initial Actual. Call these new molecular species the *Adjacent Possible*. They are the molecular species that are reachable in a single reaction step from the current actual. It is of fundamental importance that the biosphere has been evolving into the adjacent possible for 3.8 billion years, from an initial diversity of perhaps 1000 organic molecules to trillions. The biotic world advances into the adjacent possible in terms of molecules, morphologies, species, behaviours and technologically, for example, from pressure flaked stones; it lurks in everything from the global economy to the computer, and the millions of products in the current global economy.

Once at a level of complexity sufficiently above the atom, the universe, the biosphere and the technosphere can never exhaust the diversity of things and events that can happen. The evolving universe and biosphere advance persistently into the adjacent possible. This means that what comes to exist at these levels of complexity is typically unique in the universe.

Now consider a heritable variation which gives rise to a new constraint, physical biotic information, that helps cause a sequence of events in a molecular agent. If that heritable variation is to the selective benefit of the agent, the new constraint, the new biotic information, will be grafted into the organism, its progeny, and the ongoing evolution of the biosphere.

It is essential to note that in the absence of heritable variation, an increase in fitness and natural selection, this new functionality would not come to exist in the universe: but lungs and flight have come to exist. The mechanisms of heritable variation and natural selection comprise an assembly process by which propagating organization is modified in normal Darwinian adaptations and preadaptations where new functionalities arise, and these modifications are built into the ongoing evolution of the biosphere.

It is clear then that heritable variation and natural selection are sufficient mechanisms in the biosphere to build an expanding mesh of functionalities as the biosphere invades the adjacent possible.

The Evolution of the Abiotic Expanding Universe: Propagating Organization Diversifying Sources of Constraint, Free Energy and Coupling of Spontaneous and Non-spontaneous Processes

We here ask whether we can find generalizations of the above analysis of information, matter, energy, constraint, work, in the biosphere, in the abiotic expanding universe.

For some time, scholars have struggled to find the union of matter, energy and information. Cases such as James Clerk Maxwell's demon, the Bekenstein bound on the entropy of a black hole, and the holographic principle, all seem to be places in physics where matter, energy and information come together. These cases merit attention but we leave them unanalysed, except for this comment.

For information to be united with matter and energy, information must be part of the physical unfolding of the universe. Thus, consider Maxwell's demon. It has been shown that the demon cannot 'win' with respect to the *second law of thermodynamics* for a closed equilibrium system (Kauffman 2000). However, in a non-equilibrium setting, the demon can win by making measurements that reduce the entropy of the measured system, with respect to the demon, faster than the most compressed record of the measured system grows, on average, in length. Now physicists usually end their argument with a claim rather like, 'Then, in principle, work could be extracted'. Such a statement is inadequate for a theory that unites matter, energy and information. What is required is that, in the non-equilibrium setting, a displacement from equilibrium that is a source of free energy must be detected by at least one measurement; a physical system able to couple to that source of free energy must have come to exist and must actually extract free energy and must release that energy in a constrained way to carry out actual work. Thereafter, this work may propagate.

If we conceive of an abiotic physical system able to carry out these processes of measurement and work extraction in the abiotic universe, it will have to be an abiotically derived system able to perform such measurements, recording the results and employ the record of the measurements to extract actual work. Such a system will be a case of propagating organization with boundary conditions as constraints, including measurements in the record as constraints on the behaviour of the system conditional on the recorded measurements, and the constrained release of energy in work. Whether the coming into existence in the universe of such a system is plausible abiotically is certainly open to question but may be worthy of consideration. Biotically, of course, such systems abound: sources of free energy from sunlight to prey are detected and coupled to work extraction. Records of sources of free energy in the form of food are seen in ant pheromone trails. The measurement of a source of free energy and extracting that free energy typically involves thresholds and other constraints or boundary conditions. For example, ants will not follow a pheromone trail if it is below a detection threshold, and the boundaries of the trail are boundary conditions on the ants' motions.

These considerations suggest that we take information to be constraint or its physical equivalent, boundary conditions that partially cause events, where the coming into existence of the constraint is itself part of propagating organization. If we do so, the issue starts to clarify in a simple way. It is fully familiar in physics that one must specify the laws, particles, the initial and boundary conditions, then calculate the behaviour of the system in a defined state space. It is common, as noted, in physics, to 'put in by hand' the boundary conditions, as in the cylinder and piston case. But in the evolving biosphere, itself part of the evolving universe, and in the evolving universe as a whole, new boundary conditions come into existence and partially determine the future unfolding of the biosphere or the universe. These evolving boundary conditions and constraints are part of the propagating organization of the universe.

We consider a single, but complex case in cosmic evolution: molecular grains found in interstellar space. These grains aggregate up to the scale of planetesimals [(astronomical objects in, e.g. protoplanetary and debris disks)]. The grains have surfaces with complex molecular features on which complex chemistry appears to be occurring. The grains themselves act as constraints, or boundary conditions, that confine reacting substrates, hence may catalyze reactions, some of which may be endergonic, requiring, for example, photons. In some cases, the product molecules presumably are bound to the growing grain, thereby modifying the boundary conditions afforded by the grain, which in turn modifies the chemical reactions that can occur. Furthermore, the product molecules can be novel substrates – hence novel sources of free energy – which again allow novel chemical reactions to occur. In short, the grains appear to behave as constraints that can partially guide spontaneous or non-spontaneous processes; can link these processes; can create new constraints enabling such processes and linked processes; and can create novel sources of free energy in the form of novel substrates able to enter into new chemical reactions.

Assume the above account is roughly correct. Then the growing grains appear to be cases in which matter, energy, and continuously evolving boundary conditions and novel sources of free energy emerge and condition the future evolution of the grains. The grains are at levels of complexity sufficiently above atoms so that what occurs is typically unique in the universe. It seems virtually sure that no two modest size grains are molecularly identical. Here we confront a union of matter, energy, and evolving and diversifying boundary conditions linking, for example, spontaneous and non-spontaneous processes and providing diversifying sources of free energy, which alter the ever-diversifying structures that come to exist in the evolving expanding universe. If this approach has merit, it appears to afford a direct union of matter, energy and information as constraint or boundary condition.

Population Genetics and Evolution in Any Biosphere

Philosophy of biology has largely grown up in the constrained environment of current terrestrial life. Its analysis of heritable variation and selection has largely ignored the physical basis of the propagating organization and closure of tasks that achieve the living state and underlie heritable variation. Moreover, we have discussed above the fact that at levels of complexity above atoms, the universe is on a unique trajectory into the adjacent possible. These physical facts are utterly requisite to descent with heritable variation and natural selection. But these aspects are simply assumed, without deeper analysis, as available to evolution. Life would have a hard time evolving at the level of complexity of quarks, gluons and atoms.

The diversity is insufficient at least. While we do not now know the implications of the broadened view of a general biology and the evolution of biospheres in a general biology, we suspect that these issues are worth careful consideration. We will make or find life anew in the next century almost certainly. Adaptation, preadaptation, the relation between the specific physical basis of each form of life and the capacity for heritable variations will become the subjects of intense study. And meanwhile, the possibility of general laws remains open to investigation. For example, it has long been hypothesized that cells are dynamically critical, poised between order and chaos. Recent evidence begins to support this possibility (Rämö et al. 2006; Serra et al. 2004; Shmulevich et al. 2005; Kesseli et al. 2009). Since critical networks are rare in the space of dynamical systems if cells are critical, it is precisely a marriage of self-organization affording such critical behaviour, and the selective usefulness of criticality, that would account for the putative results noted above. Perhaps molecular autonomous agents in any biosphere are dynamically critical. Perhaps the hinted 'fourth law of thermodynamics' discussed in *Investigations* is true of all biospheres. We simply do not know. But that does not imply that we should not search for such laws – laws that are emergent with respect to physics and part of the emergent endlessly 'creative' universe in which we appear to live.

We end by inviting philosophers of biology, physics and others, to help think through the potential implications of a new scientific worldview that goes beyond the reductionism of the past three and a half centuries to emergence and creative evolution in biology and the human economic and cultural realms that cannot be prestated. We believe that such a change in scientific worldview, if merited, will bring with it large societal changes.

Summary

We have travelled along a new path in which we have discussed Darwinian adaptations and the non-reducibility of biology to physics, the mysterious Darwinian preadaptations which seem to preclude finite prestatement and lead to evolution where the state space cannot be prestated. This brings us to serious doubts about whether Shannon information directly applies to the evolution of the biosphere, and leads to Schrödinger's aperiodic crystal and the hypothesis that information is constraints and boundary conditions, to semiotic information and records, and to the realization that, in the biosphere, it is heritable variation and natural selection that build the intricate web of propagating organization. This provides the basis for considering a new union of matter, energy, information-constraint and work in cells. This leads to questions about the abiotic universe, where information as boundary conditions affords a simple means to unite matter, energy and information. We have been led to doubt that Shannon information is physically instantiated, whereas the evolving universe and biosphere are.

We seek a new theory of propagating organization, the unfolding of Kant's statement at the outset of this chapter. We further seek a theory of the diversifying sources of free energy and constraints that are used to couple spontaneous and non-spontaneous processes into an ever-expanding diversity of processes in the biosphere and universe. We do not believe our analysis is fully adequate, but believe it is a start.

NOTES

1. Republished with permission from Springer Nature, from Stuart Kauffman, Robert K. Logan, Robert Este, Randy Goebel, David Hobill and Ilya Shmulevich (2008), 'Propagating organization: An enquiry', *Biology and Philosophy*, 23, pp. 27–45, permission conveyed through Copyright Clearance Center, Inc.
2. Here we use the word 'constraint' in a very general sense that includes 'global constraints' (e.g. conservation of energy, symmetry conditions, etc.) and 'local constraints' or boundary conditions (e.g. initial conditions, reflection or absorption at a spatial location).

REFERENCES

Apter, M. J. and Wolpert, Lewis (1965), 'Cybernetics and development I. Information theory', *Journal of Theoretical Biology*, 8:2, pp. 244–57.

Hayles, N. Katherine (1999), 'The condition of virtuality', in P. Lunenfeld (ed.), *The Digital Dialectic: New Essays on New Media*, Cambridge, MA: MIT Press, pp. 68–94.

Hood, Leroy and Galas, David (2003), 'The digital code of DNA', *Nature*, 421:6921, pp. 444–48.

Kant, Immanuel ([1790] 1951), *Critique of Judgement* (trans. J. H. Bernard), New York: Hafner Press, p. 221.

Kauffman, Stuart A. (1993), *The Origins of Order: Self-organization and Selection in Evolution*, New York: Oxford University Press.

——— (2000), *Investigations*, Oxford: Oxford University Press.

——— (2006), 'Beyond reductionism: Reinventing the sacred', *Edge.org*, 12 November, https://www.edge.org/conversation/stuart_a_kauffman-beyond-reductionism-reinventing-the-sacred. Accessed 5 April 2023.

Kauffman, Stuart A. and Clayton, Philip (2006), 'On emergence, agency, and organization', *Biology and Philosophy*, 21, pp. 501–21.

Kesseli, Juha, Ribeiro, Andre S. and Nykter, Matti (2009), 'Measuring information propagation and processing in biological systems', in S. Niiranen, J. Yli-Hietanen and A. Lugmayr (eds), *Open Information Management: Applications of Interconnectivity and Collaboration*, Hershey, PA: IGI Global, pp. 190–226.

Lee, David H., Severin, Kay and Ghadiri, M. Reza (1997a), 'Autocatalytic networks: The transition from molecular self-replication to molecular ecosystems', *Current Opinion in Chemical Biology*, 1, pp. 491–96.

Lee, David H., Severin, Kay, Yokobayashi, Yohei and Ghadiri, M. Reza (1997b), 'Emergence of symbiosis in peptide self-replication through a hypercyclic network', *Nature*, 390:6660, pp. 591–94.

Logan, Robert K. (2006), 'The extended mind model of the origin of language and culture', in N. Gontier, J. P. Van Bendegem and D. Aerts (eds), *Evolutionary Epistemology, Language and Culture*, Dordrecht: Springer, pp. 149–67.

——— (2007), *The Extended Mind: The Origin of Language and Culture*, Toronto: University of Toronto.

Mavelli, Fabio and Luisi, Pier Luigi (1996), 'Autopoietic self-reproducing vesicles: A simplified kinetic model', *Journal of Physical Chemistry*, 100:41, pp. 16600–607.

Maynard Smith, John (2000a), 'The concept of information in biology', *Philosophy of Science*, 67:2, pp. 177–94.

——— (2000b), 'Reply to commentaries', *Philosophy of Science*, 67, pp. 214–18.

Monod, Jacques (1971), *Chance and Necessity: An Essay on the Natural Philosophy of Modern Biology* (trans. A. Wainhouse), New York: Knopf.

Rämö, Pauli, Kesseli, Juha and Yli-Harja, Olli P. (2006), 'Perturbation avalanches and criticality in gene regulatory networks', *Journal of Theoretical Biology*, 242:1, pp. 164–70.

Schrödinger, Erwin ([1944] 1992), *What is Life?: The Physical Aspect of the Living Cell*, Cambridge: Cambridge University Press.

Serra, Roberto, Villani, Marco and Semeria, Alessandro (2004), 'Genetic network models and statistical properties of gene expression data in knock-out experiments', *Journal of Theoretical Biology*, 227:1, pp. 149–57.

Shannon, Claude Elwood (1948), 'A mathematical theory of communication', *Bell System Technical Journal*, 27:3 (July and October), pp. 379–423, 623–56.

Shiryayev, Albert Nikolayevich (ed.) (1993), 'Selected works of A. N. Kolmogorov', *Information Theory and the Theory of Algorithms*, vol. III, Dordrecht: Kluwer Academic Publishing.

Shmulevich, Ilya, Kauffman, Stuart A. and Aldana, Maximo (2005), 'Eukaryotic cells are dynamically ordered or critical but not chaotic', *Proceedings of the National Academy of Sciences*, 102:38, pp. 13439–444.

Sievers, Dane and von Kiedrowski, Günter (1994) 'Self-replication of complementary nucleotide-based oligomers', *Nature*, 369:6477, pp. 221–24.

Silberstein, Michael (2003), 'Reduction, emergence and explanation', in P. Machamer and M. Silberstein (eds), *The Blackwell Guide of the Philosophy of Science*, Malden, MA: Blackwell Publishers Inc.

Smith, Eric and Morowitz, Harold J. (2004), 'Universality in intermediary metabolism', *Proceedings of the National Academy of Sciences*, 101:36 (7 September), pp. 13168–73, https://www.pnas.org/content/101/36/13168. Accessed 5 April 2023.

Sterelny, Kim and Griffiths, Paul E. (1999), *Sex and Death: An Introduction to Philosophy of Biology*, Chicago: University of Chicago.

5

Friends, Neighbours and Enemies: An Overview of the Communal and Social Biology of Plants[1]

Roza D. Bilas, Amanda Bretman and Tom Bennett

Introduction: A Communal Existence

The discoveries of the 19th Century shook apart the idea of nature as a harmonious world. As Darwin showed in his 'lawn plot experiment', a 'struggle for existence' occurs even between humble plants; it is just a very quiet and slow struggle. And yet, when an average person thinks of interactions both within and between species, plant–plant interactions are rarely at the forefront of the mind. Even among plant scientists, interactions between plants and other plants have been traditionally overlooked relative to the interactions of plants and just about any other taxon. Plants were traditionally viewed as passive, affected by their neighbours only through indirect effects on resource availability (Pierik et al. 2013). Until the early 1980s, there was little indication that plants might have interactions every bit as complex as those seen in animals (Vicherová et al. 2020). However, as an ever-increasing body of work is demonstrating, plants do detect, respond to and interact with their neighbours. These interactions are often rather subtle and slow, and as will be described, can also be unintuitive and surprising. These qualities perhaps stem inevitably from the sessile nature of plants. Being motile gives animals the great advantage of being able to decide which individuals to aggregate with, and to avoid competitors and predators. In contrast, a plant cannot choose or change its neighbours. And most likely, it cannot kill them either (not quickly, anyway). This changes the game: a plant will often be inadvertently stuck next to problematic neighbours for its entire life; in the vast majority of cases, this is simply not a feature of animal life.

Thus, for all the pests, pathogens and predators a plant might meet, the biggest and most long-lasting biotic stress it will encounter will probably be its neighbours. Neighbouring plants represent a direct threat to the resources (light, water, mineral nutrients) that a plant must access, and – regardless of how the interaction plays out – it would therefore seem imperative that plants can detect and respond to their neighbours. But can they, and do they? In this review, we aim to provide an overview of our current knowledge of plant–plant interactions, particularly focusing on the question of whether plants can detect and differentially respond to friend, neighbour and enemy.

Plant–Plant Interactions: Theory and Controversy

There is a range of interactions that could in theory occur between two neighbouring plants, with different fitness outcomes for the actor (plant A) and the recipient (plant B) (Figure 5.1). In describing these, we largely adhere to the definitions in the review by Stuart A. West, Ashleigh S. Griffin and Andy Gardner (2007) from the original ideas of William D. Hamilton (1964, 1970). Plant A could compete for resources with plant B, in the process reducing B's fitness. However, depending on the competitive ability of B, the outcome for the fitness of A might be beneficial, neutral or costly. In a 'selfish' outcome A benefits at B's expense, for instance if the competition is highly asymmetric, while in a 'spiteful' outcome there is a fitness cost to both A and B, for instance if the competitors exhaust the resource (the so-called 'Tragedy of the Commons') (Hamilton, 1970). Alternatively, plant A could cooperate with plant B, in the process increasing B's fitness. The outcome of this might also be beneficial to plant A ('mutualism') – for instance by avoiding spending resources on competition – neutral, or occur at a cost to A ('altruism'). In the case of altruism, A might still indirectly benefit if B is a close relative, since this will increase the inclusive fitness of A (i.e. kin selection). Plant A could behave without regard to plant B ('live and let live'), but a neutral strategy could still inadvertently have a range of outcomes for plant B. 'Facilitation' describes beneficial interactions where there is negligible cost or benefit to A but a benefit to B (similar to byproduct reciprocity in West et al. 2007). Facilitation has been more strictly defined as the positive effect on the environment by one species that improves the fitness of another species (Bertness and Callaway 1994; Bronstein 2009).

Two implicit assumptions when proposing the existence of these different interactions are that plants can actively detect and distinguish their neighbours, and actively respond by altering their growth or behaviour relative to a 'no-neighbour' scenario. However, in neither case is there currently conclusive

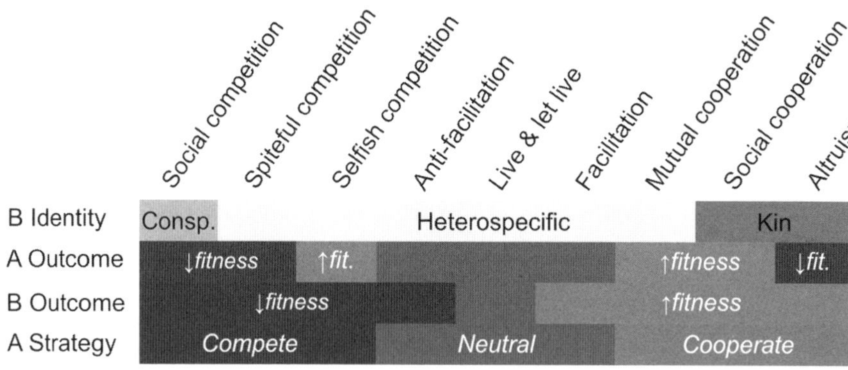

FIGURE 5.1: Plant–plant interactions. Matrix showing the possible types of plant–plant interactions, depending on the strategy of the actor (a), the identity of the recipient (b), and the actual fitness outcomes for both (a) and (b). Social interactions are those occurring with members of the same species (conspecifics – consp.), either related (kin) or unrelated. These may have different properties from equivalent interactions with heterospecifics because of the symmetry of the competition, or because the actor ultimately benefits from the increased fitness of the recipient by increased inclusive fitness (kin selection). Facilitation is an inadvertent benefit to the recipient caused by plant A's impact on the environment; in theory, plant A could also inadvertently reduce the recipient's fitness (anti-facilitation). [Colour figure at wileyonlinelibrary.com.]

supporting evidence. In part, this is because (as described above) the measurable outcome of interactions between plants does not necessarily reflect any 'intent' on behalf of the plants, making it very difficult to infer whether a plant is following a given strategy, or simply growing with no regard to its neighbours. These unproven assumptions create an interpretational minefield for the whole field of plant–plant interactions, which we refer to hereafter as the 'identity problem' and the 'response problem'. In terms of the identity problem, it is now fairly incontrovertible that plants do detect their neighbours (see Section on Mechanisms of Neighbour Detection), but it remains less clear whether they can distinguish between their neighbours. While there is certainly experimental evidence that plants can do this, it is primarily phenomenological, and therefore open to interpretation (see Section on Distinguishing Between Neighbours). In terms of response problem, there is again experimental evidence that competitive, cooperative and facilitative interactions do occur in plants (see Sections on Responding to Neighbours and Allelopathy). However, the evidence is rarely conclusive, and there are often other plausible explanations for these observations.

In writing this review, it is our general thesis that plants do distinguish between, and differentially respond to their neighbours. We will examine the evidence, and compare this to the null hypothesis that 'plants act to increase their fitness without reference to the identity of their neighbours'. Based on the evidence, we cannot currently reject this hypothesis, but we assess whether parsimony prevents us from accepting it, while also considering what evidence would be required to finally reject this hypothesis.

Mechanisms of Neighbour Detection

Traditionally, it was assumed that plants could only detect their neighbours passively, by detecting changes in resource (light, water and nutrient) availability caused by other plants. Above-ground, plants have well-described responses to poor light quality or shading, whether plant-generated or not (Roig-Villanova and Martínez-García 2016). Root growth responses to soil nutrient availability have been extensively characterized (e.g. Shahzad and Amtmann 2017), and will be inevitably be triggered if neighbouring plants deplete the environment of, for example, nitrate and phosphate (de Kroon et al. 2003; Nord et al. 2011; Schenk 2006). Furthermore, it is certainly clear that plants can, and do passively respond to the presence of neighbouring plants simply due to the reduced availability of resources in the environment (Pierik et al. 2013; Schenk 2006). However, it is also now clear that plants have multiple mechanisms by which they can actively detect neighbouring plants (Figure 5.2). Mostly, plants use 'cues' to do this – information that neighbouring plants cannot avoid making available for detection (Karlovsky 2008; Shelef et al. 2019). In this section, we introduce the main mechanisms of neighbour detection, critique the evidence for more speculative mechanisms, and examine whether these mechanisms might allow plants to distinguish between neighbours.

Light

Plant organs absorb, reflect and scatter incoming solar radiation, reducing its photosynthetically active radiation (PAR), red: far red ratio (R:FR), and blue fluence rate. This creates unique cues of neighbour presence, distinct from fluctuations in ambient light levels and quality. Plants are exquisitely sensitive to these perturbations in light quality, even in the absence of reduction in PAR, because they indicate current or future competition for light (Roig-Villanova and Martínez-García 2016). These responses to altered light quality are particularly associated with direct shading, but because of their extensive scattering of light, plants can use light cues to detect each other over significant distances in the absence of shading (Roig-Villanova and Martínez-García

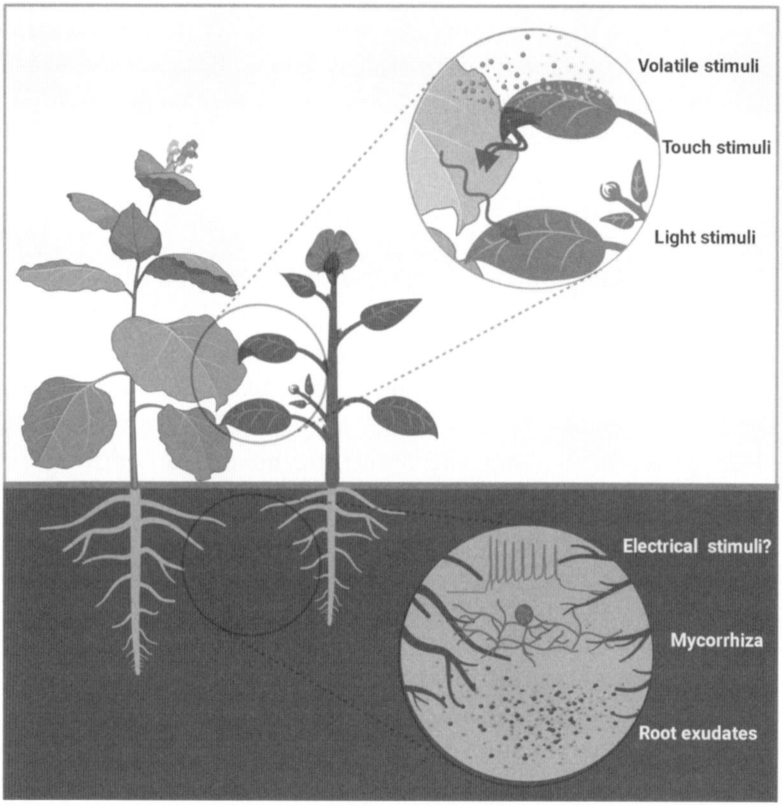

FIGURE 5.2: Mechanisms of neighbour detection. A range of above and below ground mechanisms may allow plants to detect the presence of neighbours. Whilst evidence suggests that plants do use multiple cues for neighbour detection (see Section on Mechanisms), a general understanding of which cues carry sufficient information for discrimination between species or relatives remains elusive. [Colour figure at wileyonlinelibrary.com.]

2016). Neighbour-generated light cues are detected by the well-known and highly-characterized plant photoreceptors, particular phytochromes, with their strong sensitivity to far-red light. The tips of leaves are the primary site of light cue sensing, reducing the likelihood that self-shading will trigger these responses (Pantazopoulou et al. 2017). These photoreceptor-mediated responses to neighbours represent the most unambiguous evidence of the active nature of plant–plant interactions, and have been well characterized at the ecological, physiological and molecular levels; they are reviewed in more detail elsewhere (Huber et al. 2021).

Although light quality is incontrovertibly used as a neighbour cue by plants, it is very simple in nature, and hard to envisage how it could encode specific information about the identity of neighbouring plants. In the case of Arabidopsis, arguments have been made to suggest that light signals can allow discrimination between kin and non-kin (Crepy and Casal 2015; Crepy and Casal 2016), but it would be perhaps fairer to say that light signals can allow plants to distinguish between different morphological states (Till-Bottraud and de Villemereuil 2016). Similarly, light signals can permit some level of neighbour recognition by 'phenotype matching' among tree species. For instance, sycamore trees cast deep shade, tolerate deep shade and have relatively slow growth; silver birch trees by contrast cast a weak shade, are poorly shade tolerant, and have fast growth. In the deep shade of a sycamore tree, the growth of sycamore seedlings is promoted over birch seedlings, but in the light shade of a birch tree, growth of birch seedlings is greater than sycamore seedlings (Gilbert et al. 2001). Only the 'type' of neighbour is being detected in these examples, rather than its specific identity, but this nevertheless demonstrates the potential of light signals to carry complex information.

Touch

Plants are naturally exposed to mechanical forces by a range of factors (e.g. wind, insects, physical obstacles). As such, they have mechanisms that allow them to detect and respond to these mechanical forces, which can be in part characterized as a response to 'touch' (Hamant and Haswell 2017). The effectiveness of mechanical stimulus depends on length and repetitiveness of the signal and less on the force applied (Anten et al. 2010). In the context of plant–plant interactions, plants are sensitive to even the light touch of neighbouring plants, which provides a rapid indicator of competitor presence (Markovic et al. 2016); these touch stimuli are perceived with high sensitivity by leaf trichomes and root tips (Massa and Gilroy 2003; Zhou et al. 2017). Plant-generated touch is clearly distinguishable from other mechanical stimuli, since touch, wind and mechanical damage evoke distinct molecular responses (Anten et al. 2010; Markovic et al. 2016). Responses to touch include increasing growth away from neighbouring plants, or acclimation by production of more resistant structures.

Touch by neighbouring plants seems to play a particularly important role in priming plants for further interactions, and has been found to modify both release of VOCs and of root exudates (Elhakeem et al. 2018; Markovic et al. 2019). In Arabidopsis, touch of leaf tips promotes leaf hyponasty, which move the leaves into a position in which they can better detect scattered light signals to 'confirm' the presence of neighbouring plants (de Wit et al. 2012). This highlights the interactive nature of neighbour cue-use, and the ways in which this multimodality can convey complex information. However, touch is a very simple cue and there is currently no reason to suppose that it can convey information about the identity of neighbouring plants.

Chemical Signals: VOCs

Plants emit a large range of organic chemicals into the environment, which are usually divided into volatile organic compounds (VOCs) and exudates. VOCs are typically found as free molecules in the gaseous phase, and can be released both above and below ground. In contrast, exudates would typically be secreted by the root system solubilized in water. We will cover volatiles and exudates separately, but they are perhaps best viewed as a continuum of chemical cues released into the environment by plants.

More than 1,000 different plant-emitted VOCs have been identified to date, including highly volatile compounds such as isoprene, monoterpenes, methanol and ethylene, and other more moderately volatile compounds (e.g. terpenes, methyl jasmonate, methyl salicylate and 'green leaf volatiles') (Baldwin 2010; Heil and Karban 2010; Ninkovic et al. 2019). Plants could therefore theoretically emit a very large number of VOC combinations, which might dynamically reflect both their physiological status and species (Karban et al. 2006; Pichersky et al. 2006; Ueda et al. 2012). Consistent with the ready availability of this 'information', plants can detect and respond to the presence of neighbouring plants through VOC emissions. For instance, an indication of emitter plant proximity can be 'deduced' by a receiver plant due to the fast rate of diffusion of highly volatile compounds relative to moderately volatile ones (Baldwin 2010; Heil and Karban 2010; Ninkovic et al. 2019). VOC detection and response are reviewed in detail elsewhere (Ninkovic et al. 2021).

It has been hypothesized that VOC emission evolved to quickly spread information about herbivory or pathogen attacks, although it is unclear whether their primary function is to spread this information within the same plant, or between neighbouring plants (Farmer 2001; Heil and Karban 2010; Morrell and Kessler 2017). Plants can rapidly spread information systemically to distal organs through vascular-associated calcium signalling (Toyota et al. 2018), but VOCs may act as second system to achieve the same effect. While VOC emissions would attenuate rapidly over longer distances, they would be very efficient over short distances, particularly where the corresponding vascular connection is very long (i.e. between neighbouring branches connected to a distant stem) (Heil and Adame-Alvarez 2010). Nevertheless, priming of defence responses in related neighbours could also have driven the evolution of VOC emission through effects on inclusive fitness (Karban et al. 2013; Shiojiri and Karban 2008). The simple nature of common

VOC signals would also allow eavesdropping by unrelated neighbours (Karban et al. 2003; Ninkovic et al. 2013). This unintentional sharing of information with unrelated neighbours may not be maladaptive (since reducing overall herbivory/pathogen pressure may benefit the emitting plant), but could also have led to the evolution of VOCs 'chemotypes' in a number of plant species (Gouinguene et al. 2001; Karban et al. 2014; Ninkovic 2003). There is evidence that VOC-mediated herbivory defence is more effective if signalled by related conspecific plants (Karban et al. 2013) even though the emission of VOCs may be reduced in conspecific stands (Kigathi et al. 2013).

VOCs thus seem like an excellent candidate for a mechanism by which plants not only detect, but distinguish their neighbours. As combinations, VOCs could have sufficient complexity to allow the distinction between neighbours, and there is some evidence that plants can indeed distinguish between neighbours on this basis (Karban et al. 2013). However, it is not clear exactly how plants might be able to do this; neither which VOCs might be involved in distinguishing between neighbours, nor how these volatiles are perceived. Furthermore, it should be noted that currently, VOC detection by plants has mostly been associated with defensive priming, rather than responding to the presence of neighbouring plants per se.

Chemical Signals: Exudates

Plants exude significant quantities of organic molecules into the soil (Bais et al. 2003), which play a wide range of functional roles such as conditioning soil by changing its adhesive properties or pH (Vives-Peris et al. 2020). Plants also release signalling molecules that promote the formation of beneficial symbioses with micro-organisms, that suppress pathogens, and which act allelopathically (see Section on Allelopathy) (Ehlers et al. 2020; Rolfe et al. 2019). These compounds may persist in soil due to low rates of oxidation and photodecomposition (Karlovsky 2008), and alter soil properties in a way perceptible to both contemporary neighbours and future generations, known as plant–soil feedback (Hu et al. 2018; Van der Putten et al. 2013). The soil is thus full of potential cues for the presence of neighbouring plants, but to what extent do plants actively detect and respond to these exudates?

There is reasonable evidence that plants can detect the mix of chemicals exuded by other plants, in the absence of actual neighbouring plants or any nutrient depletion, and can respond with alterations in root architecture and growth (Biedrzycki et al. 2010; Kong et al. 2018; Semchenko et al. 2014; Yang et al. 2018). Gradients of exudates in the soil might thus provide information on the proximity of neighbouring plants, as well as their physiological status, and may thus allow roots to precisely avoid neighbouring roots (Fang et al. 2013). The build-up of exudates near physical obstacles might also explain the ability of roots to avoid these obstacles without touching them (Falik et al. 2005). Since exudates possess much greater complexity than most other mechanisms of neighbour detection, root exudates have regularly been suggested to be key factors in the apparent ability of plants to distinguish self/non-self and kin/non-kin (discussed further below) (Biedrzycki et al. 2010; Semchenko et al. 2014; Yang et al. 2018).

It is certainly possible that, as combinations, exudates would have sufficient complexity to permit plants to distinguish between neighbours, but it is not currently clear which exudate compounds plants can detect. The candidates suggested so far, such as jasmonic acid and (−)-loliolide, seem likely to be generic signals, involved in triggering broad responses such as allelopathy (Kong et al. 2018). Strigolactones are a class of phytohormones that are also exuded into the soil and play a broad signalling role in the rhizosphere, including promoting formation of mycorrhizal associations (Waters et al. 2017). Since they strongly regulate plant growth, strigolactones are obvious candidates to act as plant–plant signals, but currently there is no clear evidence that this occurs, and indeed some evidence to the contrary (Kong et al. 2018). At the moment, root exudates seem the most likely candidates to act in neighbour recognition, but it remains a clear priority for the field to identify which signals are involved, whether they do encode specific information, and if so how. Root exudate detection and response is reviewed in detail elsewhere (Wang et al. 2021).

Acoustic and Electrical Stimuli

The mechanisms discussed above are unambiguously involved in neighbour detection in plants. Plants have also been proposed to use acoustic and electrical cues to detect and respond their neighbours. Plants generate acoustic vibrations by cavitation of gas bubbles in the xylem, and resulting alterations in vessel diameter (Hölttä et al. 2005; Laschimke et al. 2006), particularly in drought stressed plants (Zweifel and Zeugin 2008). Plants can also detect acoustic vibrations of sufficient magnitude via the mechanosensing pathway that integrates touch and mechanical stimuli (Ghosh et al. 2016; Ghosh et al. 2017). Acoustic vibrations can induce defence responses, suggesting a role in promoting resistance to vibration-emitting predators (Appel and Cocroft 2014; Choi et al. 2017; Kim et al. 2020), and plants can detect pollinator sound frequencies and release pollen or increase nectar in response (De Luca and Vallejo-Marin 2013; Veits et al. 2019). Thus, plants have been suggested to detect their neighbours through acoustic cues (Gagliano et al. 2012; Rodrigo-Moreno et al. 2017), but these studies are controversial, and there is currently no clear evidence for acoustic neighbour detection.

Plants generate electrophysiological action and variation potentials throughout the plant body, though their function is unclear (Davies 2006; Fromm 2006; Stahlberg et al. 2006). One recent suggestion is that continuous electrical signalling could allow communication between organs (de Toledo et al. 2019; Sukhov et al. 2019). Electric stimuli have been shown to be conducted through the soil, exchanged between neighbouring plants, and to invoke electrical signals in the receiver, whether con- or heterospecific (Volkov and Shtessel 2017; Volkov and Shtessel 2018; Volkov et al. 2019). In principle, electrical cues could be used to detect neighbours, and moreover to assess the competitive abilities of neighbours, since there is a relationship between electrical activity and the physiological state of the plant. Electric signal generation can be triggered by many external stimuli, for example wounding and touch stimuli (Degli Agosti 2014; Mousavi et al. 2013; Szechyńska-Hebda et al. 2010), and a rapidly improving understanding of electric signals has revealed specific patterns connected to water status (Comparini et al. 2020), salt stress (Wang et al. 2019) and infection by pathogens (Simmi et al. 2020). However, at the moment, there is no evidence that plants do use electrical signals for neighbour detection, though it is an interesting possibility, and more work to understand its relevance is definitely warranted.

Via Fungal Intermediaries

Symbiotic associations between plant roots and fungi are extremely common, with arbuscular mycorrhizal symbiosis present in over 70 per cent plant species (Cosme et al. 2018). As such, plants inevitably become interconnected via mycorrhizal fungi in a ‘common mycorrhizal network’ (CMN) of multiple con- and heterospecific shareholders. Therefore, it has been hypothesized that CMN might transfer interplant signals via the hyphal external surface, cytoplasmic streaming or electrical signal conduction (Barto et al. 2012; Johnson and Gilbert 2015). For instance, it has been suggested that signalling molecules, including jasmonic acid, can be transferred through CMN, allowing priming of neighbouring plants to pathogen infection (Song et al. 2010) and aphid or caterpillar herbivory (Babikova et al. 2013; Song et al. 2014). However, while there is certainly evidence that mycorrhizal colonization improves disease resistance, this might be due to nutrient support rather than priming effects (Delavaux et al. 2017). Thus far, the only well documented and widely accepted impact of CMN is the effect on plant competitiveness by nutrient transfer and growth promotion (Delavaux et al. 2017; Parniske 2008; Smith and Smith 2012). Furthermore, in general, it should be noted that CMN add no information to the system, but are simply proposed as a more efficient means of transmitting chemical cues between plants.

Distinguishing Between Neighbours

As highlighted above, a very important problem in the field is whether plants can distinguish between their neighbours. Work in this area can be broadly divided into two separate but related questions: can plants distinguish their own body from those of neighbouring plants (self/non-self recognition) and can plants distinguish between different neighbouring plants on the basis of relatedness (kin/non-kin recognition)? These areas are reviewed in detail elsewhere (Anten and Chen 2021), but we aim to provide a concise discussion here.

Self/Non-self Recognition

The self/non-self recognition question arises because all the neighbour detection cues discussed above can also be generated by the plant itself. How can a plant therefore specifically respond to the presence of a neighbour (i.e. by competing or not competing) unless it is able to distinguish between signals arising from itself and its neighbours? How do plants 'filter out' cues generated by their own body and avoid competing with themselves? Work on self/non-self discrimination has particularly focused on interactions between the root systems of plants, because that is where the most obvious evidence for (apparent) competition or non-competition between plants is found. A variety of evidence has been obtained for self/non-discrimination via the root systems, in a relatively short burst of activity.

One set of work used a system in which two plants with 'split roots' share two pots, or are grown separately in one pot (Falik et al. 2003; Gersani et al. 2001; Maina et al. 2002; O'Brien et al. 2005). Under these conditions, 'sharer' plants apparently competed with each other and over-proliferated roots relative to 'owners'. However, these results were subsequently heavily criticized, because of the failure to take into account confounding variables; although the sharer and owner plants have access to the same amount of nutrients, sharer plants have access to twice the soil volume – and soil volume in itself has a very strong effect on plant growth (Hess and De Kroon 2007; Poorter et al. 2012; Semchenko et al. 2007). A subsequent series of studies showed no evidence for over-proliferation in the presence of neighbours (Lankinen 2008; Markham and Halwas 2011; McNickle and Brown 2014; Meier et al. 2013; Nord et al. 2011; Semchenko et al. 2007), leading to suggestions that plants essentially ignored the presence of neighbours in determining root growth, and only responded to nutrient availability (McNickle and Brown 2014; Nord et al. 2011) – but again, the effect of soil volume was largely ignored. When all variables are properly accounted for, the presence of neighbouring plants reduces the growth of roots in peas - an effect in the opposite direction to that originally proposed (Chen et al. 2015).

In a second approach, clonally-propagating plants, such as strawberry, clover and buffalo grass were used to test the idea further (Falik et al. 2006; Gruntman and Novoplansky 2004; Holzapfel and Alpert 2003). These plants produce 'ramets' (new clonal individuals) which remain physiologically connected. In a two-ramet system, physiological disconnection resulted in more root growth in both ramets than if they remained connected, which was taken as evidence for the plants now recognizing each other as non-self, and competing. Again, these studies have been criticized for their experimental design (Hess and De Kroon 2007), but also defended as providing evidence for 'self/non-self recognition ... [that] cannot be denied' (Chen et al. 2012). In truth, these experiments provide no direct evidence for neighbour detection; they simply show that two small plants grow differently from one larger twin plant. The inference that this is due to self/non-self discrimination is a matter of interpretation, rather than firm experimental evidence.

Overall, perhaps the largest problem with self/non-self recognition is that no convincing mechanism has been identified that could account for it. Self/non-self recognition inevitably invokes the existence of additional contextual signals from either self or non-self that permit discrimination of the origin of cues (Chen et al. 2012). This is theoretically possible, since different species release different molecules into the environment, but as defined, self/non-self recognition requires even genetically identical plants to be able to distinguish the origin of cues. It is very difficult to see how identical plants can possibly generate distinct cues that allow unambiguous self/non-self discrimination. Thus, a variety of intricate mechanisms involving internal oscillations and/or electric signals have been suggested to explain self/non-self recognition, but there is no convincing experimental evidence for these (Chen et al. 2012; Depuydt 2014).

Kin/Non-kin Recognition

Over the last decade, interest in the self/non-self question has clearly diminished, and attention has focused more strongly on the question of whether plants can distinguish kin from non-kin (Dudley et al. 2013). Kin/non-kin recognition suggests that plants have a mechanism that allows them to distinguish closely related neighbours (kin) from all other neighbours (non-kin) (Callaway and Mahall 2007; Chen et al. 2012). Again, work on kin/non-kin discrimination has particularly focused on interactions between the root systems of plants, because that is where the most obvious evidence for competition or non-competition between plants is found. However, the phenomenon is not necessarily restricted to the roots; potential examples from the shoot system have already been mentioned (Crepy and Casal 2015; Karban et al. 2013) and there may be interactions above and below ground, with root contact being required for recognition but a response being expressed in shoots (Murphy and Dudley 2009).

As with self/non-self recognition, the initial studies which apparently demonstrated kin/non-kin recognition (Andalo et al. 2001; Donohue 2003; Dudley and File 2007) were also subject to criticism for problems in experimental design and statistical approach (Chen et al. 2012). However, subsequent studies have provided more firm evidence of kin/non-kin recognition, although again, this is primarily phenomenological (Bhatt et al. 2011; Marler 2013; Palmer et al. 2016; Takigahira and Yamawo 2019). In these experiments, plants are generally found to reduce their root growth (i.e. to 'cooperate') in the presence of kindred plants, but not (or less so) in the presence of non-kin plants. For instance, in rice plants grown together in transparent agar columns, the root systems of plants of the same variety remain completely segregated (the roots systems never touch each other), which does not occur when the plants are from different varieties (Fang et al. 2013). This is a striking result, which is very difficult to explain without the existence of some kin recognition mechanism. Further work in rice supports the existence of kin recognition across a spectrum of relatedness (Yang et al. 2018). Furthermore, plants of the same ecotype in Arabidopsis have been proposed to display reduction in root growth that is not present between ecotypes (Biedrzycki et al. 2010). This study was challenged on the basis that there is no difference in gene expression in plants grown with members of the same ecotype versus different ecotypes (Masclaux et al. 2012), but this does not preclude kin recognition in Arabidopsis; it may simply mean that the kin group operates at a different taxonomic level than the ecotype.

Although the phenomenological evidence is firmer, there is currently no clear mechanistic explanation for kin/non-kin recognition. However, it is generally easier to conceptualize how kin/non-kin recognition might work than self/non-self recognition. The work described in the previous section shows that plants can detect their neighbours through a range of different mechanisms, and that some of these have sufficient complexity to permit identity recognition. In this context, most attention has focused on root exudates, and there is certainly some evidence that these can invoke kin or non-kin responses, but it is not clear which molecules might do so (Biedrzycki et al. 2010; Semchenko et al. 2014; Yang et al. 2018). Different species certainly release different profiles of exudates, and thus the presence of 'foreign' molecules could certainly trigger a competitive response. However, an alternative model would be that plants release a molecule that positively identifies them as kin, and down-regulate their growth when exposed to this molecule (whether self-generated or not); in the absence of this signal, normal growth responses occur. Furthermore, such a molecule might be recognized with decreasing efficiency by increasingly distant kin, resulting in the observed spectrum of responses as a function of relatedness.

Although we do not yet understand its basis, the evolution of kin recognition and cooperative responses would be readily explicable via well-established effects on inclusive fitness (Hamilton 1964; Mitteldorf and Wilson 2000). Indeed, given their reproductive systems, we might expect kin cooperation to be especially strong among land plants. While animals have evolved multiple strategies to avoid inbreeding and therefore rarely have a relatedness (R) of greater than 0.5 (full-sibling) (Pusey and Wolf 1996), many plants reproduce clonally or asexually ($R = 1$), or with some degree of self-pollination ($R \sim 1$). Similarly, poor dispersal of offspring in many plants will result in local populations with a high degree inbreeding, except where self-incompatibility mechanisms are present (Barrett and Harder 2017).

Species Recognition

In addition to kin recognition, it is also possible that plants can distinguish between members of the same species and those of different species (Krannitz and Caldwell 1995; Mahall and Callaway 1991; Novoplansky 2009). Given that conspecifics usually compete for the same ecological niche, being able to detect neighbour species identity could be adaptive. For example, in experimental plots of multiple tropical tree species, coarse roots were shorter but denser in trees growing near conspecifics than heterospecifics, interpreted as a strategy to minimize competition without sacrificing nutrient acquisition (Madsen et al. 2020). However, in other experiments, there is no evidence for differential responses to conspecific versus heterospecific neighbours (Jacob et al. 2017). An intriguing (but purely theoretical) possibility is that species detection might occur by the same mechanisms as kin recognition, since conspecifics are to some extent kindred, even if only very distantly. Members of the same species might be recognized as 'distant kin', triggering weaker or different responses than close kin. Given the low dispersal of many species, plants are likely to be closely related to any conspecifics that are also their neighbours (Semchenko et al. 2013). In such populations with 'high viscosity', kin selection could occur without any requirement for a discrimination system (Hamilton 1964), though it would be highly unlikely that individuals never encountered distantly related/unrelated neighbours. Thus, a kin recognition system may not need to discriminate kin very precisely to still confer a selective benefit via inclusive fitness, which could allow the same system to be used for both kin and species recognition. This may be particularly the case in basal land plant lineages (mosses, liverworts, hornworts) which are particularly characterized by short-range dispersal through asexual reproduction, leading to continuous assemblies of closely related individuals (Frey and Kürschner 2011; Partridge and Harvey 1988; Renzaglia et al. 2000; Taylor et al. 2005). Within species, related mosses co-operatively avoid overgrowth (Proust et al. 2011), whereas heterospecific mosses seem to actively compete in response to light and VOC cues from neighbours (Vicherová et al. 2020). While species-level recognition remains largely uncharacterized, it is an intriguing prospect deserving more attention.

A Model for Neighbour Identification

If plants cannot distinguish between themselves and their neighbours, then they cannot possibly respond differentially to their neighbours. Understanding the 'identity problem' is thus at the crux of the whole field of plant–plant interactions, and its future direction. Tantalizingly, the current evidence suggests that they can indeed distinguish between their neighbours. Certainly, we cannot accept our null hypothesis that plants make no reference to the identity of neighbours, although it is perhaps not yet safe to reject it outright. But how exactly might plants distinguish between their neighbours? In our view, the phenomenological evidence for self/non-self recognition is too weak, and the mechanistic explanations too convoluted to persist with that model. Conversely, kin/non-kin recognition has better phenomenological support, and avoids some of the theoretical problems inherent in self/non-self recognition. It is also easier to see a clear selective advantage for the evolution of kin/non-kin discrimination via effects on inclusive fitness, and mechanistic explanations could be straightforward, even if currently obscure. Furthermore, kin/non-kin recognition circumvents much of the requirement for a self/non-self recognition system. The self/non-self system was invoked to allow plants to respond differently to their own cues with respect to those of their neighbours. However, in the case of kin plants, the need to respond differently to 'self' may be very much less, since plants should not compete with their kin any more than they should compete with themselves; the most important non-self plants to detect are those that are also non-kin.

Nevertheless, when responding to plant-generated cues, it is no doubt useful for plants to filter out self-generated cues. We have already encountered evidence for spatially specific cue-responses designed to filter out self-generated cues (e.g. light sensing in leaf tips), and for multimodal cue-use (e.g. touch sensitization of light, volatile and exudate signalling). We thus propose that plants use spatially restricted, multimodal processing to distinguish between self and non-self generated cues on a probabilistic basis, supported by a kin/non-kin recognition system to distinguish between the neighbours thereby identified (Figure 5.3). Whether this model is correct or not, a clear challenge for the field is to solidify our understanding of plant identity-recognition by a much deeper understanding of the mechanisms by which plants recognize the presence of other plants.

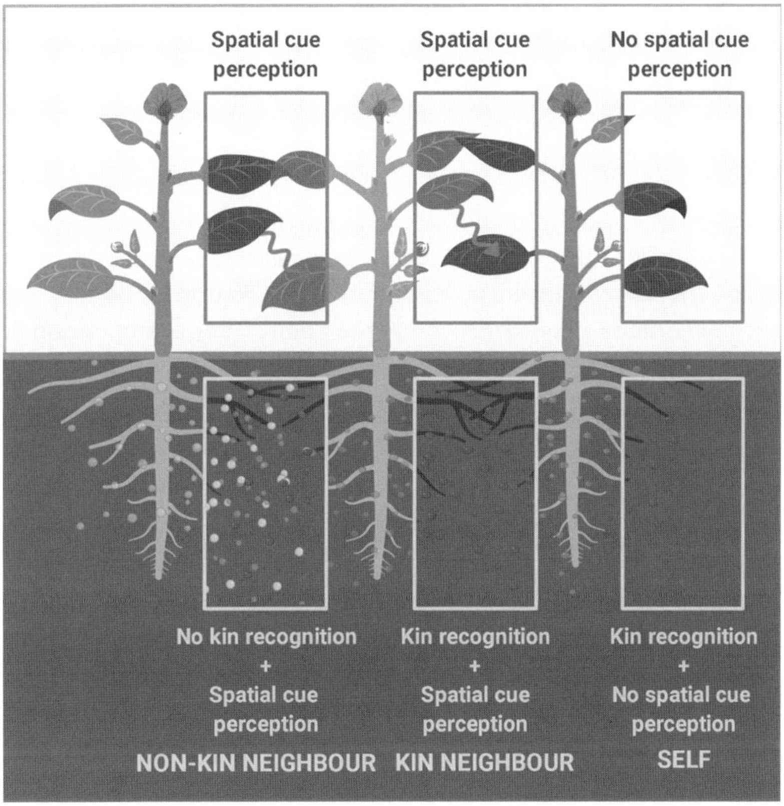

FIGURE 5.3: A possible model for neighbour discrimination with respect to kin- and self-recognition in plants. We suggest that plants could show a range of responses to neighbours based on discrimination using the integration of various cues. Plants exude a 'kin recognition' signal from their roots, but not the root tips (dark/light circles). On the right-hand side, the plant's root tips detect its own kin recognition signal at a low concentration, and detects no other spatial cues (light, touch, volatiles) for neighbour presence, thus perceiving that its roots are encountering 'self'. In the middle, the two plants detect each other's kin recognition signals at high concentration, and the presence of a kin neighbour is confirmed by above-ground spatial cues; both plants thus perceive that their roots are encountering 'kin'. On the left-hand side, the non-kin plant root tips only detect their own kin recognition signal at a low concentration. Aboveground spatial cues indicate the presence of a neighbour, so the plant perceives that its roots are encountering a non-kin neighbour. [Colour figure can be viewed at wileyonlinelibrary.com.]

Responding to Neighbours

So far, we have seen that plants can certainly detect their neighbours, and can probably distinguish between their neighbours on the basis of relatedness. But can plants use this information to differentially respond to neighbours, adopting different strategies depending on neighbour and context? In this section, we examine some of the observed outcomes of plant–plant interactions (Figure 5.4), and examine to what extent these can be viewed as deliberate strategies by plants.

Competition

Competition between organisms is a potent ecological and evolutionary driver (Aschehoug et al. 2016), and evidence for competition between neighbouring plants comes from both formal experimentation and ecological observation. Competition could theoretically occur within the shoot or root system, but experimentally most attention has been paid to the possibility of competition between root systems. Plants have been proposed to over-proliferate either roots or shoots in the presence of a neighbour, a competitive response that in theory increases resource capture at the expense of the neighbour (Gersani et al. 2001). However, if the neighbour responds in kind, both plants increase their expenditure on growth without actually gaining more resources, or may collectively exhaust resources in a 'tragedy of the commons' scenario, which is an evolutionarily stable state (Novoplansky 2009; Smyeka and Herben 2017). However the results from formal experimentation (e.g. Gersani et al. 2001) have been criticized for problems in experimental design (see above) (Hess and De Kroon 2007; Semchenko et al. 2007), and the apparent over-proliferation explained as a consequence of altered nutrient availability or increased pot volume (Hess and De Kroon 2007; Nord et al. 2011; Semchenko et al. 2007).

Whether individuals take a competitive strategy could be context dependent; the same factors may promote competition under some scenarios but not others. The stress gradient hypothesis posits that there will be greater competition in benign conditions and facilitation in stressful ones (Bertness and Callaway 1994). Various studies on plant–plant interactions have found support for this idea (e.g. meta-analysis by He et al. 2013, but see Rysavy et al. 2016). The stress gradient hypothesis might be particularly relevant to plants found in stressful environments prone to exposure, temperature variation and drought (Eränen and Kozlov 2008). Shifts may occur during the lifetime of an individual plant, for example where seedling establishment is improved by the presence of others, but adults suffer lower reproduction from competition (Lara-Romero et al. 2016). In some cases, less stress-tolerant genotypes are less negatively affected by competition

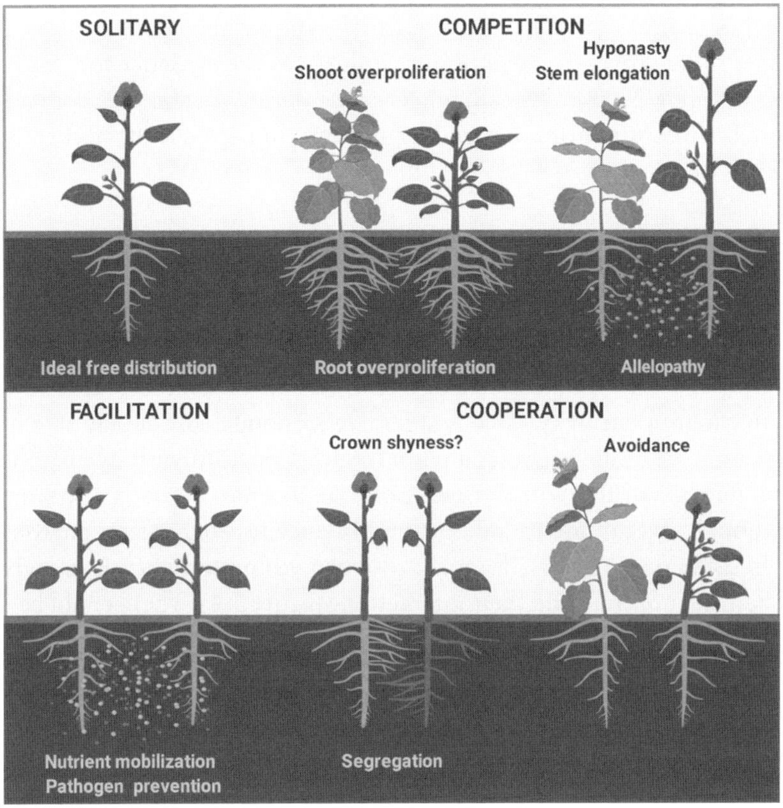

FIGURE 5.4: Responses to neighbouring plants. A summary of possible outcomes in interactions between neighbouring plants, relative to the solitary growth pattern. When unconstrained by neighbours (solitary) plants may organise their growth in relation to resource availability in an 'Ideal Free Distribution'. When faced with a near neighbour, interactions may take a number of forms. Competitive outcomes involve increased root growth, with the possible release of allelopathic exudates or volatile organic compounds, and increased stem elongation and altered organ positioning and size. Cooperation between conspecifics (perhaps often close relatives) may involve a reduction in root and shoot growth, enhanced spatial segregation of organs to avoid overlapping resource capture, or release of public goods such as nutrients or anti-pathogen molecules. Facilitation between heterospecifics may be manifested in a range of ways that are non-competitive or actively anti-competitive, and might involve avoidance or segregation of root and shoot systems, synchronous flower or seed production. [Colour figure can be viewed at wileyonlinelibrary.com.]

(e.g. Zhang and Tielbörger 2019). This trade-off between stress tolerance and competitive ability might contribute to the maintenance of polymorphisms in these traits. Theoretically, neighbour competition should be stronger between conspecifics (social competition), their niche requirements being more similar than those of heterospecific neighbours, but so far the evidence for this is mixed (Ehlers and Bilde 2019). Overall, whether competitive strategies depend on the identity of the competitors, beyond asymmetry in competitive ability, remains to be resolved.

Cooperation and Facilitation Between Heterospecifics

Avoiding unnecessary competition for resources with a neighbouring plant likely increases the fitness of many plants, even if this also inadvertently benefits neighbouring plants. The simplest strategy towards an unrelated plant would be a completely neutral strategy ('live and let live'), though this clearly might not be an active strategy. Alternatively, a plant might cooperate with its neighbours by reducing its growth to avoid competition (Semchenko et al. 2010; Wang et al. 2021), though there is not abundant firm evidence for such effects (Novoplansky 2009). In some species pairs, complex root growth patterns, such as segregation and avoidance, occur during their interaction (Figure 5.4). These could be deliberate non-competitive responses, since they are specific to the species pairs (Cahill Jr and McNickle 2011), but might also be inadvertent responses rather than an active strategy (in effect, a form of facilitation). The phenomenon of 'crown shyness', in which tree shoot systems of certain species do not spatially overlap with other members of the same species (Goudie, Polsson and Ott 2008), is also a possible cooperative response, although the genesis of this phenomenon is unclear and might not involve neighbour detection.

The evidence for facilitation is rather stronger, and there are many examples of such effects between neighbouring plants (see Li et al. 2014; McIntire and Fajardo 2014; Schöb et al. 2018). For instance, neighbours can ameliorate stressful conditions or create novel niches for other plants. For instance, facilitation is viewed as a key explanation for the over-yielding observed in intercropping systems such as maize/faba bean (Li et al. 2014). In phosphorus impoverished steppe grassland, some species can benefit from phosphorous-mobilizing heterospecific neighbours, but only if the non-mobilizing species can express sufficient plasticity in root growth (Yu et al. 2020). 'Nurse' plants enable seedling establishment, especially in very harsh conditions such as arid environments (Loayza et al. 2017). Plants could also share in 'public goods', such as attraction and maintenance of pollinator populations throughout the season (Dudley 2015). This can be particularly important for plants with low density populations in which mating can be problematic (Allee effects); neighbours can provide reproductive benefits by attraction of pollinators (Lachmuth et al. 2018). Other hypothesized examples of public goods include synchronous flower production ('co-flowering') that increases the attractiveness of displays (Torices et al. 2018) and synchronous production of seeds ('masting') which swamp predators (Dudley 2015). It should be noted that although the evidence for facilitation is generally good, it does not provide support for the concept of active plant–plant interactions, since facilitative effects can occur in the absence of neighbour detection.

Social Interactions

In cooperative and altruistic social responses, plants would be expected to down-regulate root or shoot growth in the presence of a related conspecifics, which may ultimately benefit the reproductive success of both participants (Dudley 2015). These effects could benefit fitness both directly (via the reduction of competition for both plants) and indirectly (via the increased success of both plants). Again, most of the formal experimentation has focused on root system responses, and suggests that plants respond to close kin (or their exudates) by reducing root growth, and maintaining spatial segregation (Biedrzycki et al. 2010; Fang et al. 2013; Semchenko et al. 2014). It is also worth noting that when sharer/owner experiments are repeated with all variables controlled for (see above), sharers actually reduce their root growth relative to owners (Chen et al. 2015). Kin recognition-driven cooperation has been proposed to reduce investment in competitive traits (Cahill et al. 2010; Bhatt et al. 2011; File et al. 2012; Fang et al. 2013), improve resource capture (Lepik et al. 2012) and increase fitness (Biernaskie 2011; Donohue 2003; Torices et al. 2018; Yang et al. 2018). However, a broad criticism of these studies is that they rarely measure the full extent of individual inclusive fitness (Ehlers and Bilde 2019). In general, the direct evidence for social cooperation is again rather mixed, but some of the kin-induced alterations are very difficult to explain without invoking some form of cooperation between plants.

Allelopathy

In some ways, allelopathy has the inverse problem to competition and cooperation; the experimental evidence for the phenomenon is very clear, but the benefits to plants are much less clear. Conventionally, allelopathy is defined as the release chemicals which reduce the growth, development, survival and reproduction of competitors (Figure 5.4). In plants, allelopathic compounds are exuded into the rhizosphere (Kato-Noguchi et al. 2010; Kong et al. 2018), emitted aboveground as VOCs (Kong et al. 2004), deposited in pollen (Roshchina et al. 2009) or may also be non-specifically released during litter decomposition

(Rice 2012). Allelochemicals vary in mode of uptake and mechanism of action, and their effectiveness may vary depending on spatio-temporal factors such as activation (Gaofeng et al. 2018; He et al. 2019; Rice 2012; Weston and Duke 2003). There is little debate that allelochemicals have phytotoxic potential, but the evidence that they accumulate to sufficiently high levels to cause phytotoxicity under natural conditions is not strong (Zeng 2014). Allelopathic chemicals should also have autotoxic effects on the source plant, a paradox that has been long debated (Rice 2012; Smith 1979). The highest concentration of allelochemicals should be observed close to allelopathic plants, selecting for tolerance to their own allelochemicals (Fitter 2003), which has recently been connected with suppression of defence mechanisms, implying a high cost of allelopathic behaviour (Li et al. 2019; Liu et al. 2019; Xu and Zhang 2015). Furthermore, allelopathic compounds may also have hormetic effects; at sub-toxic concentrations, they may actually induce stress tolerance and stimulate growth of neighbouring plants (Agathokleous and Calabrese 2019; Stebbing 2003). Thus, we can reasonably ask what the benefits of plant allelochemicals are – and what this might show us about plant–plant interactions.

There are two non-mutually exclusive answers to this question. Firstly, allelochemicals might be exuded for a different function, with their phytotoxicity only being a secondary and/or rarely expressed property. For instance, some allelochemicals may enhance nutrient uptake by the exuding plant (Tharayil et al. 2009; Zhang et al. 2020), or function in the recruitment or maintenance of beneficial soil microbes, although tests of this hypothesis so far have returned ambiguous results (Callaway et al. 2008; Cipollini et al. 2008; Mishra et al. 2013; Stinson et al. 2006). At natural concentrations, allelochemicals might function as signals to neighbouring plants (rather than toxins), or might allow detection of obstacles in soil, triggering root navigation (Agathokleous and Calabrese 2019; Falik et al. 2005; Fang et al. 2013; Semchenko et al. 2008).

Secondly, allelochemicals might be exuded primarily for competitive reasons, and their costs/drawbacks offset by some other mechanism. For instance, it has been suggested that allelochemicals act indirectly on competitors by preventing nutrient uptake, or altering soil microbiome composition, rather than needing to be exuded at directly phytotoxic (and autotoxic) concentrations (Zeng 2014). Another possibility is that the costs are reduced by minimizing the expression of the trait. Deployment of allelochemicals seems to be tightly connected to neighbour detection, and particularly non-kin neighbours. Active exudation can be triggered by the presence of neighbours, detected through common signals such as (−)-loliolide, jasmonic and salicylic acids (Kong et al. 2018; Li et al. 2016; Uesugi et al. 2019). Furthermore, allelopathy may only be expressed at certain developmental stages when it is the greatest advantage, particularly during seedling establishment (Kong et al. 2018). This may give plants an early competitive advantage over neighbours that can be maintained without allelochemical release; thus, the costs of allelopathy may only be transiently incurred.

The most straightforward interpretation of current evidence brings us tantalizingly close to concluding that allelopathy is not just a competitive trait, but one induced in response to the identity of the neighbouring plants. As such, allelopathy could represent the keystone in the argument that plants do differentially detect and respond to their neighbours. And yet, like other key questions in the field, the evidence is currently not conclusive enough, and other interpretations remain possible.

Conclusion and Perspectives

As the work reviewed here shows, plants are active participants in their interactions with each other, interactions that are every bit as complex as those seen between other organisms. Plants can detect their neighbours through a range of mechanisms (Section on Mechanisms), communicate with their neighbours, and attempt to manipulate their growth and function (Section on Allelopathy). We cannot yet be completely certain that plants can distinguish between the identity of their neighbours (Section on Distinguishing), and respond differentially depending on this identity (Section on Responding), but the evidence currently supports those ideas. Admittedly, the evidence for this is primarily phenomenological, and based on observation of plants growing together; as we have seen, alternative explanations of these studies are usually possible.

Critical to moving the field forward is to demonstrate the mechanistic basis by which plants may differentially detect and respond to neighbours (Figure 5.3). This requires investigating how plants integrate the potential cues and whether acting on complex social information requires a multimodal system (Dore et al. 2018). If kin recognition could be detected at the molecular level, for instance – as a consequence of ligands binding to receptors – or if different gene expression patterns could be observed in plants exposed to kin and non-kin— then much of the theoretical tension in the field would be resolved. This information would also be key to revealing whether plants are sensitive to the spectrum of genetic relatedness from self to heterospecifics, and to what extent life history and ecology have driven the evolution of these mechanisms. For example, do mating and dispersal systems that produce highly genetically structured populations correlate with more finely tuned mechanisms? We therefore advocate that the direction of future travel should certainly be towards an increased molecular understanding of plant–plant interactions, but in a holistic way that integrates observation and deeper eco-evolutionary understanding of these interactions in natural environments.

The study of plant–plant interactions has major implications for understanding and sustainably improving crop yields. Crop systems represent an obvious break from natural biological systems, particularly in their radical reorganization of plant communities into mono-cultures of closely related plants. Plant–plant interactions are hugely important in the functioning of crop species, because the drive to maximize yield from a minimum of land creates a situation where the collective yield of the group is much more important than the yield of the individual. Highly competitive plants with high individual fitness therefore likely make poorer crops than individuals with intermediate fitness (Weiner 2017). Due to modern breeding and utilization of inbred varieties, crops have effectively undergone 'group selection' for traits that enable them to yield highly at a field level, and to maintain consistency even in stressful conditions (Blum 2018; Weiner 2017). This artificial selection has likely radically altered the dynamics of plant–plant interactions in crop species compared to their wild relatives and indeed, compared to traditional landraces. For instance, competitive above- and belowground interactions were shown to be less pronounced for example within modern wheat cultivars compared to traditional landraces (Zhu et al. 2019). Since the Green Revolution, crops have also been inadvertently bred for 'density resistance', the ability to perform better under high sowing densities, by reducing inhibition that is caused solely by the presence of neighbouring plants (i.e. that is not caused by resource limitations) (Choe et al. 2016). The role of plant–plant interactions in determining optimal crop densities is reviewed elsewhere (Postma et al. 2021). Interactions between crops and weeds, and between pairs of crops in intercropping systems are further examples of the importance of plant–plant interactions in agricultural systems. Thus, by better understanding the nature of plant interactions, we can also aim to improve crop systems.

Overall, it is exciting times for the field, with a wealth of new data continuing to dispel the notion that plants are passive actors in their environment. In hindsight, it seems unlikely that plants have flourished in all terrestrial ecosystems for 500 million years without being able to actively detect and respond to other plants—whether friend, neighbour or enemy.

ACKNOWLEDGEMENTS

We thank James Cooper and Cara Wheeldon for helpful discussions, and our three anonymous peer reviewers for very helpful suggestions.

NOTE

1. Republished with permission from John Wiley & Sons Ltd., from Roza D. Bilas, Amanda Bretman and Tom Bennett (2021), 'Friends, neighbours and enemies: An overview of the communal and social biology of plants', *Plant, Cell & Environment*, 44, pp. 997–1013. Permission conveyed through Copyright Clearance Center, Inc.

REFERENCES

Agathokleous, Evgenios and Calabrese, Edward J. (2019), 'Hormesis can enhance agricultural sustainability in a changing world', *Global Food Security* 20, pp. 150–55.

Andalo, Christophe, Goldringer, Isabelle and Godelle, Bernard (2001), 'Inter- and intragenotypic competition under elevated carbon dioxide in *Arabidopsis thaliana*', *Ecology*, 82:1, pp. 157–64.

Anten, Niels P. R., Alcalá-Herrera, Rafael, Schieving, Feike and Onoda, Yusuke (2010), 'Wind and mechanical stimuli differentially affect leaf traits in *Plantago* major', *New Phytologist*, 188:2, pp. 554–64.

Anten, Niels P. R. and Chen, Bin J. W. (2021), 'Detect thy family: Mechanisms, ecology and agricultural aspects of kin recognition in plants', *Plant, Cell & Environment*, 44:4, pp. 1059–71.

Appel, Heidi M. and Cocroft, Reginald B. (2014), 'Plants respond to leaf vibrations caused by insect herbivore chewing', *Oecologia*, 175:4, pp. 1257–66.

Aschehoug, Erik T., Brooker, Rob, Atwater, Daniel Z., Maron, John L. and Callaway, Ragan M. (2016), 'The mechanisms and consequences of interspecific competition among plants', *Annual Review of Ecology, Evolution, and Systematics*, 47, pp. 263–81.

Babikova, Zdenka, Gilbert, Lucy, Bruce, Toby J. A., Birkett, Michael, Caulfield, John C., Woodcock, Christine, Pickett, John A. and Johnson, David (2013), 'Underground signals carried through common mycelial networks warn neighbouring plants of aphid attack', *Ecology Letters*, 16:7, pp. 835–43.

Bais, Harsh P., Vepachedu, Ramarao, Gilroy, Simon, Callaway, Ragan M. and Vivanco, Jorge M. (2003), 'Allelopathy and exotic plant invasion: From molecules and genes to species interactions', *Science*, 301:5638, pp. 1377–80.

Baldwin, Ian T. (2010), 'Plant volatiles', *Current Biology*, 20:9, pp. R392–97.

Barrett, Spencer C. H. and Harder, Lawrence D. (2017), 'The ecology of mating and its evolutionary consequences in seed plants', *Annual Review of Ecology, Evolution, and Systematics*, 48, pp. 135–57.

Barto, E. Kathryn, Weidenhamer, Jeffrey D., Cipollini, Don and Rillig, Matthias C. (2012), 'Fungal superhighways: Do common mycorrhizal networks enhance below ground communication?', *Trends in Plant Science*, 17:11, pp. 633–37.

Bhatt, Mudra V., Khandelwal, Aditi and Dudley, Susan A. (2011), 'Kin recognition, not competitive interactions, predicts root allocation in young Cakile edentula seedling pairs', *New Phytologist*, 189:4, pp. 1135–42.

Biedrzycki, Meredith L., Jilany, Tafari A., Dudley, Susan A. and Bais, Harsh P. (2010), 'Root exudates mediate kin recognition in plants', *Communicative & Integrative Biology*, 3:1, pp. 28–35.

Biernaskie, Jay M. (2011), 'Evidence for competition and cooperation among climbing plants', *Proceedings of the Royal Society B: Biological Sciences*, 278:1714, pp. 1989–96.

Blum, Abraham (2018), *Plant Breeding for Stress Environments*, New York: CRC Press.

Bronstein, Judith L. (2009), 'The evolution of facilitation and mutualism', *Journal of Ecology*, 97:6, pp. 1160–70.

Cahill Jr., James F. and McNickle, Gordon G. (2011), 'The behavioral ecology of nutrient foraging by plants', *Annual Review of Ecology, Evolution, and Systematics*, 42, pp. 289–311.

Cahill Jr., James F., McNickle, Gordon G., Haag, Joshua J., Lamb, Eric G., Nyanumba, Samson M. and St. Clair, Colleen Cassady (2010), 'Plants integrate information about nutrients and neighbors', *Science*, 328:5986, p. 1657.

Callaway, Ragan M. and Mahall, Bruce E. (2007), 'Family roots', *Nature*, 448:7150, pp. 145–46.

Callaway, Ragan M., Cipollini, Don, Barto, Kathryn, Thelen, Giles C., Hallett, Steven G., Prati, Daniel, Stinson, Kristina and Klironomos, John (2008), 'Novel weapons: Invasive plant suppresses fungal mutualists in America but not in its native Europe', *Ecology*, 89:4, pp. 1043–55.

Chen, Bin J. W., During, Heinjo J. and Anten, Niels P. R. (2012), 'Detect thy neighbor: Identity recognition at the root level in plants', *Plant Science*, 195, pp. 157–67.

Chen, Bin J. W., During, Heinjo J., Vermeulen, Peter J., de Kroon, Hans, Poorter, Hendrik and Anten, Niels P. R. (2015), 'Corrections for rooting volume and plant size reveal negative effects of neighbour presence on root allocation in pea', *Functional Ecology*, 29:11, pp. 1383–91.

Choe, Eunsoo, Drnevich, Jenny and Williams, Martin M. (2016), 'Identification of crowding stress tolerance co-expression networks involved in sweet corn yield', *PLoS One*, 11:1, https://doi.org/10.1371/journal.pone.0147418. Accessed 4 April 2023.

Choi, Bosung, Ghosh, Ritesh, Gururani, Mayank Anand, Shanmugam, Gnanendra, Jeon, Junhyun, Kim, Jonggeun, Park, Soo-Chul, et al. (2017), 'Positive regulatory role of sound vibration treatment in *Arabidopsis thaliana* against Botrytis cinerea infection', *Scientific Reports*, 7:1, pp. 1–14.

Cipollini, Don, Stevenson, Randall and Cipollini, Kendra (2008), 'Contrasting effects of allelochemicals from two invasive plants on the performance of a nonmycorrhizal plant', *International Journal of Plant Sciences*, 169:3, pp. 371–75.

Comparini, Diego, Masi, Elisa, Pandolfi, Camilla, Sabbatini, Leonardo, Dolfi, Marco, Morosi, Simone and Mancuso, Stefano (2020), 'Stem electrical properties associated with water stress conditions in olive tree', *Agricultural Water Management*, 234, https://doi.org/10.1016/j.agwat.2020.106109. Accessed 4 April 2023.

Cosme, Marco, Fernández, Ivan, Van der Heijden, Marcel G. A. and Pieterse, Corné M. J. (2018), 'Non-mycorrhizal plants: The exceptions that prove the rule', *Trends in Plant Science*, 23:7, pp. 577–87.

Crepy, María A. and Casal, Jorge J. (2015), 'Photoreceptor-mediated kin recognition in plants', *New Phytologist*, 205:1, pp. 329–38.

——— (2016), 'Kin recognition by self-referent phenotype matching in plants', *New Phytologist*, 209:1, pp. 15–16.

Davies, Eric (2006), 'Electrical signals in plants: Facts and hypotheses', in A. G. Volkov (ed.), *Plant Electrophysiology: Theory and Methods*, Berlin/Heidelberg: Springer, pp. 407–22.

Degli Agosti, Robert (2014), 'Touch-induced action potentials in *Arabidopsis thaliana*', *Archives des Sciences*, 67, pp. 125–38.

de Kroon, Hans, Mommer, Liesje and Nishiwaki, Aya (2003), 'Root competition: Towards a mechanistic understanding', *Root Ecology*, pp. 215–34.

De Luca, Paul A. and Vallejo-Marin, Mario (2013), 'What's the "buzz" about? The ecology and evolutionary significance of buzz-pollination', *Current Opinion in Plant Biology*, 16:4, pp. 429–35.

de Toledo, Gabriel R. A., Parise, André G., Simmi, Francine Z., Costa, Adrya V. L., Senko, Luiz G. S., Debono, Marc-Williams and Souza, Gustavo M. (2019), 'Plant electrome: The electrical dimension of plant life', *Theoretical and Experimental Plant Physiology*, 31, pp. 21–46.

de Wit, Mieke, Kegge, Wouter, Evers, Jochem B., Vergeer-van Eijk, Marleen H., Gankema, Paulien, Voesenek, Laurentius A. C. J. and Pierik, Ronald (2012), 'Plant neighbor detection through touching leaf tips precedes phytochrome signals', *Proceedings of the National Academy of Sciences*, 109:36, pp. 14705–10.

Delavaux, Camille S., Smith-Ramesh, Lauren M. and Kuebbing, Sara E. (2017), 'Beyond nutrients: A meta-analysis of the diverse effects of arbuscular mycorrhizal fungi on plants and soils', *Ecology*, 98:8, pp. 2111–19.

Depuydt, Stephen (2014), 'Arguments for and against self and non-self root recognition in plants', *Frontiers in Plant Science*, 5, p. 614.

Donohue, Kathleen (2003), 'The influence of neighbor relatedness on multilevel selection in the Great Lakes sea rocket', *The American Naturalist*, 162:1, pp. 77–92.

Dore, Alice A., McDowall, Laurin, Rouse, James, Bretman, Amanda, Gage, Matthew J. G. and Chapman, Tracey (2018), 'The role of complex cues in social and reproductive plasticity', *Behavioral Ecology and Sociobiology*, 72, pp. 1–15.

Dudley, Susan, A. (2015), 'Plant cooperation', *AoB Plants*, 7, plv113, https://doi.org/10.1093/aobpla/plv113. Accessed 4 April 2023.

Dudley, Susan A. and File, Amanda L. (2007), 'Kin recognition in an annual plant', *Biology Letters*, 3:4, pp. 435–38.

Dudley, Susan A., Murphy, Guillermo P. and File, Amanda L. (2013), 'Kin recognition and competition in plants', *Functional Ecology*, pp. 898–906.

Ehlers, Bodil K. and Bilde, Trine (2019), 'Inclusive fitness, asymmetric competition and kin selection in plants', *Oikos*, 128:6, pp. 765–74.

Ehlers, Bodil K., Berg, Matty P., Staudt, Michael, Holmstrup, Martin, Glasius, Marianne, Ellers, Jacintha, Tomiolo, Sara, Madsen, René B., Slotsbo, Stine and Penuelas, Josep (2020), 'Plant secondary compounds in soil and their role in belowground species interactions', *Trends in Ecology & Evolution*, 35:8, pp. 716–30, https://doi.org/10.1016/j.tree.2020.04.001. Accessed 4 April 2023.

Elhakeem, Ali, Markovic, Dimitrije, Broberg, Anders, Anten, Niels P. R. and Ninkovic, Velemir (2018), 'Aboveground mechanical stimuli affect belowground plant-plant communication', *PLoS One*, 13:5, https://doi.org/10.1371/journal.pone.0195646. Accessed 4 April 2023.

Eränen, Janne K., and Kozlov, Mikhail V. (2008), 'Increasing intraspecific facilitation in exposed environments: Consistent results from mountain birch populations in two subarctic stress gradients', *Oikos*, 117:10, pp. 1569–77.

Falik, Omer, de Kroon, Hans and Novoplansky, Ariel (2006), 'Physiologically-mediated self/non-self root discrimination in *Trifolium repens* has mixed effects on plant performance', *Plant Signaling & Behavior*, 1:3, pp. 116–121.

Falik, Omer, Reides, Perla, Gersani, Mordechai and Novoplansky, Ariel (2003), 'Self/non-self discrimination in roots', *Journal of Ecology*, pp. 525–31.

——— (2005), 'Root navigation by self inhibition', *Plant, Cell & Environment*, 28:4, pp. 562–69.

Fang, Suqin, Clark, Randy T., Zheng, Ying, Iyer-Pascuzzi, Anjali S., Weitz, Joshua S., Kochian, Leon V., Edelsbrunner, Herbert, Liao, Hong and Benfey, Philip N. (2013), 'Genotypic recognition and spatial responses by rice roots', *Proceedings of the National Academy of Sciences*, 110:7, pp. 2670–75.

Farmer, Edward E. (2001), 'Surface-to-air signals', *Nature*, 411:6839, pp. 854–56.

File, Amanda L., Klironomos, John, Maherali, Hafiz and Dudley, Susan A. (2012), 'Plant kin recognition enhances abundance of symbiotic microbial partner', *PLoS One*, https://doi.org/10.1371/journal.pone.0045648. Accessed 4 April 2023.

Fitter, Alastair (2003), 'Making allelopathy respectable', *Science*, 301:5638, pp. 1337–38.

Frey, Wolfgang and Kürschner, Harald (2011), 'Asexual reproduction, habitat colonization and habitat maintenance in bryophytes', *Flora-Morphology, Distribution, Functional Ecology of Plants*, 206:3, pp. 173–84.

Fromm, Jörg (2006), 'Long-distance electrical signaling and physiological functions in higher plants', in A. G. Volkov (ed.), *Plant Electrophysiology: Theory and Methods*, Berlin/Heidelberg: Springer, pp. 269–85.

Gagliano, Monica, Mancuso, Stefano and Robert, Daniel (2012), 'Towards understanding plant bioacoustics', *Trends in Plant Science*, 17:6, pp. 323–25.

Gaofeng, Xu, Shicai, Shen, Fudou, Zhang, Yun, Zhang, Hisashi, Kato-Noguchi and David, Roy Clements (2018), 'Relationship between allelopathic effects and functional traits of different allelopathic potential rice accessions at different growth stages', *Rice Science*, 25:1, pp. 32–41.

Gersani, Mordechai, Brown, Joel S., O'Brien, Erin E., Maina, Godfrey M. and Abramsky, Zvika (2001), 'Tragedy of the commons as a result of root competition', *Journal of Ecology*, 89:4, pp. 660–69.

Ghosh, Ritesh, Gururani, Mayank A., Ponpandian, Lakshmi N., Mishra, Ratnesh C., Park, Soo-Chul, Jeong, Mi-Jeong and Bae, Hanhong (2017), 'Expression analysis of sound vibration-regulated genes by touch treatment in *Arabidopsis*', *Frontiers in Plant Science*, 8, p. 100.

Ghosh, Ritesh, Mishra, Ratnesh Chandra, Choi, Bosung, Kwon, Young Sang, Bae, Dong Won, Park, Soo-Chul, Jeong, Mi-Jeong and Bae, Hanhong (2016), 'Exposure to sound vibrations lead to transcriptomic, proteomic and hormonal changes in *Arabidopsis*', *Scientific Reports*, 6:1 pp. 1–17.

Gilbert, Ian R., Jarvis, Paul G. and Smith, Harry (2001), 'Proximity signal and shade avoidance differences between early and late successional trees', *Nature*, 411:6839, pp. 792–95.

Goudie, James W., Polsson, Kenneth R. and Ott, Peter K. (2009), 'An empirical model of crown shyness for lodgepole pine (*Pinus contorta var. latifolia* [Engl.] Critch.) in British Columbia', *Forest Ecology and Management*, 257:1, pp. 321–31.

Gouinguené, Sandrine, Degen, Thomas and Turlings, Ted C. J. (2001), 'Variability in herbivore-induced odour emissions among maize cultivars and their wild ancestors (teosinte)', *Chemoecology*, 11, pp. 9–16.

Gruntman, Michal and Novoplansky, Ariel (2004), 'Physiologically mediated self/non-self discrimination in roots', *Proceedings of the National Academy of Sciences* 101:11, pp. 3863–67.

Hamant, Olivier and Haswell, Elizabeth S. (2017), 'Life behind the wall: sensing mechanical cues in plants', *BMC Biology*, 15:1, pp. 1–9.

Hamilton, William D. (1964), 'The genetical evolution of social behaviour. II', *Journal of Theoretical Biology*, 7:1, pp. 17–52.

——— (1970), 'Selfish and spiteful behaviour in an evolutionary model', *Nature*, 228, pp. 1218–20.

He, Shuliang, Wang, S. Q., Wang, Q. Y., Zhang, C. Y., Zhang, Y. M., Liu, T. Y., Yang, S. X., Kuang, Y., Zhang, Y. X., Han, J. Y. and Qin, J. C. (2019), 'Allelochemicals as growth regulators: A review', *Allelopathy Journal*, 48:1, pp. 15–26.

He, Qiang, Bertness, Mark D. and Altieri, Andrew H. (2013), 'Global shifts towards positive species interactions with increasing environmental stress', *Ecology Letters*, 16:5, pp. 695–706.

Heil, Martin and Adame-Álvarez, Rosa M. (2010), 'Short signalling distances make plant communication a soliloquy', *Biology Letters*, 6:6, pp. 843–45.

Heil, Martin and Karban, Richard (2010), 'Explaining evolution of plant communication by airborne signals', *Trends in Ecology & Evolution*, 25:3, pp. 137–144.

Hess, Linde and De Kroon, Hans (2007), 'Effects of rooting volume and nutrient availability as an alternative explanation for root self/non-self discrimination', *Journal of Ecology*, 95:2, pp. 241–51.

Hölttä, Teemu, Vesala, Timo, Nikinmaa, Eero, Perämäki, Martti, Siivola, Erkki and Mencuccini, Maurizio (2005), 'Field measurements of ultrasonic acoustic emissions and stem diameter variations. New insight into the relationship between xylem tensions and embolism', *Tree Physiology*, 25:2, pp. 237–43.

Holzapfel, Claus and Alpert, Peter (2003), 'Root cooperation in a clonal plant: Connected strawberries segregate roots', *Oecologia*, pp. 72–77.

Hu, Lingfei, Robert, Christelle A. M., Cadot, Selma, Zhang, Xi, Ye, Meng, Li, Beibei, Manzo, Daniele, Chervet, Noemie, Steinger, Thomas, van der Heijden, Marcel G. A., Schlaeppi, Klaus and Erb, Matthias (2018), 'Root exudate metabolites drive plant-soil feedbacks on growth and defense by shaping the rhizosphere microbiota', *Nature Communications*, 9:1, p. 2738.

Huber, Martina, Nieuwendijk, Nicole M., Pantazopoulou, Chrysoula K. and Pierik, Ronald (2021), 'Light signalling shapes plant–plant interactions in dense canopies', *Plant, Cell & Environment*, 44:4, pp. 1014–29.

Jacob, Cory E., Tozzi, Eric and Willenborg, Christian J. (2017), 'Neighbour presence, not identity, influences root and shoot allocation in pea', *PLoS One*, 12:3, https://doi.org/10.1371/journal.pone.0173758. Accessed 4 April 2023.

Johnson, David and Gilbert, Lucy (2015), 'Interplant signalling through hyphal networks', *New Phytologist*, 205:4, pp. 1448–53.

Karban, Richard, Maron, John, Felton, Gary W., Ervin, Gary and Eichenseer, Herbert (2003), 'Herbivore damage to sagebrush induces resistance in wild tobacco: Evidence for eavesdropping between plants', *Oikos*, 100:2, pp. 325–32.

Karban, Richard, Shiojiri, Kaori, Huntzinger, Mikaela and McCall, Andrew C. (2006), 'Damage-induced resistance in sagebrush: Volatiles are key to intra- and interplant communication', *Ecology*, 87:4, pp. 922–30.

Karban, Richard, Shiojiri, Kaori, Ishizaki, Satomi, Wetzel, William C. and Evans, Richard Y. (2013), 'Kin recognition affects plant communication and defence', *Proceedings of the Royal Society B: Biological Sciences*, 280:1756, https://doi.org/10.1098/rspb.2012.3062. Accessed 4 April 2023.

Karban, Richard, Yang, Louie H. and Edwards, Kyle F. (2014), 'Volatile communication between plants that affects herbivory: A meta-analysis', *Ecology Letters*, 17:1, pp. 44–52.

Kato-Noguchi, Hisashi, Hasegawa, Morifumi, Ino, Takeshi, Ota, Katsumi and Kujime, Hiroya (2010), 'Contribution of momilactone A and B to rice allelopathy', *Journal of Plant Physiology*, 167:10, pp. 787–91.

Karlovsky, Petr (2008), 'Secondary metabolites in soil ecology', in *Secondary Metabolites in Soil Ecology*. Berlin/Heidelberg: Springer, pp. 1–19

Kigathi, Rose N., Weisser, Wolfgang W., Veit, Daniel, Gershenzon, Jonathan and Unsicker, Sybille B. (2013), 'Plants suppress their emission of volatiles when growing with conspecifics', *Journal of Chemical Ecology*, 39, pp. 537–45.

Kim, Joo Yeol, Kang, Ye Eun, Lee, Soo In, Kim, Jin A., Muthusamy, Muthusamy and Jeong, Mi-Jeong (2020), 'Sound waves affect the total flavonoid contents in *Medicago sativa, Brassica oleracea and Raphanus sativus sprouts', Journal of the Science of Food and Agriculture*, 100:1, pp. 431–40.

Kong, Chui-Hua, Zhang, Song-Zhu, Li, Yong-Hua, Xia, Zhi-Chao, Yang, Xue-Fang, Meiners, Scott J. and Wang, Peng (2018), 'Plant neighbor detection and allelochemical response are driven by root-secreted signaling chemicals', *Nature Communications* 9:1, pp. 1–9.

Kong, Chui-hua, Hu, Fei, Liang, Wenju, Peng, Wang and Jiang, Yong (2004), 'Allelopathic potential of *Ageratum conyzoides* at various growth stages in different habitats', *Allelopathy Journal*, 13:2, pp. 233–40.

Krannitzi, Pamela G. and Caldwell, Martin M. (1995), 'Root growth responses of three Great Basin perennials to intra and interspecific contact with other roots', *Flora*, 190:2, pp. 161–67.

Lachmuth, Susanne, Henrichmann, Colette, Horn, Juliane, Pagel, Jörn and Schurr, Frank M. (2018), 'Neighbourhood effects on plant reproduction: An experimental–analytical framework and its application to the invasive *Senecio inaequidens*', *Journal of Ecology*, 106:2, pp. 761–73.

Lankinen, Åsa (2008), 'Root competition influences pollen competitive ability in *Viola tricolor*: Effects of presence of a competitor beyond resource availability?', *Journal of Ecology*, 96:4, pp. 756–65.

Lara-Romero, Carlos, de la Cruz, Marcelino, Escribano-Ávila, Gema, García-Fernández, Alfredo and Iriondo, Jose M. (2016), 'What causes conspecific plant aggregation? Disentangling the role of dispersal, habitat heterogeneity and plant–plant interactions', *Oikos*, 125:9, pp. 1304–13.

Laschimke, Ralf, Burger, Maria and Vallen, Hartmut (2006), 'Acoustic emission analysis and experiments with physical model systems reveal a peculiar nature of the xylem tension', *Journal of Plant Physiology*, 163:10, pp. 996–1007.

Lepik, Anu, Abakumova, Maria, Zobel, Kristjan and Semchenko, Marina (2012), 'Kin recognition is density-dependent and uncommon among temperate grassland plants', *Functional Ecology*, 26:5, pp. 1214–20.

Li, Bai, Li, Yu-Ying, Wu, Hua-Mao, Zhang, Fang-Fang, Li, Chun-Jie, Li, Xue-Xian, Lambers, Hans and Li, Long (2016), 'Root exudates drive interspecific facilitation by enhancing nodulation and N_2 fixation', *Proceedings of the National Academy of Sciences*, 113:23, pp. 6496–6501.

Li, Jiancai, Liu, Xiaoli, Wang, Qi, Huangfu, Jiayi, Schuman, Meredith C. and Lou, Yonggen (2019), 'A group D MAPK protects plants from autotoxicity by suppressing herbivore-induced defense signaling', *Plant Physiology*, 179:4, 1386–1401.

Li, Long, Tilman, David, Lambers, Hans and Zhang, Fu-Suo (2014), 'Plant diversity and overyielding: Insights from belowground facilitation of intercropping in agriculture', *New Phytologist*, 203:1, pp. 63–69.

Liu, Xiaoli, Li, Jiancai, Noman, Ali and Lou, Yonggen (2019), 'Silencing OsMAPK20-5 has different effects on rice pests in the field', *Plant Signaling & Behavior*, 14:9, https://doi.org/10.1080/15592324.2019.1640562. Accessed 4 April 2023.

Loayza, Andrea P., Herrera-Madariaga, Marisol A., Carvajal, Danny E., García-Guzmán, Patricio and Squeo, Francisco A. (2017), 'Conspecific plants are better "nurses" than rocks: Consistent results revealing intraspecific facilitation as a process that promotes establishment in a hyper-arid environment', *AoB Plants*, 9:6, https://doi.org/10.1093/aobpla/plx056. Accessed 4 April 2023.

Madsen, Christopher, Potvin, Catherine, Hall, Jefferson, Sinacore, Katherine, Turner, Benjamin L. and Schnabel, Florian (2020), 'Coarse root architecture: Neighbourhood and abiotic environmental effects on five tropical tree species growing in mixtures and monocultures', *Forest Ecology and Management*, 460, https://doi.org/10.1016/j.foreco.2019.117851. Accessed 4 April 2023.

McIntire, Eliot J.B. and Fajardo, Alex (2014), 'Facilitation as a ubiquitous driver of biodiversity', *New Phytologist*, 201:2, pp. 403–16.

McNickle, Gordon G. and Brown, Joel S. (2014), 'An ideal free distribution explains the root production of plants that do not engage in a tragedy of the commons game', *Journal of Ecology*, 102:4, pp. 963–71.

Mahall, Bruce E. and Callaway, Ragan M. (1991), 'Root communication among desert shrubs', *Proceedings of the National Academy of Sciences*, 88:3, pp. 874–76.

Maina, Godfrey G., Brown, Joel S. and Gersani, Mordechai (2002), 'Intra-plant versus interplant root competition in beans: Avoidance, resource matching or tragedy of the commons', *Plant Ecology*, 160, pp. 235–47.

Markham, John and Halwas, Sara (2011), 'Effect of neighbour presence and soil volume on the growth of *Andropogon gerardii* Vitman', *Plant Ecology & Diversity*, 42–43, pp. 265–68.

Markovic, Dimitrije, Nikolic, Neda, Glinwood, Robert, Seisenbaeva, Gulaim and Ninkovic, Velemir (2016), 'Plant responses to brief touching: A mechanism for early neighbour detection?', *PLoS One*, 11:11, https://doi.org/10.1371/journal.pone.0165742. Accessed 4 April 2023.

Markovic, Dimitrije, Colzi, Ilaria, Taiti, Cosimo, Ray, Swayamjit, Scalone, Romain, Ali, Jared Gregory, Mancuso, Stefano and Ninkovic, Velemir (2019), 'Airborne signals synchronize the defenses of neighboring plants in response to touch', *Journal of Experimental Botany*, 70:2, pp. 691–700.

Marler, Thomas E. (2013), 'Kin recognition alters root and whole plant growth of split-root *Cycas edentata* seedlings', *HortScience*, 48:10, pp. 1266–69.

Masclaux, Frédéric G., Bruessow, Friederike, Schweizer, Fabian, Gouhier-Darimont, Caroline, Keller, Laurent and Reymond, Philippe (2012), 'Transcriptome analysis of intraspecific competition in *Arabidopsis thaliana* reveals organ-specific signatures related to nutrient acquisition and general stress response pathways', *BMC Plant Biology*, 12, pp. 1–13.

Massa, Gioia D. and Gilroy, Simon (2003), 'Touch modulates gravity sensing to regulate the growth of primary roots of *Arabidopsis thaliana*', *The Plant Journal*, 33:3, pp. 435–45.

Meier, Ina Christin, Angert, Alon, Falik, Omer, Shelef, Oren and Rachmilevitch, Shimon (2013), 'Increased root oxygen uptake in pea plants responding to non-self neighbors', *Planta*, 238, pp. 577–86.

Mishra, Sandhya, Upadhyay, Ram Sanmukh and Nautiyal, Chandra Shekhar (2013), 'Unravelling the beneficial role of microbial contributors in reducing the allelopathic effects of weeds', *Applied Microbiology and Biotechnology*, 97, pp. 5659–68.

Mitteldorf, Joshua and Wilson, David Sloan (2000), 'Population viscosity and the evolution of altruism', *Journal of Theoretical Biology*, 204:4, pp. 481–96.

Morrell, Kimberly and Kessler, André (2017), 'Plant communication in a widespread goldenrod: Keeping herbivores on the move', *Functional Ecology*, 31:5, pp. 1049–61.

Mousavi, Seyed A. R., Chauvin, Adeline, Pascaud, François, Kellenberger, Stephan and Farmer, Edward E. (2013), 'Glutamate receptor-like genes mediate leaf-to-leaf wound signalling', *Nature*, 500:7463, pp. 422–26.

Murphy, Guillermo P. and Dudley, Susan A. (2009), 'Kin recognition: Competition and cooperation in *Impatiens* (Balsaminaceae)', *American Journal of Botany*, 96:11, pp. 1990–96.

Ninkovic, Velemir (2003), 'Volatile communication between barley plants affects biomass allocation', *Journal of Experimental Botany*, 54:389, pp. 1931–39.

Ninkovic, Velemir, Dahlin, Iris, Vucetic, Andja, Petrovic-Obradovic, Olivera, Glinwood, Robert and Webster, Ben (2013), 'Volatile exchange between undamaged plants-a new mechanism affecting insect orientation in intercropping', *PLoS One*, 8:7, https://doi.org/10.1371/journal.pone.0069431. Accessed 4 April 2023.

Ninkovic, Velemir, Markovic, Dimitrije and Rensing, Merlin (2021), 'Plant volatiles as cues and signals in plant communication', *Plant, Cell & Environment*, 44:4, pp. 1030–43.

Ninkovic, Velemir, Rensing, Merlin, Dahlin, Iris and Markovic, Dimitrije (2019), 'Who is my neighbor? Volatile cues in plant interactions', *Plant Signaling & Behavior*, 14:9, https://doi.org/10.1080/15592324.2019.1634993. Accessed 4 April 2023.

Nord, Eric A., Zhang, Chaochun and Lynch, Jonathan P. (2011), 'Root responses to neighbouring plants in common bean are mediated by nutrient concentration rather than self/non-self recognition', *Functional Plant Biology*, 38:12, pp. 941–52.

Novoplansky, Ariel (2009), 'Picking battles wisely: Plant behaviour under competition', *Plant, Cell & Environment*, 32:6, pp. 726–41.

O'Brien, Erin E., Gersani, Mordechai and Brown, Joel S. (2005), 'Root proliferation and seed yield in response to spatial heterogeneity of below-ground competition', *New Phytologist*, 168:2, pp. 401–12.

Palmer, Andrew G., Ali, Maysaa, Yang, Shukun, Parchami, Neda, Bento, Thiara, Mazzella, Amanda, Oni, Musa, Riley, Michael C., Schneider, Karl and Massa, Nicole (2016), 'Kin recognition is a nutrient-dependent inducible phenomenon', *Plant Signaling & Behavior*, 11:9, https://doi.org/10.1080/15592324.2016.1224045. Accessed 4 April 2023.

Pantazopoulou, Chrysoula K., Bongers, Franca J., Küpers, Jesse J., Reinen, Emilie, Das, Debatosh, Evers, Jochem B., Anten, Niels P. R. and Pierik, Ronald (2017), 'Neighbor detection at the leaf tip adaptively regulates upward leaf movement through spatial auxin dynamics', *Proceedings of the National Academy of Sciences*, 114:28, pp. 7450–55.

Parniske, Martin (2008), 'Arbuscular mycorrhiza: The mother of plant root endosymbioses', *Nature Reviews Microbiology*, 6:10, pp. 763–75.

Partridge, Linda and Harvey, Paul H. (1988), 'The ecological context of life history evolution', *Science*, 241:4872, pp. 1449–55.

Pichersky, Eran, Noel, Joseph P. and Dudareva, Natalia (2006), 'Biosynthesis of plant volatiles: Nature's diversity and ingenuity', *Science*, 311:5762, pp. 808–11.

Pierik, Ronald, Mommer, Liesje and Voesenek, Laurentius A. C. J. (2013), 'Molecular mechanisms of plant competition: Neighbour detection and response strategies', *Functional Ecology*, 27:4, pp. 841–53.

Poorter, Hendrik, Bühler, Jonas, van Dusschoten, Dagmar, Climent, José and Postma, Johannes A. (2012), 'Pot size matters: A meta-analysis of the effects of rooting volume on plant growth', *Functional Plant Biology*, 39:11, pp. 839–50.

Postma, Johannes A., Hecht, Vera L., Hikosaka, Kouki, Nord, Eric A., Pons, Thijs L. and Poorter, Hendrik (2021), 'Dividing the pie: A quantitative review on plant density responses', *Plant, Cell & Environment*, 44:4, pp. 1072–94.

Proust, Hélène, Hoffmann, Beate, Xie, Xiaonan, Yoneyama, Kaori, Schaefer, Didier G., Yoneyama, Koichi, Nogué, Fabien and Rameau, Catherine (2011), 'Strigolactones regulate protonema branching and act as a quorum sensing-like signal in the moss *Physcomitrella patens*', *Development*, 138:8, pp. 1531–39.

Pusey, Anne and Wolf, Marisa (1996), 'Inbreeding avoidance in animals', *Trends in Ecology & Evolution*, 11:5, pp. 201–06.

Renzaglia, Karen Sue, Duff, R. Joel, Nickrent, Daniel L. and Garbary, David J. (2000), 'Vegetative and reproductive innovations of early land plants: Implications for a unified phylogeny', *Philosophical Transactions of the Royal Society of London. Series B: Biological Sciences*, 355:1398, pp. 769–93.

Rice, Elroy L. (2012), *Allelopathy*, 2nd ed, New York: Academic Press, pp. 104–25.

Rodrigo-Moreno, Ana, Bazihizina, Nadia, Azzarello, Elisa, Masi, Elisa, Tran, Daniel, Bouteau, François, Baluška, František and Mancuso, Stefano (2017), 'Root phonotropism: Early signalling events following sound perception in *Arabidopsis* roots', *Plant Science*, 264, pp. 9–15.

Roig-Villanova, Irma and Martínez-García, Jaime F. (2016), 'Plant responses to vegetation proximity: A whole life avoiding shade', *Frontiers in Plant Science*, 7, p. 236.

Rolfe, Stephen A., Griffiths, Joseph and Ton, Jurriaan (2019), 'Crying out for help with root exudates: Adaptive mechanisms by which stressed plants assemble health-promoting soil microbiomes', *Current Opinion in Microbiology*, 49, pp. 73–82.

Roshchina, Victoria V., Yashina, Alexandra V., Yashin, Valery A. and Prizova, Nadezhda K. (2009), 'Models to study pollen allelopathy', *Allelopathy Journal*, 23:1, pp. 3–24.

Rysavy, Anne, Seifan, Merav, Sternberg, Marcelo and Tielbörger, Katja (2016), 'Neighbour effects on shrub seedling establishment override climate change impacts in a Mediterranean community', *Journal of Vegetation Science*, 27:2, pp. 227–37.

Schenk, H. Jochen (2006), 'Root competition: Beyond resource depletion', *Journal of Ecology*, 94:4, pp. 725–39.

Schöb, Christian, Brooker, Rob W. and Zuppinger-Dingley, Debra (2018), 'Evolution of facilitation requires diverse communities', *Nature Ecology & Evolution*, 2:9, pp. 1381–85.

Semchenko, Marina, Abakumova, Maria, Lepik, Anu and Zobel, Kristjan (2013), 'Plants are least suppressed by their frequent neighbours: The relationship between competitive ability and spatial aggregation patterns', *Journal of Ecology*, 101:5, pp. 1313–21.

Semchenko, Marina, Hutchings, Michael J. and John, Elizabeth A. (2007), 'Challenging the tragedy of the commons in root competition: Confounding effects of neighbour presence and substrate volume', *Journal of Ecology*, 95:2, pp. 252–60.

Semchenko, Marina, Saar, Sirgi and Lepik, Anu (2014), 'Plant root exudates mediate neighbour recognition and trigger complex behavioural changes', *New Phytologist*, 204:3, pp. 631–37.

Semchenko, Marina, Zobel, Kristjan, Heinemeyer, Andreas and Hutchings, Michael J. (2008), 'Foraging for space and avoidance of physical obstructions by plant roots: A comparative study of grasses from contrasting habitats', *New Phytologist*, 179:4, pp. 1162–70.

Semchenko, Marina, Zobel, Kristjan and Hutchings, Michael J. (2010), 'To compete or not to compete: An experimental study of interactions between plant species with contrasting root behaviour', *Evolutionary Ecology*, 24, pp. 1433–45.

Shelef, Oren, Hahn, Philip G., Getman-Pickering, Zoe and Martinez Medina, Ainhoa (2019), 'Coming to common ground: The challenges of applying ecological theory developed aboveground to rhizosphere interactions', *Frontiers in Ecology and Evolution*, 7:58, https://doi.org/10.3389/fevo.2019.00058. Accessed 4 April 2023.

Shiojiri, Kaori and Karban, Richard (2008), 'Seasonality of herbivory and communication between individuals of sagebrush', *Arthropod-Plant Interactions*, 2, pp. 87–92.

Simmi, Francine Z., Dallagnol, Leandro José, Ferreira, Arlan Silva, Pereira, Danillo Roberto and Souza, Gustavo Maia (2020), 'Electrome alterations in a plant-pathogen system: Toward early diagnosis', *Bioelectrochemistry*, 133, https://doi.org/10.1016/j.bioelechem.2020.107493. Accessed 4 April 2023.

Smyčka, Jan and Herben, Tomáš (2017), 'Phylogenetic patterns of tragedy of commons in intraspecific root competition', *Plant and Soil*, 417, pp. 87–97.

Smith, Sally E. and Smith, F. Andrew (2012), 'Fresh perspectives on the roles of arbuscular mycorrhizal fungi in plant nutrition and growth', *Mycologia*, 104:1, pp. 1–13.

Song, Yuan Yuan, Ye, Mao, Li, Chuanyou, He, Xinhua, Zhu-Salzman, Keyan, Wang, Rui Long, Su, Yi Juan, Luo, Shi Ming and Zeng, Ren Sen (2014), 'Hijacking common mycorrhizal networks for herbivore-induced defence signal transfer between tomato plants', *Scientific Reports*, 4:1, https://doi.org/10.1038/srep03915. Accessed 4 April 2023.

Song, Yuan Yuan, Zeng, Ren Sen, Xu, Jian Feng, Li, Jun, Shen, Xiang and Yihdego, Woldemariam Gebrehiwot (2010), 'Interplant communication of tomato plants through underground common mycorrhizal networks', *PLoS One*, 5:10, https://doi.org/10.1371/journal.pone.0013324. Accessed 4 April 2023.

Stahlberg, Rainer, Cleland, Robert E. and Van Volkenburgh, Elizabeth (2006), 'Slow wave potentials—A propagating electrical signal unique to higher plants', in F. Baluška, S. Mancuso, and D. Volkmann (eds), *Communication in Plants: Neuronal Aspects of Plant Life*, Berlin/Heidelberg: Springer, pp. 291–308.

Stebbing, Anthony R. D. (2003), 'A mechanism for hormesis—a problem in the wrong discipline', *Critical Reviews in Toxicology*, 33:3–4, pp. 463–67.

Stinson, Kristina A., Campbell, Stuart A., Powell, Jeff R., Wolfe, Benjamin E., Callaway, Ragan M., Thelen, Giles C., Hallett, Steven G., Prati, Daniel and Klironomos, John N. (2006), 'Invasive plant suppresses the growth of native tree seedlings by disrupting belowground mutualisms', *PLoS Biology*, 4:5, https://doi.org/10.1371/journal.pbio.0040140. Accessed 4 April 2023.

Sukhov, Vladimir, Sukhova, Ekaterina and Vodeneev, Vladimir (2019), 'Long-distance electrical signals as a link between the local action of stressors and the systemic physiological responses in higher plants', *Progress in Biophysics and Molecular Biology*, 146, pp. 63–84.

Szechyńska-Hebda, Magdalena, Kruk, Jerzy, Górecka, Magdalena, Karpińska, Barbara and Karpiński, Stanisław (2010), 'Evidence for light wavelength-specific photoelectrophysiological signaling and memory of excess light episodes in *Arabidopsis*', *The Plant Cell*, 22:7, pp. 2201–18.

Takigahira, Hiroshi and Yamawo, Akira (2019), 'Competitive responses based on kin-discrimination underlie variations in leaf functional traits in Japanese beech (*Fagus crenata*) seedlings', *Evolutionary Ecology*, 33, pp. 521–31.

Taylor, Thomas N., Kerp, Hans and Hass, Hagen (2005), 'Life history biology of early land plants: Deciphering the gametophyte phase', *Proceedings of the National Academy of Sciences*, 102:16, pp. 5892–97.

Tharayil, Nishanth, Bhowmik, Prasanta, Alpert, Peter, Walker, Elsbeth, Amarasiriwardena, Dulasiri and Xing, Baoshan (2009), 'Dual purpose secondary compounds: Phytotoxin of *Centaurea diffusa* also facilitates nutrient uptake', *New Phytologist*, 181:2, pp. 424–34.

Till-Bottraud, Irène and de Villemereuil, Pierre (2016), 'Kin recognition or phenotype matching?', *New Phytologist*, 209:1, pp. 13–14.

Torices, Rubén, Gómez, José M. and Pannell, John R. (2018), 'Kin discrimination allows plants to modify investment towards pollinator attraction', *Nature Communications*, 9:1, pp. 1–6.

Toyota, Masatsugu, Spencer, Dirk, Sawai-Toyota, Satoe, Jiaqi, Wang, Zhang, Tong, Koo, Abraham J., Howe, Gregg A. and Gilroy, Simon (2018), 'Glutamate triggers long-distance, calcium-based plant defense signaling', *Science*, 361:6407, pp. 1112–15.

Ueda, Hirokazu, Kikuta, Yukio and Matsuda, Kazuhiko (2012), 'Plant communication: Mediated by individual or blended VOCs?', *Plant Signaling & Behavior*, 7:2, pp. 222–26.

Uesugi, Akane, Johnson, Robert and Kessler, André (2019), 'Context-dependent induction of allelopathy in plants under competition', *Oikos*, 128:10, pp. 1492–1502.

Van der Putten, Wim H., Bardgett, Richard D., Bever, James D., Bezemer, T. Martijn, Casper, Brenda B., Fukami, Tadashi, Kardol, Paul, et al. (2013), 'Plant–soil feedbacks: The past, the present and future challenges', *Journal of Ecology*, 101:2, pp. 265–76.

Veits, Marine, Khait, Itzhak, Obolski, Uri, Zinger, Eyal, Boonman, Arjan, Goldshtein, Aya, Saban, Kfir, et al. (2019), 'Flowers respond to pollinator sound within minutes by increasing nectar sugar concentration', *Ecology Letters, 22:9*, pp. 1483–92.

Vicherová, Eliška, Glinwood, Robert, Hájek, Tomáš, Šmilauer, Petr and Ninkovic, Velemir (2020), 'Bryophytes can recognize their neighbours through volatile organic compounds', *Scientific Reports*, 10:1, pp. 1–11.

Vives-Peris, Vicente, de Ollas, Carlos, Gómez-Cadenas, Aurelio and Pérez-Clemente, Rosa María (2020), 'Root exudates: From plant to rhizosphere and beyond', *Plant Cell Reports*, 39, pp. 3–17.

Volkov, Alexander G. and Shtessel, Yuri B. (2017), 'Electrotonic signal transduction between *Aloe vera* plants using underground pathways in soil: Experimental and analytical study', *AIMS Biophysics*, 4:4, pp. 576–95.

——— (2018), 'Electrical signal propagation within and between tomato plants', *Bioelectrochemistry*, 124, pp. 195–205.

Volkov, Alexander G., Toole, Shannon and WaMaina, Mwangi (2019), 'Electrical signal transmission in the plant-wide web', *Bioelectrochemistry*, 129, pp. 70–78.

Waters, Mark T., Gutjahr, Caroline, Bennett, Tom and Nelson, David C. (2017), 'Strigolactone signaling and evolution', *Annual Review of Plant Biology*, 68, pp. 291–322.

Wang, Nan-Qi, Kong, Chui-Hua, Wang, Peng and Meiners, Scott J. (2021), 'Root exudate signals in plant–plant interactions', *Plant, Cell & Environment*, 44:4, pp. 1044–58.

Wang, Zi-Yang, Qin, Xiao-Huang, Li, Jin-Hai, Fan, Li-Feng, Zhou, Qia, Wang, Yong-Qian, Zhao, Xuan, Xie, Chao-Jie, Wang, Zhong-Yi and Huang, Lan (2019), 'Highly reproducible periodic electrical potential changes associated with salt tolerance in wheat plants', *Environmental and Experimental Botany*, 160, pp. 120–30.

Weiner, Jacob (2017), 'Applying plant ecological knowledge to increase agricultural sustainability', *Journal of Ecology*, 105:4, pp. 865–70.

West, Stuart A., Griffin, Ashleigh S. and Gardner, Andy (2007), 'Social semantics: Altruism, cooperation, mutualism, strong reciprocity and group selection', *Journal of Evolutionary Biology*, 20:2, pp. 415–32.

Weston, Leslie A. and Duke, Stephen O. (2003), 'Weed and crop allelopathy', *Critical Reviews in Plant Sciences*, 22:3–4, pp. 367–89.

Yang, Xue-Fang, Li, Lei-Lei, Xu, You and Kong, Chui-Hua (2018), 'Kin recognition in rice (*Oryza sativa*) lines', *New Phytologist*, 220:2, pp. 567–78.

Xu, Juan and Zhang, Shuqun (2015), 'Mitogen-activated protein kinase cascades in signaling plant growth and development', *Trends in Plant Science*, 20:1, pp. 56–64.

Yu, Rui-Peng, Li, Xiao-Xi, Xiao, Zhi-Hua, Lambers, Hans and Li, Long (2020), 'Phosphorus facilitation and covariation of root traits in steppe species', *New Phytologist*, 226:5, pp. 1285–98.

Zeng, Ren Sen (2014), 'Allelopathy-the solution is indirect', *Journal of Chemical Ecology*, 40, pp. 515–16.

Zhang, Deshan, Lyu, Yang, Li, Hongbo, Tang, Xiaoyan, Hu, Ran, Rengel, Zed, Zhang, Fusuo, et al. (2020), 'Neighbouring plants modify maize root foraging for phosphorus: Coupling nutrients and neighbours for improved nutrient-use efficiency', *New Phytologist*, 226:1, pp. 244–53.

Zhang, Ruichang and Tielbörger, Katja (2019), 'Facilitation from an intraspecific perspective–stress tolerance determines facilitative effect and response in plants', *New Phytologist*, 221:4, pp. 2203–12.

Zhou, Li Hong, Liu, Shao Bao, Wang, Peng Fei, Lu, Tian Jian, Xu, Feng, Genin, Guy M. and Pickard, Barbara G. (2017), 'The *Arabidopsis* trichome is an active mechanosensory switch', *Plant, Cell & Environment*, 40:5, pp. 611–21.

Zhu, Yong-He, Weiner, Jacob and Li, Feng-Min (2019), 'Root proliferation in response to neighbouring roots in wheat (*Triticum aestivum*)', *Basic and Applied Ecology*, 39, pp. 10–14.

Zweifel, Roman and Zeugin, Fabiene (2008), 'Ultrasonic acoustic emissions in drought-stressed trees–more than signals from cavitation?', *New Phytologist*, 179:4, pp. 1070–79

6

The Conceptual Ecology of the Human Microbiome[1]

Nicolae Morar and Brendan J. M. Bohannan

Introduction

It has become increasingly clear that humans have on them and in them a vast array of non-human organisms. The most abundant members of our resident menagerie are microorganisms. These microorganisms and their corresponding genomes are known collectively as the *Human Microbiome* (NIH 2018). The scale and complexity of our microbiome are extraordinary; for example, we are home to hundreds of species of bacteria, protists, fungi and viruses that vary in abundance across our bodies, and from individual to individual. In total, our microbiome may include three times more microbial cells than human cells (Bianconi et al. 2013; Ravel et al. 2014; Sender et al. 2016 have proposed a 1:1 ratio).

As amazing as the number of microbial cells and species in a single human may be, the genetic diversity that our microbes make available to us is even more impressive. Although human cells contain approximately 22,333 genes (Pertea and Salzberg 2010), our microbiome contains more than two million genes (Xu and Gordon 2003; Turnbaugh et al. 2007; Qin et al. 2010). Most of the genetic variation among humans is microbial, and our microbiome can even be used as an individual identifier (Franzosa et al. 2015). This kind of biological complexity poses a serious challenge to any scientist who is striving to unravel something about the fundamental nature of human organisms. To the question 'What are we?' Tom Insel replies, 'We are, at least from the standpoint of DNA, more microbial than human' (Smith 2015: n.pag.; see also Anonymous 2008; Sleator 2010; Ray 2012; Relman 2012; Brody 2014).

Not only are we host to a diversity of microbial species and their genes, but there is increasing evidence that our microbial component has an important impact on our health and well-being. Our microbes feed us through digesting foods that we cannot, they stimulate our immune system so that we can resist organisms that could cause us harm, they influence the development of our tissues when we are young and may even impact our behaviour by changing the concentration of neurochemicals in our body. Such intimate interactions between human and microbial cells can blur the lines between what is 'human' and what is 'microbial', and challenge simple definitions of an 'individual' human (Gilbert et al. 2012; Schneider and Winslow 2014; Pradeu 2016).

In part, because of the philosophical challenge posed by our microbial companions, we struggle with how to think about them. They have been referred to as a 'second genome' and as an 'additional organ'. Our microbes have been divided into 'self' and 'non-self' components as if they are human tissue. The human body has been called an 'ecosystem' with human and microbial 'communities', a 'superorganism' made up of distinctly different organisms united for a common cause, and a 'holobiont' of host and symbionts shaped collectively by evolution.

This struggle with finding the best conceptual framework is reflected even in the history of the word 'microbiome' itself. Certainly, this word has multiple origins, but one of them was conceived as a play on the word 'biome' to mean a 'convenient ecological framework ... occupying a reasonably well-defined habitat which has distinct physico-chemical properties' (Whipps et al. 1988: 176; see also Huss 2014; Eisen 2015). As a 'microbial biome', this use of 'microbiome' in reference to the collection of microbes associated with the human body explicitly invokes the idea of 'human-as-ecosystem'. Later, the word 'microbiome' was re-envisioned as a play on the word 'genome' to mean the collection of genes provided to a host by its associated microbes (see Lederberg and McCray 2001). This concept of the 'microbiome' is shorthand for the idea of human-associated microbes as a 'second genome', and it invokes explicitly the concept of 'microbes-as-organ'.

These different approaches to describing the human–microbe system are not just benign choices of words. As with all conceptual frameworks, these different views of the human–microbe system deploy different foundational metaphors, make very different assumptions, organize experimental research in specific ways and ultimately produce different views of the origin and nature of interactions between our human and microbial parts. More importantly, these views and their assumptions can lead to very different approaches to the management of the human–microbe system that, in turn, may have very important implications for human health and well-being.

Because of these implications, it is especially important to understand each of these conceptual frameworks in detail, including their origins, underlying assumptions, limitations and implications for human health. In order to explore the 'conceptual ecology' (sensu Stotz and Griffiths 2008) of the human microbiome, we have chosen the five most common of these frameworks below (the 'organ' view, the 'immune system' view, the 'superorganism' view, the 'holobiont' view and the 'ecosystem' view). For each conceptual framework, we first highlight the premises that sustain it. Second, we discuss the benefits of each concept in helping us design possible therapeutics. Finally, we assess the limitations of each by evaluating whether knowledge claims about the human microbiome made by each framework have been validated by experimental research, and we show how each framework constrains our understanding of human–microbe systems. Many of the examples we discuss are of applications of these frameworks to specific microbiomes (i.e. microbiomes at specific body sites), most often the gut microbiome (which is the best studied). However, we do not argue (nor do we believe) that the conceptual frameworks that we describe apply only to particular locations. For example, the microbiome of both the skin and the gut has been described as a 'second immune system' (Douglas 2018: 10; see also Chiu et al. 2017).

There is a tendency among some scientists to assume that there is one correct and complete account of the human–microbe self, represented by one particular conceptual framework and that the 'correct' framework will emerge if we comparatively evaluate the alternatives. We disagree with this view. Rather we show that monistic frameworks (i.e. a particular concept and the account that it represents) produce reductive (partial) explanations that ultimately limit research and shape therapeutic agendas in biased ways, and we conclude that *only* scientifically pluralist approaches can push the boundaries of our understanding of what it means to be human in the post-microbiome era.

The Organ View

A number of biologists have proposed that we think of the human microbiome as one of our *organs* (Bocci 1992; Bäckhed et al. 2004, 2005, 2007; Stainier 2005; O'Hara and Shanahan 2006; Baquero and Nombela 2012; Clarke et al. 2014; Enders 2015). In order to fully appreciate this conception of the human microbiome, we need to unpack a series of underlying assumptions.

Assumptions of the 'Organ View' of the Human Microbiome

Textbooks of human anatomy commonly define an organ as being composed of two or more tissue types that perform one or more common functions (e.g. DeWitt 1989; see also Seeley et al. 2000). Consider our heart as a paradigmatic example. Not only is a heart made of numerous tissues (muscle, fibrous and special cells that maintain the rate of our heartbeat) but it also has a very specific and constrained function – to pump blood so that it circulates throughout our body. Often, the higher the number of tissues involved in the structure of an organ, the more complex the organ and the role it plays in our body's functioning.

There are numerous references in the scientific literature to the human microbiome as such an organ. They range from comparisons in size and mass – e.g. Bocci notes that 'the mass of bacteria ... equals ... [the size] of the liver' (Bocci 1992: 251) – to more detailed arguments based on metabolic function (Frank and Pace 2008). The human microbiome is composed of multiple cellular materials (i.e. different microbial taxa) like an organ, and it performs very specific bodily functions. These functions include metabolic functions (such as digestion of food or vitamin production; Frank and Pace 2008) or even neurological functions mediated by neurochemicals altered by our gut microbiome (Turnbaugh et al. 2007: 805).

Therapeutic Applications that Emerge from the 'Organ View' of the Human Microbiome

The 'organ view' assumes that the gut microbiome is a unified system with specific inputs and outputs. Conceptualizing the human microbiome in this way influences how we measure its relevance for our health and, consequently, it frames particular therapeutic responses.

Consider, for example, infection of the gut by the bacterium *Clostridium difficile*. Infection by *C. difficile* can lead to a number of impacts on gut health, ranging from mild diarrhoea to pseudomembranous colitis (Surawicz et al. 2013). Such infections are difficult to treat, and failure rates for antibiotic therapy are high (van Nood et al. 2013: 407; see also Taur and Pamer 2014; Youngster et al. 2014), resulting in up to 14,000 deaths annually (Knight 2015: 68). Assuming the organ view, one could consider this infection as leading to 'organ failure' of the gut microbiome, because following infection the normal metabolic function of the gut microbiome is impaired. One solution that could be envisioned would be 'organ transplantation'. In fact, this is exactly what some physicians have proposed, an approach called *faecal microbiota transplantation* (FMT). Healthy volunteers, often family members, donate stool samples, which are subsequently 'transplanted' into the patient.

There is evidence that this approach can work. In one of the first randomized trials of FMT, patients with *C. difficile* infections received one of the following three therapies: antibiotic treatment alone; antibiotic treatment followed by bowel lavage (flushing of the bowel); or antibiotic treatment followed by bowel lavage and subsequent infusion of a donor's faeces. The FMT approach resulted in an 81 per cent cure rate of colitis caused by *C. difficile*, compared to 31 per cent with antibiotic treatment alone and 23 per cent with antibiotic treatment and bowel lavage (van Nood et al. 2013: 407; see also Gough et al. 2011; Youngster et al. 2014; Kelly et al. 2016). Like an organ, the microbial composition of the transplant persisted over an observational period of three months either by replacing or co-existing with the recipient's microbiome (Li et al. 2016; Kumar et al. 2017). However, unlike an organ, the recipient microbiome tends to gradually change in composition and eventually loses its similarity to its donor (Li et al. 2016). This suggests that the restoration of metabolic function is not merely a result of transplantation but also of some other processes not accounted for by the organ concept, such as colonization and niche competition (Kumar et al. 2017).

Limitations of the 'Organ View' of the Human Microbiome

If this medical procedure is so successful, why should we not simply assume that the microbes that inhabit our body (e.g. our gut) are nothing more than one of our organs? There are a number of reasons why the 'organ view' can be insufficient.

First, this view presents us with a particularly static conception of the human microbiota. Organs tend not to change substantially over time, especially after maturity is reached. Similarly, changes over time in our microbiota tend to be minimized under this framework, and they are assumed to *not* be functionally significant (Foxman et al. 2008: 3). However, there is ample evidence that changes of the microbiota over time can be associated with significant changes in organism function. Moreover, the organ view implies that the composition of the human microbiome is impermeable to inputs of cells from outside the organism. After all, we do not exchange heart or liver cells with each other or with our outside environment. However, there is evidence that host-to-host dispersal of microbes can significantly alter the composition of microbiomes (e.g. Song et al. 2013; Tung et al. 2015; Burns et al. 2017). The organ view produces a limited conception of the human microbiome by trading the complexity of the diverse microbiota for a static input–output system whose function can be measured, removed along with the 'organ' itself and ultimately restored through transplantation.

Second, from an evolutionary perspective, the organ view is grounded on particularly implausible assumptions. The evolution of organs and organ systems is understandable; the unit of selection is clearly the level of the reproducing human body. The fitness of the organ is subsumed within the fitness of the individual and any conflicts (e.g. unconstrained organ growth, as occurs in cancer) that arise are resolved at the level of the reproducing human. But how would such an evolutionary explanation work for a microbiota within a host?

Given this view, it is especially difficult to conceive of how such a dynamic microbiome along with its host could become a unit of selection. We do not assume here that more complex evolutionary models (e.g. multilevel selection models) could not provide us with such an evolutionary explanation of our microbiome (as we discuss below). But, in the case of the 'organ view', it remains unclear how the human microbiome could have sufficient heritability across generations to evolve in a manner analogous to a vertically transmitted organ, nor is it clear how fitness conflicts between microbe and host would be avoided or resolved.

The Self–Non-self (Immune System) View

In contrast to 'organs', host-associated microbiomes have also been conceptualized as a component of a host's immune system, and the host–microbe system has been thought of as consisting of two distinctive parts – 'self' and 'non-self' – defined in part by this immune system. The immune system is primarily a *functional system* whose goal is to set up *inside-out* and to provide a defensive mechanism for a self-contained system (see Tauber 2008: 225).

Assumptions of the 'Immune System View' of the Human Microbiome

The defense system of the human body is a particularly complex system. It is made up of organs (the lymphoid organs, mainly our bone marrow and thymus), vessels (lymphatic), cells (especially T cells) and proteins (antigens) – including our skin and our mucous membranes as mechanical protective walls.

There is also evidence that components of the microbiome are involved in defining the frontier between self and non-self, essentially acting as a part of the 'immune system' broadly defined (Grice and Segre 2012: 151; see also Pradeu 2012; Eberl and Pradeu 2018). This claim is based on two arguments. The first is that members of the microbiome directly prevent pathogens from establishing – a process known as 'colonization resistance' (Spees et al. 2013; Bäumler and Sperandio 2016) or 'the right of "first occupancy"' (Sansonetti and Medzhitov 2009: 416). The second is that members of the microbiome indirectly exclude pathogens by stimulating the immune system (including both stimulating the development of the immune system and through 'training' the immune system; McFall-Ngai 2007; Lee and Mazmanian 2010; Olszak et al. 2012; Sevelsted et al. 2015; Thaiss et al. 2016). In the first case, members of the microbiome act in a way analogous to a cellular component of the immune system (by directly excluding pathogens). In the second case, the essential role that members of the microbiome play in the development of the immune system justifies (at least in the minds of some) consideration of the microbiome as an essential 'component' of the immune system (which would not be the same without the microbiome). This component acts differently than pathogens (which also 'train' the immune system) because it often does not elicit an inflammatory response (they are treated like 'self'; Hooper et al. 2012; Gallo et al. 2016).

Therapeutic Applications that Emerge from the 'Immune System View' of the Human Microbiome

Conceptualizing the human microbiome as a component of the immune system emphasizes the importance of how the microbiome is acquired early in life and the role this acquisition may play in the development of the immune system. Human fetuses in *utero* develop with a reduced exposure to microbes, and their first major exposure to a wide variety of microbes is during birth. In the case of vaginal delivery, those microbes are provided during and after the passage through the birth canal. In the case of a cesarean (C-section) delivery, the newborn is exposed to microbes living on adult skin, which represent a completely different microbial community. There is some evidence that these different exposures can result in different initial microbiomes.

These differences in initial exposure could in turn alter the functioning of the immune system, both directly (e.g. reduced colonization resistance) and indirectly (e.g. a less 'well-trained' and, thus, less robust immune system). Evidence for this link between birth mode, microbiome composition and immune system function is still scant. Dominguez-Bello and colleagues (2010) have gathered preliminary data that suggest that such links are plausible. They observed that when delivered via the birth canal, the neonate is exposed to 'an ecosystem heavily colonized by a relatively limited set of bacterial taxa' (Dominguez-Bello et al. 2010: 11971), dominated by *Lactobacillus*, *Prevotella*, *Atopobium* and *Sneathia* spp. When delivered by C-section, typical skin taxa, such as *Staphylococcus* spp., tend to dominate the neonate's microbiome, and the neonate's microbiome is less diverse.

The difference in these initial microbial communities may also lead to differences in how microbiomes develop over time. Dominguez-Bello et al. (2010) suggested that cesarean-delivered infants will likely have their gut colonization by *Lactobacillus* and other microbes delayed, and such a delay will end up having an effect on the proper development of a child's immune system (for a critique, see Bäckhed et al. 2015). The higher rates of asthma, allergies and other immune disorders reported for children delivered via C-section could be explained by differences in these initial microbial communities and how they fulfil their role in the development of adaptive immunity.

The 'self–non-self view' of the human microbiome not only stresses the importance of early acquisition of particular microbes but also suggests that we need to develop procedures that ensure the exposure of newborns to a more diverse microbiota (Knight 2015: 24–26). For example, the development of the novel and controversial practice of 'vaginal seeding' (swabbing a mother's vaginal canal and transferring this microbiota to the mouth and skin of a newborn delivered via C-section) assumes in part that our microbiota is an integral part of our immune system. However, robust evidence in support of this medical practice is still lacking (Knight 2015: 25; see also Dominguez-Bello et al. 2016). Two ongoing clinical trials are further testing the efficiency of this novel therapeutic procedure (see https://clinicaltrials.gov [accessed 6 April 2023]).

Limitations of the 'Immune System View' of the Human Microbiome

If medical procedures such as vaginal seeding are proven to be successful, why should we not simply assume that our human microbiome is just a core constituent of our immune system? We argue below that this conceptualization of the human microbiome remains epistemically partial.

First, the 'self–non-self view' of the human microbiome cannot fully explain all of the different relationships between hosts and their associated microorganisms simply because it does not capture *all* of the services we receive from our microbiome (e.g. metabolic). Interestingly enough, more precise definitions of microbial types, according to their state in the host organism (e.g. 'mutualist' or 'commensal'), do not solve this difficulty. Each new category produces a new qualifier for our microbial inhabitants, and this fine-tuning of our concepts remains inadequate to capture the complexity of host–microbe interactions (see the *Aspergillus fumigatus* or *Staphylococcus aureus* examples below).

Second, if immunity is understood as a protective mechanism, based on an adversarial logic that is centred on policing the composition of the microbiome, the divide self–non-self is not true at the level of strains of bacteria. Moreover, this view fails to appreciate the importance of host–microbe interactions and the often context-dependent outcomes of such interactions. Casadevall and Pirofski claim that '[m]icrobes and hosts are variable and unpredictable' (Casadevall and Pirofski 2014: 166). This is reinforced by the fact that a number of microbes are considered 'facultative' pathogens that cause disease *only* under certain conditions. For example, *A. fumigatus* causes pulmonary diseases primarily in hosts who suffer from leukaemia (or are otherwise immunocompromised) and *S. aureus* behaves, in one out of three human hosts, more like a commensal microbe (doing little or no harm to its host) than a pathogen. It is unclear how a self–non-self conception can account for this degree of context dependence.

Finally, if the human microbiome is to be subsumed under the immunity view, providing, for example, consistent training of the immune system, the uniqueness of each individual's microbiome (Falony et al. 2016) would appear to challenge this view. In other words, no rule could (easily) emerge that would allow consistent selection between self and non-self entities, given this complexity. Our immune system exists in part to enforce such 'rules' (Sansonetti 2004), but in practice, these rules are not simple recognition rules based on taxonomic identity, but rather are driven by location (e.g. based on where a microbe is found relative to the epithelium in the gut; Hooper 2009), and by the nature and concentration of microbial-associated biomolecules (which can vary across individuals of the same microbial strain). This is one of the reasons why there is no scientific consensus over the definition of what comprises a normal and healthy microbiome (Blumberg and Powrie 2012: 6; see also Yong 2014). These points emphasize the importance of not framing research regarding health and disease only through the lens of the immunity view of the human microbiome.

The Superorganism View

Among the popular conceptions of the human microbiome, the superorganism view has found significant traction, especially in the medical literature (Lederberg 2004; Fujimura et al. 2010; Dietert and Dietert 2012; Murdoch and Detsky 2012; Dietert 2016; Lynch 2016; Lynch and Pedersen 2016). '[H]umans are in fact a vastly complex, biologic "superorganism"', claimed Kinross et al. (2008: 396) in one such example. A 'superorganism' is a collection of organisms (from the same or from different species) that interact closely and from which functions emerge that do not exist at the level of individual organisms. This concept has a long history of use in biology (Forbes 1887), for example, to explain complex systems of individuals such as colonies of social insects (Wheeler 1911; Wilson 1968; Lumsden 1982; Seeley 1989; Wilson and Sober 1989; Mitchell 1995; Hölldobler and Wilson 2009).

The original definition of superorganism sets four conditions (Wheeler 1911). The *individuality* condition posits that a group of organisms is a superorganism *if* it behaves like a unitary whole. The *uniqueness* condition states that a superorganism displays certain idiosyncrasies in behaviour, size and structure that are both group-specific (e.g. specific to a particular insect colony) and species-specific. The *growth* and *reproduction* condition requires that superorganisms undergo a cycle of growth and reproduction that is adaptive. Finally, the *composition* condition can be broadly stated that there must be multiple kinds of individuals in a superorganism. For example, these may be reproductive and non-reproductive types (e.g. workers and queens/drones in a social insect colony), similar to an organism's duality between germ and somatic cells.

These four conditions circumscribe a specific kind of functional organization. However, as Mitchell (1995: 241) rightly points out, functional organization, even if it is adaptive, is not a sufficient *evolutionary* condition for a collection of organisms to be considered a superorganism because it does not account for the misalignment of fitness among individuals. Wilson and Sober (1989) have supplemented the original conception of the superorganism with an additional requirement. Their *fitness* condition states that a group of organisms successfully survives and reproduces as a superorganism if its variation in (group) fitness is genetically heritable and there is no difference in the fitness among individuals within the group.

Given the history of this concept and the ways in which it has been applied across different biological phenomena, some of its definitions certainly overlap with other conceptual frameworks, notably the holobiont (which we discuss below). The differences between these concepts have been explicitly addressed. For example, Gordon and colleagues highlight the (potentially) *nested* nature of those conceptual frameworks: 'an ant colony [is] [...] a superorganism, but now seen as one composed of many individual ant holobionts' (Gordon et al. 2013: 153). These two concepts are distinct in other important ways; for example, the membership of a superorganism is assumed to be very tightly regulated (e.g. membership in an insect colony is regulated *genetically* for conspecific individuals and through *ventilation and temperature* for individuals of other species; Turner 2001, 2002), while some recent conceptions of the holobiont assume less stringent control on membership (e.g. Thompson et al. 2015).

Assumptions of the 'Superorganism View' of the Human Microbiome

The biological premises that support the superorganism conceptualization of the human microbiome are neatly encapsulated in the required conditions we discuss above. First, the *individuality* condition – does the host–symbiont system in the case of the human organism function like a whole? There are numerous examples of such functional integration. For instance, microbes in our gut (primarily in the large intestine), in concert with the digestive actions from our human cells, provide us with 6–10 per cent of our energy budget by helping us to break down foods that our human cells alone could not otherwise (e.g. glycan; Gill et al. 2006; see also van Duynhoven et al. 2011). Second, the *uniqueness* condition implies that each human–microbe system displays a series of biological features that are peculiar to that individual system. It has been documented that humans shelter radically different microbial communities that may include hundreds or thousands of unique species (Knights et al. 2011) and, as a consequence, such systems have individual microbial signatures (Franzosa et al. 2015).

The *growth* and *reproductive* condition requires that the host–microbe system as a whole grow (and develop), as well as reproduce, as a unit. There is increasing evidence that supports the assumption that growth and development of humans and their microbiomes can be intimately linked (Dominguez-Bello et al. 2019). However, the reproduction of the host and microbiome together requires that there be some fidelity of microbiome membership across generations, i.e. heritability in the evolutionary sense. There is very little evidence that the human microbiome as a whole has significant heritability. For example, Rothschild et al. (2018) estimated that fewer than 2 per cent of the microbial taxa in the human microbiome exhibit significant heritability, which suggests that very few members of the human gut microbiome would meet this condition.

The *composition* condition requires that superorganisms be composed of multiple types of individuals. This is clearly met; the human body is, as Gordon and Klaenhammer put it, 'a mélange of microbial and human cells and genes, with our microbial parts providing us with attributes not encoded or expressed by our strictly defined *Homo sapiens* components' (Gordon and Klaenhammer 2011: 4514).

Finally, the *fitness* condition requires that the variation in (group) fitness is heritable and that there is no difference in the fitness among individuals within the group. It is not clear how well the human–microbe system meets this condition. There is evidence that some disorders linked to the human microbiome are indeed heritable (e.g. type 2 diabetes), which suggests that at least some components of 'group' fitness may be heritable, and there is some evidence that the presence and/or abundance of some members of the microbiome may be significantly heritable (e.g. Goodrich et al. 2016). However, there is also evidence that other members of the human microbiome are not heritable.

There is no evidence that individual members of the microbiome are equivalent in terms of fitness (part of the 'fitness condition'). This would require very strong selection at the level of the group (i.e. the human–microbe system) to counter the effects of individual-level selection. It is possible for this to occur, represented by different evolutionary models that assume different relationships between individual and group fitness and reproduction. For example, the multilevel selection (MLS) type 1 model assumes that reproduction is at the level of individuals and group fitness is the total of the individual fitness of its members (i.e. group selection; see Wilson 1975), while MLS type 2 assumes that the levels of reproduction (higher level vs. lower level) are uncoupled and group fitness is not an additive function of each member's fitness (see Okasha 2006). In contrast, a 'functional agent' model assumes that the components of an individual organism, analogous to the pieces of a machine, function in a coordinated and integrated way, thus showing adaptations at the level of the organism that are not present at the component level (Folse and Roughgarden 2010). Although such models could explain the complex evolutionary interactions between the host and its microbiota, they still need to be validated through experimental data.

Researchers have also suggested that the superorganism framework could explain the phenomenon of cospeciation (Moeller et al. 2016). Cospeciation is the observation of similar patterns of diversification across species. Cospeciation has been documented for some host and microbial species; for example, Moeller and colleagues (2016) have shown how the phylogeny of *Bacteroidaceae* and *Bifidobacteriaceae* mirrors closely patterns of primate (including hominid) speciation. Cospeciation could arise from a number of evolutionary processes, including coevolution (i.e. reciprocal selection across species), and some have argued that cospeciation between hosts and microbiome members is the evidence that the host–microbe interactions that we observe today could be the result of selection at higher levels – namely, the superorganism level (Segre and Salafsky 2016).

Therapeutic Applications that Emerge from the 'Superorganism View' of the Human Microbiome

The reasons mentioned above (e.g. intimate physical and chemical association between host and microbiome; beneficial interactions between microbiomes and their hosts that may result in cospeciation) explain why some superorganism proponents endorse a robust view of coevolution that recognizes a crucial role for microbiomes in the evolution of animals and plants and vice versa (Ley et al. 2008; McFall-Ngai et al. 2013; Segre and Salafsky 2016). This thesis urges us to consider the evolutionary history that we may have shared with our microbiome when we consider human evolution and to be mindful of the health benefits that human hosts have been outsourcing to their microbiomes. The idea that humans have coevolved with their microbiomes is at the root of many proposed therapies for microbiome-mediated disorders.

For example, there is growing interest in the effects of western diets and modern hygiene on the species that inhabit our gut, represented by research projects such as the American Gut (Knight 2015: 87–88), the Ancestral Microbiome (Clemente et al. 2015) and the Human Food Project (Leach 2014). These projects are inspired, at least in part, by popular hypotheses about human health, such as the '"disappearing microbiota" hypothesis' (Blaser and Falkow 2009: 889; see also Blaser 2006, 2011; Taube and Müller 2012). Three central ideas animate such hypotheses. First, humans and the members of their microbiomes have coevolved such that the dominant interactions are beneficial (Blaser and Falkow 2009: 887; see also Dethlefsen et al. 2007). Second, western diets along with rapid changes in human ecology (e.g. hygiene and use of antibiotics) have significantly disturbed our coevolved, indigenous microbiome. Third, this disturbance has resulted in an increase in microbiome-related disorders in western countries. The 'disappearing microbiota hypothesis' has led to calls for clinical interventions that aim to restore this 'ancestral' microbiome.

Recent studies suggest that 'modern' western lifestyles may have negatively affected some of the microbes that have inhabited our gut and provided us with services for a long time. For example, there is evidence that *Helicobacter pylori*, an ancestral bacterial tenant of the human gut, may have decreased in prevalence among individuals in industrialized nations over only two or three generations, concurrently with increases in the use of antibiotics and increased hygiene (Blaser and Falkow 2009: 891). Not only may we have lost some specific members of our microbiome in response to lifestyle changes, but our gut microbiome may have decreased in overall diversity as well. For example, Clemente and colleagues reported that members of the Yanomami community, Amerindian tribes with no documented contact with western people prior to 2008, have significantly more diverse gut microbiomes than the average American or western individual (Clemente et al. 2015; Martínez et al. 2015; see also Leach 2014).

There is also evidence for a potential correlation between the loss of microbial diversity in westernized countries and the inexplicable rise of certain diseases, such as modern allergies, asthma and metabolic diseases (Blaser 2014: 167–84). Some scientists believe that 'alterations in human macroecology have progressively affected the composition of our Indigenous microbiota, which in turn has affected human physiology and, ultimately, disease risk' (Blaser and Falkow 2009: 889). Although the links between those events and the rise of modern plagues still remain to be robustly established, the idea of 'recolonizing vacated niches' as potential future therapeutics has gained significant traction (Leach 2012; Relman 2012).

We are mindful that likely not *all* superorganism advocates would necessarily agree with the 'ancestral microbiome' assertion; however, this assertion is based on the assumption of a strongly coevolved host–microbiome system and patterns of cospeciation (Moeller et al. 2016; Segre and Salafsky 2016), which are fundamental to the superorganism view. Unlike the usual case of probiotics (developed in the next section), which focuses mainly on reintroducing certain microbes for specific functional purposes, the superorganism-informed view focuses on the therapeutic benefits that emerge from restoring our ancestral microbiome state. It assumes that there is a significant health risk associated with leaving an evolutionary 'vacated niche', such as the one previously occupied by *H. pylori*, for example, unoccupied (Blaser and Falkow 2009: 892).

Limitations of the 'Superorganism View' of the Human Microbiome

Although increasing in popularity, the superorganism framework has severe limitations. As we discuss above, this framework requires a number of conditions, only a few of which are likely met. We focus below on one of these, the fitness condition, which requires that, first, there should be no difference in fitness among individuals in the group and, second, that variation in group fitness is genetically inheritable.

The first condition takes us back to an old question in evolutionary theory: can traits with public benefits for the group and private costs for each individual organism evolve by natural selection? The possibility of such biological systems is conditioned on finding a way to minimize the conflict in fitness among various organisms and, thus, prevent the emergence of cheaters. Whether we think of cheaters or pathogens, the misalignment in fitness interest between hosts and some of their microbes would significantly hinder the possibility for group traits (such as 'health') to emerge and would limit the potential for human–microbiota coevolution. Thus, such coevolution is bounded by a necessary condition for mutualism or, minimally, of commensalism between the human body and its microbial tenants. This ecologically and evolutionary informed reasoning was defended by Dethlefsen et al. (2007).

More specifically, they claim that such complex mixtures of microorganisms and human hosts could have conjointly evolved only in the presence of certain mutualistic mechanisms, such as 'by-product benefits' or 'partner fidelity feedback' (Dethlefsen et al. 2007: 811). Dethlefsen and colleagues also believe that for any mutualistic mechanism that is not cost-free, the partners can produce mechanisms to protect their relation from cheaters, such as in the case of adaptive immunity (McFall-Ngai 2007) or quorum-sensing policing (Wang et al. 2015). If the focus is only or mainly on the cooperative mechanisms that tend to sustain the human–microbe symbiosis, then selection could *plausibly* occur at the community level (Folse and Roughgarden 2010).

But this kind of cooperation-based explanation of the human–microbiota assumes a fairly stable functional structure, whose success in realization implies, in part, less dramatic changes of the microbiota. This explanation is achieved at the expense of minimizing the importance of conflict and of misalignment in fitness interest between the host and the microbiota and between the individual organisms that constitute the microbiota itself. In addition to serious concerns as to whether policing mechanisms are the exception or the rule in human–microbe interactions, there are sufficient examples where genomic conflicts are central to evolution. Whether one thinks of parent–offspring conflicts (Haig 2010, 2014), or of situations when bacteria have evolved at the expense of the host (by manipulation of host's behaviour; Ingram et al. 2013; Alcock et al. 2014) or as a vector of disease (Kodaman et al. 2014), the superorganism model fails to account for a number of situations when significant differences in fitness exist among the individuals in the group. Thus, minimizing the importance of competition in order to support the coevolution thesis could potentially give us a biased view of human–microbiota interactions.

Second, when it comes to explaining the modes of transmission/genetic heritability, even Dethlefsen and colleagues bite the bullet since they recognize that '[i]n humans, symbionts are not vertically transmitted' (Dethlefsen et al. 2007: 814). Why does this matter? If they are not vertically transmitted, it is particularly difficult to understand how the benefits of higher conjoint fitness persist through time. If a child acquires his microbiome through colonization, succession and diversification, one cannot but reflect on the highly dynamic and stochastic nature of this process of acquisition. Any evolutionary explanation of the human microbiome has to provide us with a clear response as to how organisms acquired during one's lifetime can be passed on to the next generation (Asnicar et al. 2017). Moreover, heritability estimates of the gut microbiome are particularly low in comparison with other heritable (human) traits. The estimates range from 0 to 0.4 (in the UK twin study; Goodrich et al. 2016), 0–0.46 (in the Korean twin study; Lim et al. 2017) or 0–0.67 (in the Canadian study; Turpin et al. 2016). Most recent studies similarly provide very low heritability estimates overall for the microbiome (between 1.9 and 8.1 per cent; Rothschild et al. 2018) and thus corroborate the view that given such low heritability estimates, it is safe to infer that environmental processes tend to dominate microbiota composition. So, it is unlikely that this new human–microbiota superorganism could become a unit of selection. Not impossible or even unexplainable, but unlikely as we will show also in the next section regarding the holobiont conceptual framework.

The Holobiont View

Another concept that is becoming increasingly popular (and controversial) as a way to capture our new view of the human–microbe system is the concept of the *holobiont* and its genetic component, the *hologenome* (Rosenberg et al. 2007, 2009; Zilber-Rosenberg and Rosenberg 2008; Gordon et al. 2013; Bordenstein and Theis 2015; Theis et al. 2016; see also Arnold 2013; Gissis et al. 2017: Section 4).

This concept asserts that host–microbe systems are evolutionary units (*holobionts*), coevolved to increase their inclusive (host and microbe) fitness, and that this coevolution acts to shape the combined genomes of host and microbe (the *hologenome*). The holobiont perspective focuses on the generation and maintenance of host–microbe systems through evolution. The metaphor embedded in this concept is that the microbiome can be thought of as a *symbiont* (Roughgarden et al. 2018).

The concept of the holobiont emerged from discussions concerning the consequences of symbiotic interactions and the fact that taxonomies often do not reflect the composite phylogenetic histories of organisms such as lichens (consisting of fungi and photosynthetic microbes, as well as other microbes) and corals (consisting of the coral animal and photosynthetic microbes, as well as other associated microbes; Margulis 1991; Mindell 1992; Rohwer et al. 2002; Rosenberg 2013; Singh et al. 2013).

Although some (e.g. Forest Rohwer and colleagues; cf. Skillings 2018) claim that the concept of holobiont is synonymous with 'ecological community', Rosenberg and his collaborators have defended a conception of the holobiont as an integral piece of a new theory of evolution. The ecological conception of holobiont remains marginal today and has not prevailed in medicine. In contrast, the evolutionarily informed view of the holobiont has prevailed in capturing the complexity of the human microbiome (e.g. Gordon et al. 2013; Bordenstein and Theis 2015; Chiu and Gilbert 2015; Gilbert and Tauber 2016; Shropshire and Bordenstein 2016; Theis et al. 2016).

Assumptions of the 'Holobiont View' of the Human Microbiome

The 'holobiont' and 'hologenome' concepts bring forward a number of interesting claims and novel ideas about the nature of microbes as symbionts, their intimate and interdependent relations with the human host and the mechanics of their transmission across generations. The proponents of this view infer, from a series of examples that illustrate positive interactions between microbes and hosts, the symbiotic nature of host-associated microorganisms (Zilber-Rosenberg and Rosenberg 2008; see also Lambert 2018). For humans, this is best illustrated by the gut microbiome, which provides us with a number of important functions (Ley et al. 2006a, b; Xu and Knight 2015). Proponents of the holobiont view argue that such relationships can lead to hosts and microbes acting as evolving systems (Zilber-Rosenberg and Rosenberg 2008), with the resulting *holobionts* acting as a unit of selection in evolution (Rosenberg et al. 2007, 2009; Zilber-Rosenberg and Rosenberg 2008; Gordon et al. 2013; Guerrero et al. 2013; Singh et al. 2013; Gilbert 2014; Gilbert and Tauber 2016; Richardson 2017).

Four generalizations constitute the main body of this new conception of evolution: animals and plants are constantly involved in symbiotic relations with microorganisms; there is a process of transmission of symbiotic microorganisms between generations; the fitness of the holobiont is directly impacted by host–microbe interactions; and variation in the hologenome is a function of the sum of the host's genetic information (the relatively static host genome) and that of the symbionts (the relatively dynamic polygenome of the symbionts) and, thanks to the latter, the hologenome can change rapidly (Zilber-Rosenberg and Rosenberg 2008: 723). The holobiont view has been expanded to ten principles that significantly overlap with the original four generalizations (Bordenstein and Theis 2015; see also Theis et al. 2016; for a robust critique see Skillings 2016).

Transmission of microbes across holobiont generations is central to this view. Rosenberg and Zilber-Rosenberg claim that it has been demonstrated that microbial symbionts can be transmitted between generations, from parent to offspring, by a variety of methods such as cytoplasmic inheritance, eggs, co-prophagy, direct contact during and after birth, and insect vectors (Rosenberg and Zilber-Rosenberg 2016: 2). They also believe that horizontal or environmental modes of microbial acquisition can contribute to intergenerational transmission (Shanker et al. 1990), with lower fidelity than that of vertical transmission, but 'at a high enough frequency to affect evolution' (Rosenberg et al. 2009: 2961).

In particular, this conceptual framework certainly has the advantage of recognizing the absence of strict boundaries between the human body and the microbes that inhabit it. There is a dynamic component of the composition that this view can more easily account for than the organ model, allowing one to appreciate the variability inherent to our biochemistry as well as the shared history of our interactions with our microbes. This view better captures the inherent porosity of the human–microbe system than the superorganism framework, and it raises awareness of the ways in which our microbiota could be altered and reshaped either through diet or medical interventions such as antibiotic treatment (Blaser 2014).

Therapeutic Applications that Emerge from the 'Holobiont View' of the Human Microbiome

The focus of the holobiont view on integration, function and porosity of the body leads to a number of practical benefits to human health. We have decided here to highlight the potential of *probiotics therapy*, the use of microbes artificially introduced (e.g. by voluntary ingestion of specific taxa via probiotic products and yoghurt, among others) from outside the organism to promote health.

In order to understand how probiotics function, Knight uses the metaphor of a lawn as a heuristic device. If something unexpected happens to your lawn – it gets flooded and overgrown with other grasses and weeds, what do you do? 'It could be time to reseed it selectively' (Knight 2015: 65) and reintroduce some of the plants that you favoured. Similarly, when your gut microbiota has faced a significant disturbance, it may be time to introduce back (in a sufficient quantity and diversity) some of the microbial inhabitants that benefit your health.

Probiotic therapy has been effective in treating the gut disturbances that frequently follow antibiotic therapy (Hempel et al. 2012; Shen et al. 2017). It has even been used in concert with antibiotic therapy as a treatment for certain gut dysbioses (Applegate et al. 2013). Probiotic therapy has been suggested as an alternative to antibiotics under some circumstances, for example, in situations where antibiotic treatment is not effective because of antibiotic resistance or the condition resurges after a period of stability after antibiotic treatment (Blaser 2014: 209).

In the case of probiotics, it is crucial to remain skeptical and not to fall prey to 'over-hyped' claims about the therapeutic prowess of probiotics (or gut microbiome research in general; Bourrat 2018). Although the jury is still out concerning the overall efficacy of probiotics, one of the most recent meta-analyses suggests that some effects of probiotics are well documented and their use can be considered 'evidence-based', whereas for other clinical applications, additional studies are needed to demonstrate their efficacy definitively (Rondanelli et al. 2017). When used alone or in combination with other therapies, probiotics are – for example – useful for antibiotic-associated diarrhoea (in adults and children) and *C. difficile*-associated diarrhoea (in adults and the elderly). These encouraging results demonstrate that the holobiont conception of the human microbiome can profoundly change certain therapeutic protocols (Steed et al. 2010; Applegate et al. 2013; Hungin et al. 2013; Asemi et al. 2014) and that the inherent porosity of our bodies invites us to imagine new ways to manipulate membership of the holobiont for medical purposes (Knight 2015: 64).

Limitations of the 'Holobiont View' of the Human Microbiome

Holobiont researchers are rightly fascinated by the tight connection that exists between multicellular organisms and some members of their microbiomes. Given the existence of mutualistic interactions among some microbiome members and their hosts, it is not surprising that coevolution would be assumed. However, as for the concept of 'superorganism' described above, there are reasons to be skeptical regarding whether the assumptions that undergird the holobiont view are met by most members of our microbiomes, especially regarding heritability and coevolution.

First, for hosts and their microbiomes to act 'as an evolutionary unit', microbiomes would need to be heritable. Such heritability could most easily be achieved through vertical transmission, as in the case of intracellular bacteria (e.g. of plant-sucking insects; see Tiivel 1991). This direct mode of transmission significantly reduces the possibility of 'conflict of interest between the genes of the host and the symbionts' (Maynard Smith 1993: 35) and, thus, it reduces the evolutionary opportunities for microbes to evolve into parasites (for genetic conflicts in humans, see also Haig 2014). If the level of independence is greater, then the likelihood for those microbes to evolve towards parasitism increases. A misalignment of evolutionary interests would increase the level of parasitism, virulence and competition (Maynard Smith 1993: 35–36; Moore 2002; Ley et al. 2006a: 841). Ogilvie and colleagues rightly note, 'if the unit of selection is the entity whose survival and reproduction determines how the population of units evolves, the level of the holobiont could break down in the presence of a single devastating mutant microbial species, or gene for that matter' (Ogilvie et al. 2012: 35–36). It is possible for heritability to be significant *even* without vertical transmission. For example, heritability of microbiome members could also be achieved through biased horizontal transmission, such as transmission from an environmental source common to both parents and offspring.

As we have already shown above in our discussion of the superorganism view, there is very little evidence that the human microbiome as *a whole* has significant heritability. For example, Rothschild et al. (2018) estimated that fewer than 2 per cent of the microbial taxa in the human microbiome exhibit significant heritability. This would suggest that even if a few members of the human microbiome have the potential to coevolve with their human hosts and to act in a *holobiont-like* manner (i.e. in a mutualistic fashion with the host, strengthened by coevolution), it would still be a mistake to generalize such symbiotic relations over the *entire* microbiome.

Second, most holobiont proponents support a second inference about the nature of those positive associations: the biomolecular integration shown by numerous organism–microbiome associations, while contingent, is a function of host–microbe coevolution. For example, there is some evidence that human infants may be exposed during birth to certain early microbial colonizers present in the vaginal canal that would later help the baby break down the oligosaccharides in the maternal milk that he or she could not metabolize otherwise (LoCascio et al. 2007); the intergenerational transmission of particular microbes that can use specific oligosaccharides and the presence of these otherwise indigestible oligosaccharides in breastmilk has been used as evidence for likely coevolution between humans and their microbiota.

The researchers supporting such explanations need, however, to acknowledge two important shortcomings. First, significant parts of our microbiome and their intimate connection to the host are due to processes that are not explained by shared evolutionary history (e.g. dispersal, competition, priority effects). For example, such a relationship could arise incidentally, because of selection at the level of the individual microbial population, if the production of such a benefit to the host is a 'by-product' of a process that also benefits the microbe. Host–microbe system-level selection is only needed to explain such benefits if they come at a fitness 'cost' to the microbe, and there are very few documented examples of such 'costs' in the human microbiome (Moran and Sloan 2015: 4). Second, even if this picture is true for certain microorganisms (e.g. *Lactobacillus*), it remains fundamentally host-centric, and it underplays both microbe-to-microbe interactions as well as the significant effects of microbial dispersal and ecological drift on the whole microbiome (Costello et al. 2012; Foster et al. 2017). The fact that the holobiont view leaves out these processes, processes that help us understand the dynamic nature of our microbiome, points to significant limitations of this conceptual framework in providing us with the definitive and the most complete picture of the human microbiome.

The Ecosystem View

The description of the human–microbe system as an *ecosystem* has been frequently used by microbial biologists. An ecosystem is a concept from ecology, and it consists of a community of organisms interacting with an (usually nonliving) environment. For example, a *forest ecosystem* would include the plant, animal and microbial species found in a particular forest, interacting with each other and with the soil, water and atmosphere that surrounds them. An ecosystem is assumed to have *ecosystem functions*, processes that occur only at the ecosystem scale and to provide *ecosystem services*, functions that are particularly valued by humans. For the human–microbe system, the ecosystem would include the microbial cells in and on a particular host, the host cells with which they interact, and the larger 'environment' provided by the host. With this conceptual framework, ecosystem functions might include the digestion of food and ecosystem services could include host health. Research with germ-free animal models (e.g. gnotobiotic mice or zebrafish) provide additional evidence for the importance of health services generated by our microbial biomes. In the absence of those services, these animals display a number of health deficiencies, especially with respect to their immune system (Round and Mazmanian 2009).

Numerous scientists have reconceived our definition of human organisms and our physiological functions through the lens of this basic unit of ecology (Gerber 2014; Christian et al. 2015; Douglas and Werren 2016; Mosca et al. 2016; Foster et al. 2017). This enthusiasm led Knight to emphatically claim, '[w]e are not individuals; we are ecosystems' (Knight 2015: 3).

Assumptions of the 'Ecosystem View' of the Human Microbiome

The ecosystem conceptual framework makes a number of important assumptions, the most important being that: microbe–host systems are open with respect to immigration and emigration; variation in microbiome composition within and across individual hosts is common and potentially important; and microbiome composition varies substantially over time in important ways. These assumptions also open up the possibility that the membership of microbiomes may be determined, at least in part, by dispersal among hosts and between the host and its surrounding environment, and that, like all ecosystems, stochastic processes may play an important role as well. Some microbial biologists have considered the ecosystem metaphor to be at the other end of the conceptual spectrum from the organ metaphor. The microbiota is no longer understood as 'a forgotten organ' (O'Hara and Shanahan 2006) but 'as a continuum of dynamic ecological communities living in the numerous microhabitats of the human body' (Foxman et al. 2008: 3; see also Dethlefsen et al. 2006).

The most distinctive assumption of the ecosystem metaphor is that the human–microbe system is 'open' or 'porous' to the movement of individual microbial cells. This open-system conception is similar in concept to that of the holobiont but allows for even greater porosity and for an even greater role for the immigration and emigration of microbes in the dynamics and functioning of microbiomes.

This view also highlights the variation that occurs in microbiomes within and across the human body. It emphasizes that we are more of a series of interrelated ecological communities rather than a single entity. For example, the composition of our gut microbiota tends to be quite different than the composition of our mouth or skin microbiome. And, although those localized communities interact, they nonetheless have specific microbial signatures, at least for adult organisms. In addition to this 'within-host' variation, there is also great host-to-host variation, which could be functionally important from a health perspective (Falony et al. 2016; see also Yatsunenko et al. 2012). This variation can be substantial; for example, when it comes to samples from the colon, 'the differences between individuals are greater than the differences between different sampling sites in one individual' (Turnbaugh et al. 2007: 805; see also Eckburg et al. 2005). Such complexity and diversity call into question the idea of a 'core' human microbiome as a set of microorganisms or microbial genes that a vast majority of humans would share under similar environmental conditions.

Finally, an ecosystem conception of the human microbiome also highlights the importance of dynamic processes in constituting the kind of living system that we are (Xu and Gordon 2003; Blaser 2006; Costello et al. 2012). In other words, the functions, whether metabolic or immunity based, have to be understood not as the product of static organs but rather as the effects of community structure and microbial interactions. So, any prediction about the effectiveness of such processes has to look both at *who is there* and also at *how they interact with each other*. This is what a dynamic explanation of organic communities requires. Its goal is to show how processes of community structure play a key role in assessing the etiologies of human health and diseases (Relman 2002).

Thinking about the human body in terms of ecological processes brings a new perspective to how microbe–host systems assemble, and what drives their variation over space and time (see Dethlefsen et al. 2006; Costello et al. 2012; Lemon et al. 2012; Relman 2012). Dispersal, diversification, environmental selection and ecological drift become the key mechanisms that help us explain and predict microbiome dynamics and variation (Vellend 2010). This conception also opens up the possibility that microbiomes can exist in different stable configurations, called 'alternative stable states' by ecologists (Beisner et al. 2003). Such alternative states can arise in a number of ways, including through environmental changes or through chance events during the assembly of an ecological community, or during its recovery from disturbance. This concept can be a useful tool for understanding *dysbiosis* (an alternative microbiome configuration associated with a decline in host health or functioning; Zaneveld et al. 2017) and for understanding microbiomes that cycle among multiple configurations through time (such as the vaginal microbiome; Gonze et al. 2017). It has also proven useful as a framework for understanding changes in microbiome composition as hosts develop (Marino et al. 2014). Such changes can be conceived of as examples of ecological succession, patterned changes in ecological communities following disturbance.

Therapeutic Applications that Emerge from the 'Ecosystem View' of the Human Microbiome

The 'ecosystem view' rethinks the human body not as a battleground where the antibiotic war against pathogens takes place, but rather as a habitat that we must manage (Costello et al. 2012). This has important implications for how one envisions therapeutic interventions, opening the door, for example, to thinking of such interventions through the lens of conservation biology. This conceptual framework encourages the application of concepts such as ecosystem restoration, promotion of native species, careful management of invasive species and concerns for biological diversity to the human body, suggesting new ways to manage the human–microbe system (Costello et al. 2012; Bohannan 2015; Orr et al. 2018). But, most radically, this metaphor suggests that successful interventions could be designed to take advantage of the extreme porosity of the human–microbe system.

There is growing evidence that the human–microbe system is constantly shedding and receiving potential new microbial members. Several studies have shown that cohabitating people tend to have microbiomes that are more similar in composition than those of people who are not living in the same house (e.g. Song et al. 2013; Turroni et al. 2017). Studies of laboratory animals have also shown that physical proximity (e.g. mice in the same cage or fish in the same tank) can result in an increase in the similarity of microbiomes (Burns et al. 2017). Presumably, the homogenizing effect of cohabitation occurs through the exchange of microbiome members via immigration and emigration, as the ecosystem metaphor assumes (Vangay et al. 2018).

This extreme porosity could lead to novel approaches to microbiome management or manipulation. For example, to maintain a beneficial microbe (or set of microbes) in a particular individual microbiome, this metaphor suggests that it may be necessary to have sustained exposure to these microbial communities. This sustained exposure could result not only from frequent ingestion of probiotic supplements but it could be achieved – perhaps more easily – through the 'seeding' of the external environment (a room, for example) with beneficial bacteria, which would act as the 'reservoir' of 'immigrants' that would disperse to a person's microbiome. This idea is *definitely* not just conjecture; there are products now on the market (e.g. Better Air products) that are essentially 'misters' of probiotics that inject putatively beneficial bacteria into the air. The manufacturers claim that this can have positive effects on health (e.g. reduction in allergy symptoms), although this claim is controversial.

The ecosystem view also suggests that microbiomes could be altered (to a 'healthier' state) through cohabitation with individuals with 'healthy' microbiomes. This has been shown to work in laboratory animals; for example, Bel et al. (2014) demonstrated that 'healthy' gut microbiomes (ones that conferred a reduction in susceptibility to colitis) could be transmitted among mice in the same cage. Applying a similar approach to humans could be more effective (and less invasive) than ingesting probiotics. This possibility suggests that aspects of human health could be 'transmissible' in ways analogous to the transmission of disease and that by managing this transmission, physicians could manipulate microbiome membership and ultimately human health (Browne et al. 2017).

Limitations of the 'Ecosystem View' of the Human Microbiome

Although the ecosystem view offers many advantages as a conceptual framework for the human–microbe system, it is, like all other conceptual frameworks, a (very) partial description. Its limitations stem from differences between traditional ecosystems and 'host–microbe' ecosystems, from the assumptions inherent in an ecosystem view of the world (especially assumptions about immigration and emigration), and from the problems common to all ecosystem views, such as the lack of clear boundaries, its complexity, the diffuse nature of interactions in such a complex system and the resulting difficulties with establishing causality (Koskella et al. 2017). A few examples of such limitations are discussed below.

Traditional (i.e. non-host) ecosystems are at their simplest the combination of a living community (an assemblage of different species) and a non-living environment. Members of the community not only interact with each other but importantly with their surrounding environment as well. And although these interactions involve some feedbacks (e.g. plants change the nutrient levels in soil, which alters subsequent plant growth and survival), the fact that the environment is non-living results in these feedbacks often occurring on different temporal and spatial scales than the living species with which they are interacting. For example, plants may locally change the pH of a soil but the overall pH of a soil is determined by factors (the type of rock from which the soil has developed, the age of the soil and rainfall patterns, among others) that occur at different scales of space and time than any given plant population, and over which the plant has little influence. Even more importantly, the fact that the environment is non-living prevents it from coevolving with the species that co-occur, since evolution is a process of living things.

This distinction (of living and non-living parts of an ecosystem) limits its applicability to the human–microbe system. For example, how do we think of the 'host' component in such a system? In some ways, hosts are 'environments' and although living, they may act like a non-living environment on the scales of space and time at which microbes interact. For example, the temperature of the gut of a human may be locally or temporarily altered by the microbes within (e.g. during infection by a pathogen) but, ultimately, the temperature is controlled by factors over which microbes have little influence. But, in other ways, hosts are members of the 'community', not the 'environment'. For example, immune cells may interact with members of the microbiome in ways akin to predators and prey, altering the abundance of microbial cells (and being altered in response) through direct feedbacks that are more community-like than environment-like.

Hosts and microbes have the potential to coevolve, in ways that are not easily accounted for by the ecosystem view. Communities in ecosystems can certainly evolve, and selection at the ecosystem level has the potential to drive community evolution, at least in the laboratory (e.g. Swenson et al. 2000). But coevolution between community and environment is not possible in the conventional ecosystem view (since environments are abiotic). It would be possible for communities of microbes and communities of host cells to coevolve, but this would require that microbial community composition be heritable (as we discuss for holobionts above). Such heritability is difficult to imagine under the ecosystem view, given the extreme porosity assumed of the microbiome. For coevolution to be likely, the ecosystem view would have to be modified to include some sort of biased assembly process that would allow for heritability of microbiome composition over ecosystem 'generations'.

The ecosystem view assumes that immigration and emigration are ongoing. This results in dynamic community membership: the presence of members that are somewhat stable over time (perhaps because they thrive under the conditions of any given place and time) and members who vary strongly in abundance over time (perhaps because they do not thrive under local conditions). Depending on the rates of immigration assumed, this variable component could be quite large and the impact on human health is quite significant. The assumed constant flux of individuals leads to a view that the microbiome is not only *porous* with regards to membership (as the holobiont framework also allows) but is actually *fluid*, resulting in difficulty to define *strict* boundaries. It has been argued that it is especially difficult to assign causality in complex systems with fluid boundaries (O'Malley and Skillings 2018), which could make it very difficult to use this metaphor to develop clinical applications. Thus, the ecosystem metaphor may result in a view of human–microbe systems that is too complex to be helpful. As Foxman and colleagues note, the ecosystem metaphor has 'the advantage of [an] increased realism . . . [but] is much more complex and consequently may be less useful for some purposes' (Foxman et al. 2008: 3). Consequently, it is a challenge to find ways to translate this view into clinical practice and to determine how differences in microbial diversity correlate with disease states or, how given such great variation in microbiome composition over space and time, to design effective clinical interventions.

Synthesis

The human–microbe system is complex. Fully understanding it requires that we re-envision in important ways what it means to be 'human'. We do not have language that is ideally suited to this novel and intricate task; it is not surprising that various conceptual frameworks and metaphors have become central to our explanations of our now more complicated selves. But as central and useful as they are to human cognition, these conceptual frameworks and metaphors are also sometimes dangerous for communicating difficult and abstract ideas (Lewontin 2001). Our choices of concepts come with certain embedded assumptions about the world, which if left unchallenged ultimately limit our understanding. Concepts can also become ossified if not challenged, and if used often enough can become mistaken for fact.

Such is the situation that we find ourselves in now, as we consider the human–microbe system. Each of the common conceptual frameworks we describe here captures important truths – and misses crucial aspects – of our human–microbe selves. The organ framework captures the important functional contributions that microbes make to human physiology but misrepresents the ever-changing membership of our microbiome. In contrast, the ecosystem framework emphasizes the diversity and dynamic nature of the human–microbe system, but likely overstates the role that immigration and emigration play in our microbiome. Similarly, for the superorganism, holobiont and self–non-self views, each contains some important truths about human–microbe systems, and each overstates, misrepresents or ignores other likely key aspects.

The consequences of this confusion are not solely in the realm of communication, nor are our concerns merely intellectual. Our uncritical use of these concepts has real consequences for human health and well-being. How the parts we call 'human' and the parts we call 'microbial' interact to form a human body depends in large part on the conceptual frameworks we use. Importantly, so do the clinical approaches we take when these interactions break down. Should we 'bank' our microbes as we do blood, and 'transplant' them when we get sick? Should we expose our babies to our microbiome to strengthen them, as we might expose them to vaccines? Should we manage the membership of our microbiome like we would a national park, encouraging 'natural' members and removing invaders? Is it possible to 'restore' or 'rewild' our bodies? How we consider such interventions is altered by the assumptions we make about our complex, human–microbe selves, which influences, and is influenced by, the concepts we employ.

This situation, although confusing, is not hopeless. We believe that the way out of this confusion lies not in choosing the 'correct' metaphor (Blaser et al. 2013), but rather through strongly embracing a pluralistic approach. Although there is no evidence (yet) that one conceptual framework is more accurate or more useful or somehow 'better' than another, insights emerge as we consider these metaphors together as a whole. The conceptual space that these concepts and metaphors populate, the conceptual ecology of the human microbiome, can be described with a number of different parameters and these parameters identify crucial aspects of the human–microbe system, properties of our selves that are likely important to understand in more depth (see Table 6.1).

TABLE 6.1: Characteristics of different conceptual frameworks for host–microbe systems.

Conceptual framework	Bias	Immigration	Transmission	Functional integration	Level of selection	Fidelity of reproduction	Relative strength of host selection
Organ	Host	Very low/closed	Vertical	Strong	Strongly at group	High fidelity	Very strong
Self-nonself	Host	Low/open	Vertical	Strong	Strongly at group?	?	Strong
Superorganism	~Host	Low/restricted	Both	Moderate	Weakly at group	?	Moderate
Holobiont	Host and microbe	Low/open	Both	Moderate	Weakly at group	?	Moderate
Ecosystem	Microbe and host	High/open	Horizontal	Weak	Strongly at individual	Low fidelity	Lower

For definitions of column headings, see the text.

The concepts we use differ in the assumptions they make about the degree of functional integration (i.e. the degree to which variation in microbiome composition 'maps' onto variation in host function), about whether our associated microbes are transmitted vertically (from parent to offspring) or horizontally (through contact between individuals, regardless of relatedness, or with other aspects of the external environment), and about whether membership of our microbiome is open (subject to ongoing immigration) or closed. These properties will in turn determine whether our microbiomes are individualized (unique to each individual), malleable (sensitive to manipulation) and the degree to which evolution has worked to integrate the host–microbe system (through coevolution of host and microbe, through natural selection at the scale of the human–microbe system). Each of the concepts we use combines assumptions about these properties in different ways. To illustrate this, consider just two properties: transmission mode and degree of functional integration. The organ concept assumes strong vertical transmission and strong functional integration, the ecosystem concept assumes primarily horizontal transmission and weak functional integration and the other metaphors can be arrayed between them in this two-dimensional space (see Figure 6.1). The fact that all of these concepts are commonly used strongly suggests that we do not yet know where in this space the human–microbe system lies and illustrates how important it is to better understand these two aspects of the human–microbe system.

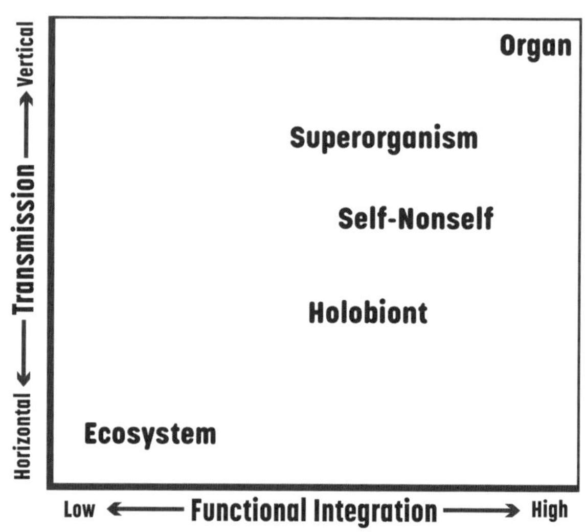

FIGURE 6.1: Microbiome conceptual frameworks vary in assumptions about transmission and functional integration.

We defend the view that it is important to maintain a pluralistic approach to our conception of the human–microbe system for two reasons. First, as we just discussed we do not yet know enough about the human–microbe system to allow us to choose among these alternative conceptions. At this stage, overuse of any one conceptual framework runs the risk of limiting our ultimate understanding of our human–microbe selves and of constraining how we approach human health and wellness. This is essentially an appeal to 'pragmatic pluralism', that arises both from a deficiency of knowledge (related to the notion of 'epistemic pluralism' as defended by Kellert et al. 2006) and from the need to judge conceptual alternatives in relation to the practical goal of developing novel therapies (Longino 2012).

Second, our suspicion is that the reason we do not yet have a clear description of the nature of the human–microbe system is that it is simultaneously many things. It is likely that some aspects of the human microbiome act 'organ-like' (stable and functionally integrated), some are more like an ecosystem (dynamic and horizontally acquired) and some take on the characteristics of a superorganism or holobiont (strongly coevolved). There is growing evidence that such variation is likely; for example, as we discussed above heritability varies greatly among microbiome members, and numerous studies have shown that functional integration also varies widely among microbiome members, with only a subset contributing strongly to human health or physiological function (Gilbert et al. 2016). Such complexity would suggest that some aspects of our microbial selves can be transplanted successfully (with subsequent transfer of function), while others must be restored gradually and carefully managed over time, and still other aspects may be highly sensitive to their initial development (and may be very difficult to subsequently change). This is an appeal to a stronger form of pluralism that arises from such biological complexity (O'Malley and Dupré 2007, 2010; Dupré 2018).

This pluralistic approach to thinking of the human–microbe system is not a novel idea. For example, this approach underlies the hunt for a 'core' microbiome (that component of the microbiome shared among all humans; see Turnbaugh et al. 2009; Shade and Handelsman 2012); the 'core' is essentially that portion of the human microbiome that is 'organ-like'. There is also evidence that the microbiome of hosts can be divided into a component that is subject to strong selection by the host, over which the host has strong control over membership and that which is present in hosts primarily through immigration from the outside (Burns et al. 2017). Conceptually speaking, one could think of this as the 'organ' component and the 'ecosystem' component. The ultimate goal then is not to avoid using various concepts and metaphors to describe the human–microbe system, nor to choose the 'best' conceptual framework, but rather to use all of these concepts to guide the exploration of our complexity as chimeric beings, and ultimately to fully understand the multitude of ways that we are human.

Acknowledgements

We thank Nelson Ting, William Cresko, Joshua Skorburg, Bryan Cwik, Jay Odenbaugh and Mark Bedau for thoughtful and encouraging comments on early drafts of this paper. Two anonymous reviewers provided critical feedback that vastly improved the manuscript. Nicolae Morar was supported by a Faculty Research Award from the Office of the Vice President for Research and Innovation at the University of Oregon. Brendan Bohannan was supported by the National Institutes of Health (Awards P50GM098911, P01GM125576 and UG30D0233989).

NOTE

1. Republished with permission of University of Chicago Press, from Nicolae Morar and Brendan J. Bohannan (2019), 'The conceptual ecology of the human microbiome', *The Quarterly Review of Biology*, 94:2, pp. 149–75, permission conveyed through Copyright Clearance Center, Inc.

REFERENCES

Alcock, Joe, Maley, Carlo C. and Aktipis, C. Athena (2014), 'Is eating behavior manipulated by the gastrointestinal microbiota? Evolutionary pressures and potential mechanisms', *Bioessays*, 36:10, pp. 940–49.

Anonymous (2008), 'Who are we?', *Nature*, 453, p. 563.

Applegate, Jennifer A., Walker, Christa L. Fischer, Ambikapathi, Ramya and Black, Robert E. (2013), 'Systematic review of probiotics for the treatment of community-acquired acute diarrhea in children', *BMC Public Health*, 13(Supplement 3):S16, pp. 1–8.

Arnold, Carrie (2013), 'The hologenome: A new view of evolution', *New Scientist*, 9 January, pp. 30–34, https://www.newscientist.com/article/mg21728992-000-the-hologenome-a-new-view-of-evolution/. Accessed 6 April 2023.

Asemi, Zatollah, Khorrami-Rad, Ashraf, Alizadeh, Sabihe-Alsadat, Shakeri, Hossein and Esmaillzadeh, Ahmad (2014), 'Effects of synbiotic food consumption on metabolic status of diabetic patients: A double-blind randomized cross-over controlled clinical trial', *Clinical Nutrition*, 33:2, pp. 198–203.

Asnicar, Francesco, Manara, Serena, Zolfo, Moreno, Truong, Duy Tin, Scholz, Matthias, Armanini, Federica, Ferretti, Pamela, Gorfer, Valentina, Pedrotti, Anna, Tett, Adrian and Segta, Nicola (2017), 'Studying vertical microbiome transmission from mothers to infants by strain-level metagenomic profiling', *MSystems*, 2:1, https://doi.org/10.1128/mSystems.00164-16. Accessed 6 April 2023.

Bäckhed, Fredrik, Ding, Hao, Wang, Ting, Hooper, Lora V., Koh, Gou Young, Nagy, Andras, Semenkovich, Clay F. and Gordon, Jeffrey I. (2004), 'The gut microbiota as an environmental factor that regulates fat storage', *Proceedings of the National Academy of Sciences*, 101:44, pp. 15718–23.

Bäckhed, Fredrik, Ley, Ruth E., Sonnenburg, Justin L., Peterson, Daniel A. and Gordon, Jeffrey I. (2005), 'Host-bacterial mutualism in the human intestine', *Science*, 307:5717, pp. 1915–20.

Bäckhed, Fredrik, Manchester, Jill K., Semenkovich, Clay F. and Gordon, Jeffrey I. (2007), 'Mechanisms underlying the resistance to diet-induced obesity in germ-free mice', *Proceedings of the National Academy of Sciences*, 104:3, pp. 979–84.

Bäckhed, Fredrik, Roswall, Josefine, Peng, Yangqing, Feng, Qiang, Jia, Huijue, Kovatcheva-Datchary, Petia, Li, Yin, Xia, Yan, Xie, Hailiang, Zhong, Huanzi, Khan, Muhammad Tanweer, Zhang, Jianfeng, Li, Junhua, Xiao, Liang, Al-Aama, Jumana, Zhang, Dongya, Lee, Ying Shiuan, Kotowska, Dorota, Colding, Camilla, Tremaroli, Valentina, Yin, Ye, Bergman, Stefan, Xu, Xun, Madsen, Lise, Kristiansen, Karsten, Dahlgren, Jovanna and Wang, Jun (2015), 'Dynamics and stabilization of the human gut microbiome during the first year of life', *Cell Host & Microbe*, 17:5, pp. 690–703.

Baquero, Fernando and Nombela, Cesar (2012), 'The microbiome as a human organ', *Clinical Microbiology and Infection*, 18, pp. 2–4.

Bäumler, Andreas J. and Sperandio, Vanessa (2016), 'Interactions between the microbiota and pathogenic bacteria in the gut', *Nature*, 535:7610, pp. 85–93.

Beisner, Beatrix E., Haydon, Daniel T. and Cuddington, Kim (2003), 'Alternative stable states in ecology', *Frontiers in Ecology and the Environment,* 1:7, pp. 376–82.

Bel, Shai, Elkis, Yoav, Elifantz, Hila, Koren, Omry, Ben-Hamo, Rotem, Lerer-Goldshtein, Tal, Rahimi, Roni, Horin, Shomron Ben, Nyska, Agraham, Shpungin, Sally and Nir, Uri (2014), 'Reprogrammed and transmissible intestinal microbiot confer diminished susceptibility to induced colitis in TMF−/− mice', *Proceedings of the National Academy of Sciences*, 111:13, pp. 4964–69.

Bianconi, Eva, Piovesan, Allison, Facchin, Federica, Beraudi, Alina, Casadei, Raffaella, Frabetti, Flavia, Vitale, Lorenza, Pelleri, Maria Chiara, Tassani, Simone, Piva, Francesco, Perez-Amodio, Soledad, Strippolli, Pierluigi and Canaider, Silvia (2013), 'An estimation of the number of cells in the human body', *Annals of Human Biology*, 40:6, pp. 463–71.

Blaser, Martin J. (2006), 'Who are we? Indigenous microbes and the ecology of human diseases', *EMBO Reports*, 7:10, pp. 956–60.

——— (2011), 'Stop the killing of beneficial bacteria', *Nature*, 476:7361, pp. 393–94.

——— (2014), *Missing Microbes: How the Overuse of Antibiotics is Fueling our Modern Plagues*, New York: Henry Holt & Co.

Blaser, Martin J., and Falkow, Stanley (2009), 'What are the consequences of the disappearing human microbiota?', *Nature Reviews Microbiology*, 7:12, pp. 887–94.

Blaser, Martin, Bork, Peer, Fraser, Claire, Knight, Rob and Wang, Jun (2013), 'The microbiome explored: Recent insights and future challenges', *Nature Reviews Microbiology*, 11:3, pp. 213–17.

Blumberg, Richard and Powrie, Fiona (2012), 'Microbiota, disease, and back to health: A metastable journey', *Science Translational Medicine*, 4:137, https://doi.org/10.1126/scitranslmed.3004184. Accessed 6 April 2023.

Bocci, Velio (1992), 'The neglected organ: Bacterial flora has a crucial immunostimulatory role', *Perspectives in Biology and Medicine*, 35:2, pp. 251–60.

Bohannan, Brendan (2015), 'Restoration of the human microbiome', in J. L. Bronstein (ed.), *Mutualism*, Oxford: Oxford University Press, pp. 279–80.

Bordenstein, Seth R. and Theis, Kevin R. (2015), 'Host biology in light of the microbiome: Ten principles of holobionts and hologenomes', *PLoS Biology*, 13:8, https://doi.org/10.1371/journal.pbio.1002226. Accessed 6 April 2023.

Bourrat, Pierrick (2018), 'Have causal claims about the gut microbiome been over hyped?', *BioEssays*, 40:12, https://doi.org/10.1002/bies.201800178. Accessed 6 April 2023.

Brody, Jane E. (2014), 'We are our bacteria', *New York Times*, 14 July, https://well.blogs.nytimes.com/2014/07/14/we-are-our-bacteria/. Accessed 6 April 2023.

Browne, Hilary P., Neville, B. Anne, Forster, Samuel C. and Lawley, Trevor D. (2017), 'Transmission of the gut microbiota: Spreading of health', *Nature Reviews Microbiology*, 15:9, pp. 531–43.

Burns, Adam R., Miller, Elizabeth, Agarwal, Meghna, Rolig, Annah S., Milligan-Myhre, Kathryn, Seredick, Steve, Guillemin, Karen and Bohannan, Brendan J.M. (2017), 'Inter-host dispersal alters microbiome assembly and can overwhelm host innate immunity in an experimental zebrafish model', *Proceedings of the National Academy of Sciences*, 114:42, pp. 11181–86.

Casadevall, Arturo and Pirofski, Liise-anne (2014), 'Microbiology: Ditch the term pathogen', *Nature*, 516:7530, pp. 165–66.

Chiu, Lynn and Gilbert, Scott F. (2015), 'The birth of the holobiont: Multi-species birthing through mutual scaffolding and niche construction', *Biosemiotics*, 8:2, pp. 191–210.

Chiu, Lynn, Bazin, Thomas, Truchetet, Marie-Elise, Schaeverbeke, Thierry, Delhaes, Laurence and Pradeu, Thomas (2017), 'Protective microbiota: From localized to long-reaching co-immunity', *Frontiers in Immunology*, 8, p. 1678.

Christian, Natalie, Whitaker, Briana Kathleen and Clay, Keith (2015), 'Microbiomes: Unifying animal and plant systems through the lens of community ecology theory', *Frontiers in Microbiology*, 6, p. 869.

Clarke, Gerard, Stilling, Roman M., Kennedy, Paul J., Stanton, Catherine, Cryan, John F. and Dinan, Timothy G. (2014), 'Minireview: Gut microbiota: The neglected endocrine organ', *Molecular Endocrinology*, 28:8, pp. 1221–38.

Clemente, Jose C., Pehrsson, Erica C., Blaser, Martin J., Sandhu, Kuldip, Gao, Zhao, Wang, Bin, Magris, Magda, Hidalgo, Glida, Contreras, Monica, Noya-Alarcón, Óscar, Lander, Orlana, McDonald, Jeremy, Cox, Mike, Walter, Jens, Oh, Phaik Lyn, Ruiz, Jean F., Rodriguez, Selena, Shen, Nan, Song, Se Jin, Metcalf, Jessica, Knight, Rob, Dantas, Gautam and Dominguez-Bello, M. Gloria (2015), 'The microbiome of uncontacted Amerindians', *Science Advances*, 1:3, https://doi.org/10.1126/sciadv.1500183. Accessed 6 April 2023.

Costello, Elizabeth K., Stagaman, Keaton, Dethlefsen, Les, Bohannan, Brendan J. M. and David, A. Relman (2012), 'The application of ecological theory toward an understanding of the human microbiome', *Science*, 336:6086, pp. 1255–62.

Dethlefsen, Les, Eckburg, Paul B., Bik, Elisabeth M. and Relman, David A. (2006), 'Assembly of the human intestinal microbiota', *Trends in Ecology & Evolution*, 21:9, pp. 517–23.

Dethlefsen, Les, McFall-Ngai, Margaret and Relman, David A. (2007), 'An ecological and evolutionary perspective on human–microbe mutualism and disease', *Nature*, 449:7164, pp. 811–18.

DeWitt, William (1989), *Human Biology: Form, Function, and Adaptation*, Glenview, IL: Scott, Foresman and Company.

Dietert, Rodney R. (2016), *The Human Superorganism: How the Microbiome is Revolutionizing the Pursuit of a Healthy Life*, New York: Dutton Press.

Dietert, Rodney and Dietert, Janice (2012), 'The completed self: An immunological view of the human-microbiome superorganism and risk of chronic diseases', *Entropy*, 14:11, pp. 2036–65.

Dominguez-Bello, Maria G., Costello, Elizabeth K., Contreras, Monica, Magris, Magda, Hidalgo, Glida, Fierer, Noah and Knight, Rob (2010), 'Delivery mode shapes the acquisition and structure of the initial microbiota across multiple body habitats in newborns', *Proceedings of the National Academy of Sciences*, 107:26, pp. 11971–75.

Dominguez-Bello, Maria G., De Jesus-Laboy, Kassandra M., Shen, Nan, Cox, Laura M., Amir, Amnon, González, Antonio, Bokulich, Nicholas A., Song, Se Jin, Hoashi, Marina, Rivera-Vinas, Juana I., Mendez, Keimari, Knight, Rob and Clemente, Jose C. (2016), 'Partial restoration of the microbiota of cesarean-born infants via vaginal microbial transfer', *Nature Medicine*, 22:3, pp. 250–53.

Dominguez-Bello, Maria G., Godoy-Vitorino, Filipa, Knight, Rob and Blaser, Martin J. (2019), 'Role of the microbiome in human development', *Gut*, 68:6, pp. 1108–14.

Douglas, Angela E. (2018), *Fundamentals of Microbiome Science: How Microbes Shape Animal Biology*, Princeton, NJ: Princeton University Press.

Douglas, Angela E. and Werren, John H. (2016), 'Holes in the hologenome: Why host-microbe symbioses are not holobionts', *MBio*, 7:2, https://doi.org/10.1128/mBio.02099-15. Accessed 6 April 2023.

Dupré, John (2018), 'Processes, organisms, kinds, and the inevitability of pluralism', in O. Bueno, R.-L. Chen and M. B. Fagan (eds), *Individuation, Process, and Scientific Practices*, Oxford: Oxford University Press, pp. 21–39.

Eberl, Gérard and Pradeu, Thomas (2018), 'Towards a general theory of immunity?', *Trends in Immunology*, 39:4, pp. 261–63.

Eckburg, Paul B., Bik, Elisabeth M., Bernstein, Charles N., Purdom, Elizabeth, Dethlefsen, Les, Sargent, Michael, Gill, Steven R., Nelson, Karen E. and Relman, David A. (2005), 'Diversity of the human intestinal microbial flora', *Science*, 308:5728, pp. 1635–38.

Eisen, Jonathan (2015), 'What does the term microbiome mean? And where did it come from? A bit of a surprise', *Microbiology of the Built Environment Network*, 8 April, https://microbe.net/2015/04/08/what-does-the-term-microbiome-mean-and-where-did-it-come-from-a-bit-of-a-surprise/. Accessed 6 April 2023.

Enders, Giulia (2015), *Gut: The Inside Story of Our Body's Most Underrated Organ*, Vancouver: Greystone Books.

Falony, Gwen, Joossens, Marie, Vieira-Silva, Sara, Wang, Jun, Darzi, Youssef, Faust, Karoline, Kurilshikov, Alexander, Bonder, Marc Jan, Valles-Colomer, Mireia, Vandeputte, Doris, Tito, Raul Y., Chaffron, Samuel, Rymenans, Leen, Verspecht, Chloë, De Sutter, Lise, Lima-Mendez, Gipsi, D'hoe, Kevin, Jonckheere, Karl, Homola, Daniel, Garcia, Roberto, Tigchelaar, Ettje F., Eeckhaudt, Linda, Fu, Jingyuan, Henckaerts, Liesbet, Zhernakova, Alexandra, Wijmenga, Cisca and Raes, Jeroen (2016), 'Population-level analysis of gut microbiome variation', *Science*, 352, pp. 560–64.

Folse, Henri J. III and Roughgarden, Joan (2010), 'What is an individual organism? A multilevel selection perspective', *Quarterly Review of Biology*, 85, pp. 447–72.

Forbes, Stephen Alfred (1887), 'The lake as a microcosm', *Bulletin of the Peoria Scientific Association*, 18, pp. 77–87.

Foster, Kevin R., Schluter, Jonas, Coyte, Katharine Z. and Rakoff-Nahoum, Seth (2017), 'The evolution of the host microbiome as an ecosystem on a leash', *Nature*, 548, pp. 43–51.

Foxman, Betsy, Goldberg, Deborah, Murdock, Courtney, Xi, Chuanwu and Gilsdorf, Janet R. (2008), 'Conceptualizing human microbiota: From multicelled organ to ecological community', *Interdisciplinary Perspectives on Infectious Diseases*, pp. 613979.

Frank, Daniel N. and Pace, Norman R. (2008), 'Gastrointestinal microbiology enters the metagenomics era', *Current Opinion in Gastroenterology*, 24, pp. 4–10.

Franzosa, Eric A., Huang, Katherine, Meadow, James F., Gevers, Dirk, Lemon, Katherine P., Bohannan, Brendan J. M. and Huttenhower, Curtis (2015), 'Identifying personal microbiomes using metagenomic codes', *Proceedings of the National Academy of Sciences of the United States of America*, 112, https://doi.org/10.1073/pnas.1423854112. Accessed 6 April 2023.

Fujimura, Kei E., Slusher, Nicole A., Cabana, Michael D. and Lynch, Susan V. (2010), 'Role of the gut microbiota in defining human health', *Expert Review of Anti-Infective Therapy*, 8, pp. 435–54.

Gallo, Richard L., Hultsch, Thomas and Farnaes, Lauge (2016), 'Recognizing that the microbiome is part of the human immune system will advance treatment of both cancer and infections', *Journal of the American Academy of Dermatology*, 74, pp. 772–74.

Gerber, Georg K. (2014), 'The dynamic microbiome', *FEBS Letters*, 588, pp. 4131–39.

Gilbert, Jack A., Quinn, Robert A., Debelius, Justine, Xu, Zhenjiang Z., Morton, James, Garg, Neha, Jansson, Janet K., Dorrestein, Pieter C. and Knight, Rob (2016), 'Microbiome-wide association studies link dynamic microbial consortia to disease', *Nature*, 535, pp. 94–103.

Gilbert, Scott F. (2014), 'A holobiont birth narrative: The epigenetic transmission of the human microbiome', *Frontiers in Genetics*, 5, p. 282.

Gilbert, Scott F. and Tauber, Alfred I. (2016), 'Rethinking individuality: The dialectics of the holobiont', *Biology and Philosophy*, 31, pp. 839–53.

Gilbert, Scott F., Sapp, Jan and Tauber, Alfred I. (2012), 'A symbiotic view of life: We have never been individuals', *Quarterly Review of Biology*, 87, pp. 325–41.

Gill, Stephen R., Pop, Mihai, DeBoy, Robert T., Eckburg, Paul B., Turnbaugh, Peter J., Samuel, Buck S., Gordon, Jeffrey I., Relman, David A., Fraser-Liggett, Claire M. and Nelson, Karen E. (2006), 'Metagenomic analysis of the human distal gut microbiome', *Science*, 31, pp.1355–59.

Gissis, Snait B., Lamm, Ehud and Shavit, Ayelet (2017), *Landscapes of Collectivity in the Life Sciences*, Cambridge, MA: MIT Press.

Gonze, Didier, Lahti, Leo, Raes, Jeroen and Faust, Karen (2017), 'Multi-stability and the origin of microbial community types', *ISME Journal*, 11, pp. 2159–66.

Goodrich, Julia K., Davenport, Emily R., Beaumont, Michelle, Jackson, Matthew A., Knight, Rob, Ober, Carole, Spector, Tim D., Bell, Jordana T., Clark, Andrew G. and Ley, Ruth E. (2016), 'Genetic determinants of the gut microbiome in UK twins', *Cell Host & Microbe*, 19, pp. 731–43.

Gordon, Jeffrey I. and Klaenhammer, Todd R. (2011), 'A rendezvous with our microbes', *Proceedings of the National Academy of Sciences of the United States of America*, 108, pp. 4513–15.

Gordon, Jeffrey I., Knowlton, Nancy, Relman, David A., Rohwer, Forest and Youle, Merry (2013), 'Superorganisms and holobionts', *Microbe*, 8, pp. 152–53.

Gough, Ethan, Shaikh, Henna and Manges, Amee R. (2011), 'Systematic review of intestinal microbiota transplantation (fecal bacteriotherapy) for recurrent *Clostridium difficile* infection', *Clinical Infectious Diseases*, 53, pp. 994–1002.

Grice, Elizabeth A. and Segre, Julia A. (2012), 'The human microbiome: Our second genome', *Annual Review of Genomics and Human Genetics*, 13, pp. 151–70.

Guerrero, Ricardo, Margulis, Lynn and Berlanga, Mercedes (2013), 'Symbio-genesis: The holobiont as a unit of evolution', *International Microbiology*, 16, pp. 133–43.

Haig, David (2010), 'Transfers and transitions: Parent-offspring conflict, genomic imprinting, and the evolution of human life history', *Proceedings of the National Academy of Sciences of the United States of America*, 10, pp. 1731–35.

——— (2014), 'Interbirth intervals: Intrafamilial, intragenomic and intrasomatic conflict', *Evolution, Medicine, and Public Health*, 1, pp. 12–17.

Hempel, Susanne, Newberry, Sydne J., Maher, Alicia R., Wang, Zhen, Miles, Jeremy, N. V., Shanman, Roberta, Johnsen, Breanne and Shekelle, Paul G. (2012), 'Probiotics for the prevention and treatment of antibiotic-associated diarrhea: A systematic review and meta-analysis', *Journal of the American Medical Association*, 30, pp. 1959–69.

Hölldobler, Bert and Wilson, Edward O. (2009), *The Superorganism: The Beauty, Elegance, and Strangeness of Insect Societies*, New York: W. W. Norton.

Hooper, Lora V. (2009), 'Do symbiotic bacteria subvert host immunity?', *Nature Reviews Microbiology*, 7, pp. 367–74.

Hooper, Lora V., Littman, Dan R. and Macpherson, Andrew J. (2012), 'Interactions between the microbiota and the immune system', *Science*, 33, pp. 1268–73.

Hungin, Amritpal Pali S., Mulligan, C., Pot, B., Whorwell, Peter J., Agréus, Lars, Fracasso, P., Lionis, Christos, Mendive, Juan M., Philipart de Foy, Jean-Michel., Rubin, Greg, Winchester, Chris and de Wit, Niek for The European Society for Primary Care Gastroenterology (2013), 'Systematic review: Probiotics in the management of lower gastrointestinal symptoms in clinical practice – An evidence-based international guide', *Alimentary Pharmacology and Therapeutics*, 3, pp. 864–86.

Huss, John E. (2014), 'Methodology and ontology in microbiome research', *Biological Theory*, 9, pp. 392–400.

Ingram, Wendy M., Goodrich, Leanne M., Robey, Ellen A. and Eisen, Michael B. (2013), 'Mice infected with low-virulence strains of *Toxoplasma gondii* lose their innate aversion to cat urine, even after extensive parasite clearance', *PLoS One*, 8, https://arxiv.org/pdf/1304.0479.pdf. Accessed 6 April 2023.

Kellert, Stephen H., Longino, Helen E. and Waters, C. Kenneth (2006), *Scientific Pluralism*, Minneapolis, MN: University of Minnesota Press.

Kelly, Colleen R., Khoruts, Alexander, Staley, Christopher, Sadowsky, Michael J., Abd, Mortadha, Alani, Mustafa, Bakow, Brianna, Curran, Patrizia, McKenney, Joyce, Tisch, Allison, Reinert, Steven E., Machan, Jason T. and Brandt, Lawrence J. (2016), 'Effect of fecal microbiota transplantation on recurrence in multiply recurrent *Clostridium difficile* infection: A randomized trial', *Annals of Internal Medicine*, 165, pp. 609–16.

Kinross, James M., von Roon, Alexander C., Holmes, Elaine, Darzi, Ara and Nicholson, Jeremy K. (2008), 'The human gut microbiome: implications for future health care', *Current Gastroenterology Reports*, 10, pp. 396–403.

Knight, Rob (2015), *Follow Your Gut: The Enormous Impact of Tiny Microbes*, New York: Simon & Schuster.

Knights, Dan, Parfrey, Laura Wegener, Zaneveld, Jesse, Lozupone, Catherine and Knight, Rob (2011), 'Human-associated microbial signatures: Examining their predictive value', *Cell Host and Microbiome*, 10, pp. 292–96.

Kodaman, Nuri, Pazos, Alvaro, Schneider, Barbara G., Piazuelo, M. Blanca, Mera, Robertino, Sobota, Rafal S., Sicinschi, Liviu A., Shaffer, Carrie L., Romero-Gallo, Judith, de Sablet, Thibaut, Harder, Reed H., Bravo, Luis E., Peek, Richard M., Jr., Wilson, Keith T., Cover, Timothy L., Williams, Scott M. and Correa, Pelayo (2014), 'Human and *Helicobacter pylori* coevolution shapes the risk of gastric disease', *Proceedings of the National Academy of Sciences*, 111, pp. 1455–60.

Koskella, Britt, Hall, Lindsay J. and Metcalf, C. Jessica E. (2017), 'The microbiome beyond the horizon of ecological and evolutionary theory', *Nature Ecology & Evolution*, 1, pp. 1606–15.

Kumar, Rajit, Yi, Nengjun, Zhi, Degui, Eipers, Peter, Goldsmith, Kelly T., Dixon, Paula, Crossman, David K., Crowley, Michael R., Lefkowitz, Elliot J., Rodriguez, J. Martin and Morrow, Casey D. (2017), 'Identification of donor microbe species that colonize and persist long term in the recipient after fecal transplant for recurrent *Clostridium difficile*', *NPJ Biofilms and Microbiomes*, 3, p. 12.

Lambert, Jonathan (2018), 'Should evolution treat our microbes as part of us?' *Quanta Magazine*, 20 November, https://www.quantamagazine.org/should-evolution-treat-our-microbes-as-part-of-us-20181120/. Accessed 6 April 2023.

Leach, Jeff (2012), *Honor Thy Symbionts*, New York: CreateSpace Independent Publishing Platform.

——— (2014), '(Re)becoming human', *Human Food Project*, 30 September, https://web.archive.org/web/20200709020624/humanfoodproject.com/rebecoming-human-happened-day-replaced-99-genes-body-hunter-gatherer/. Accessed 15 August 2023.

Lederberg, Joshua (2004), 'Of men and microbes', *New Perspective Quarterly*, 21 (Fall), pp. 92–96.

Lederberg, Joshua and McCray, Alexa T. (2001), 'Commentary: "Ome sweet" omics – A genealogical treasure of words', *The Scientist*, 15:7 (2 April), p. 8.

Lee, Yun Kyung and Mazmanian, Sarkis K. (2010), 'Has the microbiota played a critical role in the evolution of the adaptive immune system?', *Science*, 330, pp. 1768–73.

Lemon, Katherine P., Armitage, Gary C., Relman, David A. and Fischbach, Michael A. (2012), 'Microbiota-targeted therapies: An ecological perspective', *Science Translational Medicine*, 4, https://doi.org/10.1126/scitranslmed.3004183. Accessed 6 April 2023.

Lewontin, Richard Charles (2001), 'In the beginning was the word', *Science*, 29, pp. 1263–64.

Ley, Ruth E., Hamady, Micah, Lozupone, Catherine, Turnbaugh, Peter James, Ramey, Rob Roy, Bircher, J. Stephen, Schlegel, Michel L., Tucker, Tammy A., Schrenzel, Mark D., Knight, Rob and Gordon, Jeffrey I. (2008), 'Evolution of mammals and their gut microbiome', *Science*, 320, pp. 1647–51.

Ley, Ruth E., Peterson, David A. and Gordon, Jeffrey I. (2006a), 'Ecological and evolutionary forces shaping microbial diversity in the human intestine', *Cell*, 124, pp. 837–48.

Ley, Ruth E., Turnbaugh, Peter J., Klein, Samuel and Gordon, Jeffrey I. (2006b), 'Human gut microbes associated with obesity', *Nature*, 444, pp. 1022–23.

Li, Simone S., Zhu, Ana, Benes, Vladmir, Costea, Paul I., Herzog, Rajna, Hildebrand, Falk, Huerta-Cepas, Jaime, Nieuwdorp, Max, Salojärvi, Jarkko, Voigt, Anita Y., Zeller, Georg, Sunagawa, Shinichi, de Vos, Willem M. and Bork, Peer (2016), 'Durable coexistence of donor and recipient strains after fecal microbiota transplantation', *Science*, 352, pp. 586–89.

Lim, Mi Young, You, Hyan Ju, Yoon, Hyo Shin, Kwon, Bomi, Lee, Jae Yoon, Lee, Sunghee, Song, Yun-Mi, Lee, Kayoung, Sung, Joohon and Ko, GwangPyo (2017), 'The effect of heritability and host genetics on the gut microbiota and metabolic syndrome', *Gut*, 66, pp. 1031–38.

LoCascio, Riccardo G., Ninonuevo, Milady R., Freeman, Samara L., Sela, David A., Grimm, Rudolf, Lebrilla, Carlito B., Mills, David A. and German, J. Bruce (2007), 'Glycoprofiling of bifidobacterial consumption of human milk oligosaccharides demonstrates strain specific, preferential consumption of small chain glycans secreted in early human lactation', *Journal of Agricultural and Food Chemistry*, 55, pp. 8914–19.

Longino, Helen (2012), 'Knowledge for what? Monist, pluralist, pragmatist approaches to the sciences of behavior', in K. S. Plaisance and T. A. C. Reydon (eds), *Philosophy of Behavioral Biology*, New York: Springer, pp. 25–40.

Lumsden, Charles J. (1982), 'The social regulation of physical caste: The superorganism revived', *Journal of Theoretical Biology*, 95, pp. 749–81.

Lynch, Susan V. (2016), 'The lung microbiome and airway disease', *Annals of the American Thoracic Society*, 13, https://doi.org/10.1513/AnnalsATS.201605-356AW. Accessed 6 April 2023.

Lynch, Susan V. and Pedersen, Oluf (2016), 'The human intestinal microbiome in health and disease', *New England Journal of Medicine*, 375, pp. 2369–79.

Margulis, Lynn (1991), 'Symbiogenesis and symbioticism', in L. Margulis and R. Fester (eds), *Symbiosis as a Source of Evolutionary Innovation: Speciation and Morphogenesis*, Cambridge, MA: MIT Press, pp. 1–14.

Marino, Simeone, Baxter, Nielson T., Huffnagle, Gary B., Petrosino, Joseph. F. and Schloss, Patrick D. (2014), 'Mathematical modeling of primary succession of murine intestinal microbiota', *Proceedings of the National Academy of Sciences*, 111, pp. 439–44.

Martínez, Inés, Stegen, James C., Maldonado-Gómez, Maria X., Eren, A. Murat, Siba, Peter M., Greenhill, Andrew R. and Walter, Jens (2015), 'The gut microbiota of rural Papua New Guineans: Composition, diversity patterns, and ecological processes', *Cell Reports*, 11, pp. 527–38.

Maynard Smith, John (1993), 'A Darwinian view of symbiosis', in L. Margulis and R. Fester (eds), *Symbiosis as a Source of Evolutionary Innovation: Speciation and Morphogenesis*, Cambridge, MA: MIT Press, pp. 26–39.

McFall-Ngai, Margaret (2007), 'Care for the community', *Nature*, 445, p. 153.

McFall-Ngai, Margaret, Hadfield, Michael G., Bosch, Thomas C. G., Carey, Hannah V., Domazet-Loso, Tomišlav, Douglas, Angela E., Dubilier, Nicole, Eberl, Gerard, Fukami, Tadashi, Gilbert, Scott F., Hentschel, Ute, King, Nicole, Kjelleberg, Staffan, Knoll, Andrew H., Kremer, Natacha, Mazmanian, Sarkis K., Metcalf, Jessica L., Nealson, Kenneth, Pierce, Naomi E., Rawls, John F., Reid, Ann, Ruby, Edward G., Rumpho, Mary, Sanders, Jon G., Tautz, Diethard and Wernegreen, Jennifer J. (2013), 'Animals in a bacterial world, a new imperative for the life sciences', *Proceedings of the National Academy of Sciences*, 110, pp. 3229–36.

Mindell, David P. (1992), 'Phylogenetic consequences of symbioses: Eukarya and eubacteria are not monophyletic taxa', *BioSystems*, 27, pp. 53–62.

Mitchell, Sandra D. (1995), 'The superorganism metaphor: Then and now', in S. Maasen, E. Mendelsohn and P. Weigart (eds), *Biology as Society, Society as Biology: Metaphors*, Dordrecht: Kluwer Academic Publishers, pp. 231–47.

Moeller, Andrew H., Caro-Quintero, Alejandro, Mjungu, Deus, Georgiev, Alexander V., Lonsdorf, Elizabeth V., Muller, Martin N., Pusey, Anne E., Peeters, Martine, Hahn, Beatrice H. and Ochman, Howard (2016), 'Cospeciation of gut microbiota with hominids', *Science*, 353, pp. 380–82.

Moore, Janice (2002), *Parasites and the Behavior of Animals*, Oxford: Oxford University Press.

Moran, Nancy A. and Sloan, Daniel B. (2015), 'The hologenome concept: Helpful or hollow?', *PLoS Biology*, 13, https://doi.org/10.1371/journal.pbio.1002311. Accessed 6 April 2023.

Mosca, Alexis, Leclerc, Marion and Hugot, Jean P. (2016), 'Gut microbiota diversity and human diseases: Should we reintroduce key predators in our ecosystem?', *Frontiers in Microbiology*, 7, p. 455.

Murdoch, Travis B. and Detsky, Alan S. (2012), 'Time to recognize our fellow travellers', *Journal of General Internal Medicine*, 27, pp. 1704–06.

NIH (National Institutes of Health) (2018), *The Human Microbiome Project*, https://commonfund.nih.gov/hmp. Accessed 6 April 2023.

Ogilvie, Leslie A., Overall, Andrew D. J. and Jones, Brian V. (2012), 'The human-microbiome coevolutionary consortium', in L. A. Ogilvie and P. R. Hirsch (eds), *Microbial Ecological Theory: Current Perspectives*, Norfolk, VA: Caister Academic Press, pp. 25–42.

O'Hara, Ann M. and Shanahan, Fergus (2006), 'The gut flora as a forgotten organ', *EMBO Reports*, 7, pp. 688–93.

Okasha, Samir (2006), *Evolution and the Levels of Selection*, Oxford: Oxford University Press.

Olszak, Torsten, An, Dingding, Zeissig, Sebastian, Vera, Miguel Pinilla, Richter, Julia, Franke, Andre, Glickman, Jonathan N., Siebert, Reiner, Baron, Rebecca M., Kasper, Dennis L. and Blumberg, Richard S. (2012), 'Microbial exposure during early life has persistent effects on natural killer T cell function', *Science*, 336, pp. 489–93.

O'Malley, Maureen A. and Dupré, John (2007), 'Size doesn't matter: Towards a more inclusive philosophy of biology', *Biology and Philosophy*, 22, pp. 155–91.

——— (2010), 'Philosophical themes in metagenomics', in D. Marco (ed.), *Metagenomics: Theory, Methods and Applications*, Norfolk, VA: Caister Academic Press, pp. 183–208.

O'Malley, Maureen A. and Skillings, Derek J. (2018), 'Methodological strategies in microbiome research and their explanatory implications', *Perspectives in Science*, 26, pp. 239–65.

Orr, Matthew R., Kocurek, Kathryn M. and Young, Deborah L. (2018), 'Gut microbiota and human health: Insights from ecological restoration', *Quarterly Review of Biology*, 93, pp.73–90.

Pertea, Milhaela and Salzberg, Steven L. (2010), 'Between a chicken and a grape: Estimating the number of human genes', *Genome Biology*, 11:Suppl. 1, https://doi.org/10.1186/gb-2010-11-5-206. Accessed 6 April 2023.

Pradeu, Thomas (2012), *The Limits of the Self: Immunology and Biological Identity*, Oxford: Oxford University Press.

——— (2016), 'Biological individuality: Special issue', *Biology and Philosophy*, 31, pp. 761–930.

Qin, Junjie, Li, Ruiqiang, Raes, Jeroen, Arumugam, Manimozhiyan, Burgdorf, Kristoffer Solvsten, Manichanh, Chaysavanh, Nielsen, Trine, Pons, Nicolas, Levenez, Florence, Yamada, Takuji, Mende, Daniel R., Li, Junhua, Xu, Junming, Li, Shaochuan, Li, Dongfang, Cao, Jianjun, Wang, Bo, Liang, Huiqing, Zheng, Huisong, Xie, Yinlong, Tap, Julien, Lepage, Patricia, Bertlan, Marcelo, Batto, Jean-Michel, Hansen, Torben, Le Palier, Denis, Linneberg, Allan, Nielsen, H. Bjørn, Pelletier, Eric, Renault, Pierre, Sicheritz-Ponten, Thomas, Turner, Keith, Zhu, Hongmei, Yu, Chang, Li, Shengting, Jian, Min, Zhou, Yan, Li, Yingrui, Zhang, Xiuqing, Li, Songgang, Qin, Nan, Yang, Huanming, Wang, Jian, Brunak, Søren, Doré, Joel, Guarner, Francisco, Kristiansen, Karsten, Pedersen, Oluf, Parkhill, Julian, Weissenbach, Jean, MetaHIT Consortium, Bork, Peer, Ehrlich, S. Dusko and Wang, Jun (2010), 'A human gut microbial gene catalogue established by metagenomic sequencing', *Nature*, 464, pp. 59–65.

Ravel, Jacques, Glaser, Martin J., Braun, Jonathan, Brown, Eric, Bushman, Frederic D., Chang, Eugene B., Davies, Julian, Dewey, Kathryn G., Dinan, Timothy, Dominguez-Bellow, Maria,

Erdman, Susan E., Finlay, B. Brett, Garrett, Wendy S., Huffnagle, Gary B., Huttenhower, Curtis, Jansson, Janet, Jeffery, Ian B., Jobin, Christin, Khoruts, Alexander, Kong, Heidi H., Lampe, Johanna W., Ley, Ruth E., Littman, Dan R., Mazmanian, Sarkis K., Mills, David A., Neish, Andrew S., Petrof, Elaine, Relman, David A., Rhodes, Rosamond, Turnbaugh, Peter J., Young, Vincent B., Knight, Rob and White, Owen (2014), 'Human microbiome science: Vision for the future', Bethesda, MD, July 24 to 26, 2013', *Microbiome*, 2:1, p. 16.

Ray, Katrina (2012), 'Married to our gut microbiota', *Nature Reviews Gastroenterology and Hepatology*, 9, p. 555.

Relman, David A. (2002), 'The human body as a microbial observatory', *Nature Genetics*, 30, pp. 131–33.

——— (2012), 'Learning about who we are', *Nature*, 486, pp. 194–95.

Richardson, Lauren A. (2017), 'Evolving as a holobiont', *PLoS Biology*, 15, https://doi.org/10.1371/journal.pbio.2002168. Accessed 6 April 2023.

Rohwer, Forest, Seguritan, Victor, Azam, Farooq and Knowlton, Nancy (2002), 'Diversity and distribution of coral-associated bacteria', *Marine Ecology Progress Series*, 243, pp. 1–10.

Rondanelli, Mariangela, Faliva, Milena Anna, Perna, Simone, Giacosa, Attilio, Peroni, Gabriella and Castellazzi, Anna Maria (2017), 'Using probiotics in clinical practice: Where are we now? A review of existing meta-analyses', *Gut Microbes*, 8, pp. 521–43.

Rosenberg, Eugene (2013), 'On the terms holobiont and hologenome', *Microbe*, 8, p. 101.

Rosenberg, Eugene and Zilber-Rosenberg, Ilana (2016), 'Microbes drive evolution of animals and plants: The hologenome concept', *mBio*, 7, https://doi.org/10.1128/mBio.01395-15. Accessed 6 April 2023.

Rosenberg, Eugene, Koren, Omrey, Reshef, Leah, Efrony, Rotem and Zilber-Rosenberg, Ilana (2007), 'The role of microorganisms in coral health, disease and evolution', *Nature Reviews Microbiology*, 5, pp. 355–62.

Rosenberg, Eugene, Sharon, Gil and Zilber-Rosenberg, Ilana (2009), 'The hologenome theory of evolution contains Lamarckian aspects within a Darwinian framework', *Environmental Microbiology*, 11, pp. 2959–62.

Rothschild, Daphna, Weissbrod, Omer, Barkan, Elad, Kurilshikov, Alexander, Korem, Tal, Zeevi, David, Costea, Paul I., Godneva, Anastasia, Kalka, Iris N., Bar, Noam, Shilo, Smadar, Lador, Dar, Vila, Arnau Vich, Zmora, Niv, Pevner-Fischer, Meirav, Israeli, David, Kosower, Noa, Malka, Gal, Wolf, Bat Chen, Avnit-Sagi, Tali, Lotan-Pompan, Maya, Weinberger, Adina, Halpern, Zamir, Carmi, Shai, Fu, Jingyuan, Wijmenga, Cisca, Zhernakova, Alexandra, Elinav, Eran and Segal, Eran (2018), 'Environment dominates over host genetics in shaping human gut microbiota', *Nature*, 555, pp. 210–15.

Roughgarden, Joan, Gilbert, Scott F., Rosenberg, Eugene, Zilber-Rosenberg, Ilana and Lloyd, Elisabeth A. (2018), 'Holobionts as units of selection and a model of their population dynamics and evolution', *Biological Theory*, 13, pp. 44–65.

Round, June L. and Mazmanian, Sarkis K. (2009), 'The gut microbiome shapes intestinal immune responses during health and disease', *Nature Reviews Immunology*, 9, pp. 313–23.

Sansonetti, Philippe J. (2004), 'War and peace at mucosal surfaces', *Nature Reviews Immunology*, 4, pp. 953–64.

Sansonetti, Philippe J. and Medzhitov, Rusian (2009), 'Learning tolerance while fighting ignorance', *Cell*, 138, pp. 416–20.

Schneider, Gregory W. and Winslow, Russell (2014), 'Parts and wholes: The human microbiome, ecological ontology and the challenge of community', *Perspectives in Biology and Medicine*, 57, pp. 208–23.

Seeley, Rod R., Stephens, Trent D. and Tate, Philip (2000), *Anatomy and Physiology*, 5th ed., Boston, MA: McGraw-Hill.

Seeley, Thomas (1989), 'The honey bee colony as superorganism', *American Scientist*, 77, pp. 546–53.

Segre, Julia A. and Salafsky, Nick (2016), 'Hominid superorganisms', *Science*, 353, p. 350–51.

Sender, Ron, Fuchs, Shai and Milo, Ron (2016), 'Revised estimates for the number of human and bacteria cells in the body', *PLoS Biology*, 14, https://doi.org/10.1371/journal.pbio.1002533. Accessed 6 April 2023.

Sevelsted, Astrid, Stockholm, Jakob, Bønnelykke, Klaus and Bisgaard, Hans (2015), 'Cesarean section and chronic immune disorders', *Pediatrics*, 135, pp. 93–98.

Shade, Ashley and Handelsman, Jo (2012), 'Beyond the Venn diagram: The hunt for a core microbiome', *Environmental Microbiology*, 14, pp. 4–12.

Shanker, S., Lee, A. and Sorrell, Tania C. (1990), 'Horizontal transmission of *Campylobacter jejuni* amongst broiler chicks: Experimental studies', *Epidemiology and Infection*, 104, pp. 101–10.

Shen, Nicole T., Maw, Anna, Tmanova, Lyubov L., Pino, Alejandro, Ancy, Kayley, Crawford, Carl V., Simon, Matthew S. and Evans, Arthur T. (2017), 'Timely use of probiotics in hospitalized adults prevents *Clostridium difficile* infection: A systematic review with meta-regression analysis', *Gastroenterology*, 152, pp.1889–1900.

Shropshire, J. Dylan and Bordenstein, Seth R. (2016), 'Speciation by symbiosis: The microbiome and behavior', *mBio*, 7, https://doi.org/10.1128/mBio.01785-15. Accessed 6 April 2023.

Singh, Yadvir, Ahmad, Javed, Musarrat, Javed, Ehtesham, Nasreen Z. and Hasnain, Seyed E. (2013), 'Emerging importance of holobionts in evolution and in probiotics', *Gut Pathogens*, 5, p. 12.

Skillings, Derek (2016), 'Holobionts and the ecology of organisms: Multi-species communities or integrated individuals?' *Biology and Philosophy*, 31, pp. 875–92.

——— (2018), 'I, holobiont. Are you and your microbes a community or a single entity?', *Aeon*, 26 September, https://aeon.co/ideas/i-holobiont-are-you-and-your-microbes-a-community-or-a-single-entity. Accessed 6 April 2023.

Sleator, Roy D. (2010), 'The human superoganism – of microbes and men', *Medical Hypotheses*, 74, pp. 214–15.

Smith, Peter Andrey (2015), 'Can the bacteria in your gut explain your mood', *New York Times*, 23 June, https://www.nytimes.com/2015/06/28/magazine/can-the-bacteria-in-your-gut-explain-your-mood.html. Accessed 6 April 2023.

Song, Se J., Lauber, Christian, Costello, Elizabeth K., Lozupone, Catherine A., Humphrey, Gregory, Berg-Lyons, Donna, Caporaso, J. Gregory, Knights, Dan, Clemente, Jose C., Nakielny,

Sara, Gordon, Jeffrey I., Fierer, Noah and Knight, Rob (2013), 'Cohabiting family members share microbiota with one another and with their dogs,' *eLIFE*, 2, https://doi.org/10.7554/eLife.00458. Accessed 6 April 2023.

Spees, Alanna M., Lopez, Christopher A., Kingsbury, Dawn D., Winter, Sebastian E. and Bäumler, Andreas J. (2013), 'Colonization resistance: Battle of the bugs or ménage à trois with the host?' *PLOS Pathogens*, 9, https://doi.org/10.1371/journal.ppat.1003730. Accessed 6 April 2023.

Stainier, Didier Y. R. (2005), 'No organ left behind: Tales of gut development and evolution', *Science*, 307, pp. 1902–04.

Steed, H., Macfarlane, G. T., Blackett, K. L., Bahrami, B., Reynolds, N., Walsh, S. V., Cummings, J. H. and Macfarlane, S. (2010), 'Clinical trial: The microbiological and immunological effects of synbiotic consumption – A randomized double-blind placebo-controlled study in active Crohn's disease', *Alimentary Pharmacology and Therapeutics*, 32, 872–83.

Stotz, Karola and Griffiths, Paul E. (2008), 'Biohumanities: Rethinking the relationship between biosciences, philosophy and history of science, and society', *Quarterly Review of Biology*, 83, pp. 37–45.

Surawicz, Christina M., Brandt, Lawrence J., Binion, David G., Ananthakrishnan, Ashwin N., Curry, Scott R., Gilligan, Peter H., McFarland, Lynne V., Mellow, Mark and Zuckerbraun, Brian S. (2013), 'Guidelines for diagnosis, treatment, and prevention of *Clostridium difficile* infections', *American Journal of Gastroenterology*, 108, pp. 478–98.

Swenson, William, Wilson, David Sloan and Elias, Roberta (2000), 'Artificial eco-system selection', *Proceedings of the National Academy of Sciences*, 97, pp. 9110–14.

Taube, Christian and Müller, Anne (2012), 'The role of *Helicobacter pylori* infection in the development of allergic asthma', *Expert Review of Respiratory Medicine*, 6, pp. 441–49.

Tauber, Alfred I. (2008), 'The immune system and its ecology', *Philosophy of Science*, 75, pp. 224–45.

Taur, Ying and Pamer, Eric G. (2014), 'Harnessing microbiota to kill a pathogen: Fixing the microbiota to treat *Clostridium difficile* infections', *Nature Medicine*, 20, pp. 246–47.

Thaiss, Christoph A., Zmora, Niv, Levy, Maayan and Elinav, Eran (2016), 'The microbiome and innate immunity', *Nature*, 535, pp. 65–74.

Theis, Kevin, Dheilly, Nolwenn M., Klassen, Jonathan L., Brucker, Robert M., Baines, John F., Bosch, Thomas C. G., Cryan, John F., Gilbert, Scott F., Goodnight, Charles J., Lloyd, Elisabeth A., Sapp, Jan, Vandenkoornhuyse, Philippe, Zilber-Rosenberg, Ilana, Rosenberg, Eugene and Bordenstein, Seth R. (2016), 'Getting the hologenome concept right: An eco-evolutionary framework for hosts and their microbiomes', *mSystems*, 1, https://doi.org/10.1128/mSystems.00028-16. Accessed 6 April 2023.

Thompson, Janelle R., Rivera, Hanny E., Closek, Collin J. and Medina, Mónica (2015), 'Microbes in the coral holobiont: Partners through evolution, development, and ecological interactions', *Frontiers in Cellular and Infection Microbiology*, 4, p. 176.

Tiivel, Toomas H. (1991), 'Cell symbiosis, adaptation, and evolution: Insect-bacteria examples', in L. Margulis and R. Fester (eds), *Symbiosis as a Source of Evolutionary Innovation: Speciation and Morphogenesis*, Cambridge, MA: MIT Press, pp. 170–77.

Tung, Jenny, Barreiro, Luis B., Burns, Michael B., Grenier, Jean-Christophe, Lynch, Josh, Grieneisen, Laura E., Altmann, Jeanne, Alberts, Susan C., Blekhman, Ran and Archie, Elizabeth A. (2015), 'Social networks predict gut microbiome composition in wild baboons', *eLIFE*, 4, https://doi.org/10.7554/eLife.05224. Accessed 6 April 2023.

Turnbaugh, Peter J., Hamady, Micah, Yatsunenko, Tanya, Cantarel, Brandi L., Duncan, Alexis, Ley, Ruth E., Sogin, Mitchell L., Jones, William J., Roe, Bruce A., Affourtit, Jason P., Egholm, Michael, Henrissat, Bernard, Heath, Andrew C., Knight, Rob and Gordon, Jeffrey I. (2009), 'A core gut microbiome in obese and lean twins', *Nature*, 457, pp. 480–84.

Turnbaugh, Peter J., Ley, Ruth E., Hamady, Micah, Fraser-Liggett, Claire M., Knight, Rob and Gordon, Jeffrey I. (2007), 'The human microbiome project', *Nature*, 449, pp. 804–10.

Turner, J. Scott (2001), 'On the mound of *Macrotermes michaelseni* as an organ of respiratory gas exchange', *Physiological and Biochemical Zoology*, 74, pp. 798–822.

——— (2002), 'A superorganism's fuzzy boundaries', *Natural History*, 111, pp. 63–67.

Turpin, Williams, Espin-Garcia, Osvaldo, Xu, Wei, Silverberg, Mark S., Kevans, David, Smith, Michelle I., Guttman, David S., Griffiths, Anne, Panaccione, Remo, Otley, Anthony, Xu Lizhen, Shestopaloff, Konstantin, Moreno-Hagelsieb, Gabriel, GEM Project Research Consortium, Paterson, Andrew D. and Croitoru, Kenneth (2016), 'Association of host genome with intestinal microbial composition in a large healthy cohort', *Nature Genetics*, 48, pp. 1413–17.

Turroni, Silvia, Rampelli, Simone, Biagi, Elena, Consolandi, Clarissa, Severgnini, Marco, Peano, Clelia, Quercia, Sara, Soverini, Matteo, Carbonero, Franck G., Bianconi, Giovanna, Rettberg, Petra, Canganella, Francesco, Brigidi, Patrizia and Candela, Marco (2017), 'Temporal dynamics of the gut microbiota in people sharing a confined environment, a 520-day ground-based space simulation, MARS500', *Microbiome*, 5, p. 39.

van Duynhoven, John, Vaughan, Elaine E., Jacobs, Doris M., Kemperman, Robér A., van Velzen, Ewoud J. J., Gross, Gabriele, Roger, Laure C., Possemiers, Sam, Smilde, Age K., Doré, Joël, Westerhuis, Johan A. and Van de Wiele, Tom (2011), 'Metabolic fate of polyphenols in the human superorganism', *Proceedings of the National Academy of Sciences*, 108, pp. 4531–38.

van Nood, Els, Vrieze, Anne, Nieuwdorp, Max, Fuentes, Susana, Zoetendal, Erwin G., de Vos, Willem M., Visser, Caroline E., Kuijper, Ed J., Bartelsman, Joep F. W. M., Tijssen, Jan G. P., Speelman, Peter, Dijkgraaf, Marcel G. W. and Keller, Josbert J. (2013), 'Duodenal infusion of donor feces for recurrent *Clostridium difficile*', *New England Journal of Medicine*, 368, pp. 407–15.

Vangay, Palau, Johnson, Abigail J., Ward, Tonya L., Al-Ghalith, Gabriel A., Shields-Cutler, Robin R., Hillmann, Benjamin M., Lucas, Sarah K., Beura, Lalit K., Thompson, Emily A., Till, Lisa M., Batres, Rodolfo, Paw, Bwei, Pergament, Shannon L., Saenyakul, Pimpanitta, Xiong, Mary, Kim, Austin D., Kim, Grant, Masopust, David, Martens, Eric C., Angkurawaranon, Chaisiri, McGready, Rose, Kashyap, Purna C., Culhane-Pera, Kathleen A. and Knights, Dan (2018), 'US immigration westernizes the human gut microbiome', *Cell*, 175, pp. 962–72.

Vellend, Mark (2010), 'Conceptual synthesis in community ecology', *Quarterly Review of Biology*, 85, pp. 183–206.

Wang, Meizhen, Schaefer, Amy L., Dandekar, Ajai A. and Greenberg, E. Peter (2015), 'Quorum sensing and policing of *Pseudomonas aeruginosa* social cheaters', *Proceedings of the National Academy of Sciences*, 112, pp. 2187–91.

Wheeler, William M. (1911), 'The ant-colony as an organism', *Journal of Morphology*, 22, June, pp. 307–25.

Whipps, J. M., Lewis, K. and Cooke, R. C. (1988), 'Mycoparasitism and plant disease control', in N. M. Burge (ed.), *Fungi in Biological Control Systems*, Manchester: Manchester University Press, pp. 161–87.

Wilson, David Sloan (1975), 'A theory of group selection', *Proceedings of the National Academy of Sciences*, 72, pp. 143–46.

Wilson, David Sloan and Sober, Elliott (1989), 'Reviving the superorganism', *Journal of Theoretical Biology*, 136, pp. 337–56.

Wilson, Edward O. (1968), 'The superorganism concept and beyond', in R. Chauvin and C. Noiret (eds), *L'effet de Groupe chez les Animaux*, Paris: CNRS, pp. 27–39.

Xu, Jian and Gordon, Jeffrey I. (2003), 'Honor thy symbionts', *Proceedings of the National Academy of Sciences*, 100, pp. 10452–59.

Xu, Zhenjiang and Knight, Rob (2015), 'Dietary effects on human gut microbiome diversity', *British Journal of Nutrition*, 113:(Suppl. 1), https://doi.org/10.1017/S0007114514004127. Accessed 6 April 2023.

Yatsunenko, Tanya, Rey, Federico E., Manary, Mark J., Trehan, Indi, Dominguez-Bello, Maria Gloria, Contreras, Monica, Magris, Magda, Hidalgo, Glida, Baldassano, Robert N., Anokhin, Andrey P., Heath, Andrew C., Warner, Barbara, Reeder, Jens, Kucynski, Justin, Caporaso, J. Gregory, Lozupone, Catherine, A., Lauber, Christian, Clemente, Jose Carlos, Knights, Dan, Knight, Rob and Gordon, Jeffrey I. (2012), 'Human gut microbiome viewed across age and geography', *Nature*, 486, pp. 222–27.

Yong, Ed (2014), 'There is no "healthy" microbiome', *New York Times*, 1 November, https://www.nytimes.com/2014/11/02/opinion/sunday/there-is-no-healthy-microbiome.html. Accessed 6 April 2023.

Youngster, I., Sauk, J., Pindar, C., Wilson, R. G., Kaplan, J. L., Smith, M. B., Alm, E. J., Gevers, D., Russell, G. H. and Hohmann, E. L. (2014), 'Fecal microbiota transplant for relapsing *Clostridium difficile* infection using a frozen inoculum from unrelated donors: A randomized, open-label, controlled pilot study', *Clinical Infectious Diseases*, 58, pp. 1515–22.

Zaneveld, Jesse R., McMinds, Ryan and Thurber, Rebecca V. (2017), 'Stress and stability: Applying the Anna Karenina principle to animal microbiomes', *Nature Microbiology*, 2, p. 17121.

Zilber-Rosenberg, Ilana and Rosenberg, Eugene (2008), 'Role of microorganisms in the evolution of animals and plants: The hologenome theory of evolution', *FEMS Microbiology Reviews*, 32, pp. 723–35.

PART III

ENACTIONS AND VALUES

7

From Life to Mind

Mark L. Johnson

In humans, life manifests as what Maxine Sheets-Johnstone calls our 'primal animateness' (1999: 137). We are born into the world as impulsive, squirming, moving bundles of energy, exploring the perceptions and actions afforded us by our environment, the medium in which we live, move and have our being. As we grow and develop, those initially somewhat undirected impulses get channeled into specific ways of responding and valuing, which form our selfhood.

Out of all of this frenetic movement and surging childhood energy, there eventually emerges meaning, thought and mind, as functional processes that make possible our intelligent engagement with our world. Sheets-Johnstone argues that the conditions of our animate existence and action are the source of all the structures and processes by which we generate meaning and make sense of our world:

> In the beginning, we are simply infused with movement – not merely with a *propensity* to move, but with the real thing. This primal animateness, this original kinetic spontaneity that infuses our being and defines our aliveness, is our point of departure for living in the world and making sense of it.
>
> (1999: 136, original emphasis)

Other contributors to this volume provide deeply insightful accounts of the conditions for life. My project starts with some of these life-maintenance conditions – finding and ingesting nutrients, regulating internal homeostasis, finding shelter and manipulating objects – and explores how they give rise to and shape meaning and thought, which is to say, how mind arises from bodily activity. I follow John Dewey's argument that all the meaning, thought and language of which we humans are capable emerges solely from interactions of our organic bodies with the material, interpersonal and cultural environments that we inhabit (Dewey [1925] 1981). My account is, therefore, *naturalistic*, rejecting any reference to supernatural or transcendent entities, causes or processes. However, it also recognizes our fundamental human *sociality*. We depend on the continual nurturance and cooperation of others for our identity, our values and the shared meaning that binds us together in communities.

I begin with the question: how does our primal animateness serve as 'our point of departure for living in the world and making sense of it' (Sheets-Johnstone 1999: 136)? The key to an answer is to understand how the structures and processes of our sensory, motor and affective engagement with our surroundings generate our so-called 'higher' cognitive processes, actions and systems of meaning. We humans are creatures of the flesh. Our animate bodies, engaging our surroundings (physical, interpersonal and cultural) through our strivings, frustrations and satisfactions generate all the meanings of which we are capable. We can only experience, think, imagine and do what our bodies, brains and environments allow us to experience, think, imagine and do. Consequently, it is our continuous corporeal engagement with our physical and social surroundings that generates our selfhood and makes possible meaning and thought. As we will see below, the recurring patterns of our sensory and motor processes provide templates for higher-level cognitive operations of thought (Johnson 1987, 2007, 2017; Talmy 2000; Langacker [1987] 1991). For example, the center–periphery structure of our perceptual horizon (i.e. nearby focal objects fading off at the margins into a horizon of related possible experiences), gets appropriated for conceptualization and reasoning, in the form of a horizon of possible meanings within which focal meanings emerge (Merleau-Ponty 1962; Johnson 1987).

Conditions of Organism Viability

To understand how human meaning and values arise, we must begin with what it means to be a living organism in ongoing interaction with its environments. Neuroscientist Antonio Damasio observes that:

> Life is carried out within a boundary that defines a body. Life and the life urge exist inside a boundary, the selectively permeable wall that separates the internal environment from the external environment. [...] If there is no boundary, there is no body, and if there is no body, there is no organism. Life needs a boundary. I believe that minds and consciousness, when they eventually appeared in evolution, were first and foremost about life and the life urge within a boundary.
>
> (Damasio 1999: 137)

To stay alive, we have to incorporate energy, as nutrition, into our bodies and expel waste back into our surroundings. Within the internal milieu, circumscribed by our bodily boundary, we must maintain a finely tuned dynamic equilibrium, known as *homeostasis*. The parameters of this homeostatic balance operate within a fairly narrow range. If you get too hot or too cold you die. If your sodium, blood sugar or oxygen levels go too low, your brain activity shuts down and, eventually, you're a goner. The long and still ongoing evolutionary development of our species has provided our bodies and brains with exquisitely evolved mechanisms that monitor our body states as they are affected by our continuous engagement with our environments. Homeostasis thus becomes the fundamental value upon which other life values depend: '[T]he physiological state of living organism's tissues, within an optimal homeostatic range, is the deepest origin of biological value and valuations' (Damasio 2010: 48). Our many and varied needs, values and goals gravitate around our ultimate need to maintain the homeostasis necessary for life.

Homeostasis is typically construed as a return to some set-point state or pre-established value of the biological organism, in response to changes in its surroundings. However, humans, and many other animals as well, are capable of re-calibrating our expectations in anticipation of changing conditions. This process – known as *allostasis* – is not just a return to a pre-set value, but rather a proactive creation of a new dynamic equilibrium in response to changing conditions (Tucker and Luu 2012; Schulkin and Sterling 2019; Johnson and Tucker 2021; Johnson and Schulkin 2023). Life maintenance is thus an exquisitely fine-tuned back-and-forth between the organism's projected expectancies based on prior experience, and the testing of those predictions against experience arising from the organism's sensory–motor interface with its surroundings. When a discrepancy arises between what is predicted and what actually occurs, this can lead to allostatic recalibration of expectancies to meet changed conditions.

Within this allostatic process, there arise recurrent patterns of sensory and motor experience that are meaningful to us in the most intimate, visceral way. These patterns gain significance, insofar as they determine what matters for our well-being and how we make sense of things. Examples of these intrinsically meaningful patterns, called *image schemas*, are given below, in the account of embodied meaning. The basic idea is that, given the nature of our bodies and brains, and given the structure of our surroundings, patterns like up–down, near–far, center–periphery, in–out, right–left, source–path–goal, etc. establish fundamental meanings arising in our bodily experience. These patterns can then be recruited for abstract conceptualization and reasoning (Lakoff and Johnson 1999; Dodge and Lakoff 2005; Gibbs 2006).

Meanings are not mysterious abstract entities grasped by something we call mind. Instead, the experience of any object or event as meaningful is the result of *affordances* for perception and action provided by the structured energy of our bodies as they interact with the structured energy of our environments (Gibson

1979). Change our bodies and/or our environments in significant ways, and the affordances for perception and action change.

The meaning of any object, event or state of affairs is what it points to and enacts by way of experiences past, present and future, where future experience refers to anticipation of what is to come. The meaning of an object is thus the affordances for possible experiences that it provides us, given our bodily and social makeup, and the structure of the environment. An important implication is that, although meanings can be quite stable much of the time, they are always potentially subject to revision in light of changing conditions.

The Qualitative and Affective Basis for Meaning

From a naturalistic perspective, the only place to find meaning, mind and intelligence is in the structures and processes of our ongoing, patterned experiences (i.e. from organism–environment interactions). It is now a commonplace of contemporary cognitive science that almost all of this meaning-making and thinking occurs before you know it, operating beneath the level of our conscious awareness (Damasio 1999, 2010; Johnson and Tucker 2021). Most of the time, we are exploring the meaning of our situation and making non-conscious, automatic bodily adjustments, in order to remain viable and to enhance our well-being. However, for creatures like us, with evolved capacities for self-awareness, we can sometimes consciously feel these bodily homeostatic, allostatic and semantic processes. This consciousness of the meanings of qualities and relations enhances our ability to assert a modest control over the development of experience. That said, we need to keep in mind that the vast majority of our meaning and thought is carried out unconsciously, automatically and without forethought (Johnson and Schulkin 2023; Lakoff and Narayanan forthcoming).

The interdisciplinary cognitive science of the first two-thirds of the twentieth century was *cognitivist* and paid little or no attention to embodiment. It drew on information-processing psychology, artificial intelligence, Chomskyan generative linguistics and analytic philosophy of mind and language.[1] This first-generation cognitive science construed thought as manipulation of abstract symbols via formal (syntactic and logical) algorithms. Aside from supposedly providing input for sensory concepts, the body was not believed to play any major role in shaping conceptualization and reasoning. This perspective, called functionalism, argued that if humans and computers are capable of performing the same functions as part of problem-solving, then they both exhibit mind, even if they realize those functional processes in fundamentally different hardwares and wetwares (i.e. machine mechanisms, biological bodies, etc.).

In the late 1970s, as a response to the mounting failures of these disembodied approaches to explain natural language and to simulate human thinking, an *embodied* second-generation approach emerged from the intersection of phenomenology, cognitive linguistics and philosophies attentive to the importance of the body in mind, thought and language (Lakoff and Johnson [1980] 2003; Varela et al. [1991] 2016; Lakoff 1987; Langacker [1987] 1991; Gallagher 2005). This second wave recognized that human meaning, understanding and reasoning are intimately rooted in our sensory, motor and affective engagement with our environment. Meaning thus arises from recurring patterns of bodily experience, shaped by emotions, and guided by body-based and social values (Damasio 1999; Gibbs 2006; Slingerland 2008). The dominant metaphor of first-generation cognitive science – THE MIND IS A COMPUTATIONAL PROGRAM – was replaced by the guiding metaphor of second-generation approaches – THE MIND IS THE BODY. Instead of trying to impose linear formal logics on meaningless symbols to create machine grammars and algorithms that could solve certain kinds of problems, meaning and thought were seen as arising from the patterns and processes of our sensory–motor–affective interactions with the world. This second-generation cognitive science does not reject computational modelling but instead utilizes artificial neural network models that more closely capture the way humans and other animals process information and think (Feldman 2006; Johnson and Tucker 2021; Johnson and Schulkin 2023). These neural network models incorporate known neural architectures, such as center-surround structure, neural gating, spreading activation, spike timing dependence plasticity, etc. (Feldman 2006).

One of the most significant developments in recent cognitive science is the growing appreciation of the crucial role of qualities, feelings and emotions in human cognition. There is no adaptive (i.e. practical) reasoning without a functional emotional system (Damasio 1994, 2010). *Affective neuroscience* (Panksepp 1998) unmasks the mistaken over-emphasis in first-generation cognitive science on conceptual, cognitive and epistemic processes, which are assumed to be abstract and formal, independent of feelings and emotions – as if we were merely rational, dispassionate knowers. We are continuing to learn how thought and inference depend on affective processes that situate, motivate and control our thinking (Damasio 1994; Tucker and Liu 2012; Johnson and Tucker 2021).

Nearly a century ago, John Dewey ([1925] 1981, [1930] 1984) presciently highlighted the crucial but under-appreciated role of qualities in our thought processes and knowledge-seeking activities. He observed how, first and foremost, we live in a world of *qualities* – the redness of a ripe cherry, the redness of your lover's lips, the redness of a bodily wound, the redness of an Oregon coast sunset and the redness of a vermilion flycatcher. Each of these reds has its distinctive qualitative aspect and each is embedded in different fields of meaning. That's the stuff of life (and death), what attracts or repels us, rooting us in the world in the most profound visceral way.

The qualitative is not exhausted merely by discrete qualities like this pinkish-red of the gala apple, this azure blue of the April sky or this hardy pale green of the prickly pear cactus. In addition to individual qualities of objects, there is the distinctive unifying quality that demarcates a meaningful situation that we inhabit. Every quality we experience already exists within and gets its meaning from, the more comprehensive environmental situations in which we live and act. Each situation is, as an experience, a unified whole whose identity and meaning are felt as a pervasive qualitative unity that constitutes our sense of our being in this situation (Dewey [1930] 1984). Consider the teeming energy and excitement of a large athletic event. Contrast this with the very different quality that pervades the coming nightfall as you sit alone by an isolated lake high in the mountains at twilight. In either case, the distinguishing qualitative unity is *in and of the situation* itself, and only derivatively is it a feeling *in you*.

As a way of orienting oneself in one's surroundings, this felt qualitative unity is the basis for our feeling that a certain situation 'makes sense'; that is, that there are submerged meanings at work determining for us what objects, qualities and relations are relevant for our thinking. The relevance that situates us meaningfully in our world has to arise from the qualitative unity of the situation, which then guides the course of our subsequent thinking. The qualitative whole can be felt but typically not known, or fully expressed in language, in any definitive way, because it is the very qualitative and unifying background from which particular meanings emerge in the first place, as we explore our surroundings. Dewey gave the name *sense* to this mostly unconscious or marginally conscious way of meaningfully engaging one's world. It is 'an immediate and immanent meaning; it is meaning which is itself felt or directly had' (Dewey [1925] 1981: 200).

Dewey often illustrated the meaning of *sense* by using artworks, musical compositions and poems as premier examples of enacted situations with a pervasive unifying quality underlying the consummatory meaning of the artwork – *this* particular Van Gogh, *this* cubist-colorist painting by Sonia Delaunay, *this* tragic Billie Holiday love song. As Paul Ricoeur (1992) argued, an artwork provides a 'world of the work' which the perceiver can 'inhabit' or 'dwell in', in order to gather the meaning of what is happening. This *in-dwelling* is first felt and then only later explored with whatever meaning resources the audience brings to their encounter with the work. However, no system of meanings is ever fully adequate to one's sense of the situation, which is *felt*, more than *known*. To say that a given situation 'makes sense' is to say that we feel somewhat oriented and 'at home' in that situation.

It is the bodily conditions of life maintenance, motivated perception and value-based action that give rise to meaning and thought. In the terminology of what is known as '4E Cognition', our meaning and thought processes are at once *embodied* (shaped by our sensory–motor and affective systems), *embedded* (arising from the patterns of our interactions with our environments), *enactive* (creating meaning through action that transforms experience) and *extended* (incorporating structures of our external environment and storage – for cognition) (Newen et al. 2018). Following this penchant for E's, I have suggested that our 4E's might usefully be extended to 7E's by including cognition as *emotional* (rooted in and shaped by our emotions), *evolutionary* (the result of ongoing evolutionary influences on brain and body anatomy and architecture) and *exaptative* (recruiting sensory–motor structures and processes for abstract conceptualization and reasoning) (Johnson 2017).

Insofar as entire situations can make sense, that is, can *afford* us meaningful experiences, we can function more or less successfully in our surroundings. Rooted as we are in the qualitative unity of a particular situation, we can then focus on the meaning for our lives of particular qualities, objects, movements, changes and patterns of causation. Dewey called this second mode of meaning-making *signification.* This is done via the use of signs (more below) to explore the experiential implications of an object or event. On those unsettling occasions when things do not seem to make sense, we feel like aliens who are not at home in the world we are inhabiting.

There is considerable evidence that humans have evolved both core-shell and bi-lateral brain architectures that support Dewey's hypothesis that we first encounter an affect-rich, holistic situation through felt sense (processed in the limbic core and tied to our visceral needs relative to organism viability), and we then discriminate objects, qualities and relations within that situation (via the cortical shell that interfaces, through perception and action, with the external environment). (For more details, see Tucker and Luu 2012; Tucker 2007; Johnson 2007; Johnson and Tucker 2021.) The dorsal stream of motive control gives the contextual holistic setting (the *Where?* pathway) that is expansive, yet fleeting, while the ventral stream provides sustained attention to focal objects and their significance. Objects become significant and relevant within a broader experiential context established by our most basic visceral need for survival and well-being.

Dewey turned associationist psychology on its head, arguing that you don't encounter the world via fragmented sensations, which you then have to put together into objects (via an alleged process called 'association'), thereby building up experiences. Instead, you find yourself embedded in qualitatively unified, affect-rich situations within which you then articulate objects, properties and relations for focal attention, depending on your interests, values, goals and prior habits of inquiry and judgment. Objects are thus events with meaning that arise within an experiential context. In this way, we learn the meaning of some object or event – not a fixed meaning, but always only in an ongoing developing expansion of thought.

The affective dimensions of experience, meaning and thought are not limited only to feeling discrete qualities that stand out within the pervasive unifying quality of a situation. William James (1890) went so far as to observe that there are ways it feels to think. Thinking is an activity that has its own distinctive patterns of ebb and flow. When we encounter what he called 'furtherances' we are carried forward effortlessly, without critical reflection. Sometimes, however, we meet 'hindrances' that frustrate our projected expectations for future experiences. This becomes an occasion to 'stop and think', leading to a recalibration of our predictions about experience. James even claimed that logical relations – such as *if ... then*, *and*, and *but* – are rooted in basic feelings (James 1890; Johnson, 2007; Pappas 2016). For example, 'if ... then' might be felt as the experience of one thing or event as leading to an expectancy (an anticipation) of a certain related consequent event or experience (as in 'If she doesn't show up soon, then I'll go looking for her'). 'And' might be felt as the expectancy of two things being together (as in, 'She went to the library, and he went along with her'). 'But' usually follows some affirmation that then gives rise to a hesitancy that in some way challenges or alters the initial affirmation (as in, 'He may be a hunk, but he's dumb as a stone'). James liked to use the hyphenated 'thought-feeling' to remind us that what we experience is not just an intellectual thought, nor just a feeling, but a feeling-infused thought.

Another affective dimension of cognition worth mentioning is what child psychiatrist and developmental psychologist Daniel Stern (2018) identified as 'affect contours'. We describe these experiential patterns using adverbs like flowing, halting, racing, fluttering, floating, flying, dragging, surging, swelling, struggling, building up and releasing, rushing, etc., to capture distinctive qualitative aspects of what is being experienced, and how it is experienced. Stern argued that pre-linguistic children mostly experience such affect contours, which indicates the way their experience unfolds in feeling patterns, but he insisted that, as we grow into adulthood, we do not leave such affect contours behind, but rather incorporate them as basic meaning patterns that later receive explicit signification through language. These are instances of what Damasio (1999) calls 'the feeling of what happens', and which he later (2010) related to 'primordial feelings' of being alive in a body.

From Perception and Action to Meaning

Our bodies become the yardstick by which we measure the meaning and value of everything we experience – everything we think, communicate and do. One of the most mundane sources of embodied meaning shows up in the universal human practice of using body-part terms in language. We project body-part relations onto objects and events in the external world, such as the *foot* of a table, *head* of a bed, *mouth* of a river, *face* of a clock and *armpit* of the world. In Mixtec, if you want to say, 'she's sitting on a branch of the tree', you would say something like, 'she located arm tree'. 'She is sitting on the roof of the house', translates roughly as 'she located animal-back house', where an animal on four legs maps onto the roof of the house, and she is sitting on that 'animal back'. Unsurprisingly, we often anthropomorphically understand the objects and relations in the external world in terms of body part projections. Although all languages don't use the same set of body-part relations in exactly the same way, the use of body-part projections appears to be universal across cultures (Dodge and Lakoff 2005).

A second, and much more widespread and important, way that meaning grows from patterns of our bodily experience is via what George Lakoff and I (Lakoff 1987; Johnson 1987; Lakoff and Johnson 1999) called *image schemas*. An image schema is a recurring pattern of our sensory, motor and affective experience that gives order and meaning to experienced events and objects. Image schemas arise from the particular sensory and motor capacities of our bodies (e.g. seeing, touching, hearing, smelling, tasting, proprioception, kinaesthesia, manipulating objects and moving ourselves) as they encounter relatively stable and recurring structures within our environments. Image schemas are thus structural affordances that determine what and how things become meaningful to us. They operate mostly non-consciously but we can become consciously aware of them as felt patterns of our routine experiences.

Neurally, image schemas are rooted in functional brain architectures that generate perception and action. Consider, for example, what Ellen Dodge and George Lakoff (2005) call the LOCOMOTION schema. Given our animal bodies, with their evolved sensory and motor neural architectures, we are able to move our bodies through space in very specific ways, and we observe this in other animals, as well. There arises a basic general LOCOMOTION schema with the following roles and parameters recurrently present in certain types of bodily motion through space:

- A mover (person or animal),
- Body part(s) moved (e.g. legs, feet, torso, arms),
- Gait (e.g. walking, skipping, running, hopping),
- Speed (e.g. slow, medium, fast),
- Effort exerted (e.g. trudging along vs. scampering).

Different kinds of bodily movement through space are thus defined by the relevant parameter values of the roles that make up the schema. Dodge and Lakoff observe that all of these roles (and their parameters) depend on evolved brain architectures responsible for various sensory and motor functions, such as activating muscle contractions in various body parts responsible for balance, movement, increasing exertion and planning for motor actions.

For over 35 years, hundreds of articles and several books have been written exploring various image schemas that shape meaning and show up in languages the world over (Hampe 2005). Because of our bodily makeup, we experience image schemas such as OBJECT, BOUNDARY, CONTAINER, VERTICALITY, SCALAR INTENSITY, SOURCE–PATH–GOAL, FRONT–BACK, OVER–UNDER, ON, NEAR–FAR, CENTER–PERIPHERY and scores more (Johnson 2007).

Although often included among lists of image schemas, it is perhaps worth noting that certain force schemas might for certain explanatory purposes be usefully set off from object and spatial relations schemas. I originally lumped together image schemas and force schemas, such as COMPULSION, BLOCKAGE, REMOVAL OF RESTRAINT (ENABLEMENT), DIVERSION and ATTRACTION, and then analysed a generic FORCE schema with roles for *object moved, source of force, strength of force* and *direction of force* (Johnson 1987). Leonard Talmy (1985, 2000) originally developed the idea of force dynamics in linguistic meaning and grammar, exploring a large number of force-dynamic schemas in diverse languages.

The upshot of this account of body-part projections, image schemas and force schemas is to emphasize how much pre-linguistic meaning is available to us, simply by virtue of our sensory and motor capacities as they engage our surroundings at a very basic physical level of organism–environment interaction. Originally, none of this body-based meaning depends on language, but it is deepened and expanded once we acquire the syntactic, semantic and pragmatic resources of natural languages. Because human bodies share certain basic sensory, motor, and affective structures and capacities, it is to be expected that languages around the world will tend to develop linguistic constructions and terms to represent these key aspects of our perception and motor activities. Languages may sometimes vary in how they represent actions. For example, some languages code path information in verbs and manner information in adverbs, as in 'He *entered* the room *slowly*', while others encode manner information in verbs and path information in prepositions, as in 'She *strode into* the courtroom'. English uses both of these patterns, but some other languages tend to use one or the other more often (Dodge and Lakoff 2005). However, whichever option predominates in a given language concerning how manner and spatial location are represented, all languages will have ways of coding actions, their direction, and their manner, because we are beings with bodies capable of performing very specific kinds of actions that serve our viability and well-being.

From Embodiment to Abstraction

Most people don't have a problem with the idea that perception and action are embodied processes. After all, perception utilizes parts of the body (eye, head, neck, torso, hands, feet, lips, tongue, skin) and thereby parts of the brain that have evolved to carry out those perceptual activities. For actions, it utilizes all of the moveable parts of the body and the brain regions that carry out those bodily movements. However, even some who grant the obvious fact that we need our bodies to execute perception and movement still hold onto a mind/body dualism when they insist that perception and action are carried out via different functional regions of the brain than those utilized for *conceptualizing* perception and actions. They reject the idea that our perceptual and action *concepts* operate by means of the very same body and brain architectures that give rise to those perceptions and actions.

From an *embodied cognition* perspective, so-called 'higher' cognitive functions, such as conceptualization and reasoning, recruit structures and processes of sensory and motor operations. In evolutionary theory, the technical name for this is *exaptation* – the appropriation of brain architectures evolved for one type of functional process (e.g. seeing, touching and hearing) to perform a different kind of functional process (e.g. knowing, understanding, reasoning). Most of our abstract thinking, involving discriminations and integration of selected aspects of a situation, appropriates this body-based meaning from our sensory, motor, affective and interpersonal interactions. In *Metaphors We Live By* ([1980] 2003) and later in *Philosophy in the Flesh: The Embodied Mind and Its Challenge to Western Thought* (1999), Lakoff and I analysed *cross-domain mappings* of objects, properties and relations from a sensory and motor source domain to a more abstract target domain. We called these experientially based cross-domain mappings *conceptual metaphors*. For example, we analysed five major conceptual metaphors for thinking, each of which has a source domain specified for a particular type of bodily activity: perceiving, moving, manipulating objects, reproducing and eating. Take, for instance, the KNOWING IS SEEING metaphor, in which there is a concrete perceptual source domain of vision, which consists of structures like: person who sees, act of seeing, object seen, ambient lighting conditions and visual acuity. Each of these salient aspects of visual experience has a counterpart in our conceptualization of thinking-as-seeing: the person who sees maps onto the thinker, the object seen maps to the idea thought, the act of seeing maps to the act of understanding or knowing, the ambient light maps to conditions enabling knowing and visual acuity maps to intellectual ability.

In the early 1980s, when Conceptual Metaphor Theory first emerged, the primary source of evidence was linguistic. One type of linguistic evidence was polysemy generalizations, where the cross-domain mapping was used to explain why a term with a certain literal meaning (e.g. *see*, as in 'I see the forest') can also have a related meaning, via the metaphorical cross-domain mapping, in an abstract domain (e.g. 'I *see* what you mean' and 'I can't *see* the forest for the trees'). There was also evidence from inference generalizations, such as when we use an inference structure from a source domain to make inferences in the target domain (as in, 'If some object blocks your vision, you can't see what you are looking at clearly' [source domain inference] so via the *source-to-target mapping*, 'If something *obscures* an idea, you can't understand the idea clearly' [target domain inference]).

In an effort to counteract the tendency to regard metaphor as solely a linguistic phenomenon, Lakoff and I sought non-linguistic evidence (such as visual metaphor, psychological priming evidence and neural evidence) that the metaphors were both conceptual (not just linguistic) and body-based (rooted in bodily processes). Studies began to show that conceptual metaphor was pervasive in non-linguistic activities such as spontaneous gesture, music, theatre, advertising, architecture and ritual (Gibbs 2006; Johnson 2007; Kovecses 2009).

One major development supporting the bodily basis of metaphor was the theory of primary metaphor, in which Joseph Edward Grady (1997) identified a large number of basic metaphors rooted in recurring sensory–motor experiences that establish experiential correlations across different domains. To stay with my previous example, children and adults have recurring experiential correlations between seeing an object and thereby gaining understanding or knowledge of that object. You 'learn' these body-based experiential correlations just by living in the world, with the type of bodies and capacities you have, and the stable patterns you frequently encounter in daily life. In short, repeated correlations across two different domains (source and target) establish the basis for thinking of one thing (understanding) by means of a certain perceptual activity (seeing). How the source-to-target directionality of conceptual metaphors is determined requires a complex neuroscience account (Lakoff and Narayanan forthcoming; Johnson and Tucker 2021).

Over the last three decades, there is a large and mushrooming literature on how conceptual metaphors rooted in bodily and social experience are found the world over (Dancygier and Sweetser 2014; Kovecses 2009). Languages don't all necessarily share the same set of metaphors, though many metaphors appear to be nearly universal; however, there are often differences in how a particular conceptual metaphor gets elaborated in different cultures. In short, there are relatively universal source domains (with image-schematic structures and emotional dimensions) that are used around the world as the basis for common conceptual metaphors, but it is an empirical question whether there are any universal metaphors, and whether apparently similar metaphors are actually interpreted and expanded differently in different cultures. What is clearly universal (as far as we know) is that all cultures seem to use conceptual metaphors for abstract conceptualization and reasoning.

The Emergence of Language and Mind

Our survey of bodily sources for pre-linguistic meaning has already slipped over into the appearance of such bodily patterns in language proper – the use of verbal and written signs. As just noted, there is overwhelming evidence for the use of conceptual metaphor in diverse language groups around the world. However, a key claim is that the image schemas and conceptual metaphors are not, in their first appearance, dependent upon language, even though they very quickly get drawn up into the syntax and semantics of natural languages. These meaning structures and processes are embodied, and, therefore, as they work themselves up into language, linguistic meaning is embodied, too.

My hypothesis is that although there is extensive very significant pre-linguistic meaning that does not rely on language, the emergence of language dramatically enriches meaning and generates abstract thinking. So, while I submit that it is wrong to say that all meaning and thought are linguistically mediated, it is correct to say that the emergence of language is a monumental step forward in our ability to think abstractly and share meanings via cooperative communicative activities.

Because humans are qualitative creatures, it makes sense that meaning, in the first instance, is rooted in qualitatively unified situations. We begin with a grasp of the *sense* of a complex unified situation via systems of meaning that have emerged from our somatic interfacing with our surroundings, via perception and action. It is within the meaning of a complex situation (as sense) that we are then able to use signs as means for exploring the meaning of things, by focusing on their generating causes and their implications for future possible experiences. This second process, rooted in language, is what Dewey called *signification*. The meaning of any object or event is thus the experiences (past, present and future) it affords us.

In the twentieth century, there was a widespread tendency to regard all meaning and thought as mediated by language. Noam Chomsky's generative linguistics postulated a universal language (syntax) generating capacity that was not essentially dependent on the details of human embodiment. In analytic philosophy of mind and language, meaning was taken to be conceptual and propositional, with the primary unit of meaning being the proposition (expressed in a sentence). According to this view, it is language that distinguishes humans from other animals, and language is the basis for shared meaning and abstract thought. The key role of language in meaning and thought raises the question of whether sophisticated computer languages might one day equal, or even surpass, human linguistic abilities to process meanings and perform certain complex cognitive tasks. In some areas, they already have.

In contrast with this primacy-of-the-linguistic view, we have seen that language emerges and draws its life from a massive continent of pre-linguistic, non-propositional, body-based meaning that arises from our sensory, motor and affective processes. One can appreciate the wonders of human language without reducing all meaning and thought to linguistic processes. Language piggy-backs on embodied meaning structures and processes of the sort described earlier. When that happens, more abstract thought emerges, opening up new, and more complex, ways for making sense of experience. However, those who are enamored of the remarkable achievements made possible by language, too often fail to appreciate the crucial role of embodied meaning which precedes and makes possible linguistic acts. There is an unfortunate tendency to make language acquisition a rupture from pre-linguistic capacities and practices, rather than a continuation, enrichment and expansion of those body and brain-based syntactic, semantic and pragmatic activities.

One important implication of this embodied view of meaning, thought and language is that the evidence for the body-based meaning processes previously discussed extends far beyond language, to include several types of non-linguistic evidence, such as spontaneous gestures, ritual acts, psychological experiments and neuroscience research, music, theatre performance, painting, sculpture and architecture (Lakoff and Johnson 1999; Forceville and Urios-Aparisi 2009). That languages around the world should have evolved ways of representing the kinds of embodied structures of meaning and thought we have examined (plus a multitude of other syntactic and semantic patterns) is to be expected, since human perception, action and feeling manifest those very structures and relations. To recall our previous example, since there are basic structural dimensions of animal locomotion (described above as the LOCOMOTION schema), it is to be expected that most languages would have conceptual frames that include roles for *agent who acts, body part(s) used, energy exerted, objects acted upon, manner of action* and *resulting state of affairs*.

Embodied cognition theory provides numerous tools for explaining the details of languages around the world. Since it grounds meaning in pre-linguistic sensory, motor and affective experience, it has a way to explain the semantics, polysemy, inferential structure, syntactic structure and pragmatic practices that make up natural languages, as rooted in bodily processes (Feldman 2006). It explains how children can begin to learn the meaning of things they experience and do, even prior to language acquisition. Although I have not argued this here, we are beginning to see neural theories of a number of processes that underlie language (Feldman 2006; Lakoff and Narayanan forthcoming). They reveal how human brains and bodies grow, from infancy to adulthood, in an ongoing process of neural development that is learning (Johnson and Tucker 2021).

So, what is it that allows us to become aware of these embodied learning processes? We tend to say that it is our 'mind' that feels, perceives, understands, knows, judges, chooses and acts. If this mind is not disembodied, but rather profoundly embodied, then where, and how, does it exist and do what it does? What we call individual minds are relatively stable, but always developing patterns of neural activation resulting from our ongoing bodily engagements with our surroundings (physical, interpersonal and cultural). We tend to think of minds as individuated loci of 'inner' perceptions, thoughts and feelings. Dewey ([1925] 1981), however, argued that we should think of mind less individualistically and less dualistically, as continuously developing systems of shared meanings that result from conjoint cooperative communication. He then argued that each individual self gets its identity from the way it participates in cultural mind, appropriating the meanings and values developed through communicative interactions with other people. His point was that, without the systems of shared meaning that make up cultural communities, we, as individuals, would never have access to the wealth of meaning that is available to us.

Dewey's notion of mind as a communal, growing system of shared (and always developing) meanings reminds us that there is more to the meaning of things and events than what is captured in our puny individual experience. Meaning doesn't just arise from our bodily interactions with our physical surroundings but is made possible by our ability to share our intentions, experiences, thoughts and values with others in communicative practices that, while not being only linguistic, still rely heavily on shared *sense* and shared signification, the latter of which is very much dependent on language. Dewey thought that mind in this cultural sense is dependent on the emergence of language that gives us shared meanings, but he recognized that language is more than just the use of linguistic signs, since it includes 'not only gestures but rites, ceremonies, monuments and the products of industrial and fine arts' (Dewey [1925] 1981: 52). He clearly saw that there are many practices of meaning-making that are not linguistic in the narrow sense. Moreover, his notion of *sense*, as immanent meaning that is felt, is clearly non-linguistic, even as it gives rise to linguistic meaning.

Once we appreciate that meaning is embodied and extends beyond language, and once we recognize the role of qualities, feelings and emotions in meaning, we gain the ability to explain meaning in painting, architecture, music, theatre, performance, spontaneous gesture and ritual practice, and other forms of non-linguistic communicative activity (Johnson 2007). Because most of what goes on there is not conceptual and propositional meaning, it is not tied intrinsically to language in the narrow (words and sentences) sense of that term. This makes it possible for us to explain domains of meaning that too often get labelled as non-cognitive and shunted aside. We are then able to appreciate why certain mainstream views of meaning, thought and language have no resources for explaining how meaning works in these many and varied forms of symbolic interactions.

The Primacy of the Aesthetic

I want to suggest that all the processes of meaning and thought I've been discussing should be recognized as aesthetic dimensions of experience. Aesthetic theory has been profoundly shaped by the eighteenth-century view that experience can be broken down into distinct kinds (e.g. moral, political, scientific and aesthetic) and then types of judgments appropriate to those kinds of experience could supposedly be explained by the distinctive contributions of various cognitive faculties (e.g. perception, imagination, conceptualization, understanding, reason and will). According to this approach, aesthetic judgments are typically characterized as based on feelings, rather than concepts. Immanuel Kant set the stage for much subsequent aesthetic theory by asking how aesthetic judgments based on *feelings*, rather than concepts, could nevertheless claim universal validity. This perspective restricted aesthetics to a small corner of life concerned with beauty and sublimity in nature and art, which were distinguished and separated from science, morals, politics and technology. The result was a conception of aesthetics concerned primarily with formal characteristics of objects, engaged via what was called a 'disinterested judgment' and thereby divorced from the moral and practical concerns of daily life of the human condition. There are, to be sure, other historically significant theories of art and aesthetics over the last two centuries, but it was Kant who had the greatest influence in shaping subsequent theory, well into the twentieth century.

Hans-Georg Gadamer ([1960] 1989) attributed this formalist and disinterested-attitude view to Kant's 'subjectivizing' of aesthetics, which effectively removed art from the domain of cognitive meaning, and also from our practical moral and political concerns, thereby marginalizing its value for everyday life. Even worse, the assumption that all meaning is conceptual and all thought is propositional and linguistic marginalized aesthetic considerations even further. A good example of this tendency is the oft-encountered modelling of musical meaning on linguistic structures and meaning. According to the cognitivist view of meaning, we ask, 'what is it that has meaning', and the answer given is 'words and sentences'. Therefore, if there is such a thing as musical meaning, there must be structures in music analogous to words, phrases and sentences in language. Music must be a language. So, music theorists look around in a musical work for sentence-like components, which leads them to observe that we have terms like *musical ideas*, *musical questions*, *musical phrases* and *musical propositions*. Aha!, they proclaim, music *does* communicate meaning. Unfortunately, approaching musical meaning in this way captures almost nothing of the power of music or the ways it can be profoundly meaningful in our lives. Adopting such an etiolated and thin account of meaning leads us to overlook the types of embodied meaning discussed here. It robs us of the broad and rich view of meaning that we need if we hope to explain the whole range of non-linguistic meaning processes. The result is that meaning in the arts gets judged against linguistic meaning, and comes out lacking in cognitive heft.

My concern is not just with criticizing certain reductive aesthetic theories, but, more importantly, to suggest a more expansive and rich notion of aesthetics, as concerned with all of the structures and processes that go into human meaning-making. According to this conception of aesthetics, experience doesn't come demarcated into distinctive types and modes of judgment. Rather, all meaningful experience is 'aesthetic', insofar as it is shaped by the qualitative and affective dimensions we traditionally associate with the arts. Beneath and beyond conceptual and propositional structures, meaning becomes manifest through pervasive unifying qualities, particular qualities, image schemas, affect contours, patterns of feeling and emotions. This is not a willful or cavalier redefining of aesthetics, but rather an appreciation of the fact that elements typically thought of as 'merely' aesthetic, are, as we have seen, essential aspects of meaning. Consequently, aesthetics pervades and underlies *all* meaningful experience (Johnson 2007, 2018). Note that I do not say 'all' experience *simpliciter*, because some 'experience' is disorganized, loose, incoherent and undeveloped, and so is not meaningful in any significant sense.

Art, on this view, then, is an intensification and consummation of the possibilities for a meaningful experience. Art is the exemplary instance of human meaning-making, and the arts – mislabeled as practical vs. fine, since all art is practical – are seen to be pervasive in daily experience. In Dewey's terms, meaningful experience is 'art in germ' (Dewey [1934] 1987: 25).

So, following Dewey's lead, I am arguing for an aesthetics of embodied life based on the recognition of the aesthetic dimensions of all the processes that give rise to cognition and meaning. Attention to the aesthetic constituents and processes of daily life allows us to explain how both ordinary experience and also the exemplary instances of experience we call art, are meaningful for our lives.

The Values of Life – From Aesthetics to Morality

I began this essay by observing our primal animate nature as living, moving, valuing creatures. Homeostasis and allostasis are processes shaped by the fundamental biological and interpersonal values of human organisms. The fundamental conditions for life maintenance are values we have, unconsciously, automatically, without reflection and mostly beyond our control. Our interpersonal, social nature establishes another plane of values spawned by our intimate and communal engagement with other people. So, it's values – all the way down, and all the way up – from biological survival and well-being to social nurturance and care, to large-scale rights and obligations towards others.

Aesthetic elements are rooted in the value-based functioning of the organism (Johnson and Schulkin 2020; Johnson and Tucker 2021). They are not 'aesthetically disinterested', but matters profoundly relevant to our viability and well-being as living creatures. 'Better' and 'worse' enactments of meaning in so-called 'aesthetic' experience merge seamlessly with 'better' and 'worse' in morality, politics and all other value-based activities. Morals concern issues of better and worse with respect to conduct related to survival, well-being and flourishing – not just of individual agents, but of larger social groups, and even the encompassing environment. I am therefore suggesting that morality is not an abstract intellectual affair, but instead is rooted in the aesthetics of everyday experience and meaning, as developed here.

In *Morality for Humans* (2014, also see 2018), I have provided a Deweyan version, supported by contemporary neuroscience and moral psychology, of an aesthetics-based account of morality and moral deliberation. Without going into the numerous details of this aesthetics of morals, we can note two important components of such a view that give a sense of the overall project. In *Human Nature and Conduct* ([1922] 1983), Dewey opens with an account of how we are creatures of habit. Given the evolutionary establishment of our current brain and body architectures, each individual's life-long developmental shapes the patterns of neural connectivity that come to define our selfhood (Johnson and Tucker 2021; Johnson and Schulkin 2023). Simply put, through our recurring engagements with relatively stable aspects of our surroundings, we develop habits of perception, valuing, thinking and acting. A person's character is thus what Dewey ([1922] 1983) called an 'interpenetration of habits' (29). These habits and values are established in relation to the conditions for life maintenance, well-being and flourishing. Cognition is learning, carried on over the lifetime of the organism, and learning is shaped by motivational systems within the brain and body. These needs, goods and values that we have guide our perception, action and thought. Consequently, a person's moral sensibility is a complex interweaving of the values arising from the different levels or planes of their lives (biological, interpersonal and cultural). Values don't drop from the sky; rather, they emerge in our biological and social growth and functioning.

However, if all we are is bundles of habits (as David Hume suggested in his 1739–40 *Treatise on Human Nature*), then our morality would be nothing but habit-driven reproduction of values impressed on us by our biological and social situation. Dewey ([1932] 1985, [1939] 1988, [1942–48] 1989; see also Dewey and Bentley [1949] 2008) gave the name *valuing* to the mere fact that we are preferentially directed to certain objects, events or states of affairs. *Valuation*, by contrast, is our human ability to reflectively critique our received values, in light of newly emerging conditions. Valuation brings a critical perspective to bear, which was not present in mere valuing.

But what is the basis for critical appraisal? If there is no 'pure' reason, since all reasoning is motivated, value-based and rooted in emotions, then how is any critical assessment ever possible? Dewey's answer was that we are capable of 'dramatic rehearsal (in imagination)' ([1922] 1988: 132), in which we imaginatively explore how a situation would develop under the influence of different values, ends and goods. This consideration and anticipation of possible experiences is the same structural process we earlier described as the projection of expectancies for experiences that are then tested at the sensory and motor interface with our surroundings. A moral problem arises when our previous habits, formed under prior conditions, run up against new emerging conditions, for which our habitually given values are inadequate. We then have to project possible experiential scenarios, each the result of imaginatively adopting a certain set of values. The 'better' option in such inquiries is the imagined option that best reduces the indeterminacy, conflict and anxiety created by new problematic conditions. The 'better' option is recognized when it changes the pervasive unifying quality of the problematic situation into one that enhances, expands and deepens meaning, and, thereby, relative freedom of choice. It is in this sense that moral deliberation and judgment are quite analogous to aesthetic deliberation and judgment. Determining the 'better' course of action is not a matter of calculation of possible good nor is it a deduction from absolute rational principles. It is an ongoing process of exploring implications of various values, virtues and goods for growth of meaning, liberation of choice and enhanced well-being.

The problem of moral judgment, therefore, is not a matter of explaining where our moral values come from (e.g. pure reason, habit, intuition, God/gods, society and cultural institutions). We have plenty of explanations. The real problem is what we can and should do when our values conflict, either with other values we hold or with other people's values. Then the problem becomes how to harmonize values, reduce conflict and enhance meaning. Humans have a plurality of values, and the key is to develop virtues of inquiry that help us intelligently resolve conflicts among those values. Dewey calls this capacity for moral inquiry *conscientiousness* ([1922] 1988: 144). Instead of *conscience*, as a supposed source of absolute moral values and principles, what humans need most is *conscientious critical inquiry*, grounded in imaginative dramatic rehearsal. We need to cultivate intelligence – the wisdom to live rightly and well – and this, as we have seen, is a profoundly aesthetic endeavor.

From Life to the Art of Living

This chapter is an argument for a blending of science, art and the humanities, as the way to develop the intelligence we need for a critical, reflective life. The old quarrel between the sciences and the humanities is now seen to be predicated on a false methodological dualism. The traditional view is that physical nature is a deterministic system of causes and effects so that the scientific method must generate causal explanations of perceived phenomena. In contrast, since humans supposedly possess a measure of freedom, the humanities and arts require an entirely different method geared towards the interpretation of what certain objects and events mean. The result is C. P. Snow's ([1959] 1965) *The Two Cultures* – the natural sciences vs. the humanities – each claiming primacy for its distinctive method and each accusing the other of missing important factors. The sciences tout their objectivity, testability and rigor, which they accuse humanistic methods of lacking, while the humanities criticize the sciences for failure to capture the existential meaning of objects and events, and also for ignoring the aesthetic dimensions of meaning and thought.

The approach I am proposing here is an attempt to meld the sciences, humanities and arts into a new model of inquiry that appreciates the need for both criticism and testability of hypotheses but also recognizes that the human meaning of events requires not just causal explanation, but a rich and nuanced narrative understanding of the drama of life. Our best strategy is to integrate multiple approaches and then look for converging evidence. The future of research into life, meaning, thought, language, knowledge and communication lies in transdisciplinary cooperation. It requires the creation of an *artful science*, capable of both rigorous testing and humanly meaningful generalizations, and also the creation of a *scientific art*, an empirically responsible humanism that incorporates appropriate science into its understanding of the meaning and significance of objects and events. This approach also emphasizes the cultivation of moral perception and sensitivity via aesthetic bodily, affective and cognitive experiences and practices, which are arts of living.

This orientation might be thought of as an emerging third-generation cognitive science. Calling first- and second-generation cognitive science *inter*disciplinary tends to emphasize how the various disciplines both criticize and enrich one another, whereas a *trans*disciplinary orientation seeks to place all of these voices in creative synthesis, especially with an eye towards the existential, moral, political and aesthetic concerns of our lives. (For example, see Slingerland 2008; Slingerland and Collard 2011; Caruso and Flanagan 2017; Gallagher 2020.) Thus, this perspective stresses not just intellectual cross-disciplinary dialogue, but also includes modes of intelligent engagement and growth designed to cultivate values for living in a connected world facing profound issues of survival, sustainability and continued well-being.

We began with the emergence of conditions for life maintenance within an organism and eventually moved all the way to the consummatory experience of art and morality. In *The Human Eros: Eco-ontology and the Aesthetics of Existence*, Thomas Alexander (2013) suggests that our human Eros is our desire for meaning. The key point is that humans are meaning-making and meaning-seeking creatures. If there is an 'art of living', it would be the cultivation of *intelligence*, including the capacity for critical valuation, dramatic imaginative rehearsal and the appreciation of the consummatory possibilities of experience. Intelligence thus provides both means for criticizing one's (and others') habits, practices and values, and it also reveals new meanings and values that emerge from changing conditions. Yes, we first want to secure the conditions for our organismic viability in the world, but an essential part of this process is learning the meaning of things and events that helps us act intelligently in the world. An 'aesthetics of meaning and thought' (Johnson 2018; Johnson and Schulkin 2023) is the basis for an artful way of living.

NOTE

1. For an abridged account of these intertwining developments, see Lakoff and Johnson (1999), Johnson and Tucker (2021). For detailed and more nuanced histories of cognitive science, see Bechtel and Graham (1998).

REFERENCES

Alexander, Thomas (2013), *The Human Eros: Eco-ontology and the Aesthetics of Existence*, New York: Fordham University Press.

Bechtel, William and Graham, George (eds) (1998), *A Companion to Cognitive Science*, Oxford: Basil Blackwell.

Caruso, Gregg and Flanagan, Owen (eds) (2017), *Neuroexistentialism: Meaning, Morals, and Purpose in the Age of Neuroscience*, Oxford: Oxford University Press.

Damasio, Antonio (1994), *Descartes' Error: Emotion, Reason, and the Human Brain*, New York: G. P. Putnam's Sons.

——— (1999), *The Feeling of What Happens: Body and Emotion in the Making of Consciousness*, Boston, MA: Houghton Mifflin Harcourt.

——— (2010), *Self Comes to Mind: Constructing the Conscious Brain*, New York: Vintage Books.

Dancygier, Barbara and Sweetser, Eve (2014), *Figurative Language*, Cambridge: Cambridge University Press.

Dewey, John ([1925] 1981), *Experience and Nature*, in J. A. Boydston (ed.), *The Later Works Vol. 1, 1925–1953*, Carbondale, IL: Southern Illinois University Press.

——— ([1930] 1984), 'Qualitative thought', in J. A. Boydston (ed.), *The Later Works Vol. 5, 1925–1953*, Carbondale, IL: Southern Illinois University Press.

——— ([1932] 1985), *Ethics*, in J. A. Boydston (ed.), *The Later Works Vol. 7, 1925–1953*, Carbondale, IL: Southern Illinois University Press.

——— ([1934] 1987), *Art as Experience*, in J. A. Boydston (ed.), *The Later Works Vol. 10, 1925–1953*, Carbondale, IL: Southern Illinois University Press.

——— ([1922] 1988), *Human Nature and Conduct*, in J. A. Boydston (ed.), *The Middle Works Vol. 14, 1899–1924*, Carbondale, IL: Southern Illinois University Press.

——— ([1939] 1988), *Theory of Valuation*, in J. A. Boydston (ed.), *The Later Works Vol. 13, 1925–1953*, Carbondale, IL: Southern Illinois University Press.

——— ([1942–48] 1989), *Essays, Reviews, and Miscellany*, in J. A. Boydston (ed.), *The Later Works Vol. 15, 1925–1953*, Carbondale, IL: Southern Illinois University Press.

Dewey, John and Bentley, Arthur ([1949] 2008), 'Importance, significance, and meaning', in J. A. Boydston (ed.) ([1949–52] 1989), *Essays, Typescripts, and Knowing and the Known, The Later Works Vol. 16, 1925–1953*, Carbondale, IL: Southern Illinois University Press [unpublished typescript].

Dodge, Ellen and Lakoff, George (2005), 'Image schemas: From linguistic analysis to neural grounding', in B. Hampe (ed.), *From Perception to Meaning: Image Schemas in Cognitive Linguistics*, Berlin: Mouton de Gruyter, pp. 57–91.

Feldman, Jerome (2006), *From Molecule to Metaphor: A Neural Theory of Language*, Cambridge, MA: Bradford Books.

Forceville, Charles and Urios-Aparisi, Eduardo (eds) (2009), *Multimodal Metaphor*, Berlin: Mouton de Gruyter.

Gadamer, Hans-Georg ([1960] 1989), *Truth and Method*, 2nd rev. ed. (trans. J. Weinsheimer and D. G. Marshall), New York: Crossroad Publishing Company.

Gallagher, Shaun (2005), *How the Body Shapes the Mind*, Oxford: Oxford University Press.

——— (2020), *Action and Interaction*, Oxford: Oxford University Press.

Gibbs Jr., Raymond W. (2006), *Embodiment and Cognitive Science*, Cambridge: Cambridge University Press.

Gibson, James (1979), *The Ecological Approach to Visual Perception*, Boston, MA: Houghton Mifflin.

Grady, Joseph Edward (1997), 'Foundations of meaning: Primary metaphors and primary scenes', Ph.D. dissertation, Berkeley, CA: University of California.

James, William (1890), *The Principles of Psychology*, 2 vols, New York: Dover.

Johnson, Mark (1987), *The Body in the Mind: The Bodily Basis of Meaning, Imagination, and Reason*, Chicago: University of Chicago Press.

——— (2005), 'The philosophical significance of image schemas', in B. Hampe and J. E. Grady (eds), *From Perception to Meaning: Image Schemas in Cognitive Linguistics*, vol. 29, Berlin: Mouton de Gruyter, pp. 15–33.

——— (2007), *The Meaning of the Body: Aesthetics of Human Understanding*, Chicago: University of Chicago Press.

——— (2014), *Morality for Humans: Ethical Understanding from the Perspective of Cognitive Science*, Chicago: University of Chicago Press.

——— (2017), *Embodied Mind, Meaning, and Reason: How Our Bodies Give Rise to Understanding*, Chicago: University of Chicago Press.

——— (2018), *The Aesthetics of Meaning and Thought: The Bodily Roots of Philosophy, Science, Morality, and Art*, Chicago: University of Chicago Press.

Johnson, Mark L. and Schulkin, Jay (2020), 'The common currency of aesthetic sensibility', *Transactions of the Charles S. Peirce Society*, 56:3, pp. 326–48.

——— (2023), *Mind in Nature: John Dewey, Cognitive Science, and a Naturalistic Philosophy for Living*, Cambridge, MA: MIT Press.

Johnson, Mark L. and Tucker, Don M. (2021), *Out of the Cave: A Natural Philosophy of Mind and Knowing*, Cambridge, MA: MIT Press.

Kovecses, Zoltan (2009), *Metaphor: A Practical Introduction*, Cambridge: Cambridge University Press.

Lakoff, George (1987), *Fire, Women, and Dangerous Things: What Categories Reveal about the Mind*, Chicago: University of Chicago Press.

Lakoff, George and Johnson, Mark ([1980] 2003), *Metaphors We Live By, with a New Afterword*, Chicago: University of Chicago Press.

——— (1999), *Philosophy in the Flesh: The Embodied Mind and Its Challenge to Western Thought*, New York: Basic Books.

——— (2003), 'Why cognitive linguistics require embodied realism', *Cognitive Linguistics*, 13:3, pp. 245–64.

Lakoff, George and Narayanan, Srini (forthcoming), *The Neural Mind: How Brains Think*.

Langacker, Ronald ([1987] 1991), *Foundations of Cognitive Grammar*, 2 vols, Stanford, CA: Stanford University Press.

Merleau-Ponty, Maurice (1962), *Phenomenology of Perception* (trans. C. Smith), New York: Humanities Press and London: Routledge & Kegan Paul. [First published in 1945 as Phénoménologie de la perception, Paris: Éditions Gallimard.]

Newen, Albert, Leon, De Bruin, and Gallagher, Shaun (2018), *The Oxford Handbook of 4E Cognition*, Oxford: Oxford University Press.

Pappas, Gregory (2016), 'John Dewey's radical logic: The function of the qualitative in thinking', *Transactions of the C. S. Peirce Society*, 52:3, pp. 435–68.

Panksepp, Jaak (1998), *Affective Neuroscience: The Foundations of Animal and Human Emotions*, New York: Oxford University Press.

Ricoeur, Paul (1992), *Oneself as Another* (trans. K. Blamey), Chicago: University of Chicago Press.

Schulkin, Jay and Sterling, Peter (2019), 'Allostasis: A brain-centered predictive mode of physiological regulation', *Trends in Cognitive Science*, 42:10, pp. 740–52.

Sheets-Johnstone, Maxine (1999), *The Primacy of Movement*, Amsterdam: John Benjamins Publishing Co.

Slingerland, Edward (2008), *What Science Offers the Humanities: Integrating Body and Culture*, Oxford: Oxford University Press.

Slingerland, Edward and Collard, Mark (eds) (2011), *Creating Consilience: Integrating the Sciences and the Humanities*, Oxford: Oxford University Press.

Snow, C. P. ([1959] 1965), *The Two Cultures: And a Second Look*, Cambridge: Cambridge University Press.

Stern, Daniel (2018), *The Interpersonal World of the Infant: A View from Psychoanalysis and Developmental Psychology*, New York: Routledge.

Talmy, Leonard (1985), 'Force dynamics in language and thought', *Chicago Linguistics Society*, 21, pp. 293–337.

——— (2000), *Toward a Cognitive Semantics*, 2 vols, Cambridge, MA: MIT Press.

Tucker, Don M. (2007), *Mind from Body: Experience from Neural Structure*, Oxford: Oxford University Press.

Tucker, Don M. and Luu, Phan (2012), *Cognition and Neural Development*, Oxford: Oxford University Press.

Varela, Francisco J., Thompson, Evan and Rosch, Eleanor ([1991] 2016), *The Embodied Mind: Cognitive Science and Human Experience, Revised Edition*, Cambridge, MA: MIT Press.

8

Metabolism and Drift

Thomas Nail

Metabolism is the process by which organisms absorb, transform and eliminate energy. In the nineteenth century, most chemists and biologists thought of metabolism in this strictly organic sense referring to individual living creatures. However, it may surprise many that the first thinker to explicitly expand the concept of metabolism to non-living, ecological and social processes was the German philosopher, Karl Marx. This chapter shows how he did this, why he did it and the importance of this conceptual decision today.

Although historically under-appreciated by most Marxists, the significance of the idea of metabolism in Marx's work is now well-established in *ecomarxism*.[1] I will not survey this literature here but instead focus on two original contributions that highlight what I think is most important about the idea of metabolism for thinking about living and non-living systems.

First, I want to show that Marx's theory of metabolism does not appear suddenly or only in his 'mature' philosophy (Foster 2000: ix). Marx first introduced the term 'metabolism' (*Stoffwechsel*) in the *Grundrisse* (1857). However, *the* idea of a 'continuous process of nature's intra-action with itself' (MECW 3: 276) was already present in his 1841 doctoral dissertation and his 1844 manuscripts in his concepts of *zusammenhängen*, 'hanging together', *verkehren*, 'interchange' and *wechseln*, 'transformation'. Therefore, I argue that when Marx learned of the German word for metabolism, it was just one name among others for him to describe his materialist philosophy of movement and continual transformation. It was never limited to organic life for him.

Second, I want to show that Marx's theory of metabolism was not a theory about the *interaction between* humans and nature or about their *separation* by so-called metabolic 'rifts'.[2] Rather, I argue here that Marx's theory of metabolism is better understood as a process of 'metabolic drifts' or uneven/unsustainable distributions of energy.[3] Humans and nature alike can break and mend metabolic patterns of consumption, transformation and elimination. The greatest danger is pretending that there is no metabolism or that its effects are trivial compared to the circulation of economic value.

The chapter proceeds in three parts: The first shows the continuity of the concept of metabolism with other ideas of entangled transformation in Marx's work from his dissertation leading up to *Capital*. The second part describes Marx's three kinds of metabolism, and the third part describes the concept of *metabolic drift*. Finally, I conclude with some brief thoughts on how to better manage metabolic drift.

The Movement of Metabolism

The idea of metabolism was certainly connected in Marx's mind to Aristotle's use of the Greek word μεταβολή, *metabolḗ*, or change (Liddell and Scott 1940b). The term has its modern origins in the work of the Venetian physician Sanctorius (1561–1635), who described metabolism as a 'perspiratio insensibilis' ('insensible respiration'); to signify 'the realization that the living organism is undergoing continuous changes' (Bing 1971: 158–59).

However, the modern German word for metabolism, *Stoffwechsel*, did not emerge until 1815 and was not used widely until the 1830s and 1840s. One of the first to extensively use this new term was the German medical doctor, scientist and communist, Roland Daniels (1819–55), whose extraordinary book on metabolism was called *Mikrokosmos*, completed in 1850 but left unpublished due to his untimely death. However, before he passed, Daniels showed his manuscript to at least one other person: his dear friend, Karl Marx. In his manuscript, Daniels made extensive use of the concept of *Stoffwechsel* to discuss plant and animal systems, which profoundly influenced Marx, who even dedicated his book, *The Poverty of Philosophy*, to Daniels.[4] Although Marx cites the term metabolism in Justus von Liebig's *Organic Chemistry in its Application to Physiology and Pathology*, published in 1842, Daniels was his first source of influence (CIII: 949).

If we read Marx's earliest writings, it's easy to see why he fell in love with the concept of metabolism so quickly. For him, it was a way of thinking of the thoroughly entangled interconnection of natural and social patterns of motion. It was a major conceptual synthesis that united humans and nature in feedback loops of production and consumption. Humans, for Marx, are part of nature, and their social patterns take place within natural patterns. Most importantly, metabolic patterns are ones of constant transformation. The German word *Stoffwechsel* means 'material transformation' – *Stoff* means 'matter' plus *Wechsel*, which means continuous change. Its German and Greek etymology is not necessarily limited to life, and Marx knew this.

In Marx's eyes, the term *Stoffwechsel* described quite well the material and kinetic processes of flowing, cycling and circulating that Marx had been writing about since his 1841 dissertation. Marx might have even known about the idea of metabolism before Daniels and Liebig. It may have got him started thinking about things as material-kinetic processes that 'hang together', *zusammenhängen*.

Even the Greek origins of metabolism carry with them an explicitly kinetic and processual character. The Ancient Greek word μεταβολή (*metabolḗ*, 'change') comes from μετά (*metá*, 'between' or 'in the midst of') + βάλλω (*bállō*, 'I throw'). The Greek meaning here is not just 'change' but 'thrown' into a continual process of movement and intra-action within and between things (Liddell and Scott 1940c, b, a). In the Greek sense, metabolism was not unique to living organisms and was not just a change from one thing into another. It was a transformation of the whole system of nature.

When Marx wrote in *Capital* of the 'metabolism [*Stoffwechsel*] between man and nature', it echoed the description he gave in his dissertation of how form and matter continually emerge, stabilize and pass away, just as the Roman poet Lucretius described (see Nail 2020a: Chapter 1). Matter and form are two dimensions of the same dialectical process. In other words, following the insights of Marx's doctoral dissertation, which I have written about elsewhere, we should not read metabolism in *Capital* as an interaction between *separate* things such as humans and nature (Nail 2020a). This would completely undermine the dialectical and processual nature of the strictly *twofold* nature of matter, form and sensation that had guided Marx's thought up to *Capital*. In short, metabolism was the twofold movement of matter into and out of various forms. It was a *trans*-formation.

Again, we can hear echoes of metabolism in Marx's 1844 manuscripts when he writes of the 'continuous process of nature's intra-action with itself [*zusammenhängen*]' (MECW 3: 276). Thus, it is no coincidence that whenever Marx begins talking about metabolism and 'metabolic tears [*Riss*]' in the *Capital* volumes, he is almost always talking about tears *in something that hangs together* (*zusammenhängen*).[5] This is because metabolism is conceptually similar to his older and more general concept of 'intra-active hanging-together'.

This is not conjecture. Marx says almost the same thing in the line quoted above from *Capital* in the manuscripts.[6] The only difference is that in his 1844 manuscripts, he did not yet use the term *Stoffwechsel* and had to find creative ways to talk about processes of mutual interdependent transformation. By *Capital*, however, he had seen the term metabolism and used it to significant effect in various places to describe this same process of mutual change and hanging together.

Before *Capital*, though, in the *German Ideology*, Marx relied heavily on *verkehren*, or interchange. In the *Grundrisse*, the rough draft of *Capital*, he used the terms *verkehren*, *zusammenhängen*, *Stoffwechsel* and *Wechselwirkung*. The German word, *Wechselwirkung* has the word *wechsel* in it, like *Stoffwechsel*.

Instead of emphasizing changing matter (*Stoff*), though, *Wechselwirkung* emphasizes that actions or effects (*Wirkung*) fold back into one another in a kind of immanent feedback loop or continual transformation of the whole (MEGA² I.5: 126).[7]

We can also find in the *Grundrisse* a whole theory of the *zusammenhängen* of production and consumption: the 'individual produces an object and, by consuming it, returns to himself, but returns as a productive and self-reproducing individual' (G: 94). Here, Marx was already thinking of labour as a natural and social metabolic process. He even says that use-value and exchange-value 'hang together' (*zusammenhängen*).[8]

Therefore, *Stoffwechsel*, although crucial for reading Marx today, is only one among at least three other core concepts that function similarly (although not identically) throughout his work. Not only did he develop a theory of material-kinetic intra-action well before the *Grundrisse* or the term 'metabolism', but he also used other essential terms that ought to be considered alongside one another when thinking about metabolism in a general sense. Metabolism, like the idea of *zusammenhängen* in his dissertation, is the material-kinetic foundation and 'earthly basis', which he believed served as the ultimate 'source of all production and of all being' (GI: 48; G: 106). In my view, it is the single most crucial idea of Marx's philosophy. It is what makes his unique theory of materialism so exciting and promising today.

But how does metabolism work for Marx, and what does he mean when he talks about 'social metabolism'? In the next section, let's look at the three kinds of metabolism Marx distinguished.

The Threefold Theory of Metabolism

Marx uses the term metabolism in three unique senses. At the largest scale, there is a 'metabolism of nature', which includes what we call today 'earth systems'. Energy from the sun hits the earth, and energy from inside the earth comes up from the core. The rock, water, life and carbon cycle transform this energy and eventually release it into space. In a very grand sense, the entire earth has a metabolism or many intertwined metabolisms.

Within earth systems metabolisms, human animals consume energy, change their bodies and release energy. This second metabolism is what Marx calls the 'human metabolism with nature'. Third, within our metabolism with nature, humans create what he calls a 'social metabolism' based on their cycles and patterns of production and consumption. Although natural metabolism precedes human animals on the planet, once humans emerge, all three metabolic processes become aspects or folds in the same interwoven metabolic meshwork.

Central to Marx's thought was how all three metabolisms could work together in such a way to increase natural-cultural flourishing, whatever that might mean for whoever was involved. Of course, one never knows what a metabolism can do, but one can experiment and try to be sensitive to them. Ultimately, this is what capitalism, *by its nature*, cannot do.

Let's briefly look at what Marx says about each of these three kinds of metabolisms before looking at how capitalism destroys them.[9]

The Metabolism of Nature

This first kind of metabolism is what Marx calls the 'universal metabolism' or 'universal metabolic process', 'nature's metabolism' or the 'simple metabolism of nature' (MECW 30: 62, 63, 78; G: 271). This 'universal metabolic process', according to Marx, consists of 'elemental forces' that involve physical, chemical or physiological transformations such as the rusting of iron, rotting of wood, fermentation of fruits or 'transposition' of matter 'into a human organism' (MECW 30: 62; CII: 316; CI: 323).

In the *Grundrisse*, for example, Marx makes clear that his theory of production and consumption is not a strictly anthropocentric process. Nature produces and consumes itself in a continuous process of mutual metabolic transformation. He directly compares nature's metabolism to human labour.

> Consumption is also immediately production, just as in nature the consumption of elements and chemical substances is the production of the plant. It is clear that in taking in food, for example, which is a form of consumption, the human being produces his own body.
>
> (G: 90)

Even in *Capital*, Marx is quite explicit that nature is productive and metabolic without any assistance from humans. Human production and labour are subsets within the more extensive metabolic process of nature's consumption and production. This substratum is furnished by nature without human intervention. When humans engage in production, they can only proceed as nature does herself, i.e. they can only change the form of the materials. Furthermore, even in this work of modification, they are constantly helped by natural forces. Labour is therefore not the only source of material wealth, i.e. of the use-values it produces (CI: 133–34).

Marx says that nature, without human intervention, is productive. When humans make things, they are not the only source of material wealth and use-value – since they produce as nature does. This position, in my view, is irreconcilable with anthropocentrism. Nature is productive, and humans produce as nature does because they are a metabolic region of nature's production process. If this is true, humans cannot be the only meaningful metabolic process in the world.

The Human Metabolism with Nature

The second kind of metabolism is what Marx calls the 'metabolic interaction between man and nature' or the 'human metabolism with nature' (CI: 290; CIII: 959). This is the type of metabolism we find in the first chapter of *Capital*. 'It is a physiological fact that [productive activities] are functions of the human organism, and that each such function, whatever may be its nature or its form, is essentially the expenditure of human brain, nerves, muscles and sense organs' (CI: 164, cf. 283). As a natural body (*Korper*), the human body is a region or sub-cycle of natural metabolism that continually intra-acts and hangs together within nature's larger metabolic processes (CI: 283–84).

Marx often describes human activity in elemental and natural terms. For example, he says that 'Labour [is] a living agent of fermentation', the 'living, form-giving fire' that 'awakens' particular objects 'from the dead' by 'breathing life into them as the elements of a 'new formation' (CI: 292; G: 361; CI: 289; MECW 30: 63). Since humans are nature, they act as 'a force of nature', in a 'system of nature', alongside other 'elemental forces' insofar as they 'can only proceed as nature does herself, i.e. [they] can only change the form of the materials', and in their own corporeally specific way (CI: 283, 133; EPM: 181; GI: 37, 49).

In other words, it is because humans *are nature* that their activity must proceed as a metabolic interchange (*Verkehr*) with nature. Human–nature metabolism is thus a 'nature-imposed condition of human existence', in an intra-active feedback relation between production and consumption (CI: 290).

Social Metabolism

Marx describes the third kind of metabolism as an 'interdependent process of social metabolism' (CIII: 949). Just as humans emerge as a continual iteration in nature's metabolic process, so social metabolism occurs as an iterative fold in both natural and human metabolic processes. Social metabolism is not just an optional subset of human activity for Marx. It is continuous with all human activity since human action is *always social* (G: 157, 158).

All human material production also rests, 'in one way or another, on social production' and 'productive existence is possible only on this condition'. Human metabolism with nature necessarily involves a 'metabolism of social labor', and through this 'material and mental metabolism', living human individuals 'change [...] in that they bring out new qualities in themselves, develop themselves in production, transform themselves, develop new powers and ideas, new modes of intercourse [*Verkehr*], new needs and new language' (G: 489, 492, 494, 161; CII: 226).

Social metabolism, for Marx, also importantly includes non-human 'natural agents' that 'co-operate' as 'creators of wealth' (MECW 34: 32; CI: 752). Thus, the 'process of economic reproduction whatever its specific social character may be' is 'always intertwined with a process of natural reproduction'. The material production of nature continually bears, supports, maintains and gives birth to all human productions. Thus, the fermentation of alcohol, the tanning of leather, the 'production time' it takes for an animal or plant to grow and so on are all non-human forms of productive agency that participate in social metabolism.

Non-human productive agencies 'undergo physical, chemical or physiological changes while the labour process is either completely or partially suspended' (CI: 752; CII: 434, 316) – the 'ripening of corn, the growth of an oak', 'wine that ferments in the cellar' and so forth (CII: 317, 201). Nature can also suspend the labour process due to 'uncontrollable natural conditions, the seasons of the year, etc.' (CIII: 213). Nature differentiates social metabolism in regional ways related to metabolic changes. As natural and social beings, humans always hang together in natural and social coproduction.

Marx was the first to introduce the idea of 'social metabolism'. As the Marxist scholar Alfred Schmitt wrote in his book on Marx and nature, Marx 'introduced a completely new understanding of man's relation to nature', one which 'went far beyond all the bourgeois theories of nature presented by the Enlightenment' (Schmidt 2014: 78–79). In particular, Marx took the concept of metabolism from Daniels, Liebig and the sciences and then transformed it into an entire historical ontology that included nature, humans *and society*.[10]

Liebig and others only thought of metabolism in a narrow organic way, but Marx saw the importance of extending metabolism to the human relation to nature and social production. Just as plants produce and consume energy, so do patterns of social motion. In Marx's theory of metabolism, there is no ontological division between humans and nature.

But what happens when these metabolic patterns fall apart? This is what I call *metabolic drift*.

Metabolic Drift

The earth is the first agent of metabolic fracture. There have been five previous mass extinction events on our planet, and none of them required humans. In this way, living and non-living systems have a way of reorganizing and destroying themselves each time.

Humans can try to do their best to help stabilize these enormous metabolic earth systems if they want to survive. Alternatively, they can make up stories about how the earth is naturally stable or how it's here for their use alone. They can pretend that metabolic patterns are unimportant to political or economic calculations or that humans will transcend the earth with rocket ships or immaterial souls.

But for Marx, these stories are absurd. All human activity, what he called 'use-values,' comes from the Earth. 'The earth is the reservoir, from whose BOWELS the use values are to be torn' (MECW 31: 465, original emphasis). Moreover, human activity cannot leave the earth alone. Humans are constantly intervening, extracting and altering metabolic systems.

In Marx's view, it's hazardous when humans start pretending that their interventions have no metabolic effects. In its most deranged expression, some humans even believe that they deal with quality-less quantities called 'economic values'. Even the 'interdependent process of social-metabolism' is 'prescribed by the natural laws of life itself', but capitalism uses these processes as the 'sole driving force [...] to create surplus-value' to destroy the metabolic systems it relies on (CIII: 949; CI: 342).

In other words, treating humans and natural systems as quality-less quantities or 'values' independent of their metabolic conditions 'attacks the individual at the very roots of his life' and 'prevents the return to the soil of its constituent elements consumed by man' (CI: 484, 637). Capitalism systematically creates an imbalance in all three kinds of metabolism. By clearcutting forests, natural metabolic processes cannot reproduce themselves. By turning humans into workers, capitalism often forces their physiological metabolism into physical stress and poverty. By forcing vast groups of people into unnecessary labour for the profit of others, capitalism disrupts social metabolism. The free association of people to make art and society as they wish is torn apart into individuals.

Just like the process of erosion and deposition in a river, capitalism takes energy from one place and moves it to another. As water speed increases, tiny bits of the riverbank are eroded or 'torn' from their rootedness in the ground. But these bits of soil and humus from one side of the river are not destroyed nor distributed randomly. Instead, they drift through the water and are deposited and accumulated on the slower side of the river. This creates a feedback loop of faster waters, erosion and deposition until a meander pattern is visible, making sandbars of silt. Capitalism is a metabolic system like any other on Earth but one that relatively rapidly destroys other metabolic systems and thereby destroys itself.

For Marx, the 'exploitation and squandering of the powers of the earth' disrupts balanced cycles of production and consumption. This can be minor on a geological time scale, but for human life, he says, it heralds 'the coming degradation and final depopulation of the human race' (CIII: 949; CI: 381). Like a vampire, value creation 'lives only by sucking living labour, and lives the more, the more labour it sucks' (CI: 875, 342). Metabolic drift is thus not a neutral shift for us. It is a *relational and thermodynamic imbalance* such that

> the future can indeed be anticipated and ruined in both cases by premature over-exertion and exhaustion, and by disturbance of the balance between expenditure and income [... which] in the case of the worker and the land [...] exists as power and the life span of this power is shortened as a result of accelerated expenditure.
>
> (MECW 32: 442)

Capitalism creates metabolic drift by what Marx calls 'primitive accumulation'. 'So-called primitive accumulation, therefore, is nothing else than the historical process of divorcing the producer from the means of production' (CI: 874–75). Because of its 'shameless and ruthless lack of moderation, impelling it to go beyond the natural limits of labour time into the realms of madness' (MECW 33: 386), capitalism tends to self-destruct. Like cancer, it exhausts its metabolic conditions. This is the fundamental and logical paradox of primitive accumulation in capitalism. It produces a drift or imbalance into the very cycles of metabolic production and consumption that it relies on *for its own process*.

Economic value is not an immaterial or non-metabolic abstraction. In Marx's view, value cannot be *immaterial or quality-less* because nothing in nature is. That is precisely why organizing the world according to economic value has real material effects like metabolic self-destruction. The world of making and circulating something called 'economic value' is not just a social construction. It has real social effects at odds with the existing ecological and human metabolic patterns that sustain our lives. The fundamental paradox and downfall of making economic value are that it does not realize that the values it produces are not discrete or quality-less. Furthermore, its conditions of reproduction are part of the more extensive process of threefold metabolic circulation.

Capitalism is like an ouroboros: a snake that does not realize it is eating its tail until too late. Since antiquity, the image of the ouroboros has stood for the idea of eternity. It was a perfectly enclosed, discrete circle. This basic idea remains at the heart of western idealism. The snake exhausts its conditions and thus its own body by destroying them. As Marx writes, 'Nature becomes purely an object for humankind, [and as] purely a matter of utility; ceases to be recognized as a power for itself' (G: 410).

For Marx, the critical question is not how such an interwoven metabolic system was produced in the first place in nature, but rather how it became historically imbalanced into various drifts and sandbars of accumulation. As Marx writes in the *Grundrisse*:

> It is not the *unity* of living and active humanity with the natural, inorganic conditions of their metabolic exchange with nature, and hence their appropriation of nature, which requires explanation or is the result of a historic process, but rather the *separation* between these inorganic conditions of human existence and this active existence, a separation which is completely posited only in the relation of wage labour and capital.
>
> (G: 489)

Discreet economic quantities will never be able to explain entangled metabolic movements. Only by starting with the movement of metabolic production can we understand how human animals started make-believing in things like discrete economic quantities and ignoring the metabolic world in and around them.

Conclusion

The social task of communism, then, for Marx, is not to prescribe any universally correct way to organize this threefold metabolic circulation but instead simply to acknowledge the importance of material production and more carefully experiment with it. Instead of massive metabolic drifts of erosion and deposition, we could work in closer, more sensitive feedback and 'free play' with such circulation patterns.[11]

This is also what makes Marx a non-anthropocentric thinker. His philosophical framework does not assume that humans are independent or superior to nature but rather continuous with what I have called 'natural-human-social metabolic' cycles (Nail 2020a: 183). These are patterns of energy distribution that move through human bodies, ecosystems and social systems. The patterns are 'material' insofar as they are made of energy and 'kinetic' insofar as they emerge from movements of energy. These overlapping and entangled patterns are ultimately what Marx is interested in studying.

For Marx, humans would be wise to maintain an entropic 'balance between expenditure and income' of material production/consumption for the proper 'restoration, renewal and refreshment' of the 'powers of life', the 'vitality of the soil' and the 'realm of natural forces' (MECW 32: 442; CI: 376; EPM: 181; CIII: 949, 964) (see Burkett and Foster 2006). It could 'govern the human metabolism with nature in a rational way, bringing it under their collective control instead of being dominated by it as a blind power' (CIII: 959) (Mészáros 1995: 45–46).

There is much more to say about how the struggle between metabolic domination and balance has played out in history that I cannot cover here. My main point was to show that a certain kind of relational and metabolic naturalism was at the heart of Marx's philosophy since the beginning of his work. Furthermore, I think it offers a start for a much-needed theoretical synthesis for thinking about the interconnection between natural, physiological and social systems in an age of climate change and destructive capitalism. Life is only one kind of metabolism entangled with many others.

Abbreviations

Works by Marx and Engels

CI	*Capital, Volume 1* (1996), in MECW, vol. 35
CII	*Capital, Volume II* (1997), in MECW, vol. 36
CIII	*Capital, Volume III* (1998), in MECW, vol. 37
EPM	*Economic and Philosophic Manuscripts of 1844* (1975), in MECW, vol. 3
G	*Grundrisse: A Contribution to the Critique of Political Economy* (1986–87), in MECW, vols 28 and 29
GI	*The German Ideology* (1976), in MECW, vol. 5
MECW	*Marx/Engels Collected Works* (1924)
$MEGA^2$	*Die Marx-Engels-Gesamtausgabe* (II) (1975)

NOTES

1. The significance of Marx and *ecological Marxism*'s use of the concept of metabolism to designate the material-energetic interchanges between humans and the rest of nature, as mediated by the labor-process, has been well-established. See Schmidt (2014: 76–93), Fischer-Kowalski (1998: 64), Foster (1999, 2000), Dickens (2004: 62–93) and Swyngedouw (2006).
2. Foster (1999, 2000, 2021) has emphasized the 'rift' language.
3. Nor, however, is it a theory about merely metabolic 'shifts' of matters from one place to another (Moore 2017).
4. Kohei Saito (2017: 72–4, 82, 88) provides an extensive discussion of Roland Daniels and his role in introducing Marx to metabolism (*Stoffwechsel*). I thank John Foster for bringing this to my attention.
5. 'Die einen unheilbaren Riß hervorrufen in dem Zusammenhang des gesellschaftlichen und durch die Naturgesetze des Lebens vorgeschriebnen Stoffwechsels' ($MEGA^2$ II.15: 788); 'Zusammenhang endgültig zerrissen' ($MEGA^2$ II.15: 804).
6. Marx (1975) writes of the 'self-mediated [*Durchsichselbstsein*] being of nature and of man' (356).

7. 'Their own collisions with one another produce alien social power standing above them, produce their mutual interaction as a process and power independent of them' (G: 197).
8. 'Zusammenhang mit der natürlichen Daseinsform [use-value hangs together with value]' ($MEGA^2$ II.1: 81).
9. I owe a great thanks to Jason Schreiner and his excellent dissertation on Marx and metabolism for clarifying this distinction for me (Schreiner 2017).
10. For an extensive survey of the long tradition of socialist ecology, see Foster (2021).
11. This is what I call 'metabolic communism'. For a longer discussion, see Nail (2020a: 198–214). See also Nail (2020b: 242–70).

REFERENCES

Bing, Franklin C. (1971), 'The history of the word "metabolism"', *Journal of the History of Medicine and Allied Sciences*, XXVI:2, pp. 158–81.

Burkett, Paul and Foster, John Bellamy (2006), 'Metabolism, energy, and entropy in Marx's critique of political economy: Beyond the Podolinsky myth', *Theory and Society*, 35:1, pp. 109–56.

Dickens, Peter (2004), *Society & Nature: Changing Our Environment, Changing Ourselves*, Cambridge: Polity.

Fischer-Kowalski, Marina (1998), 'Society's metabolism', *Journal of Industrial Ecology*, 2:1, pp. 61–78.

Foster, John Bellamy (1999), 'Marx's theory of metabolic rift: Classical foundations for environmental sociology', *American Journal of Sociology*, 105:2, pp. 366–405.

——— (2000), 'The metabolism of nature and society', in *Marx's Ecology: Materialism and Nature*, New York: Monthly Review Press, pp. 141–77.

——— (2021), *The Return of Nature: Socialism and Ecology*, New York: Monthly Review Press.

Liddell, Henry George and Scott, Robert (1940a), 'βάλλω', in *A Greek-English Lexicon* (rev. and augmented by H. S. Jones and R. McKenzie), Oxford: Clarendon Press, https://www.perseus.tufts.edu/hopper/text?doc=Perseus:text:1999.04.0057:entry=ba/llw. Accessed 23 February 2022.

——— (1940b), 'μετά', in *A Greek-English Lexicon* (rev. and augmented by H. S. Jones and R. McKenzie), Oxford: Clarendon Press, http://www.perseus.tufts.edu/hopper/text?doc=Perseus:text:1999.04.0057:entry=meta/. Accessed 23 February 2022.

——— (1940c), 'μεταβολή', in *A Greek-English Lexicon* (rev. and augmented by H. S. Jones and R. McKenzie), Oxford: Clarendon Press, https://www.perseus.tufts.edu/hopper/text?doc=Perseus:text:1999.04.0057:entry=metabolh/. Accessed 23 February 2022.

Marx, Karl (1975), *Early Writings* (ed. Q. Hoare), New York: Vintage Books.

Marx, Karl and Engels, Friedrich (1924), *Marx/Engels Collected Works*, MECW, New York and London: International Publishers.

——— (1975), *Die Marx-Engels-Gesamtausgabe*, $MEGA^2$ II, Berlin: Akademie-Verlag.

Mészáros, István (1995), *Beyond Capital: Towards a Theory of Transition*, London: Merlin Press.

Moore, Jason W. (2017), 'Metabolic rift or metabolic shift? Dialectics, nature, and the world-historical method', *Theory and Society*, 46:4, pp. 285–318.

Nail, Thomas (2020a), *Marx in Motion: A New Materialist Marxism*, New York: Oxford University Press.

——— (2020b), *Theory of the Earth*, Stanford, CA: Stanford University Press.

Saito, Kohei (2017), 'Metabolism of political economy', in *Karl Marx's Ecosocialism: Capitalism, Nature, and the Unfinished Critique of Political Economy*, New York: Monthly Review Press.

Schmidt, Alfred (2014), *The Concept of Nature in Marx*, London, New York: Verso.

Schreiner, Jason (2017), 'The roots of life: Marx's concept of social metabolism and the dialectics of corporeal praxis', Ph.D. dissertation, Eugene, OR: University of Oregon.

Swyngedouw, Erik (2006), 'Circulations and metabolisms: (Hybrid) natures and (cyborg) cities', *Science as Culture*, 15:2, pp. 105–21.

9

From ALife to No Life: On Mediatic Contexts of Life and Death[1]

Jussi Parikka

Life and Death of Media

'What is life?' is easily an overwhelmingly grand question to tackle, but luckily we already have some possible answers. For instance, the wonderful popular book, *What is Life?* by Lynn Margulis and Dorion Sagan (1995), outlines one influential version of evolution through bacteria and symbiogenesis. It is, as Margulis' work in general, one way of thinking about alternatives to the (Neo) Darwinian focus on competition by way of providing an alternative: cooperation and symbiosis.

In his Foreword to that beautifully illustrated volume, Niles Eldredge writes:

> For microbes will not only inherit the earth (should, for example, we complex multicellular creatures fall prey to the next spasm of mass extinction); for microbes got here long before we did, and in a very real sense they already 'own', and most certainly run, the global system.
>
> (1995: xiv)

Written more than 25 years ago, the reference to humans (and other complex animals) as victims of the next mass extinction runs parallel to the increasing number of reports that warn that the coming decades are more or less decisive for the planet's future. Biodiversity loss is part of the big bundle of apocalyptic scenes that are not so much biblical as they are industrial and part of what some have coined the *capitalocene* (Moore 2015), a systematic exploitation of planetary resources with ecocidal consequences. It seems more than clear that any question about (modern) life is inscribed in questions of extinction (see Mirzoeff 2018).

Moreover, consider the opening image of Margulis and Sagan's book: two images spread across two pages as the opening gambit, on the left 'Earth in Space' and on the right '*Mycoplasma*. Phylum: Aphragmabacteria. Kingdom: Bacteria (Monera). These are some of the smallest bacteria...'. Under the images, a text in much larger font: 'Between these two realms lies all of life as we know it' (Margulis and Sagan 1995: Plate 1A, 1B). A stunning contrast, it also speaks to the decades of multiscalar imagery that starts with the earlier shots of planet Earth from space and includes also the sort of audiovisual production of telescopic and microscopic media worlds that the Charles and Ray Eames Office laid out in their short cinematic piece, *Powers of Ten* (Eames and Eames 1977).[2] It is also imagery that fits well with the discourse of the Anthropocene – as it is later coined following earlier terminologies of Gaia and such – and seems to speak to the interlinkedness of different regimes of life, from biosphere to atmosphere to mediasphere. But besides the biological and evolutionary discourse, this question shifts registers from evolution to social, cultural, political and, indeed, media theory questions. How do questions of life not only intersect with media theory interested in ecologies and ecosystems, but also the waste and politics of violence? In other words, what is life is a question that needs to be asked in the midst of the on-going financialization and militarization of environments, environmentalism and nature (Keucheyan 2016).

Where would a media theorist even start to unfold the question of life? It would surely be seen through a denial of life, life and its conditions, and life that is not merely biologically concrete but abstracted across systems and infrastructures. It is abstract only because it is not merely about living bodies as isolated entities, but multiscalar systematic relations that condition each other in strange, complex feedback loops. Abstraction then becomes a way to involve technological culture into the mix as one particular life-sustaining system. Of course, we can also ask: Is there a life of media? Mark Deuze (2012) speaks of *Media Life*, but besides our practices of living in relation *to* media, what else would there be in this topic that relates to a wider environmental and political sense, in other words, the assemblage nature of media life?

We can also narrate a longer history of life that seems parallel to the history of climate change and biodiversity loss. Artificial life (ALife) is one stable part of media histories of life with techniques of simulation of the biological world having an impact on our technological episteme since the eighteenth century (see, e.g. Riskin 2003; Bedau 2007; Bedau and Cleland 2010; Parikka 2016a). This is not to say that the biological world was not imitated or simulated earlier. One must be aware of histories of automata as a particular version of life in practice – life as practiced as serving, singing and music exemplified in the ancient automata in Greece and then especially in Middle East regions such as Baghdad and

Diyarbakir (900–1300s). From Middle East inventions to European clockworks, automata present their own sense of regularity and continuity. Questions of ALife then connect to later versions of machines that look like ducks, quack like ducks, but are not ducks. For instance, famous examples such as Jacques de Vaucanson's digesting and defecating duck, *Canard Digérateur*, presented in 1739 in France simulated some of the functions we relate to biological life. Much later, and with a much more scientific sense, was the computerized version of life, expressed in ALife simulations since the 1980s. One key question that emerged in such software-based responses to what is life was not what life is (a particular historically specific carbon-based material emergence), but life as it could be, to borrow a phrase from ALife pioneer Chris Langton in the 1980s (see Parikka 2016a).

There are many historical examples, but a crucial realization remains something expressed in more advanced philosophical notes about life and technology in the twentieth century (such as Gilbert Simondon's work): it is increasingly difficult to maintain the 'anthropological barrier' (Barthélémy 2015) that differentiates those living from those never so, those having a life from those merely inanimate and mute. Interestingly, though, the question of what constitutes life as it could be emerges at the same time when biological life as it is, is disappearing fast.

This chapter gives insight into how the life of technologies/media are not merely in the hands of the user or the assigned meanings, but the wider material history, and the wider material geography, it occupies and forms. Here the anthropological barrier needs to be discussed in terms of its more specific forms of histories such as related to colonial violence and other political conditions that specify the Anthropocene. This argument relates to environmental issues as one crucial context for understanding the impact of materiality of media (see Swartz and Wasko 2021). Technical media are increasingly recognized not only as one small waste problem but part of a planetary scale transformation that relates to both the changes in landscapes and changes in how those landscapes are seen (and in further transformation) as data. And as questions of geography and landscape, these issues are then one interface that connects to colonial histories that are constantly recycled as material infrastructures of exploitation. The life of media is also about dead waters and dead lands (see Sassen 2016a, 2016b).

Media and Waste

A Geology of Media (Parikka 2015) was an attempt to offer one approach and theoretical expression for this situation. It aimed to approach media/technologies not through their use value and practices of what we do but how they fall out of use and become environmental waste, and how before media becomes media as communication, it has already gone through multiple rounds of technological processing and has been actively and materially produced out of various geophysical materials from minerals to a range of energy transactions. It has been convincingly shown that the energy infrastructures of contemporary digital culture are heavily dependent on fossil fuels and as such, a topic also to be integrated as part of core curricula of media, visual and communication studies (see, e.g. Hogan 2015; Jacobson and Hogan 2019; Bhowmik 2019; Johnson 2019). In the book, I wanted to give a theoretical framing for this in terms of media theory as well as contemporary artistic practices. Hence, the aim was to put media technological modernity in conversation with the terminology of the Anthropocene, the particularly nasty technological industrial waste period we are living through.

But even on more local scales (which are already crystallizing the planetary scale disorders and waste), the media technological realities of disuse, of dead media persist as a materially toxic afterlife that is hard to ignore as part of the chemical legacy of technical communication. Besides *A Geology of Media*, many other researchers, artists and non-governmental organizations (NGOs) have documented the landscapes and situations where scavenging discarded media technologies for their useful components and metals is also a health hazard to be recognized as part of the broader question of media, waste and sustainability (Cubitt 2015; Taffel 2016; Starosielski and Walker 2016; Holm and Taffel 2017; Tousignant 2018; Mukherjee 2020). Such a toxic impact of media in devices and the broader context of digital culture also speak to this other reality of media technological violence (see Parikka 2016b) that is not about content but about the various levels of environmental hazards. For example, the materials contained in multiple electronic devices from smartphones to computers, devices making up the supposedly immaterial wireless culture to physical infrastructure, are also evaluated from the perspective of human labor waste that deals with these levels of media reality: lead, cadmium, mercury, barium and many more. All of these are emblematic parts of the potential damage to nervous systems and brains, kidneys and other internal organs, fetus development, etc.

Chemical health hazards of synthetic culture are hard to ignore for a media theory of life and death (see also Chen 2012). This residual side of technological culture is an odd mix of chemistry and anthropology of media disuse – of media obsolete, discarded and toxically affecting the livelihood of masses of people who deal with this sort of a legacy, not by way of accident but by way of a historical and geographical structuring of the global media economy. This is no longer media archaeology of the sort interested in the aesthetics of the obsolete but enters a different theoretical circuit that is closer to issues of postcolonial theory, critical cultural geography and other stances that map the spatial histories of dispossession and slow violence (Nixon 2013; Yusoff 2018).

The Anthropocene: Neocolonial Waste

To address ends of the world and ends of life is not merely premised on the biological inevitability of death – or even the sort of mournful metaphysics of finitude that Martin Heidegger introduced for contemporary philosophy (Braidotti 2012). In addition, one must investigate the imaginaries of ends of the world and their already existing histories, as well as Indigenous mythologies. This is what Déborah Danowski and Eduardo Viveiros de Castro (2017) set out to do in *The Ends of the World*. There is a need to map the colonial histories of the Anthropocene vis-à-vis the current imaginaries of worlds without humans so as to problematize the terms of the debate and produce a more encompassing way of addressing the issue of life. The problematization of imaginaries of the end and the problematization of generalization of Man (of the Anthropocene) go hand in hand, and over the past years, have been met with a richer sense of critical insights to their legacy. Kathryn Yusoff has argued for the necessity of understanding the Anthropocene through the excluded Blackness that is one organizing horizon for questions of human life. Hence, in Yusoff's words:

> This White Geology continues to propagate imaginaries that organize Blackness as a stratum or seismic barrier to the costs of extraction, across the coal face, the alluvial planes, and the sugarcane fields, and on the slave block, into the black communities that buffer the petrochemical industries and hurricanes to the indigenous reservations that soak up the waste of industrialization and the sociosexual effects of extraction cultures.
>
> (2018: n.pag.)

Nicholas Mirzoeff connects a critique of the Anthropocene to the specific question of life and extinction as it emerged in the nineteenth-century biology and cultural discourse. In Mirzoeff's view, this relates to the question of how to account for the racialized discourses of the human – and other species, which also justified slavery for an extended period – that then puts the Anthropocene into a different light. To quote Mirzoeff:

> In this context, moreover, the term life is key. The concept of extinction itself was part of the transformation of natural history into life science (biology) in the era of the revolutions of the enslaved and abolition (1791–1863). In the current paradigm shift to Earth systems (everything prefaced with geo-), what is the place of Black and other colonized life, human and nonhuman?
>
> (2018: 124)

Mirzoeff's effective provocation is a nod to the particularly important question of life through historically significant themes concerning race and ethnicity – a lens through #BlackLivesMatter to questions of colonization and enslavement as they persist in different guises from logistics to epistemology. The perception about the particular color lines and boundaries where nature stops, and culture begins, is one form of colonial power that is constantly inscribed across multiple practices that are also visible in a contemporary context of distributions of the effects of current waste society. Racial mattering, to refer to Mel Y. Chen's (2012) term, is present in both contemporary practices and their histories that inscribe nature as one context where politicized issues of life, death and extinction gain particular weight.

Hence, the concept of the so-called Anthropocene is also part of political and technological history, and, in this case, also part of the persisting neocolonial domination. This is a theme noted by many key writers, such as Sassen, Yusoff and others, and it has also become a viable way to frame discussions in visual and media studies. In Mirzoeff's words, there is 'a color line enshrined in natural history' (2018: 124), which comes out in rhetorics of humans as geological *masters* and in the infrastructural arrangements of how consumer products, money and waste flow across planetary geographies. Mirzoeff is able to show how natural historical marks relate to geological events, and how the knowledge about nature and particular discourses of extinction are embedded in forms of knowledge that had a strong link to colonial forms of representation.

But besides this remediation of colonial forms of division, there is also a particularly non-representational form of neocolonialism that pertains to this topic and to the weird life and death of media. Hence, consider questions of environmental waste in contexts of the Global South – western Africa, India and Pakistan – as examples of how particular intersectional concerns are inscribed into contexts of contemporary politics of the logistics of waste. Sean Cubitt (2016) articulates in more detail waste as inherently related to multiple operations of contemporary capitalism. Cubitt's environmental media message is evident from his title, *Finite Media*. It is an argument about the finite material conditions in which media culture as energy, resources and extraction is determined in relation to geographies that themselves are determined in our modern times in relation to colonial histories. In his terminology, waste is an integral part of digital media culture as it is expressed in the energy costs and in the discarded electronics that are not merely an accidental part of the cycle of media life, but integral to its existence. The creation of waste dumped on the environment and the particularly exposed human groups is the *integral waste* that is produced as a necessary externality of the global situation. This circuit is very much conditioned along historical legacies, partly related to the past decades of outsourcing production to countries of cheaper labor in different regions across Asia, and partly in the enduring histories of colonialism. As Cubitt (2016: 116) puts it, 'contemporary accumulation continues colonialism in the Global South (exocolonialism), and internally in the Global North (endocolonialism)' producing a situation that stems from a grim political history that is reproduced as a grim political present. Here, the environmental destruction picks up on the already existing multiple levels of extinction that are themselves part of the history of colonialism.

Contemporary forms of integral waste and violence have to do with geographies of resource extraction (e.g. particular regions in Africa as centers of mining as well as energy), of production (cheap labor in poorly regulated work conditions) and of waste (dumping electronics as part of the second cycle of media reuse, which unfortunately often is extraction through hazardous means by burning polyvinyl chloride [PVC] casings, etc.). Furthermore, Cubitt points out the centrality of Indigenous areas for many of these concerns. This is where the extraction industries and the logic of exploitation become clearest. Articulating this in relation to media studies, he argues that the dependence on fossil fuels and on particular minerals and related resources is the reason why an expanded look at this environmental and political condition is relevant from this media aesthetic. This perspective is radically expanded from what happens on the screen to what made the screen possible. So, whereas some software studies some years back demanded to move beyond screen essentialism to the software and other infrastructure, Cubitt takes this further to the material *and* political conditions of existence of this particular link of material production and extraction industries.

Indeed, to trace such elements as mobile digital culture to their constituent metals is one vector of this aesthetic knowledge, an articulation of scales of geography, and scales across nature cultures; for example, the lithium-mineral extraction industry's massive impact on the ecology in Chile and other parts of the world (Cubitt 2016: 64–69).

Any life of media, then, starts in the landscapes, ecologies, sites of extraction, and sites of distribution and circulation that are planetary scale conditions of contemporary technology/media. It concerns the medium of energy, geographies of those infrastructures, mining practices and their corporate routes from extraction to factories on to consumption. In extending from mere media devices to the cultural situation of life and death of technological culture, it becomes necessary to discuss the various political infrastructures put in place – also in contested areas such as the TransCanada Keystone XL (export limited) pipeline and other pipeline plans and their particular relation to Indigenous peoples and lands. Hence, the aesthetics of encampments that protests have managed to achieve by campaigns is another face of visibility when it comes to the material trails of such plans (see also Kinder 2019). As Cubitt (2016: 43–44) explains in relation to the pipeline, the plans actually involve transporting diluted bitumen (also known as dilbit).

While extracting oil from bitumen itself is environmentally costly, the substance itself is more acidic and high in sulphur, leading to increased possibilities of problems in the pipeline and hence spills (Cubitt 2016).

This itself is literally not news (although thanks to much of local and Indigenous activism, it is also forced into the mainstream media at times). As Cubitt underlines, media coverage and in general wider reporting of minor leaks does not take place. While the Deepwater Horizon offshore drilling rig was a media event (about conditions of technological culture, i.e. oil), the tens of thousands of minor leaks in the Alberta pipelines since 1990 are not. Instead of large global media events in all their weirdness (Wark 2006), the spread of these media non-events witnesses a different set of concerns: technological infrastructures failing, the mundane violence of the fossil fuel economy, the slowly creeping (Nixon 2013) extinction that defines this horizon of life of media technological culture and its energy infrastructure. This realization is particularly central to the ways in which planetary-level infrastructures move inside and outside the realms of experience and hence provides another way to think of aesthetics. It is also in this context of events and non-events that one becomes aware of the *mediations of materiality*, and *materiality of mediations* as co-determining forces. Thus, to speak of aesthetics and visibility is constantly entangled in a material history and the chemical qualities of fuels and pipelines concerned.

Conclusions

So, asked in the context of the mass-scale environmental disaster, technological culture of integral waste and political culture of neocolonialism that defines a particular angle to the Anthropocene – what constitutes life? Media as issues of waste and finite resources relates also to what can be coined as the political aesthetics of media while expanding its scope of use, as Cubitt (2016) among others has demonstrated. This aesthetics is also conceptually useful, as it concerns what scales of media use and non-use are counted as part of the analysis. Indeed, if we investigate multiscalarity of media from visual practices to infrastructure (Dvořák and Parikka 2021), the reframing of materiality is one issue that is at stake.

Consider it through the following summarizing example which again resituates questions of media and visual culture into a very different kind of material context. Scientists have observed how the so-called Anthropocene period has also been marked by an emergence of a new sort of life, a burgeoning of new geological worlds of minerals. This discovery of new minerals is related to industrialization and the new chemical reactions that these particularly important materials of technological culture when oxidized or hydrated, when in interaction with planetary processes and temperatures. To quote from a newspaper article about a nineteenth-century shipwreck carrying tin ingots:

> These human-triggered minerals include chalconatronite, a rare copper mineral that crystallises as a bright blue crust on ancient Egyptian bronze artefacts, and andersonite, a uranium-laced mineral with a fluorescent green or yellow glow that forms on the walls of mine tunnels. The bronze-hued abhurite was discovered on the wreck of the *SS Cheerful*, which sank off the coast of Cornwall, England in 1885, and only formed because of a chemical reaction between the salt water and the ship's sunken supply of tin ingots.
> (*Sciencealert* 2018: n.pag.; see Davis 2017)

Such anecdotal stories are emblematic of a bigger chemical reaction that is of a scale larger than individual remains of ships, artefacts, tin remains or other leftovers from the earthly records of culture and technology. If nothing else, this would be a rather apt way to start a science fiction novel of a synthetic ooze that starts to spread across the planet as an alien life entering not from the outer world but from the insides of current forms of production of technological culture. When it comes to the contemporary forms of waste and shipwrecks – literal and metaphorical – you can start to connect to the systematic forces of capitalism and neocolonialism. In our case, it can be connected to a different explanation or story, which has to do with the planetary scale laboratory of technological life that produces new forms of artificial life, while also accelerating the extinction of multiple existing forms of life. This is not ALife in the sense we most often understand it in computational culture, but a different set of questions of life and technology.

This laboratory of life is the other side of what comes out as waste. The emergence of a seemingly new 'life' of minerals unseen before is connected to the realities of the increasing amount of dead media and deadly technological waste. So the particular ways in which we narrate the emergence of new synthetic materials out of waste are also entangled with how we narrate the particular necropolitics of media as well. The usual media archaeological discourse about dead media and obsolescence has been helpful in expanding the scope of media studies concerns. But it is across the vast landscapes of transformation of climate and its biodiversity where media studies also has to carve out its own sense of importance in order to be able to tackle the rather ambitious and important questions of life.

NOTES

1. This research has been supported by Czech Science Foundation funded project 19-26865X 'Operational Images and Visual Culture: Media Archaeological Investigations'.
2. For updated version, also see Silleck (2002).

REFERENCES

Barthélémy, Jean Hugues (2015), *Life and Technology: An Inquiry into and Beyond Simondon,* Lúneburg: Meson Press.

Bedau, Mark A. (2007), 'Artificial life', in M. Matthen and C. Stephens (eds), *Philosophy of Biology*, Amsterdam: North Holland Publishing Co., pp. 585–603.

Bedau, Mark A. and Cleland, Carol E. (2010), 'Section III: Artifical life and synthetic biology', in *The Nature of Life*, Cambridge: Cambridge University Press, pp. 217–23.

Bhowmik, Samir (2019), 'Thermocultures of memory', in *Culture Machine*, 17, pp. 1–20, https://culturemachine.net/vol-17-thermal-objects/thermocultures-of-memory/. Accessed 7 January 2022.

Braidotti, Rosi (2012), *The Posthuman*, Cambridge: Polity.

Chen, Mel Y. (2012), *Animacies: Biopolitics, Racial Mattering, and Queer Affect*, Durham, NC: Duke University Press.

Cubitt, Sean (2016), *Finite Media: Environmental Implications of Digital Technologies*, Durham, NC: Duke University Press.

Danowski, Déborah and Viveiros de Castro, Eduardo (2017), *The Ends of the World* (trans. R. Nunes), Cambridge: Polity.

Davis, Nicola (2017), 'Minerals found in shipwreck and museum drawer "show we are living in new epoch"', *The Guardian*, 1 March, https://www.theguardian.com/science/2017/mar/01/rock-of-ages-impact-of-manmade-crystals-defining-new-geological-epoch-study. Accessed 7 January 2022.

Deuze, Mark (2012), *Media Life,* Cambridge: Polity.

Dvořák, Tomáš and Parikka, Jussi (eds) (2021), *Photography Off the Scale: Technologies and Theories of the Mass Image*, Edinburgh: University of Edinburgh Press.

Eames, Charles and Eames, Ray (dirs.) (1977), *Powers of Ten*™, Santa Monica: Pyramid Film & Video, https://youtu.be/0fKBhvDjuy0. Accessed 7 January 2022.

Eldredge, Niles (1995), 'Foreword: Undreamt philosophies', in L. Margulis and D. Sagan (eds), *What is Life?,* New York: Simon & Schuster, pp. 8–10.

Hogan, Mél (2015), 'Facebook data storage centers as the archive's underbelly', *Television and New Media*, 16:1, pp. 3–18.

Holm, Nicholas and Taffel, Sy (eds) (2017), *Ecological Entanglements in the Anthropocene*, Lanham, MD: Lexington Books.

Jacobson, Kate and Hogan, Mél (2019), 'Retrofitted data centres: A new world in the shell of the old', *Work Organisation, Labour & Globalisation*, 13:2, pp. 78–94.

Johnson, Bob (2019), *Mineral Rites: An Archaeology of the Fossil Economy*, Baltimore, MD: Johns Hopkins University Press.

Keucheyan, Razmig (2016), *Nature is a Battlefield: Towards a Political Ecology* (trans. D. Broder), Cambridge: Polity.

Kinder, Jordan B. (2019), 'Liquid ethics, fluid politics: The cultural and material politics of petroturfing', Ph.D. thesis, Alberta: University of Alberta.

Margulis, Lynn and Sagan, Dorion (1995), *What is Life?* New York: Simon & Schuster.
Mirzoeff, Nicholas (2018), 'It's not the Anthropocene, it's the white supremacy scene; or, the geological color line', in R. Griusin (ed.), *After Extinction*, Minneapolis, MN: University of Minnesota Press, pp. 123–49.
Moore, Jason (2015), *Capitalism and the Web of Life: Ecology and the Accumulation of Capital*, London: Verso.
Mukherjee, Rahul (2020), *Radiant Infrastructures Media, Environment, and Cultures of Uncertainty*, Durham, MD: Duke University Press.
Nixon, Rob (2013), *Slow Violence and the Environmentalism of the Poor*, Cambridge, MA: Harvard University Press.
Parikka, Jussi (2015), *A Geology of Media*, Minneapolis, MN: University of Minnesota Press.
——— (2016a), *Digital Contagions: A Media Archaeology of Computer Viruses*, 2nd ed., New York: Peter Lang.
——— (2016b), *A Slow, Contemporary Violence: Damaged Environments of Technological Culture*, Berlin: Sternberg.
Riskin, Jessica (2003), 'The defecating duck, or the ambiguous origins of artificial life', *Critical Inquiry,* Summer, pp. 599–633.
Sassen, Saskia (2016a), *What is Behind the New Migrations: A Massive Loss of Habitat*, re:publica, YouTube, 10 May, https://youtu.be/FPlq_ywzrQ4. Accessed 7 January 2022.
——— (2016b), 'Dead land, dead water: Saskia Sassen on loss of habitat and new migrations', *aNtiDoTe Zine*, 3 May, https://antidotezine.com/2016/06/28/dead-land-dead-water/. Accessed 7 January 2022.
Sciencealert (2018), 'An explosion of never-before-seen minerals could mark the dawn of our new geological epoch', 25 April.
Silleck, Bayley (dir.) (2002), *Cosmic Voyage*, IMAX® film presented by the Smithsonian Institution's National Air and Space Museum, USA: Warner Home Video.
Starosielski, Nicole and Walker, Janet (eds) (2016), *Sustainable Media: Critical Approaches to Media and Environment*, London and New York: Routledge.
Swartz, Jeremy and Wasko, Janet (eds) (2021), *MEDIA: A Trandisciplinary Inquiry*, Bristol: Intellect Books.
Taffel, Sy (2016), 'Technofossils of the Anthropocene: Media, geology, and plastics', *Cultural Politics*, 12:3, pp. 355–75.
Tousignant, Noémi (2018), *Edges of Exposure Toxicology and the Problem of Capacity in Postcolonial Senegal*, Durham, MD: Duke University Press.
Wark, Mckenzie (2006), 'The weird global media event and the tactical intellectual [version 3.0]', in W. H. K. Chun and T. Keenan (eds), *New Media, Old Media: A History and Theory Reader*, London and New York: Routledge, pp. 265–76.
Yusoff, Kathryn (2018), *A Billion Black Anthropocenes or None*, Minneapolis, MN: University of Minnesota Press, https://doi.org/10.5749/9781452962054. Accessed 7 January 2022.

PART IV

ECOMEDIATIONS AND EDUCATION

10

Media and Information Literacies for a Living World: Engaging with a Cyberist Era

Divina Frau-Meigs

The evidence of human impact on life has set a number of debates that currently rotate around notions of an 'Anthropocene' (Crutzen and Stoermer 2000; Randall 2022; see Subcommission on Quaternary Stratigraphy n.d). Human influence on the atmosphere and climate change, biodiversity and land mass substantiates the idea that humans (*anthropos*[1]) have significantly dominated the recent years on Earth, with geologic evidence (Zalasiewicz et al. 2010). Researchers disagree as to the beginning of an Anthropocene; some go as far back as early agricultural changes, and some contend that it started with nuclear weapons (Certini and Scalenghe 2015; also see Zalasiewicz et al. 2019). According to an article in *Nature*, 2020 marked a new threshold as human-made objects outweighed all of the living beings on Earth, an augmentation of the human-made mass that is estimated to have started over 120 years ago, at the beginning of the twentieth century (Elhacham et al. 2020).

In these heated debates, few argue about the role of information in the process because information feels immaterial and intangible. Yet looking at information could help as it is the sedimented, cumulative growth of media devices (plastrons, papyrus, parchments, papers, films, disks, etc.) that have allowed humans to archive their knowledge and transfer it to new generations in spite of daunting ecological conditions, after the Pleistocene, the last Ice Age (2.5 million to 11,700 years ago), when Homo sapiens made their first appearance on Earth, 300,000 years ago. From the early twentieth century onward, the invention of communication technologies grew at an unprecedented pace (telephone, cinema, radio, television, internet, etc.), sparking a new awareness of our place on the planet, culminating maybe when NASA published for the first time in 1967, a satellite photograph of the Earth from space, followed in 1972 by what was going to become a series of portraits of *The Blue Marble* floating in the dark void of the universe (NASA ([1972] 2001; Reinert 2011). Our singular role as a species brought about the heightened awareness of an Anthropocene, as we have by now accumulated a whole bag of blue marbles.

The turn of the twenty-first century has fostered a new augmentation, as information has moved from analogical to digital, and as paper gives way to screens in relation to the accessibility and consumability of information in all its diversity. This augmentation corresponds to a *cyberist era*, to keep the geological metaphor of our sedimented media/technologies strata. Understanding this era, its moment and momentum, can be helpful to navigate the obstacles that human-induced ecological changes (e.g. climate) are bringing to the life of plants and animals on Earth. Therefore, managing information can help find solutions to the current risks to all ecosystems, including those created by humans.

A Cyberist Paradigm

The growing role of information in the way we organize our life and culture has brought up the digital revolution. This revolution required a paradigm shift that took us beyond modernist/postmodernist views of society and mankind, as built during the nineteenth and twentieth centuries (Featherstone 2007). A cyberist view of the twenty-first century is attuned to the new sensibility that sustains our underlying social structure and frames the principles of a culture based on the complexity of information. In the modernist/postmodernist era, information was either intelligence (to be kept secret as secrecy was key to power) or news (to be made public as publicity was key to decision-making) (Habermas 1962). In a cyberist era, information is still intelligence and news, but it is also data (to be traded as intelligence) and documents (to be curated as literacy is key to knowledge formation). Secrecy, publicity and literacy are among the vital notions that vie for attention in a cyberist paradigm as their evolution is key to life on *The Blue Marble*.

The term cyberist stems from the Greek κῠβερνᾰ́ω or 'kubernáō', with four nuances, from the concrete practical notion of acting as a nautical pilot, to the more abstract notions of steering, guiding and finally governing (Liddell and Scott 1940b). All these nuances fit the current period as guiding can be fit for education and as governing shifts politics from government to multi-stakeholder governance, whereas steering and piloting are evocative of the individual control that comes with computerized information as it is secured by code (Frau-Meigs 2011a, b, c). Cyberism thus encapsulates the complexity of this integrative approach to information. It builds on and departs from 'cybernetics', a term popularized in 1948 by Norbert Wiener that lays the emphasis on automation and automatic control and contrivances ([1948] 1961) whereas cyberism considers the human pilot at the steering wheel, maintaining the ethical cape in the sea of media and data. Cybernetics thus has evolved to refer to engineering sciences that deal with machine controlled feedback mechanisms, in spite of Margaret Mead's attempt to uphold the potential cultural dimensions of cybernetics:

> [Wiener] defined it as the science of communication and control in the animal and the machine. [André-Marie] Ampère, before him, wanted cybernetics to be the science of government. For philosopher Warren McCulloch, cybernetics was an experimental epistemology concerned with the communication within an observer and between the observer and his environment. Stafford Beer, a management consultant, defined cybernetics as the science of effective organization. Anthropologist Gregory Bateson noted that whereas previous sciences dealt with matter and energy, the new science of cybernetics focuses on form and pattern.
>
> (Mead 1994: n.pag.; see also Ampère 1838: 140–41)

But this approach has remained thin and narrow, and is associated with enhanced organization and management, as well as a sense of loss of human resources and agency. Cyberism is more people-centred and recombines human energy with human patterns, giving agency back to humans' capacity for steering information and communication, the form and its formulations. It focuses on how meaning arises from the achievements of creative minds that can turn any signal into signs.

By doing so, cyberism converses with other movements that have concentrated on new forms of expression, such as modernism and postmodernism that strove to make sense of industrialization and urbanization. Cyberism does not look back on modernism the way postmodernism does but departs from both referents to consider new perspectives. It coincides with the advent of new scientific paradigms such as cognition, and its social and biocultural extrapolations provide new complex interpretations of human nature and culture in a networked environment (Castells 1997). As with all life experiments, it does not (re)solve the tensions and contradictions of mankind but displaces them with alternative value systems appearing. Cyberism corresponds to the aesthetic, ethical and political awareness of this change and plays in a counter-distinctive manner with capitalism and liberalism as they have supported modernist/postmodernist paradigms. It is based less on consumerism than on engagement, be it simple interactivity with machines or complex contributions with networked individuals, as key to an alternative value system still in the making that recognizes the need for exchange but not on the market basis of exclusive property (Bell [1970] 1996).

Redefining Information Opportunities

Cyberism prolongs Daniel Bell's *Cultural Contradictions of Capitalism* ([1970] 1996) that analyses the postmodernist twilight. He explored how the West went from an economy of production to an economy of consumption via the use of culture as a quest for 'a new sensibility' to create a new 'underlying social structure'. Bell analysed how society's culture of consumption conflicted with the mindset required for production: the economic principles of rationality, efficiency in the allocation of resources and attendant emphasis on self-discipline, delayed gratification and almost puritan restraint, while still present, clashed with the culture of consumption based on pleasure and leisure (Frau-Meigs 2011a, b, c).

With the social turn promoted by digital social media emerging between 2003 and 2006 (Myspace and LinkedIn 2003; Facebook 2004; Bebo and YouTube 2005 and Twitter 2006), the world has moved from an economy of consumption to an economy of participation, rather than an economy of attention as it is currently dubbed. The first World Summit on Information Society (WSIS) is part of this turn, as it lasted from 2003 to 2005 and is now updated (as the WSIS Forum) every year by the United Nations (UN), similar to the UN Climate Change Conferences and in compliance with the Sustainable Development Goals (SDG), set by the SDG Summit of 2015. Such events are reflections of the politics of cyberism as a form of global governance beyond the sovereignty of nations, as the fate of the Blue Marble needs new frames for collective decision-making and increased coordination (see Figure 10.1).

The current situation is also conflicted, as the mindset required for consumption clashes with the mindset of participation and is also predicated on some of the economic principles of production. Computer code increases rationality and efficiency in the allocation and management of resources, which is the business base of tech giants such as Amazon (first focused on online book distribution) and Google (first focused on online directory of businesses). Modernist pressures for conformity pushed by mass consumption patterns, labour rules and gender roles clash with cyberist pressures for creativity pushed by the fragmentation of personal experience and the elimination of physical and psychic distance in favour of expressivity, reflexivity and the blurring of gender and sex roles with new socio-cultural categories such as lesbian, gay, bisexual, transgender, queer, intersex and asexual (LGBTQIA).

In the modernist paradigm, the main fuel was physical goods; in a cyberist paradigm, the main fuel is virtual goods, information as data and documents stored in 'clouds'. The double bind of the 'information society' between consumption and participation, while it deals with the long tail of the modernism legacy and the ideology of capitalism, also generates cyberism and its budding ideology.

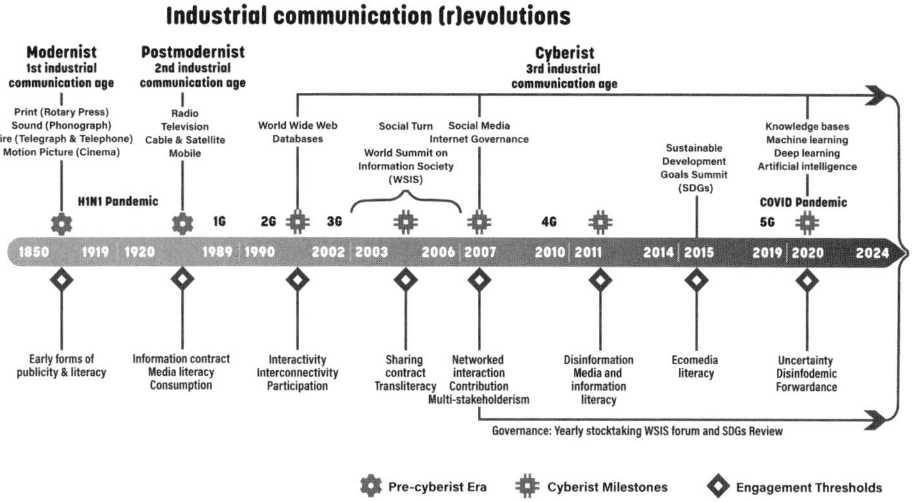

FIGURE 10.1: A Cyberist Timeline: An evolving frame of engagement (from consumption to participation to contribution).

Information is still construed as a continuation of consumption and production (data can be sold to advertisers and algorithms are considered as valuable economic secrets), but is also seen as a tool for contribution and participation (data and docs can be gathered and shared in wikis such as Wikipedia). The controlled version of information is being challenged by the chaotic and creative communication of users as they bypass established intermediaries and extoll innovative solutions for dealing with environmental and health crises. As a result, social justice seems more accessible and more democratic because the boundaries of ownership and property are being shifted, while non-proprietary code fosters solutions that promote publicity rather than secrecy, such as the Linux operating system. Additionally, virtual private networks (VPNs) can utilize secure private connections over public networks to circumvent geo-blocking, and are thriving in authoritarian regimes despite censorship controls.

Cyberism also deals with the contradictions of the transborder and transnational nature of networked culture and struggles with the distinction that Ulrich Beck (1999) makes between 'international' and 'global'. International relates to modernist views of the division of labour between raw material extraction and refined production sites with national sovereignties arbitrating trade routes, while global alludes to virtual goods and services that cross state boundaries and drive skilled labour to cosmopolitan reticular interactions that mesh together and are supported by data and media convergence, with the ideal of a single market in the horizon, as exemplified by the European policy of a 'digital single market'. None of this is neutral as it requires huge political mobilization to deal with new roles for governments and new modes of ruling, for instance, governance with new actors such as civil society forces, as exemplified in the UN's WSIS process, the SDG's yearly review and the conference of the parties (COP) process for climate change, where global treaties are continuously updated and publicly reported on a yearly basis. Cyberism is not about finite terms, as it has to deal with global uncertainties that are not finite and need to be publicly discussed and adjusted in public arenas with public consultations, hence the world summits' trend and their multi-stakeholderism.

Shifting Information Contracts

The mass and social media are not left unscathed by this cyberist moment and must revise their social contract on information. Modernist information contracts of mass media were elaborated in the 1920s on a system of publicity and veracity, relying on objectivity and investigation to distinguish news journalism from advertising and public relations focused on marketing brands and from science focused on experimentation. In terms of veracity, investigative reporting builds a particular narrative or 'story' that develops in four vertical stages: identifying a problem, making a rational analysis via expert views and proposing solutions (see Figure 10.2).

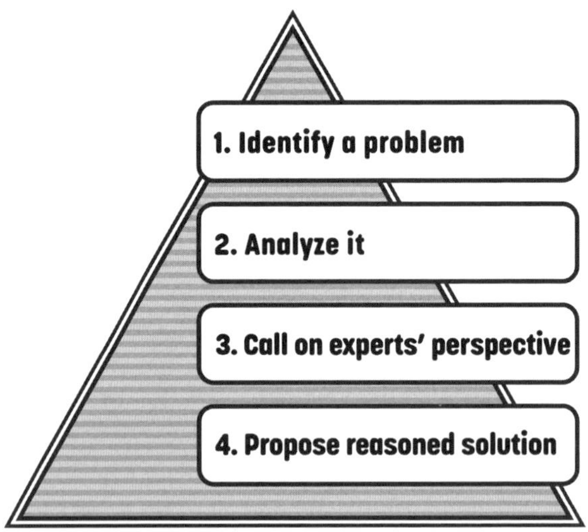

FIGURE 10.2: Information contract.

Neutrality, mentioning different points of view on an issue, and impartiality, giving roughly the same time and space to these points of view, attest to objectivity, especially since the editorial board has an agenda-setting function. It sorts the news according to importance and plays a role of guarantor and verifier before the dissemination of information. Even if the newspaper defends a certain political line, announcing such a bias strengthens, paradoxically, the norm of objectivity, where opinions and comments are often separated from facts.

Cyberist information contracts of social media disrupt modernist contracts as they add data and documents to news and are based on systems of participation related to online sharing functions. This new information-sharing contract is based on engagement in a network of peers that is amplified by reputation and recommendation, with attention and engagement metrics to monitor the whole. In terms of authenticity (not veracity), the reporting builds on a narrative or 'story' based on life experiences rather than investigation, that develops in four horizontal stages: drawing attention to a 'scandal'; arousing emotion to express the desire for change; proposing a new way of looking at the problem via a mix of expert and non-expert views and returning to solutions at the end of the process (see Figure 10.3).

No editorial board serves as a filter, though new infomediaries or influencers take on agenda-setting functions, bringing out new themes that can be picked up or not by the mass media. Alternatively, human-trained machine learning algorithms based on collaborative datasets can also decide on trends, especially the most famous recommendation algorithm 'if you liked this, then you will also like [...]'. The chronology of the media and the hierarchy of information are thus profoundly modified as exemplified by Cinematch, the algorithm used by Netflix to train viewers to see while nudging them with recommendations based on users' tastes. This may bring attention to historical documentaries such as *13th* by Ava DuVernay (2016) showing how Black people were targeted by media, government and business after the end of slavery, or *American Factory* by Steven Bognar and Julia Reichert (2019) produced by Participant Media, telling of the demise of the labour force.

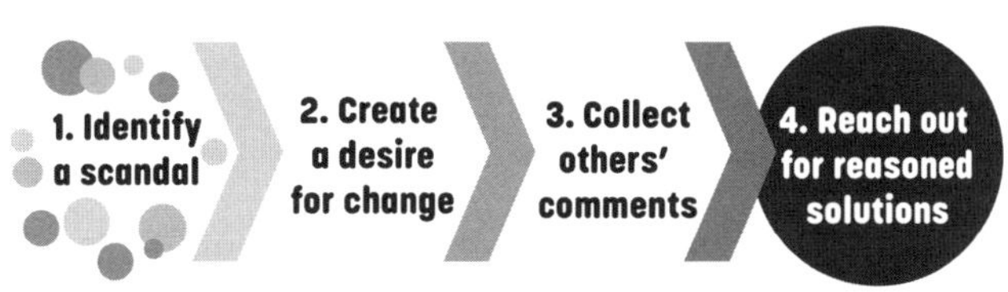

FIGURE 10.3: Sharing contract.

Cyberist contracts allow grassroot publics to express cynicism about mass media, which are perceived as elitist and attached to the establishment. Proximity, subjectivity and online and offline participation supplant objectivity and impartiality. Cyberism is disruptive as the investigative and participatory information contracts compete with each other, underlying the struggle between two cultures of control, that of the elite and of the grassroots. Theoretically, the synthesis of the publicity and participatory contracts should vouch for more transparency, more checks and balances and more democracy overall.

The sharing contract, however, can be misleading as it has been increasingly monetized, especially since 2012 when social platforms Facebook and Twitter were incorporated on the National Association of Securities Dealers Automated Quotations (Nasdaq) Stock Market and New York Stock Exchange (NYSE), respectively, and the introduction of ads to their newsfeeds, entailing the amplification of fake news. These data-driven engagements privileged 'Sponsored Stories' (Constine 2011) and therefore information that was not quality-based nor authenticity-based, but popularity-based. The algorithms for pushing these kinds of stories were set on attractiveness that could pass for agreement, assent or even mass acceptance. Some scholars have called it 'astroturfing' (Lyon and Maxwell 2004; Jacobs 2012), creating the illusion of online grassroots support with bot-generated messages. Thus, manufactured ideas and stories could find themselves on the frontline and could be monetized by their proponents in what has become a toxic industry of disinformation.

Such storytelling style has allowed some extreme and radicalized movements to carry out coordinated actions without necessarily revealing their collusion. Online polarization and radicalization are markers of the upheavals due to the contradictions between modernist and cyberist mindsets, as online participation is weaponized by disinformation to question representative democracy and challenge modernist electoral strategies. This shows how nascent cyberist ideology is, as some die-hards of the modernist era try to avoid displacement, exemplified by white nationalist movements. All the more so as, in the international mindset, non-democratic regimes feel entitled to interfere in national sovereignty issues, as revealed by such scandals as Cambridge Analytica (Wade 2018), when whistleblowers exposed how Russian trolls were bent on swaying the 2016 elections (Kertysova 2018).

The two contracts, the publicity and the participatory, seemingly evolve in two different colliding public spheres, but they are not meant to be necessarily separate. Some pure-player media co-mingle the two contracts effectively, like *Brut* and *Mediapart* in France, and *ProPublica* in the United States, where investigative journalists work with civil society and have revived citizen journalism with social media formats. The two types of publics, audiences and communities, that support these contracts, also coexist. They can be distributed along inclusive and emancipatory lines, as much as exclusive and regressive ones. They create challenges for how to sustain such diversification without undermining the collective interest.

They invite us to imagine new frameworks in the cyber era to break away from the regressive references to the modern era, without being anti-modern by going back in history to darker ages of authoritarianism. These frameworks are still unshaped for the moment, but they point to the risk of old oppositions that are less and less convincing to the new cyberist generations: right vs. left, rich vs. poor, institutionalist vs. populist, etc. Looking at the desire for change, cyberist binary oppositions, if there should be any, would rather be between short-termism and long-termism. Because in a cyberist era, it is 'the short here' and 'the long now' of the planet that are at stake due to the climate change emergency that is modifying life as we know it. No wonder that the early founders of cyberculture, like Stewart Brand, Howard Rheingold, Ted Nelson and Lee Felsenstein, who were present during the 1960s counterculture, are still active in a cyberist era, breaching the gap between modernism and cyberism, for example, in the Long Now Foundation (co-founded by Brand and Danny Hillis), sending one of their Rosetta disks on the Philae spacecraft and building the 10,000 Year Clock (Brand 2000; Turner 2006), thus setting another anthropocentric milestone.

Brand, Rheingold and other pioneers are not the only promoters of 'the long now'. Recent thinkers such as William MacAskill have been retooling their views with notions such as 'effective altruism' and suggestions for averting impending catastrophes, such as the avoidance of a 'value lock-in' that might make a system persist in spite of its toxicity (MacAskill 2022). Conversely, 'the short here' is not only associated with a modernist view and capitalism with exaggerated market profits. It can also point to the immediate care of others and/or one's own health and well-being, as well as the urgent care of the planet in the present. And long-termism does not hold a monopoly on openness and care, and can be, for some, the perpetuation of controlling current values in another guise, thus perpetuating value lock-in (e.g. eugenics and 'the plundering of our "cosmic endowment" — resources like stars, planets, asteroids and energy') (Torres 2022).

A cyberist analysis needs to keep a critical mind on the interplay between the binary opposites of 'the short here' and 'the long now'. Long-termism runs the risk of being recuperated by artificial intelligence (AI) and eugenics propounders or planet plunderers, for instance. Short-termism runs the risks of miring us in unending emergencies of continuous crises that require due diligence while preventing a distanced overview perspective. A great loss might be information, with its merging contracts, as well as useful and ethical knowledge that can potentially build sustainable solutions.

The Pandemic as Litmus Test: Information Troubles as a Cyberist Crisis

When dealing with these vying paradigms, it seems that the coronavirus disease 2019 (COVID-19) exacerbates these contradictions as it represents one of those continuous crises. It shows both a move towards a digital future of surveillance, traceability and monetization of our actions, and a future of openness, care and digital citizenship with online freedoms.

As a health crisis, the pandemic (now endemic) continues to call attention to essential workers and caretakers (such as doctors, nurses, teachers and firefighters). It has led to a general positive reassessment of public health care systems, whose public utility was very controversial before the crisis. It has also brought about decisions re-evaluating basic priorities in terms of business and commerce, with a renewed appreciation for jobs and tasks like garbage collection, as well as food and staples distribution. But this also hints at a modernist authoritarian resistance as wearing masks has been politicized and privacy further commodified as data, with (contact) tracing apps that accelerated another digital transition.

The global solution to the COVID-19 crisis was a general *e-confinement* highlighting the role of the digital revolution at the global level. People's lives moved onscreen and online. Many human functions (work, school and leisure) became mediated by digital technologies and connected networks (videoconferencing, e-learning, etc.). COVID-19 has not only been a health pandemic but also a *disinfodemic*, a disease of information contracts, be it publicity or participatory, as the value of reliable information became, literally, a matter of life or death, for instance, drinking disinfectant to combat the virus could be lethal.

'Fake news' continues to be propagated by all sorts of parties and rogue actors, who question the need for quarantining and vaccinations. It illustrates the risks of manipulating information to sow discord and increase profit. Trust in scientific inquiry and knowledge is constantly being undermined, beyond the increasing lack of trust in journalistic reporting. As such the disinfodemic appears as an unprecedented experiment into the strengths and weaknesses of information in times of uncertainty (Frau-Meigs 2022). It is an ongoing cyberist crisis that sheds light on the long now and the short here.

The disinfodemic has shown how short-termism is still attached to modernist views of capital and authority. The crisis has increased tenfold the possibility for highly concentrated digital companies such as Amazon, Apple, Facebook (parent Meta Platforms), Google (parent Alphabet) and Microsoft to collect data from large numbers of citizens with tracking apps and monitoring algorithms. They have further extended their intrusion into privacy and expanded their access to data monetization, thus consolidating their business models. The disinfodemic has revealed an interconnected patriarchy, with these companies being controlled overwhelmingly by elite male managers, connected to each other but disconnected from the unprotected cyber-working class. This patriarchy is also political and shows strands of populism and authoritarianism, with oversized egos in all types of governments, be it democracies or non-democratic regimes.

The disinfodemic also has shown how long-termism is at work with innovative attempts at care, solidarity and resolutely anti-patriarchal models, with some female heads of state prioritizing social and inter-/trans-generational solidarities. In their successful management of the crisis, they have put scientific information in the spotlight relying on medical collectives and appealing to solidarity as a vector of change and well-being, while supporting many online initiatives for the poor and the unemployed with transparent controls. They have placed an *ethics of care* into practice and have made it 'go viral' (Fine and Tronto 2020). These leaders have confirmed care theory's feminist emphases on interdependency, empathy, compassion and social justice (Gilligan 1982; Jagger 1995; Norlock 2019; Engster and Hamington 2015).

This complexified situation marks a cyberist departure from feminism as a modernist gender-based paradigm with attendant roles, as more dimensions are associated with it, such as the recognition of emotions, the expression of feelings, the acceptance of vulnerabilities and the ministrations of solidarity. It tints modernist patriarchy with other shades of masculinities that accept such emotions and vulnerabilities in both men and women, repudiating notions of dominance or submission for control (hooks 2004; Kashtan 2017).

The complexity thus generated points to an ethics of care but also underlines the 'pedagogy of cruelty' (Segato 2014: 345) that permeates the way we construct communication systems and representations, and ultimately the way we interpret information. And so in spite of the emergence of key female leaders as heads of major new media platforms and corporations over the last decade, information structures and pedagogies that we receive still impair empathy and solidarity, as well as distill violence, terror and death in 'the short here'. This systemic situation provides very little space for an information commons that counters the concentration of wealth and resources that depletes our connection with the environment, taking away our sense of agency and our potential for resilience.

The pandemic pushes us to reframe a 'new normal', accelerating the need to promote a different, non-cruel, caring pedagogy and to build informational resilience, i.e. the capacity to overcome traumatic experiences in creative ways of using information (Frau-Meigs 2023), and *forwardance*, the capacity to seize the digital media affordances to project oneself in uncertain times via cognitive strategies such as 'self-actualization, playful modelling, life-streamings and civic agency' (Frau-Meigs 2015: 18). Both informational resilience and forwardance can help us bring forth valuable lessons for the future from this global trauma, including building defences against democratic risks and expanding digital citizenship.

The Relevance of Media and Information Literacy

Cyberism by Design: Building Informational Resilience

To overcome the prevalence of screens and avoid surveillance, it is important to move towards a cyberism by design rather than by default. The short-term risk is to accept the forced evidence that e-confinement is the new normal and that digital technology is naturalized as a second nature, which it is not. This would obfuscate the fact that choice exists and that individuals can draw from a range of civic positions that relate to the type of government in which they operate. In that evaluation of the risks and opportunities, a sense of reverse engineering, including reverse surveillance (sousveillance), is essential to regain control of the narratives on screens (for instance, the 'flattening the COVID curve' slogan), and to question the hidden agendas of patriarchal governments and cyber moguls.

When considering the new normal of what to discard and what to devise, citizens need to control what is done to and with information cultures as a composite of media, docs and data. This places media and information literacy (MIL) at the core of curricula and not at the margins where it is currently. But MIL is itself caught in modernist vs. cyberist tensions, with early pioneers still supporting the focus on legacy media and new entrants proposing a transliteracy model where media and data augment each other to encompass the full scope of the current information society (Frau-Meigs 2012; also see Thomas et al. 2007). *Transliteracy* embraces the double meaning of digital convergence, both the multimedia tools for reading, writing and computing and the transdomain competencies for searching, coding and publishing online. Such convergence also introduces techno-cognitive processes such as digital smart solutions that can incorporate psychological principles guiding our consumption of information (Lewandosky et al. 2017).

To build much-needed informational resilience, MIL enables a revisitation of knowledge construction and belief systems, as well as building discernment, a deep-seated critical voice within that guides our innermost decisions. The COVID-19 crisis and its subsequent e-confinement continues to highlight MIL, much like a disease sheds light on antibodies or the need for inoculation. The disinfodemic, by contaminating democratic institutions and individual minds, has provided new insights into what constitutes healthy or reliable information in a cyberist era by drawing attention to underestimated functions that have been taken for granted: (1) the role of emotions, not merely logic, in our thought processes; (2) the power of images, not merely text, on our decision-making; (3) the influence of artificial intelligence, not merely human intelligence, to predict future decisions based on past ones; and (4) the strength of interactions within communities to validate information on the basis of values, not merely scientific proofs (Frau-Meigs 2021).

These new insights point to the dilemmas of MIL as it has to realign a modernist mindset of publicity and primacy of reason with a cyberist mindset of participation and primacy of emotion so evident in the sharing contract. The role of MIL educators and thinkers lies in being able to mitigate the negative impacts of such functions and harness their opportunities for change. Emotions alert us to the need to question how we lend authority to others. Images alert us to how we assess the authenticity and trustworthiness of our sources. Algorithms alert us to request more transparency about automated decisions based on awareness metrics (impressions, views and link clicks) and engagement metrics (likes, shares and comments). Interactions with online communities alert us to the necessary accountability and responsibility that goes along with participating and contributing in a networked society.

Paradigm Shift: MIL as Part of the Core Basics in School Programs

As notions of authority, authenticity, accountability and transparency take center stage in building informational resilience, giving publicity and secrecy a new spin, the basics and priorities of literacy also need to undergo careful scrutiny, especially with global long-term challenges of health, climate and security. The current school programs that focus mostly on text and logic relate to a modernist legacy, and are working by default, with unquestioned practices, such as school 'assignments' and 'rote learning'. They do not adequately prepare students for a cyberist era with its required attention to information as media, data and docs and how this nurtures new understandings of the complex interdependencies of body, mind and nature. The confidence crisis in education and scientific inquiry at large comes from this uncertain era where modernist illiteracy expands to the digital as a cyberist illectronism leaving entire cohorts prey to disinformation and doubt.

A full-blown cyberist core curriculum is still in the making, but its ferments are visible in calls for the *digital humanities* and the emergence of new literacies including algo-literacy, data-literacy, AI literacy (e.g. the work of projects like the Stability AI and Future of Life Institute, and for-profits like OpenAI Limited Partnership, etc.), and their array of so-called 'soft skills'. The transliteracy model tries to facilitate this transition, keeping reading and writing as fundamentals but adding coding, publishing, navigating, etc. MIL can be the core curriculum by design, with its articulated mix of info-media, info-docs and info-data that define an epistemology, its pedagogical and social uses, as well as its editorial, organizational and operational competencies (see Figure 10.4).

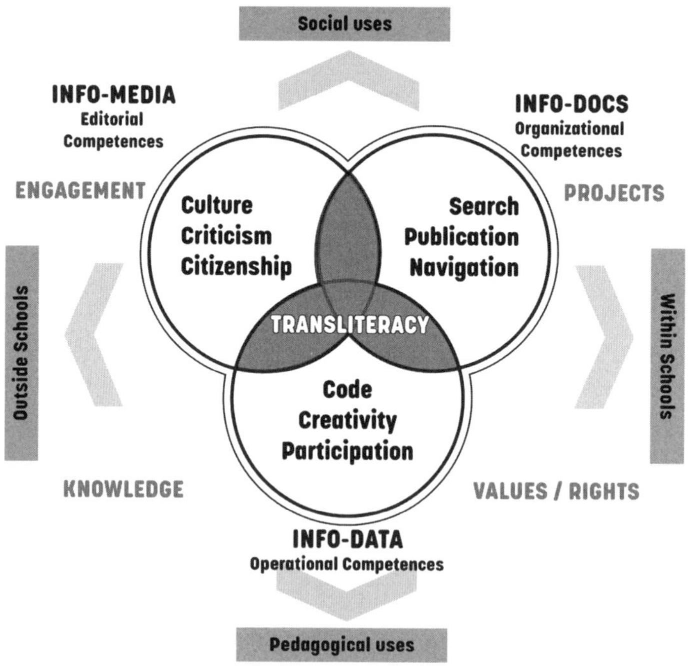

FIGURE 10.4: Transliteracy and three information cultures (Frau-Meigs 2012).

Numerous projects explore and expand this emerging curriculum, including the United Nations Educational, Scientific and Cultural Organization's (UNESCO) MIL Alliance, including Media and Information Literacy and Intercultural Dialogue (MILID), the network of MIL UNESCO chairs. Among them, the UNESCO chair 'Savoir Devenir à l'ère du développement numérique durable' (Forwardance in the era of sustainable digital development) has experimented with the implementation of such transliteracy-driven curricula, with its civil society association (Savoir*Devenir n.d.). A case in point is the YouVerify! project, funded by the European Union program, 'Media Education for all' and carried on by Savoir*Devenir along with other partners. It uses a verification plugin, InVID WeVerify, to foster critical thinking about videos and images shared on social networks and to help debunk fakes as a rapid response to combat disinformation. The project also developed a toolkit with pedagogical materials, self-paced tutorials, lesson plans and gamification (with the serious games *BotBusters* and *YouCheck Detectives*) to serve the needs of educators, students and citizens at large (YouVerify n.d.). The results showed not only increased performance but also attitude change, with students relying less on 'gut feeling' and more on 'source verification' and 'image design' (Nygren et al. 2021).

Such strategies allow forwardance to kick in, in this case, turning debunking into *prebunking interventions*, in other words, preparing the mind in advance to detect signals of disinformation and adopting reliable heuristics to counter it (Roozenbeek et al. 2020). Such strategies help delineate the new transliterate competencies needed in the twenty-first century that deal with cognitive processes such as data retrieval, fragmentation of visual docs, lateral searching, forensic filtering and comparisons across data sets. Such competencies are crucial in dealing with the consequences of a Capitalocene and especially with information related to the ecosystem. Indeed, ecological media (ecomedia) and their healthy information ecology are key to a sustainable ecosystem, especially in an endemic diseased and disinfodemic world.

Ecomedia Literacy and its Contribution to Sustainability and Democracy: Ecological Footprint and Mindprint

Drawing lessons from the disinfodemic calls for a new *information ecology*, in other words, information to help solve environmental issues and foster sustainable development. This points to an emerging trend in MIL called *ecomedia literacy* (Rust et al. 2016; López 2021). Sustainability can be construed in many different ways. In a modernist view, much attention has gone into the viability of business models from production to consumption, with an eye on key demographic changes of media audiences. But in a cyberist context, information is crucial to *environmental sustainability* (e.g. weather reports, resource management audits and air quality indices) and to public accountability via online communities organized as participatory networks. In this new context, sustainability also implies a notion of information ecology, or *ecoinformation*, and the need for open public spaces for the expression of all concerns for the future (e.g. e-waste, climate migration and cyberterrorism). This entails new attitudes to information with people sharing it and acting on it, as much as consuming it, reinforcing the need to harness the promises of a larger sustainable ecomedia and ecoinformation contract that comingles with the publicity and participatory contracts.

The disinfodemic has (de)sensitized everyone both to the market-minded responsiveness of the e-learning and social media platforms and to the lack of preparedness of school systems at local, national and global levels. The main points of vigilance around MIL were mostly dedicated to the lack of quality controls regarding commercially provided tools and content, as well as the lack of safety controls regarding the use of data and privacy protection of minors when faced with multiple screens and apps. But MIL is not just about data protection. It is also about the human-made objects and artefacts needed to construct and convey information (López 2020; Maxwell et al. 2015; Walker and Starosielski 2016).

Antonio López takes the metaphor of the cloud as a thought experiment that enables a cyberist perspective to move into forwardance. The cloud is normally connected to rain that grows trees that are in turn used to make paper, the prime modernist medium. But in a cyberist primary medium of screens, clouds today refer to 'the infrastructure of server farms that feed our gadget's apps, those servers very much depend on water for energy production and cooling' (López 2020: 383). The same could apply to the metaphor of the mine for 'data mining' that relies literally on infrastructures of cables, satellites and servers that occupy our *living* space physically, not just virtually. In other words, digital industries leave a footprint and draw on *biotic* ecosystems, thus contributing to the human-made mass and tipping the scales of environmental balance.

But ecomedia literacy is not just about drawing attention to the ecological *footprint* of the materiality of digital information, with its attendant cohort of tangible environmental issues, be it server farms CO_2 emissions or residues from mines for rare earth minerals (e.g. coltan, cassiterite and wolframite). It is also about the ecological *mindprint*, the representations and discourses about environmental issues that are mainly constructed by legacy media and are often underreported. Such constructs affect our beliefs and attitudes about the environment, and in particular, climate change. But mainstream corporate media still propagate a modernist view of the environment as a natural resource to exploit and exhaust with unlimited commercial freedoms, at the expense of – for instance – the Indigenous peoples of the Amazon (such as the Waorani in Ecuador or Kayapó Xikrin in Brazil), and the forests' flora and fauna, whose ecosystem it is. Thus, a modernist view of ecology is a parasite on the environment and does not tolerate regulatory policy. This period of mass and print media still reflects that consumption mindset by facilitating the exploitation and reinforcement of an anthropocentric view with its focus on short-term benefits.

Cyberist perspectives and social media try to emphasize alternative ecological approaches (e.g. wind and solar energies) by propagating discourses of care that show human life as co-dependent with all other living creatures on the Blue Marble. These efforts can include denouncing the carbon economy, suggesting everyday changes in support of environmental causes, as exemplified by crowdsourcing and crowdfunding for cleaning oceans or using biodegradable plastics. Though this view is gaining in currency partially due to the global e-confinement crisis, the strategy of some corporate media is to represent these issues as still open for debate, with a whole array of fake news about the climate that denies human responsibility in the process. The political economy of media needs to be scrutinized with a political economy of the environment, especially when those media rely heavily on corporate ads that promote unsustainable lifestyles.

Ecomedia and Ecoinformation Literacy

The ecomedia view is rarely taught in schools or in the perimeter of MIL. Current MIL practices still reflect that modernist mindset, acknowledging tools as empowering gadgets for students' agency and engagement, while turning a blind eye to the economics and industrial practices, thus obfuscating the ecological footprint. In terms of representations and content, the response has been a tendency to 'greenwash', that is to accept the metaphors, symbols and images that evoke *immaterial* lightness, such as 'surf', 'cloud' and 'tweets'. Greenwashing is the last attempt by polluting and waste-inducing industries to brainwash consumers into thinking that they contribute to a cleaner environment, thus deflecting attention from the underlying long-term costs of their practices.

A cyberist MIL approach requires a long view that encompasses issues of emotions, images, algorithms and interactions to foster attitude change in relation to the environment. This new ecological ideology, which could be the mark of cyberism, needs to rely on an ethics of care that includes transparency, accountability and responsibility in a networked society. To do so, an acute awareness as to how the current model of globalization impacts living systems must be fostered via education, especially as it will raise issues of *ecojustice* that may impact democratic societies and pit them against human rights or set them at war with less ecologically minded regimes. Some MIL practitioners are dealing with such issues, conjoining them to other areas of intersectional social concerns, such as racism, sexism, extremism and poverty (López 2014). But the lack of consistent curricula and shared environmental views on the Capitalocene remain, all the more so as MIL is consistently prevented from being part of the core curriculum.

Challenging the economic and political status quo will require denouncing the combined troubles brought about by patriarchy, disinformation and short-termism. We need to push for eco-ethical orientations in education and society that cannot be tackled from a modernist perspective of separate disciplinary silos. Starting with an interdisciplinary approach such as the one afforded by transliteracy and blended ecomedia and ecoinformation literacies can help change the agenda. It can be done by changing the epistemology of the current core curriculum and adding empowerment and engagement to participatory information processes, adopting a pedagogical posture that fosters political long-term thinking (see Figure 10.5). This means recovering the information commons by cleansing it of fake news pollution and pushing away from market-driven consumption, in 'the media' first as it controls the mindprint. The pedagogical is political.

The participatory contract can foster mass and social media that amplify the voices of ecojustice and the public interest, giving agency to individuals and communities that want to be included in the information commons and in the

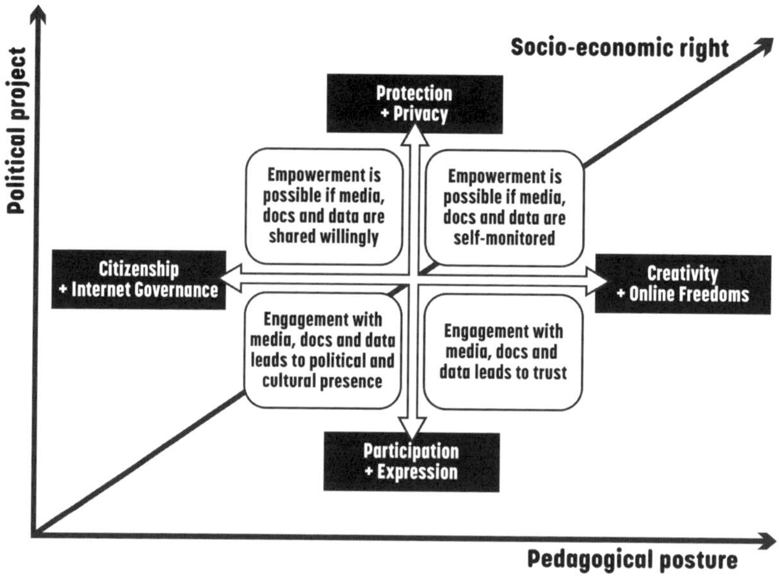

FIGURE 10.5: The epistemology of MIL (Frau-Meigs 2019).

fight for planet Earth – as it becomes clearer that there is no planet B. Ecomedia and ecoinformation literacy can develop such a sense of citizenship and belonging to a common planetary ecosystem. It can bring perspective to the short- and long-term frameworks, to the patriarchal and feminist, to the local and global and aims for an ecological common good. Environmental problems, including the digitally induced ones, will not be solved using the same kind of frameworks that created them in the first place. Critical thinking implies the invention of a new transition to ecologically minded economics and politics, with eco-ethics as part of the agenda.

As the consequences of the disinfodemic are here to stay and ring alarm bells for more troubles to come, ecomedia and ecoinformation literacy appear as a beacon of hope that casts light into the outer reaches of our lives as they interact with media, data and nature. Producing robust counter-narratives about climate change, gender injustice and an ethics of care are part of building resilience and raising preparedness to face whatever disorders might emerge in the future. Such considerations place cyberism in the logic of a Capitalocene, as the actions of humans can still be modified to maintain a whole Earth dynamic equilibrium. This widespread recognition that we are passing through a unique phase of history when individuals and nations can affect climate change or the genetic code calls attention to our larger responsibilities to the planet (Chakrabarty 2018). Cyberists who care to be humanists contemplate the role of information in all its codes (alphabetic, numeric and genetic), as well as how it affects the routine patterns of life. They are placed in the position of becoming *good ancestors* (Krznaric 2020). They can be good keepers of the Earth, stewards for the generations to come, as a sustainable democracy that thrives with the civic engagement of those who share with care.

NOTE

1. From the original Greek, ἄνθρωπος (Liddell and Scott 1940a).

REFERENCES

Ampère, André-Marie (1838), *Essai sur la philosophie des sciences, ou exposition analytique d'une classification naturelle de toutes les connaissances humaines*, Paris: Bachelier.

Beck, Ulrich (1999), *What is Globalization?* Cambridge: Polity Press.

Bell, Daniel ([1970] 1996), *The Cultural Contradictions of Capitalism* (with new Afterword), New York: Basic Books.

Bognar, Steven and Reichert, Julia (dirs.) (2019), *American Factory 美国工厂*, USA: Netflix.

Borgman, Christine L. (2000), *From Gutenberg to the Global Information Infrastructure: Access to Information in the Networked World*, Cambridge, MA: MIT Press.

Brand, Stewart (2000), *The Clock of the Long Now: Time and Responsibility*, New York: Basic Books.

Castells, Manuel (1997), *La era de la informacion: economia, sociedad y cultura, la sociedad red*, vol. 1, Madrid: Alianza.

Certini, Giacomo and Scalenghe, Riccardo (2015), 'Is the anthropocene really worthy of a formal geologic definition?', *The Anthropocene Review*, 2:1, pp. 7–80.

Chakrabarty, Dipesh (2018), 'Anthropocene time', *History and Theory*, 57:1, pp. 5–32.

Constine, Josh (2011), 'Facebook sponsored ads to appear in the Web NewsFeed in 2012', *TechCrunch*, 20 December, https://techcrunch.com/2011/12/20/sponsored-stories-news-feed/. Accessed 18 February 2022.

Crutzen, Paul J. and Stoermer, Eugene F. (2000), 'The "Anthropocene"', *Global Change Newsletter*, 41, pp. 17–18.

DuVernay, Ava (dir.) (2016), *13th*, USA: Netflix.

Elhacham, Emily, Ben-Uri, Liad, Grozovski, Jonathan, Bar-On, Yinon M. and Milo, Ron (2020), 'Global human-made mass exceeds all living biomass', *Nature*, 588, pp. 442–44.

Engster, Daniel and Hamington, Maurice (eds) (2015), *Care Ethics and Political Theory*, Oxford: Oxford University Press.

Featherstone, Mike (2007), *Consumer Culture and Postmodernism*, London: Sage.

Fine, Michael and Tronto, Joan (2020), 'Care goes viral: Care theory and research confront the global COVID-19 pandemic', *International Journal of Care and Caring*, 4:3, pp. 301–09.

Frau-Meigs, Divina (2011a), 'Attaching media education to human rights by socializing young people to ethics online', in C. Bucht, U. Carlsson and C. von Feilitzen (eds), *New Questions, New Insights, New Approaches: Contributions to the Research Forum at the World Summit on Media for Children and Youth 2010*, Gothenberg: Nordicom, pp. 173–87.

——— (2011b), 'Media matters in the cultural contradictions of the information society', *Towards a Human-Rights-Based Governance*, Strasbourg: Council of Europe Press.

——— (2011c), *Penser la société de l'écran: Dispositifs et usages*, Paris: Presses de la Sorbonne nouvelle.

——— (2012), 'Transliteracy as the new research horizon for media and information literacy', *Media Studies*, 3:6, pp. 6–16.

——— (2015), 'Augmented Media and Information Literacy (MIL)', in S. Kotilainen and R. Kupiainen (eds), *Reflections on Media Education Futures: Contributions to the Conference Media Education Futures in Tampere, Finland* 2014, Göteborg: Nordicom, pp. 13–25.

——— (2021), 'Foreword: Lasting lessons learnt from the "fake news" crisis: MIL as the 1st curriculum', in M. Hoechsmann, G. Thésée and P. R. Carr (eds), *Education for Democracy 2.0: Changing Frames of Media Literacy* , Boston, MA: Brill.

——— (2022), 'Media, information literacy and civics during the disinfodemic: The Samuel Paty Case in France', in Y. Friesem, U. Raman, I. Kanižaj and G. Y. Choi (eds), *The Routledge Handbook of Media Education Futures Post-Pandemic*, London: Routledge, pp. 455–62.

——— (ed.) (2023), 'Digital literacy, social and individual well-being', in D. Chavalarias (ed.), *Toward a Research Agenda on Digital Media and Humanity Well-Being*, France: CNRS (Centre national de la recherche scientifique), https://hal.science/hal-04091733/. Accessed 17 May 2023.

Frau-Meigs, Divina, Kotilainen, Sirkku, Hoeschmann, Michael, Shelat, Manisha and Poyntz, Stuart (eds), (2020), *Handbook of Media Education Research*, New York: Wiley.

Gilligan, Carol (1982), *In a Different Voice: Psychological Theory and Women's Development*, Cambridge, MA: Harvard University Press.

Habermas, Jurgen (1962), *The Structural Transformation of the Public Sphere: An Inquiry into a Category of Bourgeois Society*, Cambridge, MA: MIT Press.

hooks, bell (2004), *The Will to Change: Men, Masculinity, and Love*, New York: Simon and Schuster.

Jacobs, Joanne (2012), 'Faking it: How to kill a business through astroturfing on social media', *Keeping Good Companies*, 64:9, pp. 567–70.

Jagger, Alison (1995), 'Caring as a feminist practice of moral reason', in V. Held (ed.), *Justice and Care: Essential Readings in Feminist Ethics*, Boulder, CO: Westview Press.

Kertysova, Katarina (2018), 'Artificial intelligence and disinformation: How AI changes the way disinformation is produced, disseminated, and can be countered', *Security and Human Rights,* 29:1–4, pp. 55–81. Accessed 26 January 2023.

Krznaric, Roman (2020), *The Good Ancestor: How to Think Long Term in a Short Term World*, New York: The Experiment.

Lewandosky, Stephan, Ecker, Ullrich K. H. and Cook, John (2017), 'Beyond misinformation: Understanding and coping with the "post-truth" era', *Journal of Applied Research in Memory and Cognition*, 6:4, pp. 353–69.

Liddell, Henry George and Scott, Robert (1940a), 'ἄνθρωπος', in *A Greek-English Lexicon* (rev. and augmented by H. S. Jones and R. McKenzie), Oxford: Clarendon Press, http://www.perseus.tufts.edu/hopper/text?doc=Perseus:text:1999.04.0057:entry=a)/nqrwpos. Accessed 15 January 2023.

——— (1940b), 'κῠβερνᾰώ', in *A Greek-English Lexicon* (rev. and augmented by H. S. Jones and R. McKenzie), Oxford: Clarendon Press, http://www.perseus.tufts.edu/hopper/text?doc=Perseus:text:1999.04.0057:entry=kuberna/w. Accessed 25 July 2022.

López, Antonio (2014), *Greening Media Education: Bridging Media Literacy with Green Cultural Citizenship*, New York: Peter Lang.

——— (2020), 'Expanding ethics to the environment with ecomedia literacy', in D. Frau-Meigs, S. Kotilainen, M. Hoeschmann, M. Shelat and S. Poyntz (eds), (2020), *Handbook of Media Education Research*, New York: Wiley, pp. 381–397.

——— (2021), *Ecomedia Literacy: Integrating Ecology into Media Education*, New York: Routledge.

Lyon, Thomas P. and Maxwell, John W. (2004), 'Astroturf: Interest group lobbying and corporate strategy', *Journal of Economics & Management Strategy*, 13:4, pp. 561–97. Accessed 15 February 2021.

MacAskill, William (2022), *What We Owe the Future*, New York: Basic Books.

Maxwell, Richard, Raundalen, Jon and Vestberg, Nina Lager (2015), *Media and the Ecological Crisis*, New York: Routledge.

Mead, Margaret (1994), 'Cybernetics', in F. Heylighen (ed.), *Web Dictionary of Cybernetics and Systems*, http://pespmc1.vub.ac.be/ASC/CYBERNETICS.html. Accessed 19 February 2021.

NASA ([1972] 2001), 'The Blue Marble from Apollo 17', *Visible Earth*, NASA Johnson Space Center Gateway to Astronaut Photography of Earth, https://visibleearth.nasa.gov/images/55418/the-blue-marble-from-apollo-17/. Accessed 13 January 2023.

Norlock, Kathryn (2019), 'Feminist ethics', in E. N. Zalta (ed.), *The Stanford Encyclopedia of Philosophy* (Summer ed.), https://plato.stanford.edu/archives/sum2019/entries/feminism-ethics/. Accessed 29 February 2021.

Nygren, Thomas, Guath, Mona, Axelsson, Carl-Anton Werner and Frau-Meigs, Divina (2021), 'Combatting visual fake news with a professional fact-checking tool in education in France, Romania, Spain and Sweden', *Information*, 12, p. 201.

Poyntz, Stuart R., Frau-Meigs, Divina, Hoechsmann, Michael, Kotilainen, Sirkku and Pathak-Shelat, Manisha (2020), 'Introduction: Media education research in a rapidly changing media environment', in D. Frau-Meigs, S. Kotilainen, M. Hoeschmann, M. Shelat and S. Poyntz (eds), *Handbook of Media Education Research*, New York: Wiley, pp. 2–3.

Randall, Ian (2022), 'Epoch or event? Defining the Anthropocene, *Physics World*, 20 December, https://physicsworld.com/a/epoch-or-event-defining-the-anthropocene/. Accessed 13 January 2023.

Reinert, Al (2011), 'The Blue Marble shot: Our first complete photograph of Earth', *The Atlantic*, 12 April, https://www.theatlantic.com/technology/archive/2011/04/the-blue-marble-shot-our-first-complete-photograph-of-earth/237167/. Accessed 7 January 2020.

Roozenbeek, Jon, van der Linden, Sander and Nygren, Thomas (2020), 'Prebunking interventions based on "inoculation" theory can reduce susceptibility to misinformation across cultures', *Harvard Kennedy School Misinformation Review*, 3 February, http://dx.doi.org/10.37016//mr-2020-008. Accessed 6 February 2023.

Rust, Stephen, Monani, Salma and Cubitt, Sean (2016), *Ecomedia: Key Issues*, New York: Routledge.

Savoir*Devenir (n.d.), 'Ressources', website, https://savoirdevenir.net/ressources/. Accessed 18 May 2023.

Segato, Rita Laura (2014), 'Las nuevas formas de la guerra y el cuerpo de las mujeres', *Sociedade e Estado*, 29:2, pp. 341–71.

Starosielski, Nicole and Walker, Janet (eds) (2016), *Sustainable Media: Critical Approaches to Media and Environment*, London: Routledge.

Subcommission on Quaternary Stratigraphy (n.d.), *Working Group on the 'Anthropocene'*, International Commission on Stratigraphy, http://quaternary.stratigraphy.org/working-groups/anthropocene/. Accessed on 18 December 2022.

Torres, Émile P. (2022), 'Understanding "longtermism": Why this suddenly influential philosophy is so toxic', *Salon*, 20 August, https://www.salon.com/2022/08/20/understanding-longtermism-why-this-suddenly-influential-philosophy-is-so/. Accessed 27 January 2023.

Thomas, Sue, Joseph, Chris, Laccetti, Jess, Mason, Bruce, Mills, Simon, Perril, Simon and Pullinger, Kate (2007), 'Transliteracy: Crossing divides', *First Monday*, 12:12, 3 December, https://journals.uic.edu/ojs/index.php/fm/article/download/2060/1908. Accessed on 15 January 2023.

Turner, Fred (2006), *From Counterculture to Cyberculture: Stewart Brand, the Whole Earth Network, and the Rise of Digital Utopianism*, Chicago: University of Chicago Press.

Wade, Michael (2018), 'Psychographics: The behavioural analysis that helped Cambridge Analytica know voters' minds', *The Conversation*, 21 March, https://theconversation.com/psychographics-the-behavioural-analysis-that-helped-cambridge-analytica-know-voters-minds-93675. Accessed on 26 January 2023.

Weiner, Norbert ([1948] 1961), *Cybernetics or Control and Communication in the Animal and the Machine* (2nd ed.), Cambridge, MA: MIT Press.

YouVerify (n.d.), 'Resources | YouVerify', website, https://youverify.eu/en/resources. Accessed 18 May 2023.

Zalasiewicz, Jan, Waters, Colin N., Williams, Mark and Summerhayes, Colin P. (eds) (2019), *The Anthropocene as a Geological Time Unit: A Guide to the Scientific Evidence and Current Debate*, Cambridge: Cambridge University Press.

Zalasiewicz, Jan, Williams, Mark, Steffenn, Will and Crutzen, Paul (2010), 'The new world of the anthropocene', *Environmental Science and Technology*, 44:7, pp. 2228–31.

11

Journalistic Learning and Intentional Teaching with Technologies: STEM and Rural Communities

Ed Madison

The promise of new media technologies was connectivity – the notion of a 'global village' that would increase human interaction and engagement.[1] However, many millennials, along with generation Z and generation alpha students, have reported feeling disconnected, stressed and anxious about performance demands on social media, a plethora of virtual environments where they express compassion with emojis and validation by amassing 'likes', etc. (Midkiff et al. 2015; Selvi et al. 2022). Indeed, several studies indicate a correlation between mobile devices' ubiq uitous use and a rise in teen bullying, depression, anxiety, isolation, sexting and even suicide (O'Keefe and Clarke-Pearson 2011; Christakis and Moreno 2009; Selfhout et al. 2009). The list of social challenges teens face is daunting, including geopolitical instability, gun violence, climate crisis and the large-scale effects of a global pandemic. The rush to online instruction amid the coronavirus disease 2019 (COVID-19) crisis forced students to isolate themselves from engaging in face-to-face contact with friends, extended family and teachers. They had to confront jarring questions about the future they will inherit and the uncertainties it may bring (UNICEF 2021).

Simultaneously, the spread of misinformation about these grave issues compounds their influence on the quality of life and the evolution of our species, bewildering many students and the adults charged with educating them. At the height of the pandemic, institutions we once relied on for well-reasoned guidance instead gave us 'alternative facts' and dubious medical claims. This dramatic deviation from convention clouded policies and inevitably cost lives.

However, the responsibility of educating students persists beyond shifts in the political climate. We must give them tools to navigate – no matter the philosophy of leadership at any given time. Millennial and generation Z students are not disinterested in current affairs, as so often inaccurately asserted. On the contrary, as demonstrated by the above-mentioned mental health epidemic among teens, they cannot avoid them (Poindexter 2012; Coe et al. 2022).

Many students report being overwhelmed by the vast amounts of socially significant information vying for their attention and the ill-considered ways in which it is often disseminated (DeJonckheere et al. 2018). When leaders attack the fundamental role of a free press and seek to legitimize untruthful claims for political gain, they jeopardize our ability to teach young people how to distinguish facts from falsehoods and make sense of the world. According to a 2016 Stanford University study, middle school, high school and college students' inability to identify credible information is 'dismaying', 'bleak' and '[a] threat to democracy' (Domonoske 2016: n.pag.).

Social media platforms have compounded the problem previously by abdicating responsibility for moderating public posts. For example, we now know that Facebook and Twitter fell alarmingly short of detecting the malicious efforts of Russian hackers to confuse the electorate during the run-up to the 2016 United States (US) presidential election, which fueled fear and flooded Americans' news feeds with fictitious propaganda. Four years later, social media platforms were more aggressive in flagging inaccurate medical and political posts and memes, but still faced growing public and congressional concerns about their pervasive power. It remains to be seen whether regulatory agencies will eventually restrain social media's dominance. However, the public is arguably more aware of its impact.

Human ingenuity gives us the capacity to imagine a world that rises above passive and familiar dystopian views of the future. Instead, we commit to harnessing the power of technological advancements for the greater good. Conscious and intentional approaches to media and technology integration in education can empower students to meet and defeat many of society's most significant concerns head-on.

This high intention informs the work of Journalistic Learning Initiative (JLI), a standards-aligned and researched-based approach to empowering students' voices and improving educational outcomes for middle and high school students. The program expanded from two to eighteen schools in Oregon and California in its first five years, serving more than 4,500 students. JLI uses journalism education strategies to enhance learning outcomes and embeds the program in secondary English language arts and social studies courses. The approach acknowledges and honours students' intrinsic interests as a first step before introducing more challenging themes – thereby meeting students where they are (Madison 2012, 2015).

This chapter begins with a brief history of JLI and its intentional approach to teaching with technologies. It explores the JLI framework, which focuses on activating student voice and agency, and reports on how we have applied it in teaching middle and high school students. Further, it examines how JLI's approach is being used to diversify student recruitment in science, technology, engineering and math (STEM) fields through near-peer mentorship. Finally, the chapter provides a more intentional way to reconsider how we frame and speak about technology's role in our lives.

Formulating JLI

Journalism's presence in schools has a long history, most often in the form of student newspapers. Following the yellow journalism era of the late 1800s, media moguls sought to legitimize the profession and ward off congressional regulation by promoting journalism in secondary and college-level education. Academic programs also provided a reliable pipeline of prospective journalism professionals (Carey 2000). The legacy continues. Programs have expanded to include broadcasting, podcasts and a variety of digital publications.

However, in recent decades, journalism education programs have suffered from budget cuts and politicization. Instead of championing free speech, some risk-averse administrators feared transparency or conservative retribution for promoting 'fake news' defunded student media (SPLC 2023). When offered, journalism is typically an elective or an after-school program with less curricular status than traditional or Advanced Placement English (Madison 2015). While JLI student works are published, the program's emphasis differs in approach and intent. It provides teachers and their students with a journalistic lens that can be applied in English language arts courses and across the curriculum, including science and social studies courses. In its first few years, JLI recruited recent stellar journalism school graduates committed to serving their communities. We trained and placed them alongside teachers in secondary schools where their millennial sensibilities and near-peer relationship with students reinvigorated learning. The teaching team engaged students in hosting 'Interview Days', where they video conferenced or conducted in-person interviews with significant local leaders and distant experts about topics of interest. The students researched their story subjects, developed questions, wrote, revised and published their findings to the web.

We shifted our approach to supporting teachers through online training to scale the program and facilitate independent instruction. Also, rather than embed the program once or twice a week in existing English and social studies courses, JLI trains teachers to lead its stand-alone 'Effective Communicators' course. This shift provides teachers with more student-contact hours and more time to explore topics and build skills. Self-directed project-based learning is still at the heart of

JLI's work. Participants initiate and advocate for areas of inquiry and drive every aspect of the process.

JLI's inception can be traced back to 2010 when my search for a dissertation topic led me to meet Esther Wojcicki and several of her Palo Alto High School journalism students by chance at a conference at Stanford University. Their poise, professionalism and presentation skills immediately struck me. Subsequent conversations revealed that Wojcicki's scholastic journalism program was among the nation's largest, with nearly 500 of the school's 1,800 students engaged in some aspect of journalism. They generated eight distinct forms of student-produced media, including a newspaper, two feature magazines, an online news site, a sports magazine, a literary magazine, a yearbook and a daily three-camera newscast.

Of more significance were the high levels of student leadership demonstrated. During two years of study, I witnessed faculty advisors playing a secondary role to teams of student editors elected by their peers who managed their publications' day-to-day operations. I interviewed dozens of students and alumni of the program. It was evident that having authentic opportunities to express *voice* and *agency* at such a young age produced graduates who emerged as high achievers later in their professional lives.

However, significant questions remained. Were the results I witnessed an anomaly, or instead attributable to privilege? Palo Alto High School is across the street from Stanford and is located in one of the wealthiest zip codes in the US. Conversely, was it possible to replicate, reformat and scale salient elements of the school's program to benefit educators and students in far less affluent communities?

Further investigation led me to survey 664 students from ten different high schools – including Palo Alto High School – from across the US, comparing journalism students and non-journalism students (see Madison 2012, 2015). Findings indicated that social-situated learning, intrinsic motivation and skill enhancement were rated higher by journalism students than by non-journalism students after controlling for school and student demographics (community type, class standing or socioeconomic status). Journalism students in less affluent communities reported motivational beliefs and learning strategies similar to students in affluent neighborhoods. Journalism students also reported significantly higher levels of personal growth and sense of self than non-journalism students.

Nonetheless, it is fair to state that privilege affords students in affluent communities opportunities unavailable to students elsewhere. Disadvantaged students are more likely to encounter personal stressors that can impede their advancement, including domestic instability, neglect and even abuse (Becker and Luthar 2002). They are also more likely to experience food anxieties and hunger, which can adversely affect their ability to concentrate and be fully present in class (Brophy 2013).

While Palo Alto High School students are accustomed to working on the latest Apple Macintosh computers fully equipped with the same Adobe software used by professionals, a disproportionate number of schools are years behind the technology curve. Planned obsolescence is also a factor. Consumer-oriented corporations are highly motivated to create a steady flow of new products with short life cycles to drive annual sales and quarterly earnings. Conversely, school systems can be contrarian in their decision-making, are often slow to embrace change or simply do not have the budgets to stay current.

Students experience circumstances at home that often reflect the technology gaps that persist in schools. Lower-income families often cannot afford to purchase the latest devices, which can become a source of embarrassment for students whose friends have higher disposable incomes, especially given the social capital teens associate with the use of Snapchat and Instagram.

However, lack of access to the latest technologies need not impede students' classroom experiences and learning outcomes. Some educators have identified low-tech solutions they assert are more favorable than high-tech options in specific settings. These educators value *intentional teaching* with technology, a mindful approach that seeks to slow down the pace of learning rather than rush the process. The objective is to allow students to savor the experience rather than be satiated by it. Intentional teaching asks that we reconsider the value we ascribe to *multitasking* and acknowledge its potential costs. It values single-focused *sensory* learning more than simultaneous *surface-level* awareness. Consequently, learning experiences can become deeper and more meaningful.

JLI's Intentional Approach

Intentional teaching is entirely on display at Palo Alto High School. A site visit reveals pedagogical practices that surprise many first-time observers. It's not uncommon to find shiny screens and keyboards moved aside to make room for spreading out and examining printed newspapers. Dissatisfaction with the shelf life and quality of textbooks initially led Esther Wojcicki to hand her students newspapers – before desktop computers became commonplace. As digital technology emerged, she continued to use printed newspapers. Wojcicki saw opportunities to challenge students to make spatial comparisons not readily apparent in digital articles. They compared how two publications covered the same story, noting the length, placement and size of headlines and images. Observing and contrasting these editorial choices enhanced students' ability to critique what they were reading and discern biases.

Wojcicki also found that students engaged differently with screens than with printed text. Whereas screens tend to isolate a single story, newspapers can provide a broader range of discovery because they can engage your peripheral vision. Yes, hyperlinks in digital articles allow readers to click through to deeper source materials, but the material is typically related to the same narrow topic. As a practical matter, the use of newspapers also allowed Wojcicki to easily observe when students were on task as opposed to lost in their screens or surfing unrelated websites. Wojcicki and I remained mindful of the analog's attributes as we worked together to develop JLI with donor support.[2]

When defining 'technology', it is easy to forget that printed publications, much like pencils and paper, were once considered new technologies. However, it may be difficult to imagine they were met with skepticism. Plato warned against the advent of writing, stating, 'They [people] will cease to exercise memory because they rely on that which is written, calling things to remembrance no longer from within themselves, but by means of external marks' (Hackforth 1972). Not surprisingly, a common concern is expressed today about young people's reliance on mobile devices, specifically Google and Wikipedia as reference sources.

In developing JLI, our intent was never to abandon newer technologies, but to allow a student to discover that printed media remains valuable. Much like the resurgence of vinyl music, printed publications are experiencing a revival. *The Wall Street Journal* reported that book sales rose 5 per cent in 2017, compared to a 17 per cent decline in digital eBook sales (Turner 2017).

Applying JLI and Intentional Teaching in Rural Communities

Reliance on digital technologies is rarely an option for many schools in low-income communities. We intentionally sought to pilot JLI in under-resourced schools with underserved students. Oaklea Middle School, one of JLI's pilot sites, is significantly different from Palo Alto. It is in the rural community of Junction City, Oregon, a town hard hit by economic downturns and a dwindling timber industry. There are no shiny new Apple computers loaded with the latest software in their 1970s understated cinder block building. Teachers must schedule a time to use a single shared computer lab, housed with decade-old PCs running outdated Windows software. However, learning continues to thrive undaunted.

Implementing JLI at Oaklea required resourcefulness and perseverance, including lobbying the school district's Internet Technology office for special permission to use video conferencing software. Despite its obvious value as a pedagogical tool to link students with distantly located experts, Skype and FaceTime are among the online platforms that many schools typically block in their earnest effort to protect students and eliminate classroom distractions.

While JLI serves several high schools, we discovered that middle school is ideal for introducing the program – specifically in sixth grade. We have observed that 12- and 13-year-old students are less likely to have become jaded by rote traditional pedagogical practices that diminish spontaneity and one's sense of wonder.

At the same time, middle school years can be a tumultuous time for young people, marked by a deluge of psychological and physiological transformations, including growth spurts (Lee Manning 2002), hormonal changes (Stern 1999) and social awkwardness (London et al. 2007). It is a primary time for identity construction and exploration of the self, as students struggle to define and assert who they are and who they want to become (Kinney 1993; McCracken 2008). It is also a time when students may appear to rebel against parental ideals and societal norms as acts of self-discovery and identity formation (Raby 2005; Milner 2013).

In light of these factors, we observed that while digital stimuli surround many students, they express feeling grounded when working with analogue tools. At Oaklea, JLI blended the use of both in piloting its methodologies. Students researched and prepared questions for a mix of in-person community guests and distantly located experts who addressed a wide range of topics, including homelessness, animal cruelty and post-traumatic stress disorder among war veterans. Students collaborated in teams of three, using paper and pencils for ideation and note-taking. They also used the school's inexpensive Chrome tablet devices for researching information about their guests via the internet.

As the pilot progressed, we developed a JLI core framework, comprised of four pillars (see Figure 11.1), to articulate the emerging program's foundational principles better. We realized that JLI's efficacy relies on emphasizing student *voice*, *agency*, *publication* and *reflection*. *Voice* supports students in realizing their opinions and experiences matter; when students have *agency*, they learn how their voice can influence others; *publication* acknowledges the power of sharing stories with an authentic audience – not just teachers; and *reflection* allows students to broaden their awareness and explore deeper meanings (Madison et al. 2019).

Intentional teaching is how we access and implement the JLI core framework. Our emphasis is cultivating student character, irrespective of whether they have access to newer technologies. We observe that when young people are empowered to express their authentic voices, it enhances their sense of agency, referenced by other researchers as self-efficacy (Bandura 2010) or self-determination (Deci and Ryan 2004). Publication for a genuine audience places their works in the world – making learning more than an exercise. Assignments become relevant and meaningful. Reflecting on this process through introspection, discussion and analysis completes the circle.

Journalism provides a mechanism to manifest that result – and it can be just as effectively achieved with old newspapers as it can with electronic devices. Over

FIGURE 11.1: The JLI core frameworks.

the pilot academic year, we observed a significant shift in student engagement levels – exemplified by straighter postures, longer attention spans and eagerness to participate. Writing skills also improved in terms of clarity, detail and length.

Diversifying STEM through Digital Storytelling and Near-Peer Mentoring

JLI's commitment to increasing underserved students' opportunities extends beyond alignment with English language arts and social studies curricula. National Science Foundation (NSF) grants have allowed our research team to investigate ways to use journalistic digital storytelling to motivate underrepresented students to consider STEM career pathways.

'My STEM Story', our three-year project, uses digital storytelling to capture salient interactions between six high school-curious mentees we match with six science-proficient undergraduates – all students of color. While mentoring-based approaches have shown some promise in ameliorating these issues, many mentoring programs are academically oriented, and mentors are often distant graduate students or professors (Pluth et al. 2015; Tenenbaum et al. 2014).

Near-peer mentoring has emerged as an effective alternative to traditional forms of mentoring. In contrast to conventional mentors, near-peers are just a few years older than mentees and thereby enjoy a shared language and a greater degree of closeness (Zaniewski and Reinholz 2016).

Our research team designed and facilitated a three-day summer program at the University of Oregon, which assigned documentary video crews to trail the pairs as mentees shadowed their mentors in labs, dined together and discussed STEM pathway benefits and challenges.

Classrooms of high school students will view edited video highlights and complete reflection exercises designed to measure each digital story's impact. Our study framework is *identity-based motivation* (IBM) theory (Oyserman 2015), which posits that individuals interpret situations in ways that are congruent with their currently active identities. When a task feels identity-congruent, struggle is assessed as appropriate and meaningful. However, when a task feels identity-incongruent, the same struggle can be evaluated as senseless and 'not for people like me'. As such, IBM theory informed our exploration of how STEM-related perceptions of 'struggle' and self-efficacy experienced by high school students of color can be influenced in new ways, acknowledging intersectionality factors of race, ethnicity and gender.

Mentees reported that the high point of their mentoring experience was shadowing near-peers in their respective labs. Whether wearing goggles and white coats to measure aluminum gallium minerals or peering through a microscope lens to sort worm embryos, the mentees described the experiences as immersive and memorable.

For example, Navaeh, an African American high school junior, had the opportunity to shadow Martha, a Latinx undergraduate. She studies human development's biological processes at the cellular level on her path to becoming a medical doctor and Ph.D. Their bond strengthened after discovering they were the first in their respective families to pursue college. In the lab, Navaeh was initially frustrated by her inability to sort and isolate worm embryos. Further, she reported, 'At first, it was kind of gross. I was just kind of oooh like it's moving. After I finally got over it, I was really fascinated, and I wanted to keep looking'.

If the edited videos prove effective in motivating underserved students to pursue STEM-related career paths, it could support efforts to diversify science and society may benefit from contributions that would otherwise remain marginalized; gifts that may prove to be significant innovations that affect the quality of our lives.

The Future of Educational Technologies: Potential and Pitfalls

As we evolve as humans, we will undoubtedly invent new and perhaps unimaginable ways to engage with technology. Implants, sensors and wearable devices are already emerging, and we can anticipate they will have some educational applications. A remaining question is: to what end?

In his classic book, *Understanding Media*, theorist Marshall McLuhan argued that a medium is 'any extension of ourselves', or more broadly 'any new technology' (McLuhan [1964] 2003: 19). However, debates continue as to whether our advanced technological devices liberate and enhance human capacities or whether they can mechanize and thwart otherwise organic and authentic human interactions.

Curiously, we now know that Steve Jobs prohibited his children from using Apple's iPad, and Bill Gates denied his three kids mobile phones until age 14 (Weller 2018). Similarly, many Silicon Valley tech executives send their children to low-tech Waldorf Schools where pens, paper and scissors are abundant, and no computers are found (Richtel 2011). Did these technology titans have unspoken fears about a future too heavily reliant on machines?

I'm not suggesting any form of conspiracy. Perhaps, they simply recognize the value of a more measured and monitored approach to introducing young people to technology. There is a sensible approach to negotiating young people's screen time and thoughtful use of technology in schools. It begins by realizing that we invent technology to ease and enhance our existence – not burden it.

NOTES

1. Interestingly, Marshall McLuhan's characterization of the global village has been widely misinterpreted. He clarified his position as follows: 'The global village absolutely insures maximal disagreement on all points. It never occurred to me that uniformity and tranquillity were the properties of the global village. It has more spite and envy' (McLuhan 1967: 272). He later updated the concept to a 'global theater', in which all 'become actors and there are few spectators. The population of the world is both the cast and content of this new theater' (McLuhan and Nevitt 1972: 145).
2. See John S. Hutton et al. (2020) for an interesting discussion that 'suggest[s] a need for further study into the association between screen-based media use and the developing brain, particularly during early childhood.'

REFERENCES

Bandura, Albert (2010), 'Self-efficacy', in I. B. Weiner and W. E. Craighead (eds), *The Corsini Encyclopedia of Psychology*, 4th ed., Hoboken, NJ: John Wiley & Sons.

Becker, Bronwyn E. and Luthar, Suniya S. (2002), 'Social-emotional factors affecting achievement outcomes among disadvantaged students: Closing the achievement gap', *Educational Psychologist*, 37:4, pp.197–214.

Brophy, Jere E. (2013), *Motivating Students to Learn*, New York: Routledge.

Carey, James W. (2000), 'Some personal notes on U.S. journalism education', *Journalism*, 1:1, pp. 12–23.

Christakis, Dimitri A. and Moreno, Megan A. (2009), 'Trapped in the net: Will Internet addiction become a 21st-century epidemic?' *Archives of Pediatrics & Adolescent Medicine*, 163:10, pp. 959–60.

Coe, Erica, Cordina, Jenny, Enomoto, Kana, Jacobson, Raelyn, Mei, Sharon and Seshan, Nikhil (2022), 'Addressing the unprecedented behavioral-health challenges facing Generation Z', *McKinsey & Company*, 14 January, https://www.mckinsey.com/industries/healthcare-systems-and-services/our-insights/addressing-the-unprecedented-behavioral-health-challenges-facing-generation-z. Accessed 25 January 2023.

Deci, Edward L. and Ryan, Richard M. (eds) (2004), *Handbook of Self-determination Research*, Rochester, NY: University of Rochester Press.

DeJonckheere, Melissa, Fisher, Andre and Chang, Tammy (2018), 'How has the presidential election affected young Americans?', *Child and Adolescent Psychiatry and Mental Health*, 12:8, 13 February, https://doi.org/10.1186/s13034-018-0214-7. Accessed 6 April 2023.

Domonoske, Camila (2016), 'Students have "dismaying" inability to tell fake news from real, study finds', *National Public Radio*, 23 November, https://www.npr.org/sections/thetwo-way/2016/11/23/503129818/study-finds-students-have-dismaying-inability-to-tell-fake-news-from-real. Accessed 2 February 2023.

Hackforth, Reginald (ed.) (1972), *Plato: Phaedrus* (No. 119), Cambridge: Cambridge University Press.

Hutton, John S., Dudley, Jonathan , Horowitz-Kraus, Tzipi, DeWitt, Tom and Holland, Scott K. (2020), 'Associations between screen-based media use and brain white matter integrity in preschool-aged children', *JAMA (Journal of the American Medical Association) Pediatrics* 174:1, https://jamanetwork.com/journals/jamapediatrics/fullarticle/2754101. Accessed 16 January 2023.

Kinney, David A. (1993), 'From nerds to normals: The recovery of identity among adolescents from middle school to high school', *Sociology of Education*, 66:1, pp. 21–40.

London, Bonita, Downey, Geraldine, Bonica, Cheryl and Paltin, Iris (2007), 'Social causes and consequences of rejection sensitivity', *Journal of Research on Adolescence*, 17:3, pp. 481–506.

Madison, Ed (2012), 'Journalistic learning: Rethinking and redefining language arts curricula', Ph.D. doctoral thesis, Eugene, OR: University of Oregon.

——— (2015), *Newsworthy: Cultivating Critical Thinkers, Readers and Writers in Language Arts Classrooms*, New York: Teachers College Press.

Madison, Ed, Anderson, Ross C. and Bousselot, Tracy (2019), 'Self-determined to write: Leveraging interest, collaboration and self-direction through a journalistic approach', *Reading and Writing Quarterly*, 35:5, pp. 473–95.

Manning, M. Lee (2002), 'Revisiting developmentally appropriate middle level schools', *Childhood Education*, 78:4, pp. 225–27.

McCracken, Grant D. (2008), *Transformations: Identity Construction in Contemporary Culture*, Bloomington, IN: Indiana University Press.

McLuhan, Marshall ([1964] 2003), *Understanding Media: The Extensions of Man, Critical Edition* (ed. T. W. Gordon), Corte Madera, CA: Gingko Press.

——— (1967), 'Part six: A dialogue', in G. E. Stearn, *McLuhan: Hot and Cool. A Primer for the Understanding of and a Critical Symposium with Responses by McLuhan*, Toronto: Signet Books, pp. 259–91.

McLuhan, Marshall and Nevitt, Barrington (1972), *Take Today: The Executive as Drop Out*, New York: Harcourt Brace Jovanovich, Inc.

Midkiff, Madison, Wheeler, Jessica, Baker, Emily and Steinberg, Kristin (2015), 'The relationship between social media use and stress in college students', *18th Annual Georgia College Student Research Conference*, Milledgeville, GA, Georgia College & State University, 10 April, https://kb.gcsu.edu/src/2015/friday/28/. Accessed 2 Februrary 2023.

Milner, Murray (2013), *Freaks, Geeks, and Cool Kids*, New York: Routledge.

O'Keefe, Gwenn Schurgin and Clarke-Pearson, Kathleen (2011), 'Clinical report – The impact of social media on children, adolescents, and families', *Pediatrics*, 127:4, pp. 800–04, https://doi.org/10.1542/peds.2011-0054. Accessed 2 February 2023.

Oyserman, Daphna (2015), 'Identity-based motivation', in R. Scott and S. Kosslyn (eds), *Emerging Trends in the Social and Behavioral Sciences*, Hoboken, NJ: John Wiley & Sons.

Pluth, Michael D., Boettcher, Shannon W., Nazin, George V., Greenaway, Ann L. and Hartle, Matthew D. (2015), 'Collaboration and near-peer mentoring as a platform for sustainable science education outreach', *Journal of Chemical Education*, 92, pp. 625–30.

Poindexter, Paula M. (2012), *Millennials, News, and Social Media: Is News Engagement a Thing of the Past?*, New York: Peter Lang.

Raby, Rebecca (2005), 'What is resistance?', *Journal of Youth Studies*, 8:2, pp. 151–71.

Richtel, Matt (2011), 'A Silicon Valley school that doesn't compute', *New York Times*, 22 October, https://www.nytimes.com/2011/10/23/technology/at-waldorf-school-in-silicon-valley-technology-can-wait.html. Accessed 2 February 2023.

Selfhout, Maarten H. W., Branje, Susan J. T., Delsing, M., ter Bogt, Tom F. M. and Meeus, Wim H. J. (2009), 'Different types of Internet use, depression, and social anxiety: The role of perceived friendship quality', *Journal of Adolescence*, 32:4, pp. 819–33.

Selvi, V., Bhuvaneswari, M. and Sandra, S. (2022), 'Stories: A tool to rejuvenate the culture of Generation Z and Generation Alpha', *YMER Digital*, 21:1, pp. 267–76, https://ymerdigital.com/uploads/YMER210152.pdf. Accessed 16 January 2023.

Stern, Robin (1999), 'Hormone-driven kids: A call for social and emotional learning in the middle school years', *Voices from the Middle*, 7:1, pp. 3–8.

Student Press Law Center (SPLC) (2023), 'High school press freedom & censorship', Washington, DC: SPLC, https://splc.org/high-school/. Accessed 2 February 2023.

Tenenbaum, Laura S., Anderson, Margery K., Jett, Marti and Yourick, Debra L. (2014), 'An innovative near-peer mentoring model for undergraduate and secondary students: STEM focus', *Innovative Higher Education*, 39, pp. 375–85.

Turner, Zeke (2017), 'Book publishers go back to basics', *Wall Street Journal*, 14 October, https://www.wsj.com/articles/book-publishers-go-back-to-basics-1507983856. Accessed 2 February 2023.

United Nations Children's Fund (UNICEF) (2021), *Preventing a Lost Decade: Urgent Action to Reverse the Devastating Impact of COVID-19 on Children and Young People*, New York: UNICEF, https://www.unicef.org/eap/media/10306/file/UNICEF 75 report.pdf. Accessed 16 January 2023.

Weller, Chris (2018), 'Five tech industry moguls who raise their kids tech-free', *Business Insider*, 2 February, https://www.businessinsider.com/silicon-valley-moguls-who-raised-their-kids-nearly-tech-free-2018-1. Accessed 2 February 2023.

Zaniewski, Anna M. and Reinholz, Daniel (2016), 'Increasing STEM success: A near-peer mentoring program in the physical sciences', *International Journal of STEM Education*, 3, p. 14.

12

'Dirtying' Ecocinema Studies[1]

Salma Monani and Stephen Rust

Framed in the centre of a lush green forest, two men, one brown-skinned and one paler, occupy the screen in a medium-long shot. Shirtless and wearing terrycloth loincloths, briefs and sneakers, they sit on a log and confer in what sounds like a Native language.[2] Subtitles reveal nonsensical banter about male and female physical prowess. As they swat mosquitoes, the browner man, Chingachgook, offers to take his pale friend, Hawkeye, hunting. When Hawkeye agrees, the two hop off their log and charge offscreen into the woods. For the next four minutes, the camera tracks the men through a full spectrum of shot distances, pans, tilts, high and low angles, front and back, and slow motion – as they journey to the accompaniment of an epic orchestral score. Their hunt takes on a comedic spin as the montage cutting reveals them to be dramatically crisscrossing the same small space while jumping, falling and tripping over each other. Their antics continue until sunset and they stand bent over, breathing hard and loudly dry heaving from the exertion. Undaunted, they carry on as the scene cuts to show them stepping out of the forest to the door of a fast food joint. After stoically contemplating the menu, they order 'number 4' meals (with cheese) in English and stand with looks of satisfaction while the music triumphantly swells and the scene ends.

This short video, 'Hunting' (The 1491s 2011), is posted by the Native sketch comedy group, the 1491s, on their YouTube channel. Troupe members Dallas Goldtooth (Diné/Dakota) and Ryan Red Corn (Osage Nation) perform as the characters Chingachgook and Hawkeye from James Fenimore Cooper's 1826 novel, *The Last of the Mohicans* ([1826] 1983), about the French and Indian Wars. Released in September 2011, the short has (as of 8 August 2023) nearly 300,000 views and over 5,500 likes. As one of many accompanying comments below points out, 'If you know what movie this music is from, the clip is twice as funny'. The source is Trevor Jones and Randy Edelman's acclaimed score for Michael Mann's 1992 Hollywood blockbuster movie adaptation of the novel. Cooper's novel was immensely popular at the time of its publication, and continues to provide Euro-western white culture with a foundational romanticized image of Indigenous peoples[3] as ecological Indians – noble savages living in harmony with Nature in a time and place far removed from present concerns. 'Hunting' clearly pokes fun at the ecological Indian trope and the ever-circulating imagery of this magical-mythical figure by mainstream media texts like the 'Crying Indian' public service announcement (Ad Council 1971), *Dances with Wolves* (Costner 1990) and *Twilight* (Hardwicke 2008), as well as *Avatar* and its sequel (Cameron 2009, 2022) that perpetuate this stereotype (see, e.g. Marrubio and Buffalohead 2012; Diamond 2009; Takeuchi 2010; Coles 2020; Asenap 2022). By using slapstick comedy generated through the emotional and cognitive dissonance occurring between the acting and the music, 'Hunting' critiques the highly constructed nature of this mainstream media representation.

'Hunting' uses irony to laugh at the myth of the stereotypical 'ecological Indian'. Hawkeye and Chingachgook are contemporary characters who put on a noble face but struggle with mosquitoes and do their hunting at a modern fast food restaurant. Though only five-minutes long, 'Hunting's' comedy, thus, also makes space to express complex environmental realities as experienced and negotiated by Native people who subvert the cultural logic of capitalism through their decolonial use of the modern digital space of the internet. In a 2016 interview with the *Los Angeles Times* during the Dakota Access Pipeline protests, Goldtooth explains the 1491s' larger goal:

> The world is so lazy, society is so lazy, that it continuously tries to minimize the complexities of entire cultures and people in the simplest of terms [...] Our job is to push back, to make it complex. Don't be afraid to dirty the narrative.
>
> (Yardley 2016: n.pag.)

'Hunting' makes space for viewers to complicate and 'dirty the narrative' of Eurocentrism. While the film does not provide explicit political or ideological commentary about its juxtaposition of the forest and the fast food restaurant, it implicitly exploits the connection for humour and, when read through contexts of Indigenous experiences, it functions as a window (as we will soon demonstrate) into specific ecological relations with land, the disruption settler colonialism brings to Indigenous food sovereignty and ways that Indigenous peoples work to redress these disruptions.

In beginning with 'Hunting', we consider how the field of ecocinema studies, which attends to cinema through an *eco*critical lens, has been self-reflexively 'dirtying' narratives within the broader fields of cinema and media studies and within its own ranks. Ecocinema studies' ecocritical lens draws attention to environmental issues and ecological thinking, which was considered fringe thinking in cinema studies when we first began our work together in 2009, as noted in earlier overviews (Ivakhiv 2008; Willoquet-Maricondi 2010; Rust et al. 2013). This is no longer the case as the crises of climate change and a broader turn towards *environmental* humanities, an interdisciplinary humanities initiative that 'envisions ecological crises fundamentally as questions of socioeconomic inequality, cultural difference, and divergent histories, values, and ethical frameworks' (Heise 2017: 2), propels attention towards media's role in these matters.

The international organization, the Society for Cinema and Media Studies (SCMS), now hosts a growing Environment and Media special interest group, founded by scholars such as Janet Walker and Hunter Vaughan (who are, respectively, editors of two new high-profile journals in the field, *Media + Environment* and *Journal of Environmental Media*). Across the field (and the humanities in general), environment is no longer siloed as 'nature out there' or a symbolic abstract to denote 'human nature'. Instead, the intimate and complex ways in which environment and media are socially constructed and materially manifested make them subjects of interdisciplinary interest (Rust et al. 2023). Understanding media and environment's entangled material and cultural relations is increasingly urgent in the twenty-first century as variants of audiovisual (i.e. cinematic) media production, distribution, consumption and post-consumption undeniably affect *all* planetary life, human and non-human .

Below, we consider questions included in this collection through an auxiliary question: '*whose* lives are mediated, and how?' By bringing the question of 'whose' to life, we deliberately move away from a Eurocentric sense of life as a 'what' (an object) to instead suggest a flow of diverse *subjective* encounters now giving meaning to media and environmental relations. These subjective encounters of life open space to trouble and collapse traditionally erected boundaries between media and environment, human and non-human, as well as centre and margin, power and disenfranchisement. 'Hunting' is one such example that does this work, subverting simplistic narratives of '*what* is *the* "Indian"?' (a vanishing, noble eco-savage sidekick to the white settler) to '*who* are these individuals, and how are they mediating their *own* lives?' Framing our discussion of such subjectivities as 'Whose Visions and Voices?' and 'Whose Feelings and Emotions?', below we discuss ecocinema studies' engagements with two cutting-edge areas of ecocritical research – decolonial studies and affect studies. By elaborating these fields and their overlaps, as well as pointing in our conclusion to potential points of connection to developments in materiality studies, another important area of environmental studies research, we demonstrate how ecocinema studies has begun to decenter prior universalizing and Eurocentric trajectories of cinema scholarship.

Whose Visions and Voices?: Decentering the Eurocentric Camera and Shifting the Intellectual Focus

Many overviews of the field (including ours in *Ecocinema Theory and Practice* [2013]) look to the late 1990s and early 2000s as the period when scholars with interests in various modes of film – from wildlife television documentaries to Hollywood and independent cinema – began to produce monographs with attention to environmental themes (Mitman 1999; Bousé 2000; Ingram 2000; MacDonald 2001). Perhaps not surprisingly, all of these studies focus on films made by Euro-western filmmakers. While many were notably silent on the presence of non-western representations, others served to remind us of how western cinema constructed 'nature' in specifically western ways (e.g. as wilderness separate from humans) but tended to replicate troubling stereotypes of non-white peoples. Many studies began to question the 'Eurocentric gaze', which is a point of view that privileges western ideas and ideologies. David Ingram, for example, notes that Hollywood films such as Mann's *The Last of the Mohicans* (1992), suggest 'an idealization of Native cultures as living in a pristine, unfallen Eden' as 'a product of white liberal guilt and wishful thinking' (2000: 46). Such an idealization masks histories of colonial genocide and makes invisible Indigenous realities of struggle and survival.

Despite these useful critiques, very little of this early scholarship in the field turned directly to cinema *made by* non-white people or asked what it might mean to present such cinema as a corrective to the Eurocentric gaze. It wasn't until a decade later that such work began to permeate ecocinema studies, through articles such as Jennifer Machiorlatti's on activist documentaries made by North American Native communities in the collection *Framing the World* (2010) and Sheldon Lu and Jiayan Mi's anthology *Chinese Ecocinema* (2011). These moves have been furthered through studies of transnational cinema, which as Pietari Kääpä and Tommy Gustafsson note 'may expand producers and audiences' ecological perception and cognitive abilities', to interrogate 'the contradictions, complexities, obstacles and opportunities emerging in the clash of the local and the global' (2013: 4, 19). Some of this work was also occurring at the margins of cinema studies, for example, in collections focused more broadly at the intersections of Indigenous and literary ecocritical studies (Monani and Adamson 2017).

The push to interrogate ecocinema's 'contradictions, complexities [...] in the clash of the local and the global' has brought ecocritical attention to non-western cinema while simultaneously obscuring its ecocritical potential (Kääpä and Gustafsson 2013: 19). This is because Hollywood dominates the global film industry, and thus, even with this move to think transnationally, it continues to be the central focus of ecocritical research. Hollywood's presence is hard to ignore. For example,

James Cameron's blockbuster movie *Avatar* (2009) might be the most ecocritically analysed film of all time, and as numerous articles on it attest, to their credit, ecocritics admirably problematize and 'dirty' the Hollywood industry's capitalist narrative (Vaughan 2019). In contrast, non-western cinema *still* remains relatively marginalized in ecocinema studies. This lack of attention is slowly shifting, propelled by a broader imperative within the environmental humanities to seek alternatives to Eurocentric capitalism, which has disproportionately harmed non-western humans and non-humans alike (Rust et al. 2023). The symposium, *A Clockwork Green: Ecomedia in the Anthropocene* (2018),[4] showcased this diversifying focus, with dedicated panels on cinema made by Indigenous peoples, as well as examples of Pakistani, Japanese and Nigerian cinema. For example, from Twitter feeds that draw attention to Indigenous film to curated pedagogical compilations of such films, scholars are drawing attention to thriving and burgeoning archives of cinema that subverts the Hollywood centre (BlackDeer 2022; Lempert [2015] 2023). However, even as ecocinema scholars become more invested in non-western cinema, there are additional ways that such scholarship has yet to reach its potential.

Ecocinema studies, like much environmental studies scholarship, is rooted in western canons and most scholars are white. These factors contribute two compounding challenges that are deeply structural. First, canonical adherence guides us to speak with and to Eurocentric intellectual and linguistic traditions, and not necessarily to non-western counterparts. While such Eurocentric readings are useful, they miss opportunities to consider non-western lenses, and in some cases, do injustice to texts and contexts.

Take our own work, for example. Earlier in our careers, we each wrote essays about *Avatar* that addressed its symbolic representations of Indigenous peoples. Salma critiqued *Avatar* and turned her praise instead to *Indigenous-made* cinema while using the concept of the 'ecological Indian' as a representational frame that evokes sympathy and empathy (Monani 2014). Looking back, Salma's reading makes blinkered assumptions about the audiences from whom sympathy and empathy are expected. The article's assumptions about assumed audiences are problematic in light of subsequent work by Glen Coulthard (Yellowknife Dene) and others who point out that Indigenous peoples are not necessarily seeking a 'politics of recognition' from white settler society (Coulthard 2014). Steve examined *Avatar* as a progressive mega-blockbuster that disrupted business as usual in the Hollywood film industry through its emotionally charged interrogation of western culture's extractivist ideologies (Rust 2014). Writing at a critical historical moment in the debate over climate change, land use practices in Amazonia and Indigenous sovereignty, he connected *Avatar*'s symbolic depictions of Native genocide to writings about the past but failed to put *Avatar* into direct conversion with contemporary Indigenous cinema. Neither of us are Indigenous, and we wrote papers seeped in the Eurocentric canonical media scholarship that preceded us, but should have gone further to incorporate the voices of Indigenous people speaking for themselves.

Such work of bringing Indigenous perspectives and agency to the fore can be understood through Salma's own evolving work on Indigenous cinema within a frame she calls *d-ecocinema criticism*, which brings decolonial methodologies to ecocinema criticism. Emerging from the field of Indigenous studies, decolonial methodologies allow us to rethink traditional Eurocentric research processes. For example, as Kim Tallbear (Sisseton Wahpeton Oyate) writes, 'It is helpful to think of the research process as a professional networking process with colleagues (not subjects), as an opportunity for sharing knowledge, not data gathering' (Tallbear 2014: 2). This understanding rejects a Eurocentric research mode that erects an 'objective' boundary between 'those who know [the researcher] and those from whom the raw materials of knowledge production are extracted [the subject]'. Building on this idea, as non-Indigenous scholars reading texts that are specific to each filmmaker's particularized Indigenous world-views, we can't claim to be the 'one who knows'. We definitely bring 'outsider' approaches to reading the films but can aspire to decolonial readings that account for the 'insider' insights of Indigenous filmmakers and communities. As Indigenous cinema scholar Davinia Thornley suggests, such collaborative readings are not 'fawning or uncritical', but instead account for the epistemes of the culture producing it, not the 'outside' culture watching it, as far as possible (Thornley 2014).

As we become more aware of how skewed towards privileging the Eurocentric gaze our academic practices are, and the inherent problems of such limited points of view, d-ecocinema criticism helps productively reframe our work. However, here, a second challenge arises. If most scholars are white (or, not a cultural insider to a particular community), the stakes for ethical engagement with Indigenous and non-western scholars and communities and the risk of failure or virtual signalling may seem so daunting that such terrain is too fraught to enter. Yet, if we are to do a better job in attending to the visions of such filmmakers, we must also do better listening to intellectual voices from their communities. Ecocinema studies has been, and always will be, a self-reflexive field, and this process of rethinking whose voices and visions we engage, *and how*, animates those drawn to the field. To say more, we turn now to ecocinema's intersections with affect studies.

Whose Feelings and Emotions? Diversifying the Affective Possibilities of Ecocinema

Affect theory is a field of study attentive to the role of emotions in society and cultures. It is informed by developments in psychology, neuroscience, as well as cultural epistemologies of experiential knowledge. The affective turn has informed some of the most dynamic conversations unfolding in the humanities (see, e.g. Tompkins 1962; Nathanson 1992; Ahmed 2004; and Rifkin 2012). For ecocinema scholars, the search to better understand affect has meant investigating 'how films portray human emotional relationships with the more-than-human world and how such films act upon viewer's emotions' (Weik von Mossner 2014: 1). There are two overlapping yet distinct strands to this work – audience reception studies, with attention to methods in the empirical social sciences, and a more rhetorical and philosophical approach to affect theory, attentive to ecocritical readings.

Among the most influential essays in the field's development is one that sits squarely in the latter category – Scott MacDonald's 'Toward an Eco-Cinema' (MacDonald 2004). A manifesto of sorts published in the leading environmental humanities journal, *Interdisciplinary Studies in Literature and Environment (ISLE)*, sparked the field's formation through its argument that particular camera, editing, sound and other production techniques can be deployed by filmmakers (and teachers) to help 'retrain' viewers' perceptions of the natural world and 'counter the damaging psychic and environmental effects of the commercial cinema' (2004: 112). As an approach embraced by many, MacDonald's attempt to define ecocinema as something akin to slow cinema (i.e. cinema of long takes and little camera movement) prompted scholars to ask what is 'ecocinema' and for *whom* is it intended?

In his touchstone essay, 'The aesthetics and ethics of eco-film criticism' (Ingram 2013), and subsequent articles, David Ingram turned to cognitive film studies and empirical communication studies to demonstrate how different social and behavioural factors can nuance audiences' responses to different modes of cinema and that slow cinema techniques – which reject plot-driven narratives for long, lingering takes – may not always be effective in inspiring environmental engagement. And, in fact, empirical studies demonstrate that mainstream environmental movies, like *The Day After Tomorrow* (Emmerich 2004) and *Avatar*, which are often dismissed by intellectuals, can inspire viewers to make perceptual connections between their lives, society and the planet (see, for example, Leiserowitz 2004; Reusswig et al. 2004; Aoyagi-Usui 2004; Balmford et al. 2004; Brereton and Hong 2013; Gould et al. 2013; Haluza-Delay et al. 2013).

Ecocinema scholars have generated a sizable archive of audience reception articles by collaborating with and building on the work of colleagues in environmental communications – a vibrant interdisciplinary realm of communication studies which is 'attentive to diverse forms of interpersonal, group, public, organizational and mediated communication that make up the social debate about environmental issues and problems, and our relationship to the rest of nature' (Meisner 2020: n.pag.). Audience research surveys such as those conducted by the Yale Center for

Climate Change Communication on *The Day After Tomorrow*, inspired a special issue of the journal *Interactions* published in 2014, edited by Pietari Kääpä (2014). Championing the emerging subfield of *empirical ecocriticism*, scholars like Alexa Weik von Mossner and Pat Brereton endorse quantitative evaluation of the viewer/text relationship as methods of backing up anecdotal claims of cinematic affects and effects (Schneider-Mayerson et al. 2020). However, even as ecocinema studies turn to empiricism, much of this work is confirming both 'the important power of film's visual appeals in framing [environmental issues like] climate change' and also film's general inability on its own to produce a 'long term effect [...] on participant understanding' (Bondi et al. 2021: 221).

In all, the complications of cinema's affects and effects are not lost on ecocinema scholars. Cinema scholars have long debated cinema's potential to directly and indirectly affect individuals and societies by reshaping viewers' perceptions of the world outside the theatre and inviting/persuading/triggering them to change their actions as a result. Founded in 1919, during the Russian Revolution, the Moscow Film School became one of the earliest incubators of such theories through the writing, research and filmmaking of faculty like Lev Kuleshov, Sergei Eisenstein and Vsevolod Pudovkin. Eisenstein in particular advocated strongly for filmmakers to deploy a cinema of attractions and montage to mobilize the masses in support of the Bolshevik cause (Eisenstein 1968). Eisenstein made several internationally famous movies – notably *Strike* (1925), *Battleship Potemkin* (1925) and *October* (co-directed with Aleksandrov1928) – informed by his theories about the power of images and editing but his career took a dramatic turn when Josef Stalin came to power and he found his work manipulated and his life at risk. As Eisenstein's montage techniques were subsequently put into the service of capitalism by western filmmakers to lure ever larger audiences into theatres, Eisenstein spent the latter part of his life serving the Stalinist state as essentially a prisoner.

Fortunately, ecocritics doing the important work of affect and audience studies have generally heeded the advice of theorist Andrew Hageman to avoid the dream of a clockwork green. Referencing the well-known novel (Burgess 1962) and film, *A Clockwork Orange* (Kubrick 1971), Hageman's provocation for scholars to remain relentlessly open-minded and self-critical has been a rallying cry of sorts in the field as folks recognize that movies alone cannot lead to social change. Hageman's work, much like MacDonald's, is not based on empirical research but turns to philosophical and rhetorical theories to speculate on film's (and audiences') perceptual possibilities and limitations. In this vein, a significant individual contribution is Adrian Ivakhiv's *Ecologies of the Moving Image: Cinema, Affect, Nature* (Ivakhiv 2013). Drawing on western philosophers such as Gilles Deleuze, Félix Guattari, Alfred North Whitehead and Charles Sanders Pierce, Ivakhiv constructs a comprehensive eco-philosophy of cinema. While Ivakhiv's theories and cinematic examples are primarily Eurocentric, his theory of cinema, which is attentive to the essential *dynamics* of cinema's 'film-world' with the 'extra-film-world', makes room to think beyond western spaces. Such openings are even more evident in two well-received collections, *Moving Environments: Affect, Emotion, Ecology, and Film* and *Affective Ecocriticism*, which include non-western texts (Weik von Mossner 2014; Bladow and Ladino 2018). Both these collections also include work by Nicole Seymour, whose extensive attention to cinema at the intersections of affect studies and queer ecologies deserves mention for its wise council to diversify our understanding of audiences and affects and not take everything in life quite so seriously.

Seymour encourages us to queer our understandings of and attachments to mainstream environmentalist discourse. She reminds us that mainstream environmental discourse is not only problematic in its elitist white and heteronormative rhetoric but often incorporates a stifling sense of martyrlike 'gloom and doom'. Calling for 'an ironic affective stance', she invites us to consider how films that use comedy, camp and satire and thus, not commonly understood as environmental, can nonetheless 'foment environmentalist agendas, while encouraging us to question both the "eco" and the "cinema" in ecocinema' (Seymour 2014: 74). Her call to embrace irony and the unexpected has inspired an outpouring of affective ecocritical work examining modes of cinema (such as comedy and horror) traditionally ignored in environmental thought (Rust and Soles 2014; Murray and Heumann 2016; Tidwell and Barclay 2018). Again, though most of this work continues to focus on western cinema, it also engages non-western cinema. Seymour herself intentionally turns to Black, Indigenous and People of Color (BIPOC) queer scholars to illuminate the environmental justice aspects of what she tongue-in-cheek calls 'Bad Environmentalisms' (Seymour 2018).

Such openings to consider how cinematic encounters can engender diverse affects from diverse audiences is vital, and perhaps more so is recognizing that feelings and emotions are complex expressions that embody cultural learning (Monani 2016; Bohr 2015). In the case of the 1491s' 'Hunting' short, such culturally attuned engagements reveal layers of enriched understandings of how the film encodes messages and for whom. At its most obvious, the short pokes fun at Hollywood's stereotypes. Yet, the spoof's power takes on added value when approached from the perspective of Indigenous audiences. For example, scholar Kyle Powys Whyte (Potawatomi) chuckles at the short, noting how *The Last of the Mohicans* is a common butt of native humour. He notes that Hawkeye's heartfelt words to Cora: 'I will find you! Wherever you are, and, however long it takes; I will find you!' are often a playful refrain among Indigenous friends when they part ways, even if they know they'll see each other in a short while. The melodrama of the film is a popular subject of ridicule, as suggested by another endearing three-minute short created by Darryl Nepinak (Salteaux/Cree), 'The Last of the Nepinaks' (2005).

At a deeper level, the short's humour invites conversations about specific ecological relations with land, the ruptures settler colonialism makes to these relations and moves to re-suture such relations. Most noticeably, 'Hunting' points to how Indigenous food sovereignty – in this case, hunting – is constrained by settler colonialism. The short film was shot on sovereign lands, the Menominee reservation, yet the forest space seems small and the characters ultimately resort to eating generic fast food. Histories of Indigenous peoples' dispossession and settler colonial control contextualize these images. Forced onto reservations, constrained by punitive hunting laws and restricted agricultural practices, Indigenous peoples' ability to engage their food systems, as Whyte writes, for its 'collective capacities' – its ability to facilitate 'valuable goods like political sovereignty, nutrition, and spirituality' – has been severely hampered (Whyte 2018). Policed by systems of violence that dehumanize Indigenous agency, Indigenous communities often find themselves instead with fragmented and precarious food options systemically manufactured and characterized as 'food deserts' (see, e.g. Krogstad 2014; Jernigan et al. 2017; Canada Without Poverty 2023).

While 'Hunting' alludes to these realities, its primary tone is deliberately humorous, rather than of despair. It gives little more than a minute to the easily identifiable symbol of food insecurity – a fast food joint. Instead, it devotes the majority of its time to negotiating the forest. These onscreen choices take on added affective significance when viewed from the epistemes of many North American Indigenous cultures that valued land as life and humour as 'healing medicine' long before colonial disruptions (Taylor 2006). They evoke the activist work that Goldtooth, Red Corn and the Menominee nation are all doing offscreen to counter food insecurity and promote instead what Whyte describes as 'collective continuance' – 'a society's overall adaptive capacity to maintain its members cultural integrity, health, economic vitality, and political order into the future' (Whyte 2018: 355). Collective continuance recognizes the imbrication of individual social elements (like food procurement) as entangled in larger contexts of community resilience, and was a guiding principle of many traditional Indigenous societies that did not, as capitalism does, silo land relations. Goldtooth is part of the leadership team of the Indigenous Environmental Network (IEN), an organization at the forefront of grassroots efforts within Indigenous communities; Red Corn is involved with the Red Corn Native Foods company that was established by his grandparents on the Osage reservation; and the Menominee reservation (where the film is shot) has developed educational programs dedicated to reclaiming food sovereignty.[5] Given these insights, 'Hunting's' attention to the characters' valiant but fruitless attempt to actually go hunting is both a cheeky thumbing at settler systems that dictate rules of food acquisition and a subversive attempt to portray the very real challenges of healing such disrupted ecological relations.

Ultimately, 'Hunting's' comedic and light-hearted affect when approached through lenses that allow for Indigenous perspectives reveal an enriched reading of the short, one that is simultaneously ecological and decolonial. Such readings are imperative at a time when we are increasingly faced with the limits of the Eurocentric gaze and its entwined relations to capitalism's material violence on human and non-human lives alike.

Conclusion

When reflecting on the motivations and intentions of scholars in a research field that is growing as rapidly as ecocinema studies, any critique short of a monograph can provide but a snapshot. We have framed our analysis intentionally around the question of whose lives, voices and feelings matter in ecocinema studies to allow sufficient space to give 'Hunting' the attention it deserves while widening our scope to examine the field's purview and to constantly be self-reflexive in 'dirtying' our narratives. Even as we write, there is self-examination of the field's whiteness and what that means. For example, in the wake of African American George Floyd's killing by police officers in Minneapolis and the protests in support of the Black Lives Matter movement that grabbed the nation's attention in the summer of 2020, the Advisory Board of the journal *Media+Environment* reflected on its own whiteness and the gatekeeping consequences that shape its practices. The *Journal of Environmental Media*, focusing on new media similarly seeks voices and intellectual insights from beyond the western frame.

How we grow our ecomedia community deserves continued alertness, and while in this essay we outline how research on whose voices and feelings are foregrounded, other areas of rapid growth in the field, particularly investigations of cinema and media infrastructures and residual materialities, also invite an interrogation of whose bodies are impacted. Since the publication of the UCLA Institute for the Environment's comprehensive study *Sustainability in the Motion Picture Industry* in 2006, scholars like Richard Maxwell and Toby Miller (2012), Lisa Parks (2012 and in progress), Nicole Starosielski (2015) and Hunter Vaughan (2019) have contributed to understanding the environmental impacts of media on material bodies and beings – human and non-human.

Within the motion picture and television industry and in cinema production programs around the world, a new era of experimentation is poised to unfold as filmmakers, students and production companies with shared interest in sustainability and justice collaborate to use new digital technologies to their fullest advantage while remaining ever-mindful of the pollution they are contributing to earth, air and water. Those looking to explore 'the inextricable relationship between moving images and the natural resources that sustain them' by collaborating to bring new voices and visions to the foreground can take inspiration from Nadia Bozak's monograph *The Cinematic Footprint* (Bozak 2012: 1). An early effort in the field to blend the study of media's ecological materiality with its aesthetics and affects, Bozak looked outside of Hollywood (and ends with a chapter on Isuma TV, an Inuit initiative) to raise critical questions about the aesthetics of low-budget digital cinema, carbon-neutral production, and whether a filmmaker who makes intentional efforts to reduce their carbon footprint lends themselves creatively to discovering new kinds of ecocinema techniques that will engage audiences and ecocritics alike. We hope these developments prompt us to revisit this conversation in another decade to reflect on and rethink the issues we've raised, and we look forward to the possibility that life allows us both the opportunity to make good on our intentions.

NOTES

1. The authors want to thank the editors of this collection for their persistence in following up with us regarding working our *What is Life?* conference presentations into publication. The process sparked an opportunity for Salma and Steve to work together and consider the state of our field since the publication of *Ecocinema Theory and Practice* (2013). Salma is also very grateful to the co-participants of the Andrew Mellon Summer Institute in Environmental Humanities hosted by Colby College in 2019 for reading and commenting on drafts – a special thanks to Kyle Powys Whyte and to Priscilla Ybarra for their encouragement to continue this work on 'Hunting'. Also grateful to Nicole Seymour and Sarah Wald for being the first to help Salma think through the idea of d-ecocinema criticism.
2. The dialogue is actually gibberish with a few Ojibwe and Navajo words sprinkled in.
3. Please note that we capitalize Indigenous very intentionally throughout this essay. Following the suggestion of the Associated Press, this move is attentive to current awareness of existing imbalances in how we identify groups. European and Americans are capitalized, but other groups (Indigenous, in particular) have not been given the same privilege (Scire 2020; see https://www.niemanlab.org/2020/06/ap-style-is-now-to-capitalize-black-and-indigenous/. Accessed 27 January 2023).
4. 'A Clockwork Green: Ecomedia in the Anthropocene' was billed as a virtual 'Nearly Carbon-Neutral Symposium' at the University of California, Santa Barbara (UCSB), sponsored by the Association for the Study of Literature and Environment (ASLE) and UCSB, 10–14 June 2018. See http://ehc.english.ucsb.edu/?page_id=17731/. Accessed on 27 January 2023.
5. See Indigenous Environmental Network (https://www.ienearth.org. Accessed on 20 July 2023), Red Corn Native Foods (http://redcorn.com/. Accessed on 27 January 2023)

and Menominee Rebuilders (https://www.menomineerebuilders.org/food-sovereignty-initiative/. Accessed on 27 January 2023).

REFERENCES

The 1491s (2011), 'Hunting', Canada, YouTube, 11 September, https://youtu.be/7oKtyYIIcaQ. Accessed 4 January 2021.

Ad Council and Keep America Beautiful (1971), 'The Crying Indian', Marstellar, Inc. [ad agency], https://www.adcouncil.org/our-story/our-history/the-classics/. Accessed 17 January 2023.

Ahmed, Sara (2004), *The Cultural Politics of Emotion*, 2nd ed., Edinburgh: Edinburgh University Press.

Aoyagi-Usui, Midori (2004), '*The Day After Tomorrow*: A study on the impact of a global warming movie on the Japanese public', *National Institute for Environmental Studies (NIES) Working Paper* (unpublished), pp. 41–3.

Asenap, Jason (2022), '*Avatar: The Way of Water*, or how not to make Indigenous futurism movies', *Salon*, 22 December, https://www.salon.com/2022/12/22/avatar-the-way-of-water-or-how-not-to-make-indigenous-futurism-movies_partner/. Accessed 17 January 2023.

Balmford, Andrew, Manica, Andrea, Airey, Lesley, Birkin, Linda, Oliver, Amy and Schleicher, Judith (2004), 'Letters to the Editor: "Hollywood, climate change, and the public"', *Science*, 305:5691, 17 September, p. 1713.

BlackDeer, Autumn Asher (@DrBlackDeer) (2022), 'Don't wanna watch the colonial glorifying blue people movie? Check out these sci-fi films by actual Indigenous people telling our own stories instead', Twitter, 16 December, https://twitter.com/DrBlackDeer/status/1603884556020391936. Accessed 27 January 2023. [Note: Login for list.]

Bladow, Kyle and Ladino, Jennifer (eds) (2018), *Affective Ecologies*, Lincoln, NE: University of Nebraska Press.

Bohr, Marco (2015), 'Cinema of emancipation and Zacharias Kunuk's *Atanarjuat: The Fast Runner*', in S. MacKenzie and A. W. Stenport (eds), *Films on Ice: Cinema of the Arctic*, Edinburgh: Edinburgh University Press, pp. 84–96.

Bondi, Brittany A., Monani, Salma B., Principato, Sarah and Barlett, Christopher (2021), 'Examining the impact of climate change film as an educational tool', *Applied Environmental Education and Communication*, 20:3, pp. 221–37.

Bousé, Derek (2000), *Wildlife Films*, Philadelphia, PA: University of Pennsylvania Press.

Bozak, Nadia (2012), *The Cinematic Footprint: Lights, Camera, Natural Resources*, New Brunswick, NJ: Rutgers University Press.

Brereton, Pat and Hong, Chao-Ping (2013), 'Audience responses to environmental fiction and non-fiction films', *Interactions: Studies in Communication and Culture*, 4:2, pp. 171–99.

Burgess, Anthony (1962), *A Clockwork Orange*, London: Heinemann.

Cameron, James (dir.) (2009), *Avatar*, USA: 20th Century Fox.

——— (2022), *Avatar: The Way of Water*, USA: 20th Century Fox.

Canada Without Poverty (2023), 'Fact about poverty', Canada Without Poverty, http://www.cwp-csp.ca/poverty/just-the-facts/. Accessed 17 January 2023.

Coles, Lily (2020), 'Indigenous representation in *The Twilight Saga* film *New Moon*, by Lily Coles', Dalhousie University Cinema and Media Studies Blog, 7 May, https://blogs.dal.ca/cinema/2020/05/07/indigenous-representation-in-the-twilight-saga-film-new-moon-by-lily-coles. Accessed 15 December 2020.

Cooper, James Fenimore ([1826] 1983), *The Last of the Mohicans: A Narrative of 1757*, Albany, NY: State University of New York.

Coulthard, Glen (2014), *Red Skin, White Masks: Rejecting the Colonial Politics of Recognition*, Minneapolis, MN: University of Minnesota Press.

Costner, Kevin (dir.) (1990), *Dances with Wolves*, USA: Orion Pictures.

Diamond, Neil (2009), *Reel Injun*, Canada: Resolution Pictures and National Film Board of Canada.

Eisenstein, Sergei (dir.) (1925) *Strike*, Soviet Union: Goskino.

——— (1925) *Battleship Potemkin*, Soviet Union: Goskino.

Eisenstein, Sergei (1968), *Film Essays and a Lecture* (ed. and trans. J. Leyda), New York: Dobson Books.

Eisenstein, Sergei and Aleksandrov, Grigori (dirs.) (1928), *October: Ten Days That Shook the World*, Soviet Union: Sovkino.

Emmerich, Roland (dir.) (2004), *The Day After Tomorrow*, USA: 20th Century Fox.

Gould, Rachelle K., Ardoin, Nicole M. and Hashimoto, Jennifer K. (2013), 'Transposing the conversation into popular idiom: The reaction to *Avatar* in Hawai'i', in B. Taylor (ed.), *Avatar and Nature Spirituality*, Waterloo: Wilfrid Laurier University Press, pp. 95–122.

Haluza-Delay, Randolph, Ferber, Michael P. and Wiebe-Neufeld, Tim (2013), 'Watching *Avatar* from "AvaTar Sands" Land', in B. Taylor (ed.), *Avatar and Nature Spirituality*, Waterloo: Wilfrid Laurier University Press, pp. 123–41.

Hardwicke, Catherine (dir.) (2008), *Twilight*, USA: Summit Entertainment.

Heise, Ursula (2017), 'Planet, species, justice—and the stories we tell about them', *Routledge Companion to the Environmental Humanities*, New York: Routledge, pp. 1–9.

Ingram, David (2000), *Green Screen: Environmentalism and Hollywood Cinema*, Exeter: University of Exeter Press.

——— (2013), 'The aesthetics and ethics of eco-film criticism', in S. Rust, S. Monani and S. Cubitt (eds), *Ecocinema Theory and Practice*, New York: Routledge, pp. 63–86.

Ivakhiv, Adrian (2008), 'Green film criticism and its futures', *Interdisciplinary Studies in Literature and Environment*, 15:2, pp. 1–28.

——— (2013), *Ecologies of the Moving Image: Cinema, Affect, Nature*, Waterloo: Wilfrid Laurier University Press.

Jernigan, Valarie Blue Bird, Huyser, Kimberly R., Valdes, Jimmy and Watts Simonds, Vanessa (2017), 'Food insecurity among American Indians and Alaska Natives: A national profile using the current population survey – Food security supplement', *Journal of Hunger & Environmental Nutrition*, 12:1, pp. 1–10.

Jones, Trevor and Edelman, Randy (1992), *The Last of the Mohicans: Original Motion Picture Soundtrack*, Los Angeles, CA: Morgan Creek Records.

Kääpä, Pietari (ed.) (2014), 'Ecocinema audiences', *Special issue of Interaction: Studies in Communications and Cultures*, 4, p. 2.

Kääpä, Pietari and Gustafsson, Tommy (2013), *Transnational Ecocinema: Film Culture in an Era of Ecological Transformation*, Bristol: Intellect.

Krogstad, Jens Manual (2014), 'One in four Native Americans and Alaska Natives living in poverty', Pew Research Center, 13 June, https://www.pewresearch.org/short-reads/2014/06/13/1-in-4-native-americans-and-alaska-natives-are-living-in-poverty/. Accessed 18 July 2019.

Kubrick, Stanley (dir.) (1971), *A Clockwork Orange*, USA: Warner Bros.

Leiserowitz, Anthony A. (2004), 'Before and after *The Day After Tomorrow*: A U.S. study of climate risk perception', *Environment*, 46:9, pp. 22–37.

Lempert, William. ([2015] 2023), 'Navajos on Mars: Native Science Fiction Futures', *Space and Anthropology*, [21 September] January, https://medium.com/space-anthropology/navajos-on-mars-4c336175d945. Accessed 31 January 2023.

Lu, Sheldon and Mi, Jiayan (2009), *Chinese Ecocinema: In the Age of Environmental Challenge*, Hong Kong: Hong Kong University Press.

MacDonald, Scott (2001), *The Garden in the Machine: A Field Guide to Independent Films about Place*, Berkeley, CA: University of California Press.

——— (2004), 'Toward an eco-cinema', *Interdisciplinary Studies in Literature and Environment*, 11:2, pp. 107–32.

Mann, Michael (dir.) (1992), *The Last of the Mohicans*, USA: Morgan Creek Entertainment and Twentieth Century Fox.

Marrubio, Elise and Buffalohead, Eric (eds) (2012), *Native Americans on Film: Conversations, Teaching and Theory*, Lexington, KY: University of Kentucky Press.

Maxwell, Richard and Miller, Toby (2012), *Greening the Media*, London: Oxford University Press.

Meisner, Mark (2020), *Environmental Communication: What It Is and Why It Matters*, IECA.org, https://web.archive.org/web/20220104020947/https://www.theieca.org/resources/environmental-communication-what-it-and-why-it-matters. Accessed 15 December 2020.

Mitman, Greg (1999), *Reel Nature: America's Romance with Wildlife on Film*, Cambridge, MA: Harvard University Press.

Monani, Salma (2014), 'Evoking sympathy and empathy: The ecological Indian and Indigenous eco-activism', in A. W. von Mossner (ed.), *Moving Environments: Affect, Emotion, Ecology, and Film*, Waterloo: Wilfrid Laurier University Press, pp. 225–46.

——— (2016), 'Feeling and healing eco-social catastrophes: In the "horrific" slipstream of Danis Goulet's *Wakening*', *Global Weirding: Special Issue of Paradoxa*, 28, pp. 192–213.

Monani, Salma and Adamson, Joni (eds) (2017), *Ecocriticism and Indigenous Studies: Conversations from Earth to Cosmos*, New York: Routledge.

Murray, Robin L. and Heumann, Joseph K. (2016), *Monstrous Nature: Environment and Horror on the Big Screen*, Lincoln, NE: University of Nebraska Press.

Nathanson, Donald L. (1992), *Shame and Pride: Affect, Sex, and the Birth of the Self*, New York: W.W. Norton.

Nepinak, Darryl (2005), 'The Last of the Nepinaks', Vimeo, https://vimeo.com/38899868. Accessed 4 January 2021.

Parks, Lisa (2012), *Down to Earth: Satellite Technologies, Industries and Cultures*, New Brunswick, NJ: Rutgers University Press.

——— (forthcoming), *Mixed Signals: Media Infrastructures and Cultural Geographies.*

Reusswig, Fritz, Schwarzkopf, Julia and Pohlenz, Philipp (2004), 'Double impact: The climate blockbuster *The Day After Tomorrow* and its impact on the German cinema public', *Potsdam Institute for Climate Impact Research (PIK) Report No. 92*, Potsdam: PIK, https://www.pik-potsdam.de/en/news/latest-news/archive/2004/pr92.pdf. Accessed 15 December 2020.

Rifkin, Mark (2012), *The Erotics of Sovereignty: Queer Native Writing in the Era of Self-Determination*, Minneapolis, MN: University of Minnesota Press.

Rust, Stephen (2014), '*Avatar*: Ecorealism and the blockbuster melodrama', in B. Taylor (ed.), *Avatar and Nature Spirituality*, Waterloo: Wilfrid Laurier University Press, pp. 23–36.

Rust, Stephen, Monani, Salma and Cubitt, Sean (eds) (2013), *Ecocinema Theory and Practice*, New York: Routledge.

——— (eds) (2023), *Ecocinema Theory and Practice 2*, New York: Routledge.

Rust, Stephen and Soles, Carter (eds) (2014), 'Ecohorror special cluster: "Living in fear, living in dread, pretty soon we'll all be dead"', *ISLE: Interdisciplinary Studies in Literature and Environment*, 21:3, pp. 509–12.

Scire, Sarah (2020), 'AP style is now to capitalize Black (and Indigenous)', *NiemanLab*, 22 June, https://www.niemanlab.org/2020/06/ap-style-is-now-to-capitalize-black-and-indigenous/. Accessed 27 January 2023.

Schneider-Mayerson, Matthew, Weik von Mossner, Alexa and Małecki, W. P. (2020), 'Empirical ecocriticism special cluster: "Environmental texts and empirical methods"', *ISLE: Interdisciplinary Studies in Literature and Environment*, 27:2, pp. 327–36.

Seymour, Nicole (2014), 'Irony and contemporary ecocinema: Theorizing a new affective paradigm', in A. W. von Mossner (ed.), *Moving Environments: Affect, Emotion, Ecology and Film*, Waterloo: Wilfrid Laurier Press, pp. 61–78.

——— (2018), 'The queerness of environmental affect', in K. Bladow and J. Ladino (eds), *Affective Ecocriticism: Emotion, Embodiment, Environment*, Lincoln, NE: University of Nebraska Press, pp. 235–55.

Starosielski, Nicole (2015), *The Undersea Network*, Durham, NC: Duke University Press.

Starosielski, Nicole and Walker, Janet (eds) (2016), *Sustainable Media: Critical Approaches to Media and Environment*, New York: Routledge.

Tallbear, Kim (2014), 'Standing with and speaking with faith: A feminist Indigenous approach to inquiry', *Journal of Research Practice*, 10:2, pp. 1–7.

Takeuchi, Craig (2010), '*The Twilight Saga: Eclipse*: University professors analyze Indigenous content', *The Georgia Straight*, 29 June, https://www.straight.com/blogra/twilight-saga-eclipse-university-professors-analyze-indigenous-content. Accessed 15 December 2020.

Taylor, Drew Hayden (ed.) (2006), *Me Funny*, Vancouver: Douglas and MacIntyre.

Thornley, Davinia (2014), *Cinema, Cross-Cultural Collaboration and Criticism*, London: Palgrave Macmillan.

Tidwell, Christy and Barclay, Bridgitte (eds) (2018), *Gender and Environment in Science Fiction*, Lexington, KY: Lexington Books.

Tompkins, Sylvan (1962), *After Imagery Consciousness: Vol. 1: The Positive Affects*, New York: Springer Publishing.

UCLA Institute for the Environment (2006), *Sustainability in the Motion Picture Industry*, Contractor's Report to the Board, California Integrated Waste Management Board, https://web.archive.org/web/20211128090218/https://www.ioes.ucla.edu/wp-content/uploads/mpisreport.pdf. Accessed 15 April 2021.

Vaughan, Hunter (2019), *Hollywood's Dirtiest Secret: The Hidden Environmental Costs of the Movies*, New York: Columbia University Press.

Weik von Mossner, Alexa (ed.) (2014), *Moving Environments: Affect, Emotion, Ecology, and Film*, Waterloo: Wilfrid Laurier University Press.

Whyte, Kyle P. (2018), 'Food sovereignty, justice, and Indigenous peoples: An essay on settler colonialism and collective continuity', in A. Barnhill, T. Doggett and M. Budolfson (eds), *Oxford Handbook of Food Ethics*, London: Oxford University Press, pp. 345–66.

Willoquet-Maricondi, Paula (ed.) (2010), *Framing the World: Explorations in Ecocriticism and Film*, Charlottesville, VA: University of Virginia Press.

Yardley, William (2016), 'An environmental activist who uses comedy to help stop oil pipelines', *Los Angeles Times*, 12 September, https://www.latimes.com/nation/la-na-sej-dallas-goldtooth-activist-comedian-20160902-snap-story.html. Accessed 15 December 2020.

PART V

SYNTHESES AND BIODESIGNING

13

System Dynamics, Machine Learning and Structural Validation[1]

William Schoenberg and Jeremy Swartz

Today human, non-human and more-than-human societies face a multitude of complex problems that require understanding, not merely as individual relationships, but as entire systems to recognize unintended consequences and develop effective policy solutions. For instance, we need to understand the value of *system dynamics* (SD) from historical, educational and practical perspectives, and how it is commonly utilized to make sense of the world. This chapter presents an overview of SD, from modeling to critiques, as well as an explanation of the structural origins of behaviors. The discussion includes white box vs. black box models, what makes models useful, and how to improve the quality of models via structural validation, as well as how to make use of the power and adaptability of machine learning (ML) to improve the quality of models. We argue that feedback loop dominance (LD) analysis illuminates the reasons for observed behaviors and thus allows for better structural validation of models. Methods of feedback system neural networks and feedforward artificial neural networks (ANNs) are offered as ways to do exploratory data analyses which can help to ease model conceptualization processes. We conclude with attention to *systems education, public policymaking and regulation*, especially pertaining *to* SD *participatory* approaches in relationship to ML, deep learning (DL) and artificial intelligence (AI).

Modeling, Systems and Dynamics

The underlying language of science is mathematics, which codifies science's understanding of the universe in the forms of axioms, causal relationships and natural laws. The organization of related sets of mathematically codified understandings defines a *model*. The scientific method is constructed around the development and testing of models, and only those models that prove useful in generating understanding of real-world phenomena that they represent survive.

Colloquially, *systems* primarily are the natural phenomena modeled, although the term has a wider meaning when applied in the context of models. For instance, our solar system is a system, a set of processes, which act on our sun, the planets and their moons, as well as all the asteroids, planetoids, gases and other matter contained within our celestial neighborhood. That (natural) system is modeled using equations built from scientific laws. But the colloquial language we've developed is not specific enough between the natural world and our human representations and understandings of it. Therefore, by taking a wider, holistic view, the meaning of the word system becomes a set of interrelated processes that come together to produce *dynamics*. The individual behaviors that celestial mechanics describe in the context of our solar system include the spin of the Earth around its axis, or the movement of Venus and Mars through space as they orbit the Sun. The instantaneous and individual behaviors of the bodies within the solar system which happen in each moment looked at together over a continuum of time and across a set of interacting bodies are celestial dynamics. Dynamic behaviors do not occur in isolation from time or by single elements alone; dynamics are the outcomes of a system composed of at least two elements observed over time. They apply broadly across systems, both natural and mathematical, and as long as there is interaction between the elements producing behavior over time, there are dynamics. All of the work that science does to model systems is centered around creating an understanding of dynamics or the cause-and-effect interactions which exist between things.

The current treatment of modeling used thus far has been exceedingly broad in scope, but the underlying practices and disciplines are relatively easy to categorize. This chapter focuses on mathematical models, specifically differential equation models, as opposed to physical models. Within the domain of mathematical modeling, there are two general categories: white box and black box.

White box models are the oldest and most widespread; all the examples referred to thus far are white box models based on differential equations, where the meaning of each variable and of those equations which house them, have direct and interpretable meaning. In white box models, each variable (or node) within the network of model equations that represent a system has a known and agreed-upon real-world analogue from the natural system being modeled.

On the other hand, *black box models* are contemporary tools originating in the 1950s with the advent of artificial neurons, and later on ANNs, random forests, etc. Black box models are directly and fully based on experimental data, where the understanding they encode in the form of their mathematical structure is not directly interpretable. Each of the nodes within their network of model equations does not have a direct known analogue to the natural system they represent.

Both white and black box models have an underlying mathematical structure. The key difference is the encoding of the understanding of the natural system that those models represent within their mathematical structure. White box models present that understanding in the form of interpretable causal relationships, whereas black box models present more opaque mathematical structures, which do not provide any specific meaning when studied element by element.

Within the realm of white box models, which are the primary focus of this chapter, there are further sub-categories, which are useful for understanding the breadth of modeling techniques. There are models constructed of systems of differential equations (analytical or simulated), or models constructed by the interactions of agents (actors, particles, etc.), or models which create dynamics based upon the generation of a discrete sequence of events, among others. The ways to construct, interpret and analyse white box models are nearly as boundless as science itself, but that doesn't mean there aren't some useful abstractions that can apply across wide swaths of white and black box models to help better integrate them into our understanding.

The first of these useful abstractions across all systems and models are *states*, which pertain to the memory of a system (natural or mathematical) and provide the system with continuity across time as it creates dynamics. In a model, states are designed to represent the same quantities they do in their natural system in an attempt to mimic the natural system the model purports to represent. States have direct analogues to their natural equivalents. Examples are the number of people in a city, jelly beans in a jar or the location of a planet in space. This doesn't mean that states have to be physical quantities, but can be used to track soft concepts (e.g. the amount of knowledge that a person or group of people have about an idea).

A state represents a fundamental element of a system whose value at any given moment in time is dependent upon its previous value, and any new changes brought about by the other elements within the system between the current moment and the next. For instance, if time within a system were to stop, the states are the parts of the system that would still be measurable and have a value. Consider a basic system of a bathtub filling with water where a spigot releases water to fill the tub and a drain at the bottom empties it. The amount of water in the bathtub is a state of that system; other states would be the size of the spigot or the drain. The amount of water which flows into the tub via the spigot or leaves via the drain are the *rates* which change the value of the states over time. Therefore, the rates (of change of the state) are the key property which defines a state. States are the fundamental elements of systems whose value can only be changed over time. The water in the bathtub cannot rise or lower in a single instant without the passage of time. Water must be added over some interval of time, no matter how infinitesimal that interval is.

The second useful abstraction is *feedback*, or an abstraction used to represent processes which self-modify their own states over time without continued action from a source of action. For example, take a simple population model of planet Earth. In this overly simplistic model, there is one state which represents the number of people, and two rates which modify the value of that state over time. First, the birth rate which represents new arrivals added to the population, and second, the death rate, which represents existing members of the population who depart through death. There are two feedback processes. The first is the more people in the population, the more births there are, which means over time (notwithstanding anything else), there will be more people. This is a *positive* or *reinforcing* feedback loop which leads to, if left unchecked, exponential behaviors of either growth or decay, or in other words, dramatic changes to the states of the system. The second feedback process in this system says that the more people there are in a population, the more people die, which means fewer people in the population. This is a *negative* or *balancing* feedback loop which leads to logarithmic stabilizing behaviors in the states of the system.

How can we apply this set of abstractions to the popular modern-day white box modeling paradigms? First, *agent-based* models use states to capture the attributes of actors (individuals) or items according to their location in space, their colour, etc. The interactions between agents can be looked at through the lens of a feedback loop – any action that an agent takes which modifies one of its states that later comes back to affect that same state. Also, the actions of agents or the impacts of other agents on an agent of interest are often delayed in time. *Discrete event* models can be looked at the same way as they are structures composed of states linked by feedback, whose rates rather than being computed in a continuous fashion over a series of uniform time intervals, change their states over a discrete continuum of time or in non-uniform intervals of time. Lastly, SD models are composed of *stocks* or states and *flows* or rates, as well as feedback loops which connect the stocks and often create time delays. Each of these abstractions can be recognized in the network of model equations generated by all of these modeling paradigms.

In the context of black box models, e.g. ML (specifically ANNs), a subfield of AI,[2] states can be thought of as *nodes*. In *recurrent neural networks* where nodes' values are dependent upon their previous states, it would be fair to call each recurrent node a state. In supervised and reinforcement learning, feedback exists in the context of training, but not in the mathematical structure of the final model. Training connects the feedforward relationships from the input nodes to the output nodes using weights and biases set during backpropagation that depends upon the outputs of the neural network to re-adjust those weights and biases. This feedback loop between outputs, and weights and biases back to outputs, does not exist in the final form of the network of equations of a standard *multilayer perceptron* neural network, a class of ANNs. In fact, for the very large majority of ANNs, no feedback at all is present in the final structure.

Now that we've laid the groundwork for understanding systems and models, we turn to the problem of defining and measuring what makes a model useful. It is often noted that 'All models are wrong but some are useful' (Box 1979: 202). This begs the questions: what makes a model useful, how can we measure a model's utility, and how can we be confident that the ways in which a model is wrong don't affect the learning we are trying to do using the model?

SD and its Modeling Process

SD[3] as a field seeks to understand via *simulation models*[4] the underlying *nonlinear*[5] behavior of systems using accumulations (states), feedback and time delays. It grew out of Jay W. Forrester's research in the 1950s at the Massachusetts Institute of Technology (MIT). In the early 1960s, SD was focused primarily on modeling and understanding problems in business or management as Forrester was a professor at the recently formed Sloan School of Management at MIT and published *Industrial Dynamics* in 1961. Forrester's (1969) *Urban Dynamics* demonstrated the beginnings of the broad utility of the SD methodology with the study of the management of a city. Branching out from businesses and cities ultimately led him to the Club of Rome in 1970, where he became deeply interested in global socio-economic modeling and managing global systems, publishing *World Dynamics* in 1971. Subsequently, *World Dynamics* laid the groundwork for *The Limits to Growth* (Meadows et al. 1972), which became a seminal publication because it was one of the first to take the position that humanity could grow to surpass the physical carrying capacity of its habitat, planet Earth. (Also see Meadows et al. 2004; Randers and Goluke 2020; Herrington 2020.)

Since the 1970s, there has been a continuing history of good modeling work coming out of the SD field, including John D. Sterman (2000) and Peter M. Senge ([1990] 2006), as well as the recent work of Prevention Impacts Simulation Model (PRISM) for chronic disease policy-making, climate interactive and others.[6] As a method, SD is very broad and generally applicable to a wide variety of systems, deriving power and utility from its laser focus on the understanding of systems regardless of the context of those systems. Generally, the purpose of the SD method is to help people make better decisions in the face of complexity (Sterman 2000). To better understand the critical discourse around SD, it is important to understand how it is practiced and implemented.

In 1994, Forrester described the practice of SD in general as the process of understanding and improving systems. In the same paper, Forrester laid out the general steps of the modeling process shown in Figure 13.1. The process is clearly not linear and exposes many opportunities for iteration and revision.

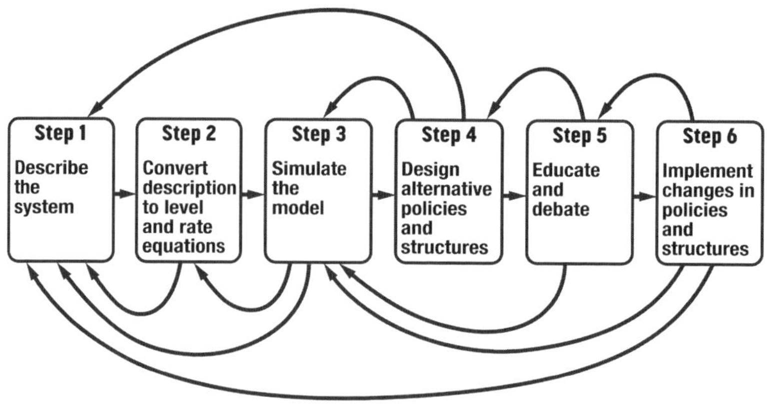

FIGURE 13.1: The SD modeling process (Forrester 1994).

Figure 13.1 illustrates how SD at its core generalizes real-world observations into complex (computational) mathematical models, which can be seen in steps one and two. It then uses those mathematical models as a basis for developing, then sharing, inferences or policies which are then used as the basis for either education and/or making decisions in the real-world, which can be seen in steps three through six. As an example, we can apply this process to understanding a typical business issue of declining revenues, or an environmental problem such as climate change. Step one would be to describe how the business operates, or how products are demanded, produced and sold. For the climate system, it would be to enumerate how the Earth's various natural systems work together to regulate climate. The second step would be to encode that understanding into a series of equations (a model), making sure to get more descriptions of the system as questions about the specifics of the problem arise. Step three involves the modeler simulating that model (which can require that more equations be written, or more information be solicited). If done properly, this results in a depiction of the original problem by the model. In the fourth step, the modeler looks for solutions to the problem in the model. For the business example of declining revenue, the solutions may be adding more salespeople, lowering the price, etc., which may influence each of the previous steps. For the climate system, it may include adjusting human greenhouse gas (GHG) emissions, or land use rates. Step five is to use the model to educate the key stakeholders about what the model says and why, which includes the possibility of learning more and therefore adjusting previous steps. Thus, in the sixth step, the stakeholders can implement the change and either 'fix' the original problem which would end the process, or go back to any of the other steps and understand other ways to improve the model.

Critical Discourse Around SD

Early criticisms of the work of Forrester and other system dynamists are typified by economist, William D. Nordhaus's (1973) response to *World Dynamics* (Forrester 1971). Nordhaus took issue with Forrester's neo-Malthusian position and criticized the method of SD and computer simulation by saying 'without an accurate model there is no assurance that systems dynamics is better than mental models; the main result is a spurious and misleading precision' (1157). He argued that the model was 'measurement without data' and that none of the relationships between the variables in the model are couched in empirical studies or observed data. Essentially, Nordhaus was claiming that by abstracting to such a degree as Forrester did, all meaning and connection with the real world was lost and that inferences drawn from this model could not in any way contribute to the understanding of that real-world system, a reductionist/logical positivist perspective.

Nordhaus continued by pointing out that there was insufficient validation done through empirical testing. Applying a reductionist/logical positivist perspective to this work would result in an argument that insufficient sensitivity analysis was performed on this model yielding an incorrect perception of accuracy and precision. In fact, one of the things Nordhaus did in his paper was to perform a *sensitivity analysis* on the parameters and relationships in the world dynamics model to demonstrate the possible range of outcomes and policies that could be produced. As a key proponent of a holistic worldview, we believe that Forrester would argue against Nordhaus by saying that the model has value because it informs discussion and thinking on the topics of boundless economic and population growth.

Recently, SD has been expanded with a few steps added to the process shown in Figure 13.1. Validation and uncertainty testing now are done on any model before it can be realistically used to provide policy recommendations or serve as a basis for education in response to the criticisms of subscribers to the reductionist/logical positivist worldview. Barlas (1996) described the process for model validation placing emphasis on the modeler to ensure that their models are structurally and behaviorally valid between simulation and policy analysis. At a high-level, behavioral validation means that the model is generally capable of reproducing historical data with a high degree of accuracy without the model having foreknowledge of that history. And at the same level, *structural validation* refers to the concept that the model is producing that behavior for the right reasons (e.g. that the system of equations that the model is composed of is logically and empirically valid).

Building upon the previous examples of declining revenues or changing GHG emissions, between the simulation and policy steps (three and four), as well as between the policy construction and the education steps (four and five), the modeler now must explicitly focus on validation, where it was more implicit before. For instance, in the business example, the modeler must check that their model reproduces the decline in revenue with a high level of accuracy. In the climate case, they must confirm that their model accurately portrays historical climate data. They may confirm with subject matter experts and existing bodies of literature that the system of equations used to encode the knowledge that they gathered about the system is logical and borne out by data. Other tests include assessing their model's sensitivity to parameters to make sure point estimates are not the primary cause for behavior or to identify specific assumptions which must be further tested and validated. They typically also subject their model to extreme conditions tests to make sure that, for instance in the business case, when there are no salespeople, there are no sales. Barlas' addition is to make sure that the modeling process emphasizes exposing a theory to data to try to find cases where the model doesn't explain the data.

Barlas produced a list of all the validation tests a model must go through before being considered useful for policy recommendations. He breaks these tests down into three categories: direct structure tests, structure-oriented behavior tests and behavior pattern tests. He demonstrates how these tests confirm and validate structure and behavior, as well as (quite importantly) cover model purpose and problem identification (the issue of boundaries).

Addressing Criticisms of Modeling via Automated Feedback LD Analysis

SD models of consequence to policy can contain as few as three to five feedback loops or as many as tens of millions, as was the case for the model underlying *Urban Dynamics*. By studying the relationships between the feedback loops in an SD model, we can understand where model behavior comes from, and via the process of induction, relate that back to real-world processes that are responsible for the problem being modeled.

This process requires significant experience, knowledge and practice and sets a high barrier to entry into the field of useful SD modeling in a policy-making context, which acts in part as a safeguard for policy-makers, who must ultimately undertake any responsibility for making actionable decisions based upon the recommendations from SD models. On the other hand, this process is error-prone and may not yield a proper understanding of model behavior and therefore may lead to real-world determinants of system behavior which can drive policy-makers to make sub-optimal decisions based on improperly understood stories and narratives of system behavior.

Feedback LD Analysis

The current methods (both mathematical and analytical) relate model behavior back to the underlying feedback loop structure. This process, whether done by hand or algorithm, is called *feedback loop dominance* (LD) analysis. Performing LD analysis relies on either practitioner intuition and experience (the art of modeling and model analysis) or complex algorithmic LD analysis. The most popular technique of the latter in the literature is eigenvalue elasticity analysis, but other notable techniques include the pathway participation metric, or loop impact analysis. The former is taught as part of the methodology of model building, while the latter comes from 40 years of work on techniques to derive and explain model behavior based on the analysis of structure (for example, see Graham 1977; Forrester 1982; Eberlein 1984; Davidsen 1991; Mojtahedzadeh 1996; Ford 1999; Saleh 2002; Mojtahedzadeh et al. 2004; Gonçalves 2009; Saleh et al. 2010; Kampmann 2012; Hayward and Boswell 2014; Moxnes and Davidsen 2016; Oliva 2016; Sato 2016; Hayward and Roach 2017; Naumov and Oliva 2019; Oliva 2020).

LD analysis sheds light on the origins of behavior in SD models by relating observed behavior back to the feedback process(es) that created it (Forrester 1961; Richardson 1991). LD analysis is concerned with the discovery of the strength and polarity of the key feedback loops existent within models as time progresses within those models (Richardson 1995). Over any period of time in a simulating model, some feedback loops are most important to the expressed dynamics of the model above all others, and those feedback loops are referred to as the dominant feedback loops. LD is a concept measured at each and every specific point in time when the model simulates. This means during some points in time (for instance, in a simple susceptible-infectious epidemic model), the behavior of the model early on is dominated by a reinforcing feedback loop which spreads a disease to new susceptible suitable hosts, but later on the behavior of the model may come to be dominated by a balancing feedback loop which limits the spread of the disease due to running out of susceptible, suitable hosts. This continuum or pattern of shifting LD is called the *LD profile*.

By identifying the LD profile of a model, we better understand the causes for the observed behavior over that period of time from a mathematical perspective. The explanation of the causes for the dynamics exhibited by the model changes as the structure or inputs into that model change. However, it must be noted that the LD profile on its own says nothing about whether or not the model is structurally valid and relevant to the problem being studied. It still requires expert assessment to validate if the measured LD profile captures the essence of the underlying system being modeled. To a traditionalist, the ability to identify the LD profile of a model (assuming it is a perfect representation of the underlying system) is the key to the development of robust policy options, performing impact assessment by way of simulation, and the formulation of policy recommendations. Understanding which feedback processes are most meaningful and how they contribute to the dynamics being changed (again only if the model is a perfect representation of reality), demonstrates complete mastery of the system and the ability to intervene safely and productively, knowing everything that will happen is a consequence of your intervention.

Moving past the issues typified by Nordhaus and addressed by Barlas' emphasis on model validation (both behavioral and structural), we're left with just a single strong criticism of the SD approach, and of modeling as a general practice. The criticism is that modelers cannot provably demonstrate that their models are representative of the systems they purport to embody and/or understand. *Validation* is a process of elimination, subjecting the model to batteries of tests aimed at finding flaws, where the model clearly misrepresents reality. Each of the validation practices in use today – confirming that models are able to replicate historical dynamics without foreknowledge, or confirming that the equations within a model are each logical and valid in isolation – helps to build confidence that the model is an accurate representation of the system. The paradigm here, and of science in general, is that you can never prove a model right, only wrong. Another important pursuit within SD is a mathematical system capable of performing LD analysis algorithmically using a mathematical process to understand which feedback processes are responsible for the behavior of the model at each and every point in time (Sterman 2000).

Automated Analysis: Loops That Matter

David Ford (1999: 4, 5) stated, 'system dynamicists need at least two things: (1) automated analysis tools applicable to models with many loops and (2) a clear and unambiguous understanding of LD and how it impacts system behavior'. Recently, there has been a breakthrough in automating model analysis via an easy-to-use and widely applicable *automated* method for objectively measuring LD called 'Loops That Matter' (LTM) depicted in Figure 13.2 (Schoenberg et al. 2019).

LTM tells the modeler which feedback loops are responsible for producing the dynamics generated by the model at each point in time while it simulates. It does this by analyzing each equation in the model to determine the importance of each independent variable to its dependent variable(s) and uses that information to generate a measurement of 'feedback loop importance', which is compared across all feedback loops in a model (Schoenberg et al. 2020; Schoenberg et al. 2023). By knowing the importance of each link and feedback loop in the model, modelers can speak

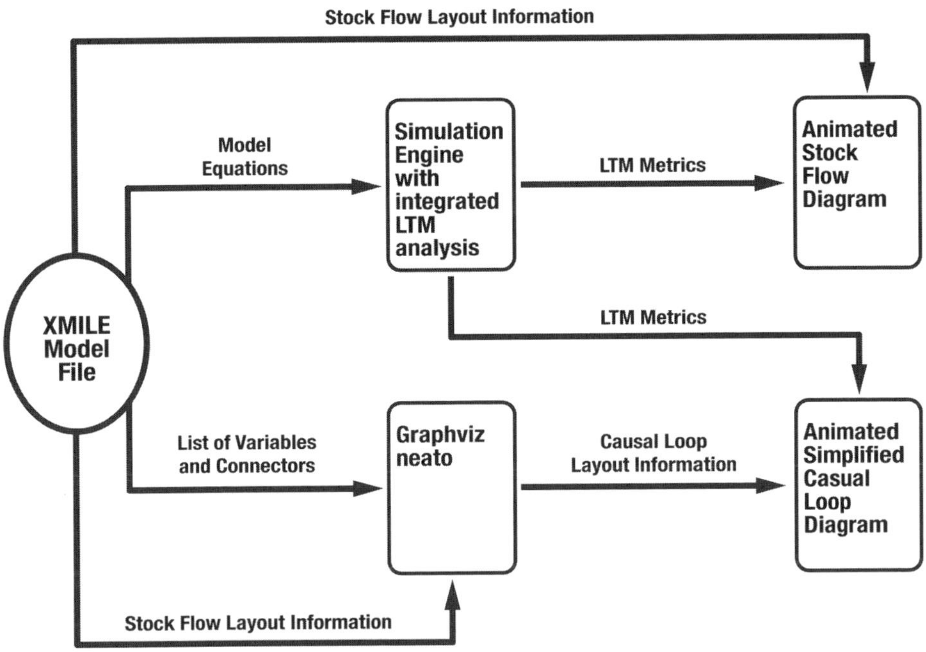

FIGURE 13.2: Diagram of how Loops That Matter (LTM) process works.

with confidence about where specifically in their model behavior is being generated. The theory underlying LTM is very broad in its scope, even while the implementation of the algorithm is currently limited in its application to SD-style models. The primary reason that LTM represents a major breakthrough is because it can be used as a powerful new structural validation test measuring whether or not the model produces behavior for the right reasons, or in other words, that the feedback LD of the model matches with an empirically derived understanding of the system which the model represents. For instance, we could rule out the susceptible-infectious-recovered (SIR) epidemic model as useful if during the exponential spread of the disease, the model's behavior is being dominated by a feedback loop (or set of loops) which has nothing to do with disease spread. By algorithmically calculating the LD profile, LTM makes it abundantly clear why a model produces the behavior it does, which allows the modeler and other subject matter experts to validate whether the model provided reasonable explanations of the underlying system.

Therefore, to address the aforementioned philosophical criticism of SD – that models may not properly represent reality – the SD modeling process needs to be once again amended. The change adds an LTM analysis to the model validation step so that the generated explanation for dynamics, the LD profile, can be validated against both the pre-existing understanding of the system being modeled and the empirical data gathered about that system. Thus, modelers can be assured that if their model passes the existing battery of validation tests, it is much less likely to be a random coincidence. The model is a much more useful tool for answering questions where we do not understand how the system will respond to perturbations. So even though 'all models are wrong', we'll at least know which models are useful and that those will be the models in which descriptions of where behavior comes from match reality in all cases where we currently understand the system under study.

Turning a Black Box Model into a White Box Model

Modern-day ML methods, including probabilistic modeling, kernel machines or DL,[7] have arisen from a strong, almost single-minded, focus on empiricism making use of the extraordinary amounts of observational data available from a plethora of sources (Schölkopf and Smola 2008; Ghahramani 2015; Goodfellow et al. 2016). When viewed through the lens of accurate prediction power, ML has proven to be quite successful, but present-day techniques typically fail to reveal the fundamental causal mechanisms driving behavior. Although it cannot be ignored that interpreting the structure behind black box, DL models is an active research area (Montavon et al. 2018). To make full use of the technological advancements in ML, significant emphasis must be placed on finding a valid and interpretable causal understanding of the underlying real-world system (Runge et al. 2019).

The study of observational causal inference exists alongside the field of ML and is focused on drawing conclusions about causal connections between variables by studying the response in an effect variable when a cause is being changed (Pearl 2009). Largely based on statistics, observational causal inference started with the seminal works of Norbert Wiener and Clive W. J. Granger and has been growing and developing since the 1950s (Wiener et al. 1956; Granger 1969). The most well-known method is *Granger causality*,[8] which tests whether omitting the past of a time series X in a time series model including Y's own and other covariates' past increases the prediction error of the next time step of Y (Granger 1969). Granger causality is useful in discovering specific causal links in a system, but it fails to generalize to complex nonlinear systems, typically failing to identify all the links in the networks of feedback relationships which govern these systems from a holistic perspective (Spirtes and Zhang 2016).

Observation causal inference methods which are not based on Granger causality have been categorized into the following three broad frameworks: nonlinear state-space methods, causal network learning algorithms and structural causal models (Runge et al. 2019). These approaches are designed to discover causality in directed acyclic graphs and are incapable of operating in complex dynamic feedback-rich environments, which have been argued to be at the core of the most intractable and relevant problems (Sterman 2000). Considering the current state of the art in ML and automated observational causal inference, it becomes clear that the ML field lacks a direct understanding of the relevance of the causal structures underlying their models to real-world systems.

The power of the LTM method for discovering the origins of behavior in models has direct utility outside of the field of SD. By combining states, feedback loops and feedforward ANNs, we're now able to do automated causal inference using an ML approach and relying on LTM to turn the generated black box model into a white box model in all the ways that matter, namely, the ability to generate a feedback LD profile (Schoenberg 2019). This approach, called a *feedback system neural network* (FSNN), constructs a model with one state for each dimension in the dataset. Each state is connected to each and every other state (including itself) using a standard *multilayer perceptron artificial neural network* (MLP-ANN) forming a system of ordinary differential equations. This is an SD model with a factorial of the number of states (stocks) feedback loops, typically an amazingly large number (trillions, quintillions, a googolplex or more). That model is trained (parameterized) via back-propagation (calibration) to fit the empirical data. That parameterized black box model is then analysed using LTM to discover the origins of behavior via its feedback LD profile ultimately turning the black box model transparent enough to reason about, visualize and understand what the model has learned about the data, and therefore to validate the generated model's mathematical structure.

The FSNN method (presented in Figure 13.3) acts as a machine that produces behaviorally accurate, feedback-rich structural hypotheses directly from data where the polarity and contribution of each link between all states in the system is known. The machine takes time series data measured from the real world as its input, and, with no additional input, produces a behaviorally relevant causal dynamic hypothesis, which is much more easily validated by subject matter experts than standard neural nets and other ML technologies. The validation performed also is of a much higher quality because it is a structural validation of the machine-learned understanding of the system rather than a behavioral validation of the outputs of the model.

By combining SD and ML, FSNNs demonstrate a powerful new ML technique for finding the causality in arbitrary datasets. By using MLP-ANNs to represent the relationships between the states, the universal approximation theorem (Cybenko 1989) ensures that each relationship is capable of reproducing the real-world function which links the states together. By linking all states together in a directed cyclic graph, this new ML method becomes much more likely to find accurate causal models for the data. As Forrester might say, this is because nonlinear feedback systems are at the heart of complex dynamic real-world problems and those

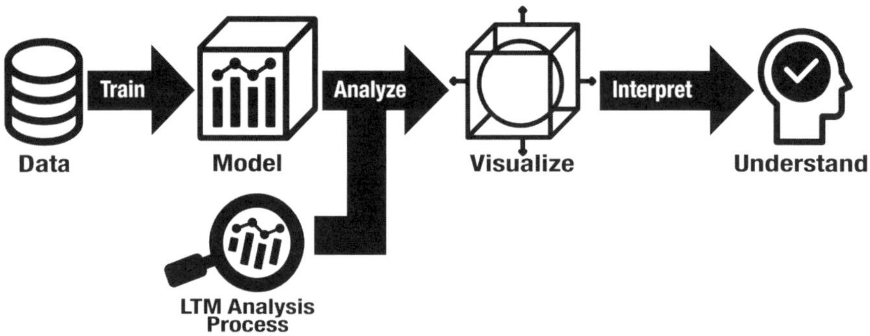

FIGURE 13.3: High-level overview of the ML approach to causal inference including LTM.

systems are best represented in mathematics as systems of ordinary differential equations containing feedback, which this new method does as a matter of purpose.

FSNNs are not cure-alls which replace the existing SD modeling process. They are most useful when you're trying to describe and understand the potential causal relationships in the system in the early phases of the modeling process. FSNNs are useful for exploratory data analysis but are not meant to be the end of the process, but rather the beginning. They are especially useful for intractable problems which exist in inscrutable systems where there is a lot of data, but precious little understanding of the high-level relationships which govern the processes responsible for generating the observed dynamics. An example of such a case may be in genomics where there is a glut of empirical evidence on gene expression, but much less understanding of how those gene expressions are related.

Using Modeling to Solve Problems and General Conclusions

Human and non-human communities face many complex problems, including environmental change, food insecurity, worsening inequality, etc. Solving these problems requires understanding, not merely of individual relationships, but entire systems, so as to recognize unintended consequences. The tools, techniques and problem-solving, or in other words, *technologies* (Swartz et al. 2019) necessary to make progress on the development of society may be based in part on models, whether white box or black box. Following the processes laid out by Forrester and amended by Barlas is a start to developing models, but as previously suggested, there are at least two steps forward that modelers and systems scientists can embrace.

First is to improve the quality of models by doing better structural validation using LTM to produce a feedback LD profile that illuminates the underlying causal structure, clarifying the feedback-based explanation of dynamics. Whether the model is SD, agent-based, or black/white box, it is no longer sufficient to state that a model represents a system because it reproduces behavior or has some pieces of logical structure which can be verified as existing in the natural system. Modelers must now demonstrate that their models reproduce the behavior of the systems they purport to represent for the right reasons, by demonstrating the feedback LD profile of their models matches what is known about the systems in reality. Without this, those models are objectively less useful because there is no assurance that they will accurately respond to non-historically observed inputs.

Second, modelers can use techniques like FSNNs to make use of the power and adaptability of ML models to quickly build hypotheses to do exploratory data analysis which ought to ease the model conceptualization process. The major overarching emphasis here is on structural validation and ensuring that models are useful because they demonstrably capture the essence of reality for more reasons than just behavior matching in the past.

Some keys to developing robust societies capable of tackling systemic issues noted in this chapter, are a deeper integration of systems education, public policymaking and regulation. For instance, there has been research into 'the benefits of using small system dynamics models to address public policy questions' with a special focus on urban planning and social welfare (Ghaffarzadegan et al. 2010: 1).

Similarly, other work considers participatory approaches and applications of SD in relationship to ML, DL, AI and various societal contexts, including fostering increased attention to bias, fairness, inclusivity and participation of vulnerable stakeholders, as well as with wider sociotechnical frameworks (Hovmand 2014; Martin Jr. et al. 2020a, b; Király and Miskolczi 2019; Gullett et al. 2022).

Organizations that work to get systems thinking, sciences and philosophy integrated into the K-12 education curriculum are vital to the future abilities of societies to mitigate and adapt to these challenges and potentially avoid future intractable problems.[9] Current SD research presents insights into disaster management, education, epidemiology and public health, environmental science, public policy, restoration ecology, systems biology and more (e.g. see Payne-Sturges et al. 2023).

Future research is needed around the benefits and more importantly the consequences of SD in relationship to ML, DL and AI, cf. from merely '[m]aking the invisible visible' to something qualitatively different in '[g]athering strength, gathering storms' (Stone et al. 2016; Littman et al. 2021) and human-centered design (Nestor et al. 2023). But it's not just scientists, modelers or analysts who need to be trained to think systemically. It is also educators, policymakers, stakeholders, artists and everyone living, breathing and functioning as parts of systems – all of whom can bring forth *caring* interventions to solve difficult problems.

NOTES

1. This chapter has been significantly revised from earlier versions. See Schoenberg and Swartz (2022).
2. Machine learning refers to computer models or algorithms which self-modify their state through trial and error (optimization, backpropagation) on a given set of empirical data.
3. Developed in the late 1950s at MIT, the earliest SD modeling language was DYNAMO (DYNAmic MOdel), which has a programming language like interface. The first of the visual SD tools based around programmable stock and flow diagrams was STELLA (Systems Thinking, Experimental Learning Laboratory with Animation), developed by high-performance systems (now isee systems) and was first released in 1985 for the Macintosh. Other popular visual SD modeling tools are Vensim (produced by Ventana), and Powersim, which was initially called Constructor, now Studio. Other tools exist, but these four are the most notable.
4. As opposed to systems of differential equations which are analytically solved which are typically referred to as dynamic systems.
5. Nonlinear behavior is when a change in the inputs to the system does not create a proportional change in the outputs.
6. The System Dynamics Applications Award is presented for the best 'real world' application. See https://systemdynamics.org/system-dynamics-applications-award. Accessed 14 February 2023.
7. Machine learning algorithms are typically broken down into three categories:
   - *Supervised learning* is when the input data to the algorithm contains paired sets of inputs with outputs, and the algorithm develops a model using that input data which can map any arbitrary input to its specific and correct output.
   - *Unsupervised learning* is when the input data to the algorithm does not contain any outputs. It creates a model which classifies and groups the given data as it sees fit without any preconceived notion for how the inputs should be arranged into outputs.
   - *Reinforcement learning* is when an algorithmically generated model receives feedback about its output as it navigates the problem space. Rather than pairing each input with a specific output (supervised learning), the algorithm trains the model by giving it rewards as it produces certain outputs.

   Deep learning refers specifically to ANNs with multiple layers where a deep neural network is one which allows for universal approximation as defined by the universal approximation theorem.
8. Granger causality is a statistical concept which uses prediction to measure if a change in an independent variable causes a change in a dependent variable. The technique is based on linear regression, although there are nonlinear methods which are difficult to apply.

9. For instance, the Creative Learning Exchange, Acton, MA, http://www.clexchange.org/cle/about.asp; Waters Center for Systems Thinking, Tucson, AZ, https://waterscenterst.org. Both accessed 14 February 2023. Also see Forrester (2016).

REFERENCES

Barlas, Yaman (1996), 'Formal aspects of model validity and validation in system dynamics', *System Dynamics Review*, 12:3, pp. 183–210.

Box, George E. P. (1979), 'Robustness in the strategy of scientific model building', in R. L. Launer and G. N. Wilkinson (eds), *Robustness in Statistics*, New York: Academic Press, Inc., pp. 201–36.

Cybenko, George (1989), 'Approximation by superpositions of a sigmoidal function', *Mathematics of Control, Signals and Systems*, 2:4, pp. 303–14.

Davidsen, Pål I. (1991), *The Structure-Behavior Graph*, Cambridge, MA: MIT Press, The System Dynamics Group.

Eberlein, Robert Larry (1984), 'Simplifying dynamic models by retaining selected behavior modes', Ph.D. dissertation, Cambridge, MA: Massachusetts Institute of Technology.

Ford, David N. (1999), 'A behavioral approach to feedback loop dominance analysis', *System Dynamics Review*, 15:1, pp. 3–36.

Forrester, Jay W. (1961), *Industrial Dynamics*, Cambridge, MA: MIT Press.

——— (1969), *Urban Dynamics*, Cambridge, MA: MIT Press.

——— (1971), *World Dynamics*, Cambridge, MA: Wright-Allen Press.

——— (1994), 'System dynamics, systems thinking, and soft OR', *System Dynamics Review*, 10:2&3, pp. 245–56.

——— (2016), 'Learning through system dynamics as preparation for the 21st century', *System Dynamics Review* 32:3&4, pp. 187–203.

Forrester, Nathan Blair (1982), 'A dynamic synthesis of basic macroeconomic theory: Implications for stabilization policy analysis', Ph.D. dissertation, Cambridge, MA: Massachusetts Institute of Technology.

Ghaffarzadegan, Navid, Lyneis, John and Richardson, George P. (2011), 'How small system dynamics models can help the public policy process', *System Dynamics Review*, 27:1, pp. 22–44.

Ghahramani, Zoubin (2015), 'Probabilistic machine learning and artificial intelligence', *Nature*, 521:7553, p. 452.

Gonçalves, Paulo (2009), 'Behavior modes, pathways and overall trajectories: Eigenvector and eigenvalue analysis of dynamic systems', *System Dynamics Review*, 25:1, pp. 35–62.

Goodfellow, Ian, Bengio, Yoshua and Courville, Aaron (2016), *Deep Learning*, Cambridge, MA: MIT Press.

Graham, Alan Karl (1977), 'Principles of the relationship between structure and behavior of dynamic systems', Ph.D. dissertation, Cambridge, MA: Massachusetts Institute of Technology.

Granger, Clive W. J. (1969), 'Investigating causal relations by econometric models and cross-spectral methods', *Econometrica: Journal of the Econometric Society*, 37, pp. 424–38.

Gullett, Heidi L., Brown, Gregory L., Collins, Delores, Halko, Martha, Gotler, Robin, Stange, Kurt C. and Hovmand, Peter S. (2022), 'Using Community-Based System Dynamics to Address Structural Racism in Community Health Improvement', *Journal of Public Health Management and Practice* 28: 4(Supp.), pp. S130–37.

Hayward, John and Boswell, Graeme P. (2014), 'Model behaviour and the concept of loop impact: A practical method', *System Dynamics Review*, 30:1, pp. 29–57.

Hayward, John and Roach, Paul A. (2017), 'Newton's laws as an interpretive framework in system dynamics', *System Dynamics Review*, 33:3&4, pp. 183–218.

Herrington, Gaya (2020), 'Update to limits to growth: Comparing the World3 model with empirical data', *Journal of Industrial Ecology*, 25:3, pp. 614–26.

Hovmand, Peter S. (2014), *Group Model Building and Community-based System Dynamics* Process, New York: Springer Science+Business Media.

Kampmann, Christian Erik (2012), 'Feedback loop gains and system behaviour', *System Dynamics Review*, 28:4, pp. 370–95. [First published in *Proceedings of the 1996 International System Dynamics Conference*, Cambridge, MA: Systems Dynamics Society, pp. 21–25.]

Király, Gábor and Miskolczi, Péter (2019), 'Dynamics of participation: System dynamics and participation—An empirical review', *Systems Research and Behavioral Science*, 36:2, pp. 199–210.

Littman, Michael L., Ajunwa, Ifeoma, Berger, Guy, Boutilier, Craig, Currie, Morgan, Doshi-Velez, Finale, Hadfield, Gillian, Horowitz, Michael C., Isbell, Charles, Kitano, Hiroaki, Levy, Karen, Lyons, Terah, Mitchell, Melanie, Shah, Julie, Sloman, Steven, Vallor, Shannon and Walsh, Toby (2021), 'Gathering strength, gathering storms: The one hundred year study on artificial intelligence (AI100)', 2021 Study Panel Report, Stanford: Stanford University, September, https://ai100.stanford.edu/sites/g/files/sbiybj18871/files/media/file/AI100Report_MT_10.pdf. Accessed 11 May 2023.

Martin Jr., Donald, Prabhakaran, Vinodkumar, Kuhlberg, Jill, Smart, Andrew and Isaac, William S. (2020a), 'Extending the machine learning abstraction boundary: A complex systems approach to incorporate societal context', *arXiv preprint*, 2006:09663.

——— (2020b), 'Participatory problem formulation for fairer machine learning through community based system dynamics', *arXiv preprint*, 2005:07572.

Meadows, Donella H., Meadows, Dennis L., Randers, Jørgen and Behrens, William W., III (1972), *The Limits to Growth: A Report for the Club of Rome's Project on the Predicament of Mankind*, New York: Universe Books.

Meadows, Donella H., Randers, Jørrgen and Meadows, Dennis L. (2004), *Limits to Growth: The 30-Year Update*, White River Junction, VT: Chelsea Green Publishing.

Mojtahedzadeh, Mohammad T. (1996), *Structural Analysis of the URBAN1 Model*, Working Paper, Albany, NY: University at Albany State University of New York (SUNY).

Mojtahedzadeh, Mohammad, Andersen, David and Richardson, George P. (2004), 'Using Digest to implement the pathway participation method for detecting influential system structure', *System Dynamics Review*, 20:1, pp. 1–20.

Montavon, Grégoire, Samek, Wojciech and Müller, Klaus-Robert (2018), 'Methods for interpreting and understanding deep neural networks', *Digital Signal Processing*, 73, pp. 1–15.

Moxnes, Erling and Davidsen, Pål I. (2016), 'Intuitive understanding of steady-state and transient behaviors', *System Dynamics Review*, 32:2, pp. 128–53.

Naumov, Sergey and Oliva, Rogelio (2019), *Structural Dominance Analysis Toolset*, System Dynamics Group, Cambridge, MA: Massachusetts Institute of Technology.

Nestor, Maslej, Fattorini, Loredana, Brynjolfsson, Erik, Etchemendy, John, Ligett, Katrina, Lyons, Terah, Manyika, James, Ngo, Helen, Niebles, Juan Carlos, Parli, Vanessa, Shoham, Yoav, Wald, Russell, Clark, Jack and Perrault, Raymond (2023), 'The AI Index 2023 Annual Report', AI Index Steering Committee, Institute for Human-Centered AI, Stanford: Stanford University, April, https://aiindex.stanford.edu/wp-content/uploads/2023/04/HAI_AI-Index-Report_2023.pdf. Accessed 11 May 2023.

Nordhaus, William D. (1973), 'World dynamics: Measurement without data', *The Economic Journal*, 83:332, pp. 1156–83.

Oliva, Rogelio (2016), 'Structural dominance analysis of large and stochastic models', *System Dynamics Review*, 32:1, pp. 26–51.

——— (2020), 'On structural dominance analysis', *System Dynamics Review*, 36:1, pp. 8–28.

Payne-Sturges, Devon C., Ballard, Ellis, Cory-Slechta, Deborah A., Thomas, Stephen B. and Hovmand, Peter (2023), 'Making the invisible visible: Using a qualitative system dynamics model to map disparities in cumulative environmental stressors and children's neurodevelopment', *Environmental Research*, 21:115295.

Pearl, Judea (2009), 'Causal inference in statistics: An overview', *Statistics Surveys*, 3, pp. 96–146.

Randers, Jorgen and Goluke, Ulrich (2020), 'An earth system model shows self-sustained thawing of permafrost even if all man-made GHG emissions stop in 2020', *Scientific Reports*, 10:1 , pp. 1–9.

Richardson, George P. (1991), *Feedback Thought in Social Science and Systems Theory*, Philadelphia, PA: University of Pennsylvania Press.

——— (1995), 'Loop polarity, loop dominance, and the concept of dominant polarity', *System Dynamics Review*, 11:1, pp. 67–88.

Runge, Jakob, Bathiany, Sebastian, Bollt, Erik, Camps-Valls, Gustau, Coumou, Dim, Deyle, Ethan, Glymour, Clark, Kretschmer, Marlene, Mahecha, Miguel D., Muñoz-Marí, Jordi, van Nes, Egbert H., Peters, Jonas, Quax, Rick, Reichstein, Markus, Scheffer, Marten, Schölkopf, Bernhard, Spirtes, Peter, Sugihara, George, Sun, Jie, Zhang, Kun and Zscheischler, Jakob (2019), 'Inferring causation from time series in Earth system sciences', *Nature Communications*, 10:1, p. 2553.

Saleh, Mohamed M. (2002), 'The characterization of model behavior and its causal foundation', Ph.D. dissertation, Bergen: University of Bergen.

Saleh, Mohamed, Oliva, Rogelio, Kampmann, Christian Erik and Davidsen, Pål (2010), 'A comprehensive analytical approach for policy analysis of system dynamics models', *European Journal of Operational Research*, 203:3, pp. 673–83.

Sato, Jeremy B. (2016), 'State space analysis of dominant structures in dynamic social systems', Ph.D. dissertation, St. Louis, MO: Washington University in St. Louis.

Schoenberg, William ([2019] 2020), 'Feedback system neural networks for inferring causality in directed cyclic graphs', *arXiv* preprint, 1908:10336, https://arxiv.org/abs/1908.10336. Accessed 24 March 2021.

Schoenberg, William, Davidsen, Pål and Eberlein, Robert (2020), 'Understanding model behavior using the loops that matter method', *System Dynamics Review*, 36:2, pp. 158–90.

Schoenberg, William, Hayward, John and Eberlein, Robert (2023), 'Improving loops that matter', *Systems Dynamics Review*, 39:2, pp. 140–51.

Schoenberg, William and Swartz, Jeremy (2022), 'Building more robust system dynamics models through validation', in *International Conference of the System Dynamics Society 2021*, vol. 2, Red Hook, NY: Curran Associates, Inc., pp. 923–39.

Schölkopf, Bernhard and Smola, Alexander J. (2008), *Learning with Kernels: Support Vector Machines, Regularization, Optimization and Beyond*, Cambridge, MA: MIT Press.

Senge, Peter M. ([1990] 2006), *The Fifth Discipline: The Art & Practice of the Learning Organization, Revised and Updated Edition*, New York: Doubleday.

Spirtes, Peter and Kun Zhang (2016), 'Causal discovery and inference: Concepts and recent methodological advances', *Applied Informatics*, 3:1, p. 3.

Sterman, John (2000), *Business Dynamics: Systems Thinking and Modeling for a Complex World*, New York: Irwin/McGraw-Hill.

Stone, Peter, Brooks, Rodney, Brynjolfsson, Erik, Calo, Ryan, Etzioni, Oren, Hager, Greg, Hirschberg, Julia, Kalyanakrishnan, Shivaram, Kamar, Ece, Kraus, Sarit, Leyton-Brown, Kevin, Parkes, David, Press, William, Saxenian, AnnaLee, Shah, Julie, Tambe, Milind and Teller, Astro (2016), 'Artificial intelligence and life in 2030: The one hundred year study on artificial intelligence', Report of the 2015–2016 Study Panel, Stanford: Stanford University, September, https://ai100.stanford.edu/sites/g/files/sbiybj18871/files/media/file/ai100report10032016fnl_singles.pdf. Accessed 11 May 2023.

Swartz, Jeremy, Wasko, Janet, Marvin, Carolyn, Logan, Robert K. and Coleman, Beth (2019), 'Philosophy of technology: Who is in the saddle?', *Journalism & Mass Communication Quarterly*, 96:2, pp. 351–66.

Wiener, Norbert (1956), 'The theory of prediction', in E. F. Beckenbach (ed.), *Modern Mathematics for the Engineer: First Series*, New York: McGraw-Hill, pp. 165–90.

14

Life from the Edge of Synthetic Biology

Pier Luigi Luisi

This chapter explores life from the edge of *synthetic biology* (SynBio), expanding on accounts of systems, autopoiesis and cognition noted earlier in this volume. Research is presented from the laboratory on biological structures that do not exist in nature, especially experiments into minimal cells and spontaneous overcrowding. I speculate on how a systemic view integrates bottom-up and top-down approaches, and briefly discuss multiple ways of knowing.

Generally, SynBio is part of the natural sciences and stands within the broader field of biotechnology. There are basically two main camps of interest[1] and by far the most popular is *bioengineering*,[2] essentially genetic engineering, mostly of bacteria and other microorganisms with the aim of producing, for example, new pharmaceuticals. The other camp of non-genetic or *chemical synthetic biology* is where new structures are built by the classic methods of organic chemistry.[3]

The most reductive way to see life in terms of chemistry claims that life can be reduced to a unique chemical reaction. For example,[4] the energy required for motion and movements comes from the combustion of sugar in the body through cellular respiration:

$$C_6H_{12}O_6 + 6O_2 = 6CO_2 + 6H_2O + \text{energy}$$

But where does the sugar come from? From starch and polysaccharides, which are polymers of sugar or various plant products, such as rice. And how is all this produced? Chlorophyll synthesis, a part of photosynthesis, whereby plants produce sugar and oxygen at the same time:

$$6CO_2 + 6H_2O = C_6H_{12}O_6 + 6O_2$$

This is the same equation written in the opposite way but this process requires a lot of energy. And where does this energy come from? The sun.

This reductionist view argues that life is a machine that works by the force of the sun. The sun gives the energy to produce plants and their products, like starch and rice; humans eat all that, acquire oxygen and produce carbon dioxide. Thus, plants and humankind are unified in a wholeness, with the sun as the main driving force. Indeed, the two equations above are actually only one, and it is assumed that this is all you need to know about life. We can agree about the importance of that basic equation, but we certainly need to understand much more about life.

Autopoiesis

We can start noticing that these two equations work in living organisms, thanks to mitochondria or chloroplasts in biological cells – all life on our planet is cellular life. Thus, to get to a higher dimension of the chemistry of life, we must look at the cell. This was the approach that Humberto Maturana and Francisco Varela proposed with their theory of *autopoiesis* in the 1970s (Maturana and Varela [1987] 1992; Varela et al. 1974). The simplest way to introduce the concept is to observe that all living organisms on our planet implement life in the same way. For simplicity, let us look at our own body. The hemoglobin in our body lasts only a few days, and then it degrades and disappears. But it is immediately replaced and regenerated from the inside of the body. And the same is true for most of our cells, which do not last more than a couple of months; they die but are replaced by new ones.

This continuous replacement gives an impression of continuity: the liver cell remains a liver cell and yesterday's cat remains the cat of today. There is *self-maintenance* due to the fact that the living organism is capable of regenerating the components that are used and consumed during the life cycle. We can therefore propose, with the words of Maturana and Varela, a kind of definition of life based on the behaviour of the living cell: 'A living system is [...] capable of self-production and self-maintenance through a regenerative network of processes which takes place within a boundary of its own making' (Damiano and Luisi 2010: 147).

In other words, the cell maintains its own identity despite an enormous number of transformations and does so thanks to a process of regeneration from within. This is also true for every macroscopic organism: microbe, cat or elephant.

The simplest living cell is an immensely complex organism, consisting of thousands of compounds, all interconnected and mutually dependent, as illustrated in Figure 14.1. This is only part of the metabolic map of the cell of *Escherichia coli*, but the picture is generally valid for any kind of cell.

And from this, we can learn some interesting things about the life of a cell or the life of an elephant. For example, if we ask the question: where is the life of

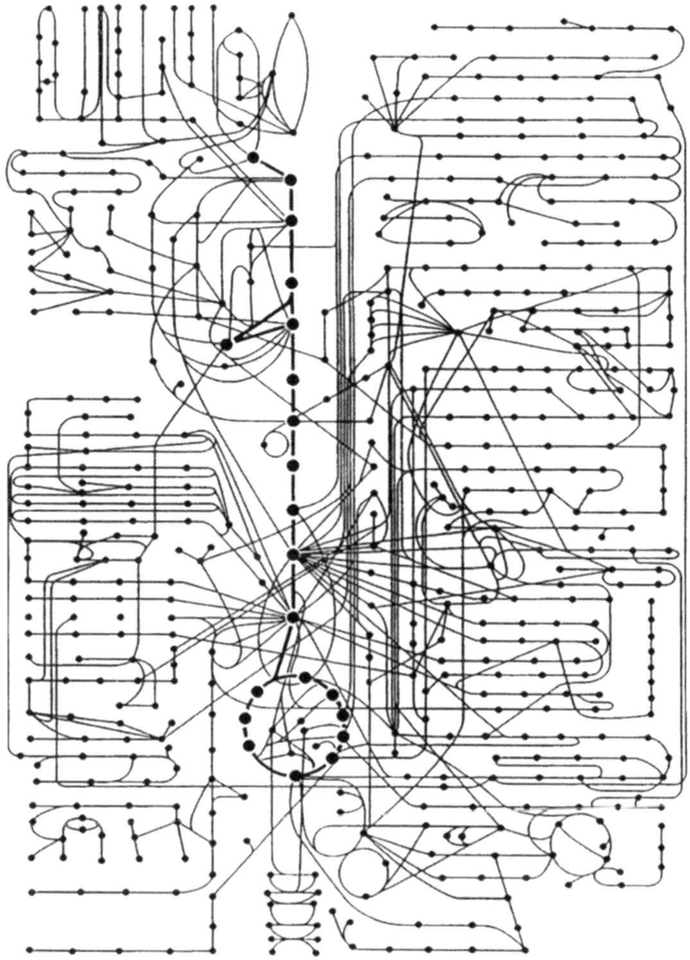

FIGURE 14.1: Part of the metabolic pathway/network of a bacterium (*E. coli*). Note that each point is a chemical compound, each line is a chemical (enzymatic) reaction leading from one compound to the next one, each reaction being catalyzed, i.e. made possible at normal conditions, by an enzyme, a large specialized protein. Indeed, a very complex system (Luisi 2016: 374).

the cell (or elephant) localized? Is there a particular spot or reaction center that is dominant or that is the director of the entire enterprise? The answer is that life is not localizable. The life of the cell is the whole network of reactions; there are no compounds which stand alone and direct the scene. DNA (the code-script) is certainly very important, but this is the product of dozens of chemical reactions, each of which in turn being due to other reactions. Life can thus be understood as a network of bodily reactions, experiences, memory, emotions and environments, as a systemic dimension, where each of these aspects is linked with and dependent on all the others (see Luisi 2016).[5]

Cognition

Obviously, regeneration is due to energy and food coming from the environment. One cannot talk about life without considering at the same time its environment. This introduces the notion of *cognition* to the previously noted definition of autopoiesis. Cognition is, according to Maturana, the 'doing' of the organism in order to maintain its own autopoiesis, and, in other terms, is basically the capability of the organism to recognize, interact and adapt with its environment (Maturana and Varela [1973] 1980, [1987] 1992; Varela et al. 1974). Accordingly, Luisa Damiano and Luisi (2010) proposed a reformulated description of an autopoietic system as:

> a network of production processes which produces the components which, through their interactions and transformations, permanently regenerate the network of processes constituting the system itself as a concrete topological unit, separated from its *medium* by a boundary and related to it through cognitive or adaptive coupling.
> (149, emphasis added)

It is important to note that life and cognition are not necessarily taken to be strictly equivalent but rather as closely related phenomena that share essential (sensorimotor) characteristics (Van Duijn et al. 2006: 159; see also Varela 2000; Bechtel and Bich 2021). Some authors like to make a distinction between *minimal cognition* or 'sensorimotor coordination abilities' (Van Duijn et al.: 167)[6] and increasingly complex cases of cognition, 'such as reasoning, explicit problem-solving or symbolization' (166).

Autopoietic Systems in the Laboratory: Are They Living?

Now let us consider whether a minimal autopoietic cell can be realized in a laboratory. One simple pictorial way to represent a working cell is presented in Figure 14.2. Notice that all chemistry takes place within the borders, and therefore the system can be seen as an autopoietic system.

The question that then comes to mind: is it possible to realize experimentally something like this in the laboratory?

One such system made in the early 1990s by my team at the ETH (Eidgenössische Technische Hochschule) Zurich in Switzerland, used as a spherical component the simplest possible system, the aqueous *micelle*. We started from a biphasic system,

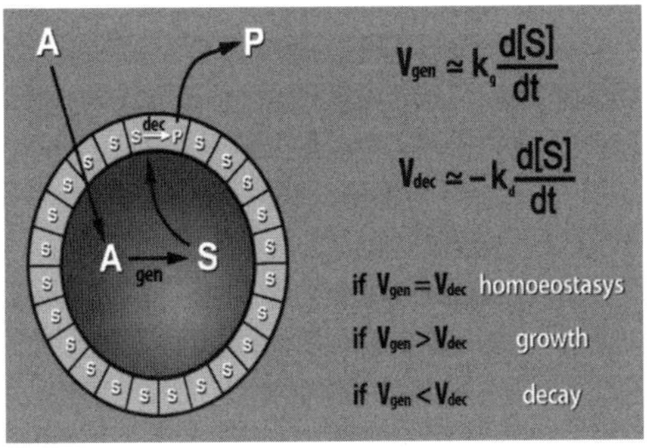

$$V_{gen} \simeq k_g \frac{d[S]}{dt}$$

$$V_{dec} \simeq -k_d \frac{d[S]}{dt}$$

FIGURE 14.2: The minimal autopoietic cell: a cell schematized by a spherical structure formed by the surfactant S, while nutrient A is entering and is transformed by the internal mechanisms of the structure into a cellular component. There is also a destructive reaction which transforms S into P. There are only two basic chemical reactions, and depending on the relative velocity of the two, we can have three different modes of existence of the cell, homeostasis, growth/self-reproduction and decay/death (Luisi 2016: 125; see also Bich and Green 2018: 3926–27).

in which the water-insoluble ethyl caprylate, $CH_3(CH_2)_6COOCH_2CH_3$, lies over an alkaline water phase (Bachmann et al. 1991, 1992). The hydrolysis of ethyl caprylate at the macroscopic interface produces aqueous micelles in the water phase, but then these micelles attract and hydrolyze further ethyl caprylate, and then we have a beautiful exponential increase of the concentration of micelles, as shown in Figure 14.3.

This is an autopoietic system. If life and autopoiesis are equivalent, then the question whether the micellar system illustrated in Figure 14.3 is living becomes a proper question. However, with a deeper analysis, one can ask: is autopoiesis the minimal and sufficient condition for life? Or is it only a necessary, but not also the necessary condition? This question has been debated, and in my opinion needs further scrutiny (see Bitbol and Luisi 2004).

Vesicles to Minimal Cells: From Autopoiesis to Synthetic Biology

Vesicles appear to be the proper candidates to model a biological cell, as they can reach cellular dimensions (up to 100–200 microns). In particular, liposomes, which are vesicles constituted by double layers of lipid surfactants, are more structurally related to biological cells. And it is well known that liposomes can incorporate proteins and other macromolecules (Luisi 2016), thus more closely simulating a cell. Another important feature of vesicles is that they can be self-reproducing, as shown by our group long ago (Walde et al. 1994). The incorporation of proteins into vesicles takes us into the realms of *bioengineering* and *chemical synthetic biology*.

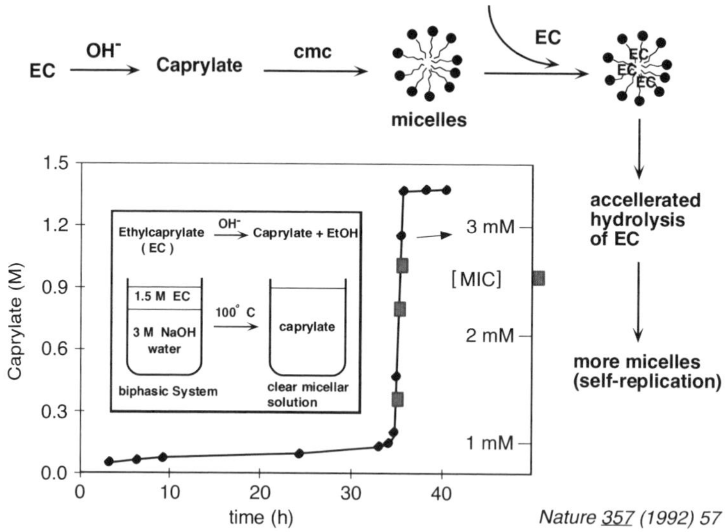

FIGURE 14.3: The time course of the production of ethyl caprylate (EC), until the critical micelle concentration is reached, whereby the formation of micelles becomes exponential (Bachmann et al. 1991).

One beautiful example of the latter is reported in Figure 14.4, where a self-replicating ribonucleic acid (RNA) is incorporated into self-reproducing vesicles (Oberholzer et al. 1995).

But is it possible to synthesize by this way a real biological cell in the lab? We know that the simplest modern biological cells consist of at least a couple of hundred genes (Gil et al. 2004). However, as one of the arguments goes, most likely the first biological cells in the evolutionary pathway were much simpler, perhaps consisting of only ten to twenty genes.

Can then one attempt the construction of a *minimal cell*, defined as a biological cell which has the minimal number of components and still acts as a biological cell? I believed that such a question makes sense and my group invested much time and effort in the study of protein-containing liposome (Schmidli et al. 1991; Murtas et al. 2007; Stano et al. 2011). Such work was successful in terms of SynBio, in the sense that a series of novel biological structures were obtained and characterized, which were previously not present in nature.

However, the gap between what we could design in the lab (one, two or maximal three macromolecules inside the liposomes), and the complexity of real biological cells is so large that I recognized later that this 'bottom-up' approach (in other words, synthesizing components from very simple molecules under prebiotic conditions)[7] was not the way to arrive at the laboratory synthesis of living cells.

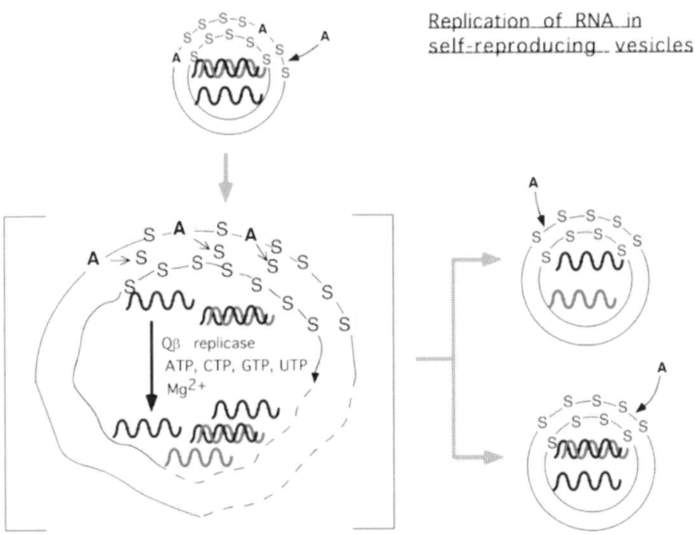

FIGURE 14.4: Self-replication of RNA in self-reproducing vesicles (Oberholzer et al. 1995).

Spontaneous Macromolecular Overcrowding

Let me add that these studies on liposomes have added some basic and important new knowledge about the physical chemistry of vesicles themselves. One example, which will be useful also for our later discussion, is the phenomenon of *spontaneous overcrowding* (Luisi et al. 2010; de Souza et al. 2012; de Souza et al. 2014; Stano et al. 2015). Figure 14.5 shows the procedure used to incorporate proteins into vesicles, which involves adding a vesicle-forming surfactant to the protein solution; the vesicles form and by closing they incorporate the protein. Very simple, except that you would then expect a Poisson distribution of the solute inside the compartments.

We initially did this experiment with ferritin, a protein which is large enough to be seen as a single spot in electron microscopy, and the surprising result was a very anomalous distribution of the solute macromolecules inside the vesicles. This was in fact a kind of 'all or nothing' distribution, diagrammatically shown in Figure 14.5, and from images of the actual experiment in Figure 14.6 (Luisi et al. 2010; de Souza et al. 2012; de Souza et al. 2014).

The fact that macromolecules can accumulate spontaneously into self-forming liposomes has been utilized also to incorporate an entire kit for the biosynthesis of green fluorescent protein (GFP) with the advantage that you can see the progress

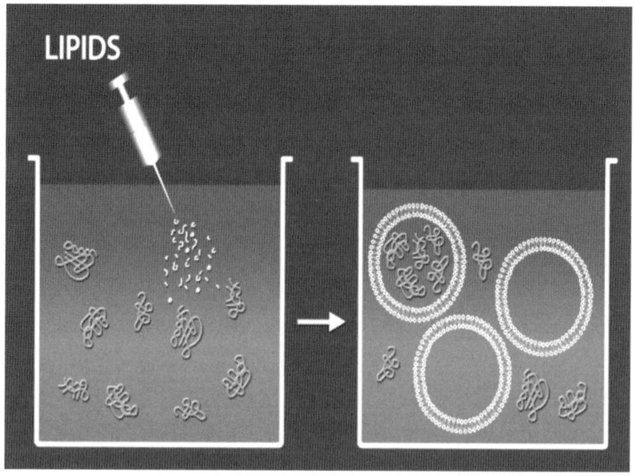

FIGURE 14.5: Formation of protein-containing protein by injecting a vesicle-forming surfactant into the protein solution. The surprising result is that, instead of a Poisson normal distribution of the solute into the liposomes, we have a case of quasi 'all or nothing' incorporation.

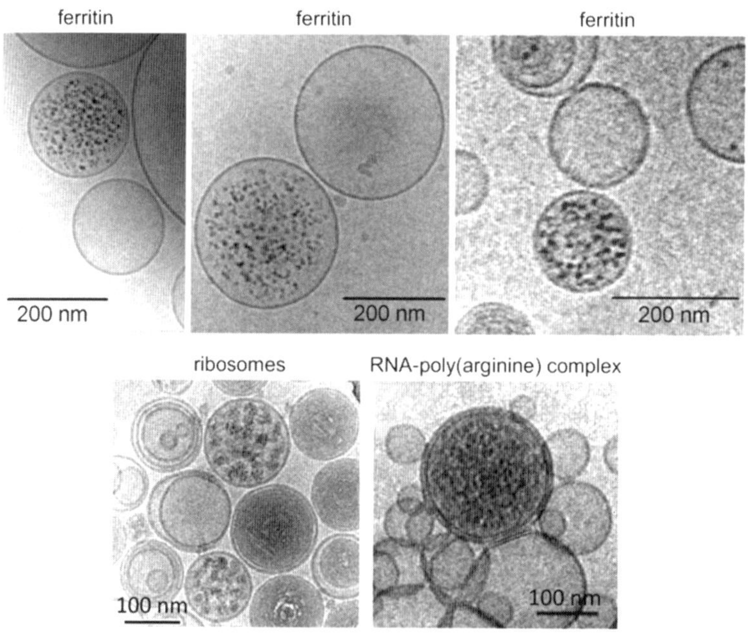

FIGURE 14.6: Actual transmission electron microscopy (TEM) micrographs of ferritin containing POPC liposomes, showing that indeed there are a few over-filled liposomes and a lot of 'empty' ones (Luisi et al. 2010; de Souza et al. 2012; de Souza et al. 2014).

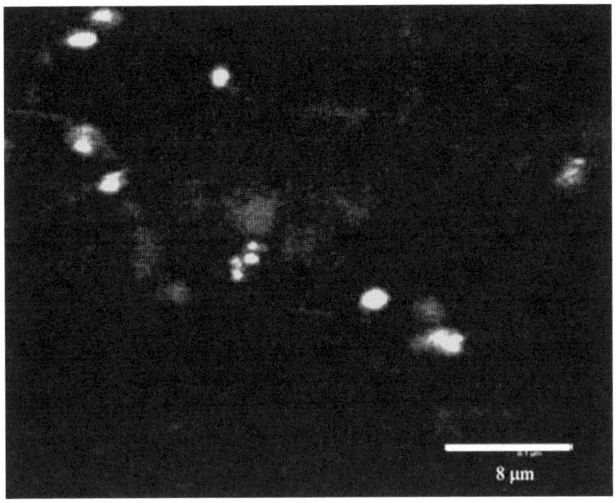

FIGURE 14.7: Biosynthesis of GFP inside liposomes (Murtas et al. 2007).

of the reaction, thanks to the formation of the green color inside the liposomes (Murtas et al. 2007), as shown in Figure 14.7.

Beyond Bottom-up Synthetic Biology: A Systemic Approach

Over recent years, I have mentioned my skepticism towards the bottom-up approach to the construction of the minimal living cell (or making 'new forms of life from scratch', also known as *wet artificial life*; see Bedau and Cleland 2010: 218). This is an interesting point also from a general, philosophical view.

According to Alexander Ivanovic Oparin, life on Earth originated from prebiotic molecules, which underwent a kind of spontaneous 'molecular evolution' by which the molecular structures and functionality, in due course, increased up to the formation of the first proto-cellular structures, which then, eventually, became our first cells. Here the term spontaneous means determined by physicochemical laws, in other words, no divine intervention. There is no demonstration that life really started that way, as nobody has been able to replicate the process in the lab.

Oparin's way of thinking had its origin in the Darwinian view, whereby, in general, complexity evolves from simpler structures (e.g. DNA and enzymes). This way of thinking is the classic bottom-up approach[8] and seems indeed to be the most logical way to build complexity. However, this elicits the question of whether this way of thinking is one of the main reasons for the impasse in research on the origin of life, with the general question: why is the origin of life on Earth still a mystery for science?[9] (see also Luisi 2020).

But are there alternative ways of thinking? If we return to the experiment in Figure 14.6, the net result is rather clear and important: given a diluted solution of biopolymers, the production of vesicles *in situ* is going to produce spontaneously, cell-like compartments having a high concentration of reagents – so as to possibly permit interactions and reactivity which are not possible in a diluted solution. Although the number of overcrowded vesicles is much smaller than the empty ones, even if there are only 0.1 percent of them, we will still have billions. And they will differ from each other in terms of stochastic distribution and relative concentration of the reagents.

We might first speculate that the overcrowded vesicles may be able to interact with each other forming a dynamic system or consortium. Another daring speculation is that, given their extremely large number of continuously changing structures, there may be at least one which eventually will have the right ingredients and concentration of proteins and nucleic acids to start life, or more simply, to start the first dynamic steps towards life (Luisi 2020). This scenario also appears in keeping with some views of the physics of self-organized criticality (Luisi 2018: 7, 2020). 'The combination of dynamic minimal stability and spatial scaling leads to a power law for temporal fluctuations' (Bak et al. 1987: 381).

Thus, we would have with an initial mutual fusion, re-equilibration towards a system with quasi-homogeneous vesicles in terms of content and mutual reactivity, and selection to arrive at an aggregate (quasi-colony) of vesicles as viable cell-like structures.

This then is a systemic view: what is essential is the *multiplicity*, the great number of components in dynamic equilibrium that permits the assumption that at least one of them can be or become the 'right one' to start. Aside from the speculative character of such an argument, there is an important general point: that nature, and probably also evolution, does not like to produce one single object or pathway at a time, but rather a great multiplicity. This can be understood as a contingency network with no predetermined logical connection between one step and the next that could then work with selection. All this has not been thoroughly studied yet and requires confirming experiments.

NOTES

1. See Chiarabelli and Luisi (2014), Luisi and Stano (2011) and Luisi et al. (2006).
2. 'The application of engineering principles to investigate biological systems creates synthetic systems or develops materials or devices that supplement or replace natural components' (Hine 2019).

3. It is not the place here to review all the works on SynBio. For a few typical examples, see Chiarabelli and Luisi (2014), Luisi and Stano (2011), Luisi et al. (2006), Forster and Church (2007), Benner and Sismour (2005) and Pfeifer et al. (2001) or the extensive reference list in Luisi (2016: 419–56).
4. Example adapted from Premio Basile finalist children's book (Luisi 1996).
5. See Green (2022): 'Systems are in this context generally understood as large networks of integrated components exhibiting non-linear dynamics. However, there is no unified view of what constitutes a systems-approach'.
6. For example, bacteria 'can use internal models of the world to make decisions' (Bechtel and Bich 2021: 8). Also see Bich and Moreno (2016) and Bich and Frick (2018).

   > [T]here are different possible interpretations of the terms 'top-down' and 'bottom-up' in the literature, and we believe that, to avoid confusion, the term 'reconstruction' is perhaps more appropriate in minimal cell studies, making it clear that, in this procedure, one does not necessarily reach the construction of an extant cell or something that exists on Earth. Since they do not exist in our biological life, the term 'artificial cell' may be used.
   >
   > (Luisi et al. 2006: 7)

   Ruiz-Mirazo et al. (2004: 329) similarly point to 'the tension between the individual-metabolic and the collective-ecological sides of the phenomenon'. Meanwhile, Bedau et al. (2009: 66) note that 'In the long run, bottom-up and top-down synthetic biology will increasingly blend and become harder to distinguish'.
7. More contemporarily, this view of life is included in NASA's definition and assumed by many RNA researchers: 'Life is a self-sustained chemical system capable of undergoing Darwinian evolution' (Joyce 1994 noted in Luisi 1998; also see Hays 2015). This 'reflects the obvious prejudice that the molecular mechanism of nucleic acids must be the main basis for defining life [...] bring[ing] life and evolution to equivalence' (Luisi 2017: 7).
8. 'In his book of 1936, Oparin expressed his views in the form of a dialectical materialist analysis explicitly citing Friedrich Engels. He described life as a naturally emergent stage in the evolution of matter, one in which physicochemical laws had been supplemented by the "purely biological" laws of natural selection and metabolism' (Luisi 2016: 5).
9. Oparin ([1938] 1965: 33) noted:

   > Engels (1933) shows that a consistent materialistic philosophy can follow only a single path in the attempt to solve the problem of the origin of life. Life has neither arisen spontaneously nor has it existed eternally. It must have, therefore, resulted from a long evolution of matter, its origin being merely one step in the course of its historical development.

REFERENCES

Bachmann, Pascale Angelica, Walde, Peter, Luisi, Pier Luigi and Lang, Jacques (1991), 'Self-replicating micelles: Aqueous micelles and enzymatically driven reactions in reverse micelles', *Journal of the American Chemical Society*, 113:22, pp. 8204–09.

Bachmann, Pascale Angelica, Luisi, Pier Luigi and Lang, Jacques (1992), 'Autocatalytic self-replicating micelles as models for prebiotic structures', *Nature*, 357, pp. 57–59.

Bak, Per, Tang, Chao and Wiesenfeld, Kurt (1987), 'Self-organized criticality: An explanation of the 1/f noise', *Physical Review Letters*, 59:4, pp. 381–84.

Bechtel, William and Bich, Leonardo (2021), 'Grounding cognition: Heterarchical control mechanisms in biology', *Philosophical Transactions of the Royal Society of London. Series B, Biological Sciences*, 376:1820.

Bedau, Mark A. and Cleland, Carol E. (2010), *The Nature of Life: Classical and Contemporary Perspectives from Philosophy and Science*, Cambridge: Cambridge University Press.

Bedau, Mark A., Parke, Emily C., Tangen, Uwe and Hantsche-Tangen, Brigitte (2009), 'Social and ethical checkpoints for bottom-up synthetic biology, or protocells', *Systems and Synthetic Biology*, 3:1, pp. 65–75.

Benner, Steven A. and Sismour, A. Michael (2005), 'Synthetic biology', *Nature Reviews Genetics*, 6, pp. 533–43.

Bich, Leonardo and Frick, Ramiro (2018), 'Synthetic modelling of biological communication: A theoretical and operational framework for the investigation of minimal life and cognition', *Complex Systems*, 27:3, pp. 267–87.

Bich, Leonardo and Green, Sara (2018), 'Is defining life pointless? Operational definitions at the frontiers of biology', *Synthese*, 195:9, pp. 3919–46.

Bich, Leonardo and Moreno, Alvaro (2016), 'The role of regulation in the origin and synthetic modelling of minimal cognition', *Biosystems*, 148, pp. 12–21.

Bitbol, Michel and Luisi, Pier Luigi (2004), 'Autopoiesis with or without cognition: Defining life at its edge', *Journal of the Royal Society Interface*, 1:1, pp. 99–107.

Chiarabelli, Christiano and Luisi, Pier Luigi (2014), 'Chemical synthetic biology', *Science Progress*, 97:1, pp. 48–61.

Damiano, Luisa and Luisi, Pier Luigi (2010), 'Towards an autopoietic redefinition of life', *Origins of Life and Evolution of Biospheres*, 40:2, pp. 145–49.

de Souza, Tereza Pereira, Fahr, Alfred, Luisi, Pier Luigi and Stano, Pasquale (2014), 'Spontaneous encapsulation and concentration of biological macromolecules in liposomes: An intriguing phenomenon and its relevance in origins of life', *Journal of Molecular Evolution*, 79:5&6, pp. 179–92.

de Souza, Tereza Pereira, Stano, Pasquale, Steiniger, Frank, D'Aguanno, Erica, Altamura, Emiliano, Fahr, Alfred and Luisi, Pier Luigi (2012), 'Encapsulation of ferritin, ribosomes, and ribo-peptide complexes inside liposomes: Insights into the origin of metabolism', *Origins of Life and Evolution of Biospheres*, 42:5, pp. 421–28.

Engels, Friedrich (1933), *Dialectics of Nature (in Russian)*, Moscow: Partizdat.

Forster, Anthony C. and Church, George M. (2007), 'Synthetic biology projects in vitro', *Genome Research*, 17, pp. 1–6.

Gil, Rosario, Silva, Francisco J., Peretó, Juli and Moya, Andrés (2004), 'Determination of the core of a minimal bacteria gene set', *Microbiology and Molecular Biology Review*, 68:3, pp. 518–37.

Green, Sara (2022), 'Philosophy of systems and synthetic biology', in E. N. Zalta (ed.), *The Stanford Encyclopedia of Philosophy*, Stanford, CA: The Metaphysics Research Lab, Center for the Study of Language and Information (CSLI), Stanford University (Summer), https://plato.stanford.edu/archives/sum2022/entries/systems-synthetic-biology/. Accessed 31 May 2023.

Hays, Lindsey, L. (editor-in-chief) (2015), *NASA Astrobiology Strategy. National Aeronautics and Space Administration*, Washington, DC: NASA Astrobiology Institute.

Hine, Robert (ed.) (2019), 'Systems biology', in *A Dictionary of Biology*, 8th ed., New York: Oxford University Press.

Joyce, Gerald F. (1994), 'Foreword', in D. W. Deamer and G. R. Fleischaker (eds), *Origins of Life: The Central Concepts*, Boston, MA: Jones & Bartlett, pp. xi–xii.

Luisi, Pier Luigi (1996), *Remedius l'alchimista*, Firenze: Giunti Editore.

——— (1998), *About Various Definitions of Life: Origins of Life and Evolution of the Biosphere*, 28, pp. 613–22.

——— (2016), *The Emergence of Life: From Chemical Origins to Synthetic Biology*, 2nd ed., Cambridge: Cambridge University Press.

——— (2018), 'The system view of the origin of cellular life', *Archives of Life Science and Nutritional Research*, 1:1, pp. 1–10.

——— (2020), 'Why the origin of life is still a mystery?', *Journal of Systems Chemistry*, 8, pp. 1–8.

Luisi, Pier Luigi, Allegretti, Matteo, de Souza, Tereza Pereira, Steiniger, Frank, Fahr, Alfred and Stano, Pasquale (2010), 'Spontaneous protein overcrowding in liposomes: A new vista for the origin of cellular metabolism', *ChemBioChem*, 11:14, pp. 1989–92.

Luisi, Pier Luigi, Ferri, Francisca and Stano, Pasquale (2006), 'Approaches to semi-synthetic minimal cells: A review', *Naturwissenschaften*, 93:1, pp. 1–13.

Luisi, Pier Luigi and Stano, Pasquale (eds) (2011), *The Minimal Cell: The Biophysics of Cell Compartment and the Origin of Cell Functionality*, Dordrecht: Springer.

Maturana, Humberto R. and Varela, Francisco J. ([1973] 1980), *Autopoiesis and Cognition: The Realization of the Living*, Dordrecht: D. Reidel Publishing, pp. 59–145.

——— ([1987] 1992), *The Tree of Knowledge: The Biological Roots of Human Understanding, Revised Edition*, Boulder, CO: Shambhala Publications.

Murtas, Giovanni, Kuruma, Yutetsu, Bianchini, Paolo, Diaspro, Alberto and Luisi, Pier Luigi (2007), 'Protein synthesis in liposomes with a minimal set of enzymes', *Biochemical and Biophysical Research Commununications*, 363:1, pp. 12–17.

Oberholzer, Thomas, Wick, Roger, Luisi, Pier Luigi and Biebricher, Christof K. (1995), 'Enzymatic RNA replication in self-reproducing vesicles: An approach to a minimal cell', *Biochemical and Biophysical Research Commununications*, 207, pp. 250–59.

Oparin, Aleksandr Ivanovich ([1938] 1965), *The Origin of Life*, 2nd ed. (trans. S. Morgulis), New York: Dover Publications, Inc. [First published in Russian as *Vozniknovenie zhizni na zemle*, Moscow: Izd. Akad. Nauk SSSR, 1936].

Pfeifer, Blaine A., Admiraal, Suzanne J., Gramajo, Hugo, Cane, David E. and Khosla, Chaitan (2001), 'Biosynthesis of complex polyketides in a metabolically engineered strain of *E. coli*', *Science*, 291:5509, pp. 1790–92.

Ruiz-Mirazo, Kepa, Peretó, Juli and Moreno, Alvaro (2004), 'A universal definition of life: Autonomy and open-ended evolution', *Origins of Life and Evolution of the Biosphere*, 34:3, pp. 323–46.

Schmidli, Peter Kurt, Schurtenberger, Peter and Luisi, Pier Luigi (1991), 'Liposome-mediated enzymatic synthesis of phosphatidylcholine as an approach to self-replicating liposomes', *Journal of the American Chemical Society*, 113:21, pp. 8127–30.

Stano, Pasquale, Carrara, Paolo, Kuruma, Yutetsu, de Souza, Tereza Pereira and Luisi, Pier Luigi (2011), 'Compartmentalized reactions as a case of soft-matter biotechnology: Synthesis of proteins and nucleic acids inside lipid vesicles', *Journal of Materials Chemistry*, 21, pp. 18887–902.

Stano, Pasquale, de Souza, Tereza Pereira, Carrara, Paolo, Altamura, Emiliano, D'Aguanno, Erica, Caputo, Margherita, Luisi, Pier Luigi and Mavelli, Fabio (2015), 'Recent biophysical issues about the preparation of solute-filled lipid vesicles', *Mechanics of Advanced Materials and Structures*, 22:9, pp. 748–59.

Van Duijn, Marc, Keijzer, Fred and Franken, Daan (2006), 'Principles of minimal cognition: Casting cognition as sensorimotor coordination', *Adaptive Behavior*, 14:2, pp. 157–70.

Varela, Francisco J. (2000), *El Fenómeno de la Vida*, Santiago de Chile: Dolmen Ediciones.

Varela, Francisco J., Maturana, Humberto R. and Uribe, Ricardo B. (1974), 'Autopoiesis: The organization of living system, its characterization and a model', *Biosystems*, 5:4, pp. 187–96.

Walde, Peter, Wick, Roger, Fresta, Massimo, Mangone, Annarosa and Luisi, Pier Luigi (1994), 'Autopoietic self-reproduction of fatty acid vesicles', *Journal of the American Chemical Society*, 116:26, pp. 11649–54.

15

Templating Life: DNA as Nature's Hard Drive, Version 2.0[1]

Mél Hogan and Tessa J. Brown

Given an ever-expanding ecosystem of digital data generated and circulated online since the late 1990s, a supporting material infrastructure has been continuously updated to accommodate growth and volatility of data storage spurred by global internet communications (Hu 2015). By 2025, predictions suggest we will be producing 463 exabytes of data globally, every day – one exabyte being 1000^6 bytes of data (Desjardins 2019). Most of the data we currently produce is stored on servers in industrial data center complexes. Our ability to store this much data is strained and pushes us to re-conceptualize archives as a political, material, social and environmental space, as well as one increasingly entangled with biology and the human body (Ceze and Strauss 2017).

Our digital data storage solution, which is to store data on computer/server hard drives, is susceptible to disruption and destruction by exposure to floods, magnetic fields and extreme heat, among other events (Starosielski 2014). The data centres housing servers also require tremendous amounts of upkeep, energy, land and water to operate and maintain systems for backup connectivity (via generators and batteries). However, innovations in storage solutions attempt to anticipate and circumvent these threats, leading technologists and scientists to consider archival modalities that are environmentally friendly, durable, dense, efficient and robust. Tape cartridges, quartz disks and now synthetic deoxyribonucleic acid (DNA) seem most promising. Storing data – text, audio, software and video – onto synthetic DNA has been successfully accomplished and is touted as one of the most promising storage technologies, especially for fixed, long-term storage (Hogan 2020). As a biological molecule, DNA remains forever legible even without special equipment (because the equipment to code/decode DNA can always be recreated). DNA's four types of nucleobases, A (adenine), T (thymine), C (cytosine) and G (guanine), can be found time and again in any living thing.

For as long as there is life on the planet, the code can be retrieved and the mechanisms to decode can be recreated. And without life – presumably – the need for such an archive disappears.

DNA editing and storage are remarkable in their potential. Using the letters ACTG (assigned to DNA for encoding data) as binary code (see Figure 15.1) has the potential to store all of the world's digital data in a container the size of a shoebox. This decade-old idea was tested in July 2017, when Harvard researchers encoded/decoded a short movie or Graphics Interchange Format (GIF) – five frames of Eadweard Muybridge's 1887 galloping horse – onto living bacteria.[2] The process involved assigning individual pixels a DNA code based on their shade of grey and encoding it onto the bacteria, *Escherichia coli*, using the clustered regularly interspaced short palindromic repeats (CRISPR) gene editing tool. The bacteria multiplied and created perfectly intact copies of the short film. Moreover, the sequence was retrieved in chronological order (Shipman et al. 2017; Kolata 2017). This innovation is significant even as DNA has effectively always already been a storage medium. As science writer Ed Yong explains:

> [V]iral DNA always gets inserted in the same place, and new sequences get added after old ones, as if the bacteria were stacking books on a shelf. They use these archives to guide an enzyme called Cas9 [CRISPR associated protein 9], which cuts and disables any viral DNA that matches the stored sequences. So, first and foremost, CRISPR is a kind of genetic memory – a system for storing information.
>
> (Yong 2017: n.pag.)

Other scientists have also considered DNA as a living archive. In particular, Karin Ljubič Fister and Iztok Fister Jr. invite us to see this living archive's future materialities:

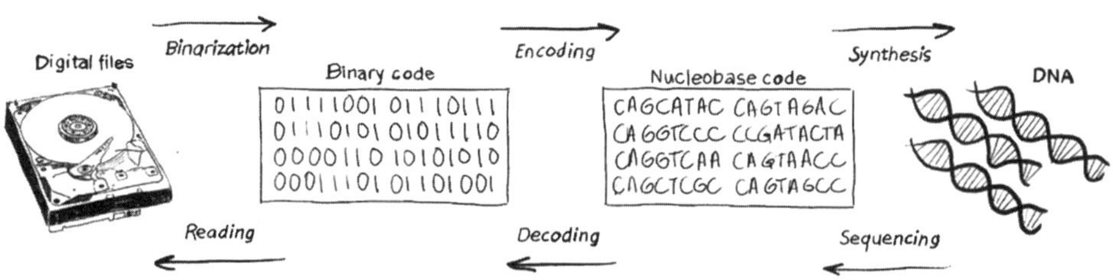

FIGURE 15.1: This image is adapted by Mél Hogan from 'dna-storage-2.jpg' in Viramgama (2018). Digital files can be rendered in binary code, which is then recoded in nucleobases and written into synthetic DNA – when the data needs to be retrieved, that synthetic DNA can be sequenced, rewritten as binary code, and read as a digital file.

> And just imagine how this could change the way we browse, access and store digital data. Imagine having your own, small, handy DNA sequence reader...you go to a park, take a leaf, put it inside...and yes, you can listen to the Rolling Stones or browse Wikipedia directly from a leaf.
>
> (Fister and Fister n.d.: n.pag.; see also Shankland 2019)

With the new research, scientists continue to consider DNA as an archive that would end external storage:[3] why store 'onto' anything when DNA is itself programmable? Why not instead program bacteria to record cellular life? The idea of DNA storage is already, in this sense, old. As Seth Shipman, a geneticist at Harvard Medical School, explained:

> We encoded images and a movie into DNA in a living cell which is fun, but it's not really the point of the system [...] What we're trying to develop is a molecular recorder that can sit inside living cells and collect data over time.
>
> (Sample 2017: n.pag.; see also Yim et al. 2021; Machemer 2021)

Scientists are now trying to conceive of the best way to maximize what DNA does best – reproduction. However, Robert F. Service (2021) points out that '[R]esearchers will also need to come up with ways to prevent their messages from degrading as the bacteria mutate as they replicate. But [...], it may give James Bond a new tool for hiding messages in plain sight'.

Instead of conceiving the archive of the future as one that requires constant conversions to better and newer media for storage, and for longer and more stable preservation, DNA could mean that memory remains stored in memory itself – at a cellular level. Faced with a seeming inability to stem the rising tide of data, DNA-based storage also appears to be a viable path to mitigating ecological crises, despite its unclear energy demands. At the very least, it conserves biodiversity and preserves the remnants of humanity (Hogan 2021b; Mendell et al. 2022). But rather than externalizing our lives onto records, we could *embody* those records. Whose bodies would be tasked with carrying memory? Would this be a burden or an honour? A service or a sacrifice?

While there are now successful experiments encoding/decoding data onto DNA, a little reflection is being done about this kind of innovation in relation to DNA's colonial legacy: it is linking of race and genetics, forensic anthropology, DNA profiling, tracking migrational history and so on. As such, we explore the promise of DNA as a medium for data storage by questioning both its historical complicity and its ongoing conquest and/or decolonization. We do this foremost by building from the work of those noted in this chapter, as well as Lily Kay (2000), Evelyn Fox Keller (2002), Yuval Noah Harari (2016), Alondra Nelson (2016), Kaushik Sunder

Rajan (2006, 2017) and Robert M. Sapolsky (2017), who together complicate the social history of DNA to address the specificity of an embodied archive. We use their insights to imagine how DNA storage is at once a distinct concept, situated outside the body, and also a scientific endeavor for understanding human heredity from within. How these two mechanisms come together is by way of increased computational power and its infrastructural logics, or in other words, computational biology (including bioinformatics), forever merging binary and DNA codes, memory and archive (see Park and Allaby 2017; NIH IRP 2023).

Nature's Hard Drive

To understand the impetus to use DNA as a storage mechanism, it is important to first return to the notion of 'discovery' by scientists. Like many such discoveries, DNA has contested origins and emerges from deep-seated European and Enlightenment-era ideals. These ideals have become largely normalized – science is the main mode of scrutiny for a rational understanding of the human body and mind.[4]

The discovery of DNA in the 1950s was attributed to American biologist James Dewey Watson, English physicist Francis Harry Compton Crick, New Zealand-born British biophysicist, Maurice Hugh Frederick Wilkins (NobelPrize.org 2021) and later on to English chemist and X-ray crystallographer, Rosalind Elsie Franklin (Science History Institute [2016] 2017; Smith 2018).[5] However, the nuclei had been spotted under the microscope of Swiss physiological chemist Friedrich Miescher as early as 1869 (Pray 2008; see also Miescher 1871: 456). Despite Miescher's amazing sighting, and his careful documentation of the chemical nature of the molecule, his findings remained unacknowledged for 50 years. Only then in 1919 did Russian biochemist Phoebus Levene continue Miescher's work and go on to identify ribonucleic acid (RNA) as containing ribose instead of deoxyribose in DNA.[6] While understated at the time, genetic code would eventually take on meaning alongside Boolean algebra, as binary code (0s and 1s) was being refined for computing and telecommunications to store and manage data. Only in 2017, however, would the two truly merge.

While Levene's preliminary model for DNA is overly simple in contrast to what we know today, it paved the way for Erwin Chargaff, an Austro-Hungarian-born biochemist, to connect DNA to hereditary units or genes. He identified the number of purines (A and G) in a DNA molecule as equal to the number of pyrimidines (T and C). In other words, A = T and C = G. It is from here that Watson, Crick and Wilkins (as well as Franklin) assembled the sum of the evidence and were (eventually) acknowledged with discovering DNA. By that time, in 1953, they had access to better equipment (such as X-rays and model building), which helped them to present DNA's structure as a three-dimensional double-helix.

Since these scientists' important contributions, other variations have been found, emphases placed elsewhere, and research expanded upon by science and technology studies, art and design, the natural, cognitive and social sciences, the digital humanities and critical code and data studies around the world.[7] DNA's history has been one of collective conceptualization, imagination, innovation and exploration. However, this history remains a paradigm focused on and giving credit to those who have worked to render visible and visual DNA's molecular structure. It is perhaps this visual dynamic that inspired the specific ways that we would later imagine cutting into it, replacing segments and editing it like any other medium. The stories surrounding the discovery of DNA have overlooked many social, cultural, political and economic conditions, motivations and contexts that surround quests to discover and visualize molecules and render them intelligible by and for western scientific ends (TallBear 2013; Wray 2015).

DNA has not meant the same thing to all people at all times. In its current and dominant formation, and as a western apparatus, it is foremost a code. By definition, code is 'any system of symbols, together with the rules for their association, that can be used to represent or transfer information' (Cammack et al. 2008), e.g. a system of signals or symbols for communication, instructions for a computer or the genetic code (*OED Online* 2023; see also Hine 2019). However, owning the code, controlling its uses and framing the way it serves societal ideals is something scholars have tried to grapple with – the shifting interpretations and specificity of contexts shaping DNA's meaning.

In particular, Jenny Reardon and Kim TallBear put into question the ways in which DNA molecules have been harvested by the western scientific community as almost always separate from the body (Reardon and TallBear 2012). This disconnect has meant a lack of attention paid to issues that pertain to belonging and ownership in its embodied sense. Drawing on the example of Native Americans and the Havasupai people, Reardon and TallBear point to the paradoxical relationship of science and property. On the one hand, whiteness and property have long been recognized as central to western nation-building. On the other hand – and through the promotion of science as neutral, objective and of universal interest – it has negated those very ideals of property for 'others'. By failing to consider divergent property interests in this way, the western scientific paradigm sees any other modality of identity construction as impeding on it (and on the so-called universal progress and knowledge) (Reardon and TallBear 2012: 243).[8] Science is powerful, but it cannot be both objective and imposed without undoing itself – and without becoming ideological.

This point about DNA's connection to ancestral stories is made also, though very differently, in *The Social Life of DNA: Race, Reparations, and Reconciliation After the Genome* by Alondra Nelson (2018). Nelson's argument is that for many African Americans, DNA testing can hold a special appeal despite being part of a larger technological surveillance apparatus that has historically proven to be discriminatory and racist (such as dataveillance on prisoners, medical testing, and video recordings of police killings). Given the complicated genealogy of slavery, and the lack of records created or recovered as a result, African Americans can use DNA to access stories that speak to their identities *before* the slave trade. This shifts identity categories in very important ways – it puts into question which moments in our shared lineages and timelines mark who we are or understand ourselves to be culturally. These are but two interventions among many that question DNA's significance. However, taken together they clearly make the point about the need to create space for many and sometimes contradictory narratives about the social power of genetic code and its deployments.[9]

Eugenics before Genomics

The quest for something gene-like, however, began as a cultural project rather than a scientific one. Predated by eugenics – another colonial 'science' that studies hereditary improvements of the human race by controlled selective breeding as well as sterilization and euthanasia – DNA emerged from a violent, highly racist, xenophobic framework, which scientists would struggle to mitigate in their work thereafter. Those advancing the science also could not always anticipate nor foretell the devastating repercussions of how genetic theory in concert with novel ideas from psychology and biology were to be interpreted and used for political control.

In 1859, Charles Darwin published *The Origin of Species*, which challenged the commonly held idea at the time that God had created the world in seven days and suggested instead that humans shared a common ancestor with apes. This did not mean that humans had evolved from the animals, but rather shared a lineage. However, the theory was (and remains) largely misunderstood as humans being descendants of apes, which translated into the notion that humans could therefore be more 'primitive', 'authentic' or 'evolved'. These were the seeds of a movement that attempted to value some people over others based on biological markers that anchored people both geographically and temporally. This mindset was further amplified by the fact that Darwin was Francis Galton's cousin – a familial connection that unwittingly contributed further to the birth of eugenics (rather than, for instance, deepening and challenging ideas of human exceptionalism, or of animal–human relationships) (Haraway 2016).

Eugenics was coined by Galton in 1883. By then he had already published an article and a book on the subject of heritability in which he put forward the idea that certain traits might be better for humanity than others. In 1865, he wrote a piece suggesting that civilization had diminished man's inherent ability to improve because it had thwarted the process of natural selection (Galton 1865). A few years later, he followed with a book called *Hereditary Genius*, which laid the groundwork for a quantitative understanding of human traits and the heritability of human intelligence from a social sciences perspective (Galton 1869). In the time between these two publications, several other key texts came out and many policies were enacted, including Gregor Mendel's 1866 paper 'Versuche über Pflanzenhybriden' ('Experiments on Plant Hybridization') containing his findings on pea plant heredity, which laid the foundations for modern-day genetic theory; the 1867 US-based 'Ugly Laws' to prevent vagrancy and poverty, or public confrontation with any kind of so-called bodily deformity; and the Indian Act in Canada in 1876. Other violently racist colonial policies were also instated against First Nations and Indigenous peoples in what was deemed Canadian territory. Residential schools in Canada began to be enforced in the 1870s and were kept in place until 1996, to then be replaced by child and family service agencies, which destroyed families, traditional knowledge, communities and cultures, the detrimental effects of which remain in full effect (see Marshall and Gallant [2012] 2021; Cowan 2020). While in the early 1900s, the legitimacy of eugenics was starting to fade in favor of behaviorism, its institutionalized and industrialized effects had spread and garnered a momentum that had become difficult to halt. Eugenics thinking had a grip on society.

Early research into human classification in the West worked to dismantle the perceived inconveniences of diversity and difference, be it to address immigration, mental illness, the uneven distribution of wealth, deviations from social norms, or cultural practices and languages. Together, policies (such as the Insane Asylum Act (1873), the Indian Act, the US Act to regulate immigration (1882), etc.) and infrastructure (schools, asylums, prisons, etc.) became a foundation of modernity and the conditions for colonial science to flourish. The first quarter of the twentieth century saw an onslaught of research (largely misguided inspirations from Darwin and Mendel) dedicated to the growing obsession with bodily classification and control. For decades – and despite significant resistance to attributing scientificity to the concept – eugenics was validated, serving as a basis for policy-making in the western world. In Canada, for example, the Immigration Act of 1910 was instated to make sure that 'only the best individuals entered and remained in Canada' and to protect against 'prohibited classes' that would, by that logic, pollute and dilute the population of its 'best traits' over time and for future generations of Canadians (Parmelee 1910). While the focus on genes as social markers changed over time, some of the ideals from eugenics took hold in other ways. They never entirely vanished (see NHGRI n.d.; Veit et al. 2021). In fact, there is a very real possibility that biologically deterministic technologies will be weaponized against the already-oppressed, forming the backbone of 'datafied eugenics'. This is a process by which 'traits determined to be more or less desirable determine one's path forward within algorithmically set limits' (Hogan 2021b: 20).

In 1990, the Human Genome Project (HGP) began (Lewis 2021); 'Dolly the sheep' was cloned in 1996; the genetic code of the fruit fly was decoded in 2000; the mouse had its genome decoded in 2002; the HGP's first draft was declared complete in 2003 and subsequently corrected and 'finished' (Zimmer 2021); since 2006, you can have your beloved dog cloned (Hickman 2013); in 2008, scientists completed the genome-wide sequence of an extinct animal, the woolly mammoth (Greshko 2021); in 2015, scientists edited the DNA in human embryos (with much controversy) (Stein 2015) and in 2017, DNA was used for data storage and gene therapy.

A future emphasis of DNA as research, discourse and object of interest is on getting people to playfully volunteer their DNA to lab testing in hopes of revealing data on health and ancestry, presumably to better understand one's past and/or future 'potential', i.e. that fit into popular self-enhancement and 'true self' narratives. DNA services that are offered include commercial applications of gene sequencing, such as FamilyTreeDNA, AncestryDNA and 23andMe, although arguably these are not foremost about health or ancestry as advertised. Like most social media sites that portend to be for community-building (e.g. Alphabet, Meta), DNA service sites are in fact yet another mode to collect data from people for profit and, ultimately, to surveil them. Those are the pillars of our current digital communications industry (profit and surveillance-for-profit) which cannot be disentangled from the motives of how DNA, as a concept, is being appropriated, normalized and deployed (Lanchester 2017).

Further, biocapital has become a case study that allows us to observe the birth of new systems and logics in capitalism, as well as a form of capitalism that emerges along with new technologies and epistemologies (Rajan 2006). In the case of genomics, we see both of these forms of biocapital at work, as companies begin to co-opt the human genome, turning DNA into profit, using it to reinforce neoliberal ideologies and commitments, and building upon it a new form of capitalism, one in which biotechnology firms (and subsidiaries), such as Roche (Genentech), Bayer (Monsanto) and other biopharmaceutical companies imagine their never-ending quest for profit will be realized. DNA as data storage, then, creates – or rather, promises to create – a system wherein we can continue to produce data at exponentially increasing rates, where our profligacy need not be curbed, because DNA will save us from ecological disaster and preserve important traces of humanity in itself as well (see also Haraway 1997; Helmreich 2008; Birch and Tyfield 2012: 303–04; Molteni 2018). As Rajan has observed: 'Life, for biocapital, is a business plan' (2006: 144).

In July 2017, a U.S. Food and Drug Administration (FDA) panel 'opened a new era in medicine' by approving 'the first-ever treatment that genetically alters a patient's own cells to fight cancer, transforming them into what scientists call "a living drug" that powerfully bolsters the immune system to shut down the disease' (Grady 2017). This was the first gene therapy ever to reach the market. However, as the science of DNA and genomics replaced the pseudoscience of eugenics, remnant problematics continue to influence many aspects in the quest to map the human, both in and as DNA. While gene therapy is certainly altering the ways in which people understand their bodies and mortality, gene testing already has as well in other ways. In *The Gene*, Siddhartha Mukherjee, warns against a few of these effects. One is an understanding of DNA as a 'manual of pre-vivorship' (e.g. preventative double mastectomies). The second is the serious impact of gender diagnosing in cultures – such as India and China – where historically, it has been deemed favorable to have male children (Mukherjee 2016). As such, humans continue to be classified and ranked based on biomarkers and genetic markers that suggest rather than dictate potential outcomes (see also Sapolsky 2017).

With the first draft of the human genome mapped (a project led by J. Craig Venter and Francis Sellers Collins) and a better sense of the ways in which gene clusters function, a desire to alter and edit genes became part of the scientific imaginary. This was especially true in work that sought to cure genetic diseases or to correct gene mutations taken up by researchers in countries like Japan, the Netherlands and Spain, who were instrumental in the discovery of clustered DNA repeats that decades later would prove to be the entry point into DNA editing. For example, conceptualized by Jennifer Doudna and her team (Doudna and Sternberg 2017), CRISPR is the first lab-tested gene editing tool. CRISPR allows for the manipulation of DNA in the nucleus of cells.[10] This places evolution more firmly than ever before in the hands of scientists, at least insofar as they consider themselves masters of the concept and code. Doudna and Sternberg reflect on the editing process, claiming that 'the genome – an organism's entire DNA content, including all its genes – has become almost as editable as a simple piece of text' (2017: xiii). However, while the CRISPR tool is impressive, the risks have yet to be fully assessed both in terms of its over-idealized technical precision and in terms of the aesthetics guiding the editing process (see also Greely 2021).

The Dataist's Archive

With science, we continually shift and re-establish boundaries between internal and external, and between nature and environment. We continue to progress towards the goal of recording our cells' memories,[11] which would reinstate a new kind of binary. However, for the moment we remain in this strange liminal space where data is stored in DNA, kept in vials and suspended according to their status as living or as life. As noted by several other authors in this volume, life has been understood and engineered increasingly from a mechanistic view – something that we imagine can be broken down into code, recorded and replayed. And yet, even though the initial draft map of the human genome was announced in 2003, it was not reported completed until twenty years later in August 2023 (NIH 2023). Over these two decades, numerous claims have surrounded this project, notably that genomics promises to render human life in a sense legible, notably that it would cure diseases like cancer and finally eradicate racism once and for all (Reardon and Tallbear 2012). To capture significantly more population diversity, the human genome references sequences have been an initial step to comprise the human *pangenome* reference. In May 2023, the first draft pangenome included sequences from forty-seven people of diverse ancestries with more to come (NHGRI 2023). As science seeks control over life's blueprints and building blocks, we enter the era of natural mutability in an unpredictable environment. Whereas nature was fixed and nurture was manipulatable – now, these are reversed (Keller 2010). We construct human nature, but the environment is out of our control. An ultimate technological fix in fact might be dependent on a global environmental catastrophe (Hogan 2017). Indeed, the coronavirus pandemic has 'demonstrate[d] the interconnectedness of all things – covid, care, community, nature, ewaste, racism, greed – in both the making and undoing of our modern communication systems' (Hogan 2021a). Perhaps this intensifying scrutiny at the cellular scale is a response to the global scale of our communications infrastructure, which is too big, too dispersed and off-limits. Our new media infrastructure is built as an environment unto itself, more resilient to environmental disasters than other industries and more invested in making surveillance a question of convenience.

It is not a coincidence, then, that with the rise of automation and machine learning (see Schoenberg and Swartz 2023), there is also growing xenophobia (as further evidenced by issues on university campuses, as well as political extremism) (Subbaraman 2021). Technology not only relies on second-class citizens to run the chain of supply at both ends (from mining rare earth minerals, to assembly lines, to dissecting disused technology/e-waste) but is programmed by humans according to their prejudices and biases (Bivens and Haimson 2016; Essex 2021). Even though programmers may not be intentionally coding the apparatus with mal intent, the infrastructure is made manifest by the ideals of the culture that builds it (Alang 2017; Davey 2016; Simonite 2017; Morrison 2021).

The infrastructure, therefore, maintains a disequilibrium between people through pervasive technological logic that justifies their own ends and by generating new ways to measure and control outcomes. One has allowed the other based on both computational power and the move towards probabilities that these logics carry – to collect and aggregate all data. The progress in DNA research is also tied to growing computational power and data storage that allows DNA sequencing and the comparison of large samples. In this way, the coevolution of big data and DNA research cannot be understated (Harari 2016; Reardon 2017; Mendell et al. 2022). Increasingly, humans think in ways that enable ratings and rankings because communications infrastructures feed off of easy and often binary classifications. This contributes to the condition of 'dataism', the idea that data is more valuable than humans or that humans are somewhat useless without data (Lohr 2015). This is how we become dependent on the infrastructure and, eventually, risk *becoming* the infrastructure.

What we know now is that DNA research that relies on massive internet infrastructures and computational power is imbricated in material, ecological and conceptual logic. Our global communications infrastructure is built on social inequalities that allow for cheap labor, cheap nature, expensive products, proprietary software and hardware, and quick distribution and sales. As such, we need to stay attuned to these logics and their normalizing effects. Just as we have produced artificial intelligence and machine learning only to be later confronted with the fact that these algorithms have appropriated our sexist and racist biases, DNA, as a technological apparatus, runs similar (if not graver) risks. Like binary code, the meaning of DNA code is in its context and must be recognized as the kind of data that wields tremendous power by those in control of its concept, discourse, tools and infrastructure. That is, big data could become a renewed tool for eugenics if using mass aggregation helps draw conclusions about people that are said to be more accurate than their own ideas of self or experiences. Eugenics, in this sense, remains part of the discussion – and is closer than ever to the science of DNA – as both move towards universal and totalizing narratives. As an archive, the story of DNA reveals to us our anxieties and priorities – those that have shifted humankind and those that anchor anew a sense of perpetual loss that is to be mitigated.

NOTES

1. This article was revised from Mél Hogan's (2018) piece, 'Templating life: DNA as nature's hard drive', published in *Public*, vol. 29:57. It was updated for this volume by Tessa J. Brown and published with permission from Intellect Ltd.
2. Another example is digital photos stored as DNA. See Trafton (2021).
3. It is also represented in the concept of the 'DNA-of-things [DoT] information storage framework [...] for storing other types of data – from biological information that is not naturally recorded in DNA, such as signaling dynamics and lineage histories, to non-biological

content like books, pictures and movies' (Farzadfard 2020: 31). Meanwhile, Koch et al. explain: 'The DoT approach [...] could also be used for integrating fabrication plans and conditions into complex products such as houses and art installations, as a way of preserving information for later generations' (2020: 43; see also ETH Zürich 2019). In the early 2020s, a DNA trade group formed, representing 'a coalition of computing and biotech firms including Microsoft, Twist Bioscience, Illumina and Western Digital announced that they were forming the DNA Data Storage Alliance (DDSA)' (Vitak 2021; see also Coughlin 2021).

4. A distinction further reinforced by western science.
5. 1953 marked

> the publication of the famous Watson and Crick paper – and that of two other papers on DNA that appeared in the same issue. Neither was so high profile, but each was essential to the structure's discovery. Both were written by scientists at King's College London: one by Maurice Wilkins and his colleagues Alec Stokes and Herbert Wilson, and the other by Gosling and his PhD supervisor, Rosalind Franklin.
>
> ('Due credit' 2013)

Then, Watson, Crick and Wilkins received Nobel Prizes in 1962. 'Since the Nobel Prize committee doesn't confer awards posthumously, it means that Franklin will never share in the scientific community's highest honour for her work [... due to her] early death [in 1958]' (Smith 2018).

6. Alberto Ascoli discovered in 1900 that RNA has three of the same nucleobases as DNA, except that the fourth nucleobase is U or uracil, not T or thymine.
7. For example, the Environmental Media Lab (EML), https://environmentalmedialab.com; and the *Heliotrope Journal*, https://www.heliotropejournal.net; Department of Communication, Media and Film, University of Calgary.
8. Reardon and TallBear call on the implementation of 'intercultural frameworks' that include tribally controlled biobanks and DNA loan systems. TallBear asks us to consider what forms of relations count/have counted historically and how western science came to be the adjudicator to truth rather than a way of looking at the world, given evidence of science's partiality and often oppressive practice.
9. Interestingly, in regards to Indigenous peoples' ancestry, 'DNA technology has further strengthened the claim of a familial relationship between a living Native American and a historical figure: Tatanka Iyotake [... aka] Sitting Bull' (Dasgupta 2021).
10. While it has been touted as easy and accurate in theory, in the lab the process remains relatively cumbersome.
11. Historically, this builds on Norbert Wiener's comments postulating 'genetic memories' (Wiener 1964).

REFERENCES

Alang, Navneet (2017), 'Turns out algorithms are racist', *The New Republic*, 31 August, https://newrepublic.com/article/144644/turns-algorithms-racist. Accessed 31 January 2023.

Birch, Kean and Tyfield, David (2012), 'Theorizing the bioeconomy: Biovalue, biocapital, bioeconomics or . . . what?', *Science, Technology, & Human Values*, 38:3, pp. 299–327.

Bivens, Rena and Haimson, Oliver L. (2016), 'Baking gender into social media design: How platforms shape categories for users and advertisers', *Social Media + Society*, 2:4, pp. 1–12.

Cammack, Richard, Atwood, Teresa, Campbell, Peter, Parish, Howard, Smith, Anthony, Vella, Frank and Stirling, John (eds) (2008), 'Code', *Oxford Dictionary of Biochemistry and Molecular Biology*, 2nd ed., Oxford: Oxford University Press, https://www.oxfordreference.com/view/10.1093/acref/9780198529170.001.0001/acref-9780198529170-e-3752. Accessed 30 November 2021.

Ceze, Luis and Strauss, Karin (2017), 'Storing data in DNA brings nature into the digital universe', *Scientific American*, 29 July, https://www.scientificamerican.com/article/storing-data-in-dna-brings-nature-into-the-digital-universe. Accessed 31 January 2023.

'code, n.' (2023), *OED Online*, Oxford: Oxford University Press, https://www.oed.com/dictionary/code_n. Accessed 24 July 2023.

Coughlin, Tom (2021), 'DNA storage update', *Forbes*, 28 October, https://www.forbes.com/sites/tomcoughlin/2021/10/28/dna-storage-update/. Accessed 31 January 2023.

Cowan, Katie (2020), 'How residential schools led to intergenerational trauma in the Canadian Indigenous population to influence parenting styles and family structures over generations', *Canadian Journal of Family and Youth/Le Journal Canadien de Famille et de la Jeunesse*, 12:2, pp. 26–35, https://journals.library.ualberta.ca/cjfy/index.php/cjfy/article/download/29511/21481. Accessed 31 January 2023.

Dasgupta, Alakananda (2021), 'DNA of Native American leader Sitting Bull matched to living relative', *New Scientist*, October 28, https://www.newscientist.com/article/2295264-dna-of-native-american-leader-sitting-bull-matched-to-living-relative/. Accessed 31 January 2023.

Davey, Alba (2016), 'Clarifai wants you to correct AI's biggest gaffes', *WIRED*, 5 October, https://www.wired.com/2016/10/clarifai-wants-correct-ais-biggest-gaffes. Accessed 31 January 2023.

Desjardins, Jeff (2019), 'How much data is generated each day?', *World Economic Forum*, 17 April, https://www.weforum.org/agenda/2019/04/how-much-data-is-generated-each-day-cf4bddf29f/. Accessed 31 January 2023.

Doudna, Jennifer A. and Sternberg, Samuel H. (2017), *A Crack in Creation: Gene Editing and the Unthinkable Power to Control Evolution*, Boston, MA: Houghton Mifflin Harcourt.

'Due credit' (2013), *Nature*, 496, 270, 17 April, https://doi.org/10.1038/496270a. Accessed 31 January 2023.

Essex, David (2021), 'Why algorithmic auditing can't fully cope with AI bias in hiring', *TechTarget*, 26 August, https://searchhrsoftware.techtarget.com/feature/Why-algorithmic-auditing-cant-fully-cope-with-AI-bias-in-hiring. Accessed 31 January 2023.

ETH Zürich (2019), 'Storing data in everyday objects', YouTube, 10 December, https://youtu.be/yo5Q1WfpMtw. Accessed 31 January 2023.

Farzadfard, Fahim (2020), 'DNA storage in everyday objects', *Nature Biotechnology*, 38:1, pp. 31–32, https://www.nature.com/articles/s41587-019-0376-8. Accessed 31 January 2023.

Fister, Karin Ljubič and Fister, Iztok, Jr. (n.d.), 'Storing data into living plants', http://www.storing-data-into-living-plant.net/experiment. Accessed 16 February 2018.

Galton, Francis (1865), 'Hereditary character and talent', *Macmillan's Magazine*, vol. 12, pp. 157–66, http://www.galton.org/essays/1860-1869/galton-1865-macmillan-hereditary-talent.html. Accessed 16 February 2018.

——— (1869), *Hereditary Genius: An Inquiry Into Its Laws and Consequences*, London: Macmillan and Company.

Grady, Denise (2017), 'F.D.A. panel recommends approval for gene-altering leukemia treatment', *New York Times*, 12 July, https://www.nytimes.com/2017/07/12/health/fda-novartis-leukemia-gene-medicine.html. Accessed 31 January 2023.

Greely, Henry T. (2021), *CRISPR People: The Science and Ethics of Editing Humans*, Cambridge, MA: MIT Press, https://mitpress.mit.edu/books/crispr-people.

Greshko, Michael (2021), 'Mammoth-elephant hybrids could be created within the decade. Should they be?', *National Geographic*, 13 September, https://www.nationalgeographic.com/science/article/mammoth-elephant-hybrids-could-be-created-within-the-decade-should-they-be. Accessed 31 January 2023.

Harari, Yuval Noah (2016), *Homo Deus: A Brief History of Tomorrow*, Toronto: Signal.

Haraway, Donna J. (1997), 'Gene: Maps and portraits of life itself', in *Modest_Witness@Second_Millennium.Femaleman©_Meets_OncoMouse™: Feminism and Technoscience*, New York: Routledge, pp. 131–72, https://books.google.com/books?id=ftO4jLQ2RM8C.

——— (2016), *Staying with the Trouble: Making Kin in the Chthulucene*, Durham, NC: Duke University Press.

Harden, Kathryn Paige (2021), *The Genetic Lottery: Why DNA Matters for Social Equality*, Princeton, NJ: Princeton University Press.

Helmreich, Stefan (2008), 'Species of biocapital', *Science as Culture* 17:4, pp. 463–78.

Hickman, Leo (2013), 'If your dog is about to die, why not clone it?', *The Guardian*, 11 March, https://www.theguardian.com/science/shortcuts/2013/mar/11/dog-about-to-die-clone-it. Accessed 31 January 2023.

Hine, Robert (ed.) (2019), 'Genetic code', in *A Dictionary of Biology*, 8th ed., Oxford: Oxford University Press, https://www.oxfordreference.com/view/10.1093/acref/9780198821489.001.0001/acref-9780198821489-e-1838. Accessed 31 January 2023.

Hogan, Mél (2017), 'Servers', in J. Wenzel and P. Yaeger (eds), *Fueling Culture: 101 Words for Energy and Environment*, New York: Fordham University Press.

——— (2018), 'Templating life: DNA as nature's hard drive', *PUBLIC*, 29:57, pp. 145–53.

——— (2020), 'Genomic media', *Transmediale Marshall McLuhan Lecture 2020, Expanded Networks*, Berlin: Embassy of Canada, 29 January, https://youtu.be/zoQRASLqENo?t=3234. Accessed 31 January 2023.

——— (2021a), 'The pandemic's dark cloud', *A Nourishing Network*, 22 January, https://a-nourishing-network.radical-openness.org/the-pandemics-dark-cloud.html. Accessed 31 January 2023.

——— (2021b), 'The data center industrial complex', in M. Jue and R. Ruiz (eds), *Saturation: An Elemental Politics*, Durham, NC, London: Duke University Press.

Hu, Tung-Hui (2015), *A Prehistory of the Cloud*, Cambridge, MA: MIT Press.

Kay, Lily E. (2000), *Who Wrote the Book of Life?: A History of the Genetic Code*, Stanford, CA: Stanford University Press.

Keller, Evelyn Fox (2002), *The Century of the Gene*, Cambridge, MA: Harvard University Press.

——— (2010), *The Mirage of a Space Between Nature and Nurture, Durham, NC*: Duke University Press.

Koch, Julian, Gantenbein, Silvan, Masania, Kunal, Stark, Wendelin, J., Erlich, Yaniv and Grass, Robert N. (2020), 'A DNA-of-things storage architecture to create materials with embedded memory', *Nature Biotechnology*, 38:1, pp. 39–43, https://www.gwern.net/docs/genetics/editing/2019-koch.pdf. Accessed 31 January 2023.

Kolata, Gina (2017), 'Who needs hard drives? Scientists store film clip in DNA', *New York Times*, 12 July, https://www.nytimes.com/2017/07/12/science/film-clip-stored-in-dna.html. Accessed 31 January 2023.

Lanchester, John (2017), 'You are the product', *London Review of Books*, 17 August, https://www.lrb.co.uk/v39/n16/john-lanchester/you-are-the-product. Accessed 31 January 2023.

Lewis, Ricki (2021), 'Looking back 20 years after the unveiling of the first human genome sequence', *PLoS* [Public Library of Science] *Blogs*, 28 October, https://dnascience.plos.org/2021/10/28/looking-back-20-years-after-the-unveiling-of-the-first-human-genome-sequence/. Accessed 31 January 2023.

Lohr, Steve (2015), *Data-ism: The Revolution Transforming Decision Making, Consumer Behavior, and Almost Everything Else*, New York: HarperCollins.

Machemer, Theresa (2021), 'Scientists write "Hello World" in bacterial DNA with electricity and CRISPR', *Smithsonian Magazine*, 15 January, https://www.smithsonianmag.com/smart-news/scientists-write-hello-world-bacterial-dna-electricity-and-crispr-180976763/. Accessed 31 January 2023.

Marshall, Tabitha and Gallant, David ([2012] 2021), 'Residential Schools in Canada', (10 October) 1 June, https://www.thecanadianencyclopedia.ca/en/article/residential-schools. Accessed 31 January 2023.

Mendell, Madeleine, Hogan, Mél and Verhoeven, Deb (2022), 'Matters (and metaphors) of life and death: How DNA storage doubles back on its promise to the world', in N. Shuurman (ed.), *The Canadian Geographer/Le Géographe Canadien*, 11 January, pp. 37–47.

Miescher, Friedrich (1871), 'Ueber die chemische Zusammensetzung der Eiterzellen', *Medicinisch-chemische Untersuchungen*, 4, pp. 441–60.

Molteni, Megan (2018), 'These DNA startups want to put all of you on the blockchain', *WIRED*, 16 November, https://www.wired.com/story/these-dna-startups-want-to-put-all-of-you-on-the-blockchain/. Accessed 31 January 2023.

Morrison, Sara (2021), 'A disturbing, viral Twitter thread reveals how AI-powered insurance can go wrong', *Vox*, 27 May, https://www.vox.com/recode/22455140/lemonade-insurance-ai-twitter. Accessed 31 January 2023.

Mukherjee, Siddhartha (2016), *The Gene: An Intimate History*, New York: Scribner.

Nelson, Alondra (2016), *The Social Life of DNA: Race, Reparations, and Reconciliation After the Genome*, Boston, MA: Beacon Press.

NHGRI (National Human Genome Research Institute) (n.d.), 'Fact Sheet: Eugenics and Scientific Racism', NHGRI, https://www.genome.gov/about-genomics/fact-sheets/Eugenics-and-Scientific-Racism. Accessed 31 January 2023.

——— (2023), 'A new human "pangenome" reference', National Human Genome Research Institute, 1 June, https://www.genome.gov/about-genomics/new-human-pangenome-reference. Accessed 7 September 2023.

NIH (National Institutes of Health) (2023), 'Researchers assemble the first complete sequence of a human Y chromosome', NIH Press Releases, 23 August, https://www.nih.gov/news-events/news-releases/researchers-assemble-first-complete-sequence-human-y-chromosome. Accessed 5 September 2023.

NIH (National Institutes of Health) IRP (Intramural Research Program) (2023), 'Computational biology', NIH IRP, 8 March, https://irp.nih.gov/our-research/scientific-focus-areas/computational-biology. Accessed 7 September 2023.

NobelPrize.org (2021), 'Speed read: Deciphering life's enigma code', The Nobel Prize in Physiology or Medicine 1962, https://www.nobelprize.org/prizes/medicine/1962/summary/. Accessed 30 November 2021.

Parmelee, Charles Henry (1910), *An Act Respecting Immigration*, Ottawa: Canadian Research Knowledge Network, pp. 205–39, https://www.canadiana.ca/view/oocihm.9_07184/. Accessed 31 January 2023.

Pray, Leslie A. (2008), 'Discovery of DNA double helix: Watson and Crick', *Nature Education*, 1:1, p. 100, https://www.nature.com/scitable/topicpage/discovery-of-dna-structure-and-function-watson-397. Accessed 31 January 2023.

Rajan, Kaushik Sunder (2006), *Biocapital: The Constitution of Postgenomic Life*, Durham, NC and London: Duke University Press.

——— (2017), *Pharmocracy: Value, Politics, and Knowledge in Global Biomedicine*, Durham, NC: Duke University Press.

Reardon, Jenny (2017), *The Postgenomic Condition: Ethics, Justice, and Knowledge after the Genome*, Chicago: University of Chicago Press.

Reardon, Jenny and TallBear, Kim (2012), '"Your DNA is our history": Genomics, anthropology and the construction of whiteness as property', *Current Anthropology*, 53:S5, S233–45.

Sample, Ian (2017), 'Harvard scientists pioneer storage of video inside DNA', *The Guardian*, 13 July, http://www.theguardian.com/science/2017/jul/12/scientists-pioneer-a-new-revolution-in-biology-by-embeding-film-on-dna. Accessed 31 January 2023.

Sapolsky, Robert M. (2017), *Behave: The Biology of Humans at Our Best and Worst*, New York: Penguin Press.

Schoenberg, William and Swartz, Jeremy (2023), 'System dynamics, machine learning and structural validation', in J. Swartz and J. Wasko (eds), *LIFE: A Transdisciplinary Inquiry*, Bristol: Intellect, pp. 289–308.

Science History Institute ([2016] 2017), 'James Watson, Francis Crick, Maurice Wilkins and Rosalind Franklin', Philadelphia: Science History Institute, (June) December, https://www.sciencehistory.org/historical-profile/james-watson-francis-crick-maurice-wilkins-and-rosalind-franklin. Accessed 31 January 2023.

Service, Robert F. (2021), 'Scientists "program" living bacteria to store data', *Science*, 11 January, https://www.science.org/content/article/scientists-program-living-bacteria-store-data. Accessed 31 January 2023.

Shankland, Stephen (2019), 'Startup packs all 16GB of Wikipedia onto DNA strands to demonstrate new storage tech', *CNET*, 29 June, https://www.cnet.com/tech/computing/startup-packs-all-16gb-wikipedia-onto-dna-strands-demonstrate-new-storage-tech/. Accessed 31 January 2023.

Shipman, Seth L., Nivala, Jeff, Macklis, Jeffrey D. and Church, George M. (2017), 'CRISPR–Cas encoding of a digital movie into the genomes of a population of living bacteria', *Nature*, 547:7663, pp. 345–49, https://www.ncbi.nlm.nih.gov/pmc/articles/PMC5842791/. Accessed 31 January 2023.

Simonite, Tom (2017), 'Machines taught by photos learn a sexist view of women', *WIRED*, 21 August, https://www.wired.com/story/machines-taught-by-photos-learn-a-sexist-view-of-women. Accessed 31 January 2023.

Smith, Kiona N. (2018), 'Rosalind Franklin died 60 years ago today without the Nobel Prize she deserved', *Forbes*, https://www.forbes.com/sites/kionasmith/2018/04/16/rosalind-franklin-died-60-years-ago-today-without-the-nobel-prize-she-deserved/. Accessed 30 November 2021.

Starosielski, Nicole (2014), 'Media, hot and cold: The materiality of media heat', *International Journal of Communication*, 8:5, pp. 2504–08.

Stein, Rob (2015), 'Critics lash out at Chinese scientists who edited DNA in human embryos', *NPR.org*, 23 April, http://www.npr.org/sections/health-shots/2015/04/23/401655818/critics-lash-out-at-chinese-scientists-who-edited-dna-in-human-embryos. Accessed 31 January 2023.

Subbaraman, Nidhi (2021), 'Caltech confronted its racist past. Here's what happened', *Nature*, 599:7884, pp. 194–98, https://www.nature.com/articles/d41586-021-03052-x. Accessed 31 January 2023.

TallBear, Kim (2013), *Native American DNA: Tribal Belonging and the False Promise of Genetic Science*, Minneapolis, MN: University of Minnesota Press.

Trafton, Anne (2021), 'Could all your digital photos be stored as DNA?', *MIT News Office*, 10 June, https://news.mit.edu/2021/dna-data-storage-0610. Accessed 31 January 2023.

Viramgama, Ashutosh (2018), 'DNA Data Storage – Synthetic DNA – Future of Storage', ashutoshviramgama.com, 9 December, https://ashutoshviramgama.com/dna-data-storage-synthetic-dna-future-of-storage/. Accessed 31 January 2023.

Veit, Walter, Anomaly, Jonathan, Agar, Nicholas, Singer, Peter, Fleischman, Diana S. and Minerva, Francesca (2021), 'Can "eugenics" be defended?', *Monash Bioethics Review*, 39, pp. 1–8, https://link.springer.com/article/10.1007/s40592-021-00129-1. Accessed 31 January 2023.

Vitak, Sarah (2021), 'Technology alliance boosts efforts to store data in DNA', *Nature*, 3 March, https://www.nature.com/articles/d41586-021-00534-w. Accessed 31 January 2023.

Wiener, Norbert (1964), 'Machines smarter than men?: Interview with Dr. Norbert Wiener, noted scientist', *U.S. News & World Report*, 24 February, pp. 84–87, https://collections.nlm.nih.gov/catalog/nlm:nlmuid-101584906X7699-doc. Accessed 31 January 2023.

Wray, Britt (2015), 'What is the world really like?: Andrew Pickering explains why modern science isn't the only – or even right – way to figure that out', *Method*, November, http://www.methodquarterly.com/2015/11/what-is-the-world-really-like. Accessed 16 February 2018.

Yim, Sung Sun, McBee, Ross M., Song, Alan M., Huang, Yiming, Sheth, Ravi U. and Wang, Harris H. (2021), 'Robust direct digital-to-biological data storage in living cells', *Nature Chemical Biology*, 17:3, pp. 246–53, https://www.ncbi.nlm.nih.gov/pmc/articles/PMC7904632/. Accessed 31 January 2023.

Yong, Ed (2017), 'Scientists can use CRISPR to store images and movies in bacteria', *The Atlantic*, 12 July, https://www.theatlantic.com/science/archive/2017/07/scientists-can-use-crispr-to-store-images-and-movies-in-bacteria/533400. Accessed 31 January 2023.

Zimmer, Carl (2021), 'Scientists finish the human genome at last', *New York Times*, 23 July, https://www.nytimes.com/2021/07/23/science/human-genome-complete.html. Accessed 31 January 2023.

PART VI

ARTFUL LIVES AND METALIVING

16

Aqueous Mediums, Urban Architectures, Anadromous Beings[1]

Brook Muller

We are constantly discovering that water is our best teacher.

—Herbert Dreiseitl, Deiter Grau and Karl Ludwig[2]

Introduction: Water Carrying Urban Meaning

We learn water's many lessons, pure and clear to Dreiseitl, Grau and Ludwig, when we pay careful, sustained attention to it, and as we allow it to settle and clarify, induce reflection and make its mark. Building on this collection's focus on communication, media and nature, how might water, a medium for transporting beings, energy and thoughts, guide us in imagining a more ecologically responsive city and reworked interactions between buildings and urban landscapes?

I consider the fate of salmon, an anadromous species that uses this fluid medium to complete its great life circuit (although numerous others could be studied for their indicator – and intrinsic – value) so as to better appreciate connections we might otherwise overlook.[3] The ensuing discussion also emphasizes the complexities and functional attributes of pre-development hydrological conditions in comparison to current paths and states of water as it courses through the city. This preliminary work provides a script for reconceptualizing architectural and urban design undertakings such that projects deliver net positive watershed impact.[4]

The emphasis on water in this chapter complements the exploration in this collection of the ways communication constitutes and permeates all avenues and forms of life. This theme furnishes an opportunity for reflecting on the intersection of language, water as universal solvent, aqueous beings and an expansion of sustainable architectural discourse that at present pays insufficient heed to the multiple, interrelated water crises that cities confront: climate change, ageing water infrastructures and the associated need for a staggering level of investment, undependable and poor quality supply, inequalities of availability, downstream ecological impacts, disconnection between urban dwellers and the precious resource they rely upon, and the enormous amount of energy consumed transporting water and making our urban and regional water systems run. Simply put, the language is dry, water is omitted or treated secondarily and near exclusively technically, and this impoverished and typically ethically and aesthetically hollow discourse has profound material and cultural ramifications.[5] I seek to enter the domain of sustainable design differently, to probe aqueous depths in order to enrich design narratives and material outcomes. With water as a carrier of meaning, we become better equipped to *enlist the hard and the grey*, that is to say, the currently toxic, impervious mediums of the city, *in the cause of the soft and the blue* for the benefit of lifeforms of all kinds.

Long Haul Carrier

The environmental historian Richard White provides a helpful point of departure for these reflections. In *The Organic Machine: The Remaking of the Columbia River*, White enlists energy as a medium to frame and draw correspondences between human labour, human constructs (specifically dams such as the Bonneville), the power of the river and the expenditures of salmon and other species who swim its swift waters. In the great metabolic exchange that defines the life cycle of salmon, nutrients that they spent years in the Pacific Ocean consuming are delivered to the upper watersheds where they once were born and where they return to spawn and die. As White says, 'Salmon thus are virtually a free gift to the energy ledger of the Columbia. They bring energy garnered from outside the river back to the river' (White 1995: 15).

When millions of Chum, Pink, Chinook and Coho salmon make their way up the Pacific Northwest's riverways, hungry bears preparing to hibernate await the plenty. The parts of salmon that bears do not consume and leave behind on the forest floor decompose and provide a critical source of nitrogen and other marine-rich nutrients for trees growing tall in the region's poor volcanic soils. With respect to The Great Bear Rainforest of British Columbia, for example, it has been estimated that 'eighty percent of the nitrogen in the forest's trees comes from the salmon' (Rahr 2018). An incredible set of coevolutionary adaptions and transactions can be found here: salmon carry energy; water carries salmon; bears carry salmon from streams to land; trees take up nutrients; trees cool streams so they carry higher levels of dissolved oxygen to the benefit of salmon; and on it goes.

As trees depend on salmon, salmon depend on trees. For example, before much of the landscape was converted to settlements and farms, forested riparian corridors and Oregon White Oak prairie communities characterized much of Willamette River Valley in the western part of the state. The rain landing on leaves would travel down rivulets of bark to leaf litter and then slowly make its way to the river, becoming cooler and more oxygen-rich along the way. Rivers and floodplains connected, and wetland complexes furthered processes of filtration and cleansing. Rivers overstepped their banks during the major rain events of winter. Side channels would form, with eddies or 'refugia' providing ideal spots for fish to wait out storms and surges. 'Coarse woody debris' – fallen dead trees and the remains of large branches along the edges of rivers and side channels – helped establish favourable conditions for salmon spawning in gravel beds by reducing streamflow velocities while allowing sufficient flow to deliver abundant oxygen. This woody debris also provided shade to keep water temperatures cooler and served as a critical component in the nutrient cycle. As hydrologist Ken Vigil claims 'the food chain begins with bacteria' made available by this debris and other sources.[6] Bart R. Johnson, a landscape architecture professor at the University of Oregon, maintains that, in a context of trees downed by storms, salmon were 'not just adapting to but thriving off disturbance'.[7]

A more complex spatial/habitat structure, characterized by 'stream sinuosity' (meanders) and 'natural hydrological shadows' (eddies and calm streamside channels) along with a landscape characterized by a high 'coefficient of friction' (rough, vegetated and variegated terrain that decelerates flows), supported diverse riparian biotic communities. It was a landscape both dynamic and indeterminate, with profoundly different seasonal conditions and ever-fluctuating boundaries supportive of specialized forms of life. Pulsing and coursing, stepping and bending, flooding and eroding, water was the medium making the system go.

The Urbanized Watershed Response

People almost always impact the hydrology first.

(Sarah Hinners 2017: n.pag.)[8]

This country's waterways have been transformed by omission.

(Alice Outwater 1996: 175)

With waterways serving as primary conduits of travel prior to the advent of trains, planes and automobiles, many of our cities built up along riparian corridors, atop creeks and streams and on fill where wetland complexes once existed. Urbanization dramatically alters and simplifies hydrological conditions, with dire consequences for aquatic beings. Today we are left with the damaging legacy of 'urbanized watershed response', or as Walsh et al. (2005) would characterize it, 'The Urban Stream Syndrome'. According to Andrew Karvonen (2001), we have 'created impervious landscapes of transportation networks and building rooftops, reorienting stormwater flows from predominately vertical to horizontal pathways' (vii). Solving one significant urban problem – preventing flooding and damage to buildings by removing water as quickly as possible – created myriad others.

Marc Leisenring, an engineer with the firm Geosyntec, breaks down causes of loss of stream function as a result of urbanization. These are:

- Modified hydrology
- Eroding banks and headcuts
- Channelization
- Floodplains and wetlands disconnected
- Limited upstream sediment supply
- Lack of vegetated buffers
- Nutrient enrichment[9]

By draping asphalt, concrete and rubberized membranes over large portions of the urban landscape, we have ensured that rainwater flows much more quickly and at a higher temperature – and therefore with less oxygen – than in pre-development circumstances. As it makes its way along hard surfaces where wetland complexes may have once existed, this water flows into riverways that have been channelized or 'entrenched' to prevent flooding and promote more opportune and dependable conditions for settlement (see Figure 16.1). With greater volumes of water entering a constrained river, the energy exchanges, loads and demands of which Richard White speaks are altered dramatically.[10]

Perilous and energy consuming, to begin with, the journey for salmon swimming upstream through a city in a fast-flowing river with no refugia becomes life-threatening.[11] And not only is the stormwater warm and oxygen-poor — it is toxic. The first fall rains in the Pacific Northwest pick up an elixir of heavy metals, motor oils and other pollutants that have accumulated over the dry summer months and deliver these to urban waterways at the very time many salmon species, such as the highly sensitive Coho, are making their runs upstream. The impacts are deadly, as research by Nat Scholz of the National Oceanic and Atmospheric Administration (NOAA) and Jen McIntyre of Washington State University Puyallup (WSU-Puyallup) so vividly demonstrates.[12]

Copper and zinc, commonly used materials in the building industry, are especially harmful. While copper is a nutrient for plants and animals, in higher

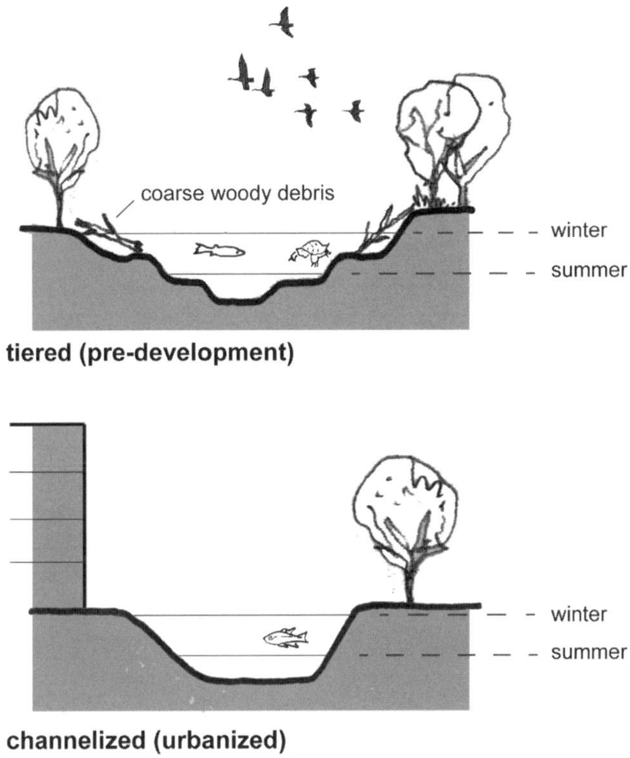

FIGURE 16.1: Intact or historic (pre-development) stream with a tiered profile vs. a simplified and channelized urban one.

concentrations it can lead to acute toxicity, chronic or long-term toxicity, diminished growth, decreased olfactory response (smell is a primary form of a salmon's defence against predation, providing warning signals), impaired swimming capability, and organ and cell damage.[13] With respect to runoff, Alan Fleming, an engineer and certified professional in stormwater quality with Kennedy/Jenks Consultants of Portland, Oregon, goes so far as to claim: 'If you fix your zinc and copper problem, everything else is taken care of'.[14]

What Inheres in the Hard and the Grey?

If we ever figure out how to make urban hydrology mimic pre-development conditions, we are well over half way there to restoring urban ecosystems.

(Sarah Hinners 2017: n.pag.)[15]

What does it mean to mimic pre-development conditions in an endeavour to remove pollutants and reintroduce functional ecological integrity in a dense urban context? For Maria Cahill, Program Manager of Recode – a Portland-Oregon-based non-profit dedicated to advancing next-generation water infrastructures with strong commitments to equity and sustainability at the heart of their mission – mimicking pre-development conditions would recognize that 'in Western Oregon, 50% of water was stored in the air' and that urban runoff has changed this dynamic and has dramatically reduced infiltration and shallow base flows. For Cahill, then, we are called upon to 'engage in practices that get 50% into the air and 50% into the ground'.[16] In response to this, movements such as Green Infrastructure (GI), Low Impact Development (LID) and Water Sensitive Urban Design (WSUD) respond to the need to redirect the flows of urban waters and show the benefits of softening the urban landscape, increasing the coefficient of friction, reducing runoff volume and improving water quality. As NOAA Fisheries and WSU-Puyallup have demonstrated, biofiltration features in the urban landscape such as green roofs, bioswales, rain gardens and in-flow through planters are highly effective mediums for increasing residence time for treatment, filtering out contaminants and slowing down the rate of stormwater flows back into the system.

Given the need to promote more ecologically responsive urban hydrological dynamics, how is it possible to aggressively accelerate positive ecological change in a context of rapid urbanization, ever greater population densities, and rising real estate costs? How are we to reconcile competing demands for limited space? For example, it has become an increasingly common situation for state departments of environmental quality to call for on-site stormwater infiltration (or reduced rates of flow if on-site treatment is not possible). Meanwhile, cities are implementing compact growth policies so as to meet affordability and transit-oriented development (TOD) goals; simply put, these efforts, one focused on more porosity in the urban landscape and the other on higher levels of impermeability, operate at cross-purposes. In addressing these goals that seem to be in conflict, in attempting to reconcile the pressures of development and the need for urban waterways to function in a manner akin to their pre-development states, Randy Hester, in his great work *Design for Ecological Democracy*, would ask the architect to embrace this very tension. He argues that the greatest opportunities for creative thought and exciting forms of design expression are to be found (Hester 2006). What sorts of inflections and multi-dimensional synergies between buildings and landscapes might be derived by confronting the three-dimensional puzzle of greater porosity and improved urban hydrology?

China's 'Sponge City' initiative provides one innovative model for infrastructures of the blue and the green to assume key functions and important spatial presences in contemporary urban design alongside the customary grey. And yet, what of existing cities composed largely of grey, hard surfaces? How can urban environments be retrofitted to perform in a manner akin to landscapes prior to urbanization? What potential inheres in the grey such that it is less disruptive and might even contribute to the cause of the blue and the green? The basic geology and morphology of certain pre-development landscapes hold clues.

Declivities

On a mild afternoon in the winter of 2001, my ex-wife Cathy and I boarded the *Indian Pacific* in Sydney, Australia. The train, beloved by railroad enthusiasts around the world, wound through the ridge of the Blue Mountains and climbed a gap before continuing west on the plateau towards the great dry interior. In the next morning's bright light, we arrived at the mining town of Broken Hill, 'Capital of the Outback', in the westernmost portion of New South Wales.[17] From Broken Hill, we proceeded northeast in a rented four-wheel vehicle along straight dirt roads amid flat terrain covered in grey green scrub, passing flocks of strutting emus and periodically descending to cross dry creek beds where white posts with black tick marks had been placed to measure water levels during flash flood events. So works the desert. Given the flatness of the landscape, we could see from a great distance the low hills of our destination, Mutawintji National Park, a place of deep significance to Aboriginal peoples (see Figure 16.2). After we had arrived,

FIGURE 16.2: The hills of Mutawintji National Park appear on the distant horizon.

we hiked up one of the hills – no more than a couple of hundred feet above the plain – and looked in all directions to the distant arcing horizon of the great red Australian desert.

Mutawintji is an island of ancient uneroded geography amidst a sea of granular soils. In the presence of the pools of water we came upon in the shaded declivities of the smooth rocky hills, we found abundant vegetation, with some shrubs a deep dark shiny green, others a chalky grey (see Figure 16.3). Where vegetation was found, we saw prints in the sand, evidence of fauna; as we listened at dawn the next day, a kangaroo scratched at our tent, looking for a handout. People coming to the area inevitably followed these signs of life and found hospitable the combination of rocks and pools, plants and animals. They did not live here year-round – the desert demanded nomadism, continual adaptation and the gathering of foodstuffs across a great expanse – but arrived seasonally from all directions to congregate and engage in rituals in this sacred rise of hills.

Our journey bore some affinity, albeit far less perilous, to those of the eighteenth-century Jesuit missionary, Eusebio Kino. In 'Maps of Water Holes', a chapter in his wonderful book *The Secret Knowledge of Water* (Childs 2000), Craig Childs speaks of Kino's ability to make sense of the hydrological reality of the

FIGURE 16.3: The cliffs of Mutawintji: where there is rock there is shade; where there is shade there is water; where there is water there is vegetation; where there is vegetation there are animals; where there are animals there are people.

landscapes that surrounded him in his efforts to secure reliable sources of water as he travelled by foot and mule from the mission in Tucson, where he was stationed, to Los Angeles and other locations for the purposes of trade, exploration and the gathering of supplies. Making his way through the Cabeza Prieta in the Sonoran Desert, one of the most desolate, arid landscapes of North America, Kino avoided the flats of the desert floor when possible and instead travelled along ridgetops, where he found shaded, water-filled pocket-like depressions (called *tinajas altas* or 'high tanks') – mediums of life amidst arid, mountainous, hard rock desolation.

The Hard Makes Possible the Soft?

Bart Johnson succinctly describes cities as in effect geological formations of rock outcrops, not unlike Mutawintji or the mountains of the Cabeza Prieta. Hard urban surfaces configured for functions such as quick conveyance to storm drains (with significant downstream effects) can be reconfigured to intercept and decelerate flows for the purposes of floodplain storage and to perform other beneficial work (taking place before the water flows to bodies such as rivers and lakes or infiltrates the ground and eventually evaporates, returning to the sky to continue the great cycle). A city has the potential to be a geological structure of water-scooped hollows and cool shady slot canyons, a structure that reinforces, and not compromises, ecological integrity.

For the American West, cities could become surrogate mountain ranges in a time of reduced snowpack, with their buildings and urban landscapes formed as at-grade storage depressions and above-grade declivities. Whether at the site or district scale, a goal becomes that of optimizing the harvesting, use, recycling and treatment of water, and, in doing so, instantiating micro-hydro loops that ultimately benefit the parent system in which projects are situated. With designers carefully following flows, buildings can achieve the next positive watershed impact in ultra-urban contexts.

An evolving nomenclature anticipates this conceptual and material transformation and offers a new way of seeing water and the city. Water resources specialist, Roy Iwai, calls the urban environment 'the pipeshed'.[18] Like a sub-watershed, a pipeshed establishes a boundary, with a rooftop and gutter constituting the edge of a portion of a catchment and the beginning of the tributary. Adopting a similar framing, Kevan Moffett, who runs the Moffett Research Lab with the Washington State University EcoHydrology Group, speaks of 'streets as streams and street trees as BMPs' (best management practices): 'Each urban street acts like an ephemeral headwater stream, and the street trees serve as that stream's riparian zone'[19] (see Figure 16.4).

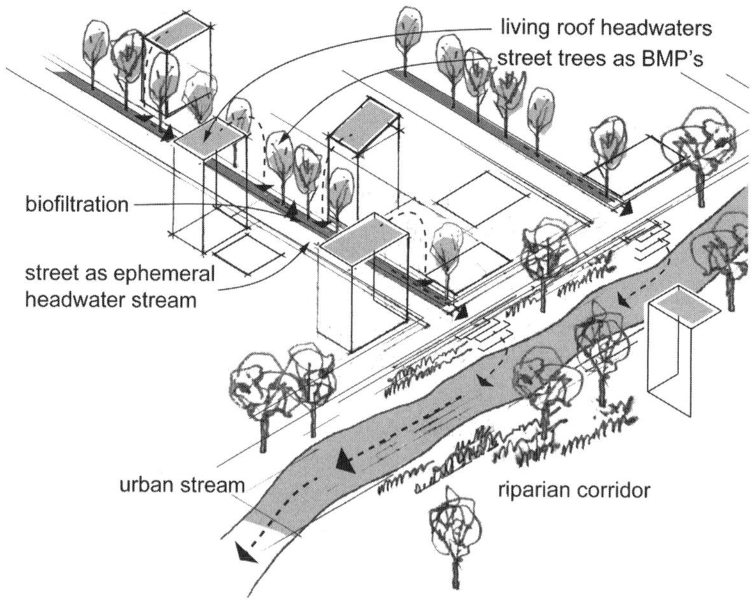

FIGURE 16.4: Gutter as the beginning of the urban pipeshed (terms from Kevan Moffett, Moffett Research Lab, Washington State University EcoHydrology Group).

Proceeding from this reorientation, and recognizing the multiple demands on urban space, designers can seek in their proposed design interventions combinations of the hard (rock-like built structures) and soft (sponge-like biological matrices) in the cause of greatly improved hydrological function and a city supportive of life.

Hydro-Urban Formations

Might urban spaces of water collection during major rain events become sites and stages of gathering during times of favourable weather? Who would comprehend this better than the Dutch, having learned over centuries to manage and prosper in their watery landscape, a constructed, constrained and densely populated space? De Urbanisten's Water Square Benthemplein, built in Rotterdam in 2013, and other civic infrastructural design projects by the firm, combine water storage with recreational spaces and other forms of community amenity (see Figure 16.5). Pickup basketball games and other public activities take place in sunken courts that at other times gather heavy rains. The temporal assumes a heightened role in

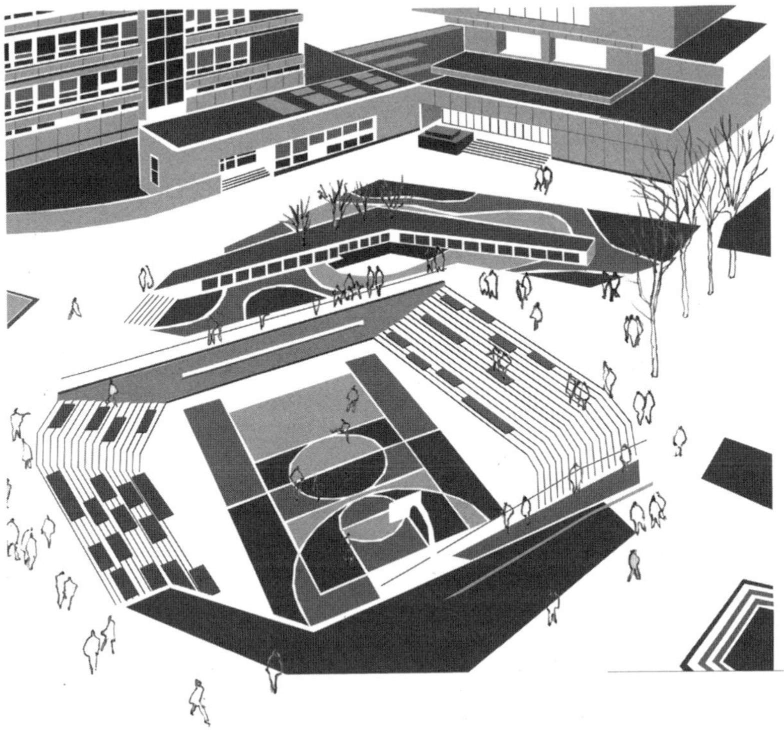

FIGURE 16.5: Water Square Benthemplein, Rotterdam by De Urbanisten.

urban placemaking; with so little slack and competing pressures on limited space, a hollowed volume assumes stacked and sequenced functions.[20]

This manner of thinking about the city guides Oakland-based Hyphae Design Laboratory's groundbreaking approach and work. Hyphae analyses flows of different water sources through a site and its environs to balance project demand with broader-scale hydrological surpluses and deficiencies. For its adaptive reuse of the Waterman Gardens multifamily residential complex, a postwar development in San Bernardino, California, Hyphae evaluated the city's hydrological conditions, recognizing that paving over a floodplain to build its downtown led to exceptionally high groundwater levels near the Santa Ana River, heightened flood risk and liquefaction of soils.[21] They also identified a corresponding lowering of groundwater at the base of the foothills several miles to the north. Instead of spending millions of dollars on an engineered solution that would involve pumping and work against the natural hydrology, Hyphae asked instead whether projects like Waterman Gardens up and down the watershed could in aggregate perform this work more resourcefully as part of a larger process of urban redevelopment, through a green and blue infrastructural approach that would capture rain and intercept and decelerate stormwater during storm events.

Serving as an ecological and water consultant for the Charles David Keeling Apartments Project, a LEED (Leadership in Energy and Environmental Design) Platinum residential complex on the campus of the University of California at San Diego (designed by Kieran Timberlake Architecture, Planning, and Research and Spurlock Poirier Landscape Architects), Hyphae utilized a manner of holistic thinking in which builtscapes and landscapes are inflected. Because the complex was to be located in a floodplain and subject to flash floods, the city of La Jolla required the incorporation of enormous and expensive underground tanks – on the order of millions of dollars. As consistent with Hyphae's commitment to 'move dollars from utilities to the landscape',[22] the design team lifted the building up off the ground and created a series of shaded and sunny green spaces for students that would also provide storage and flood mitigation during infrequent and yet at times intense rains. In lieu of tanks, the project establishes a dry creek riparian patch of landscape akin to pre-development conditions in a dense urban setting (see Figure 16.6).

FIGURE 16.6: Keeling apartments.

Hyphae demonstrates that an integrated approach to site planning, building massing and configuration can lead to projects that connect people to broader hydrological functions. With this objective in mind, several years ago I taught an advanced level architectural design studio focused on a mixed-use development on the shores of the Willamette River in Milwaukie, Oregon, immediately to the south of Portland (the current site of a wastewater treatment plant that is scheduled to go off-line sometime in the future). We made our first visit to the site by boat, travelling upstream and south on the Willamette in the company of Travis Williams, executive director of Willamette Riverkeeper. The journey provided an opportunity for Travis to share his expertise on the current state of the river in comparison to what it might have been centuries ago, and for my students to gain the perspective of looking up at our site, not down upon a river from it (Duncan 2001) (See Figure 16.7).[23] This initial approach prompted in subsequent design investigations a sectional stance that emphasized systems, interactions and flows, as opposed to a more conventional and commanding, plan-like orientation.

For his proposal, Orrin Goldsby embraced the idea of improving riparian habitat and the need for greater buffer between the river and the development. He concentrated construction along a 'great street' at the eastern edge of the site paralleling and yet at a distance from the river. Goldsby incorporated ground-floor commercial uses, upper-level residential units and an environmental education centre anchoring the north edge. This compact site planning provided the

FIGURE 16.7: Skyline of the city of Rashid (Rosetta), Egypt, from the Nile River. The vantage of a landscape from a river is vastly different than viewing a river from a bluff.

open space necessary to soften and make more complex, variable and bench-like the shore of the river and strip of land immediately adjacent to it. In Goldsby's proposal, reintroduced riparian tree and shrub communities interlaced with seasonal side channels. Provocatively, lantern-like spaces on the upper levels of the residences, facing the reestablished riparian zone, jutted into the trees, in contrast to the more brightly lit portions of the dwellings fronting the great street. An idea of improved urban habitat and hydrology informs the making of architecture down to the placement of rooms, the design of facades – screenlike for the benefit of sensitive bird species – and lighting choices. The result: riparian lantern architectures.

Samantha Rusek pursued an entirely different and yet equally water-aware approach. In her proposal, Rusek deployed parallel wings of residential units perpendicular to the south-to-north flowing river, with permeable alleys between wings. This allowed for optimal solar access for dwellings as well as an elongation of the path of water from roofs to river. Rainwater harvested from the roof is directed to large cisterns located under stairs that provide access to upper-level units. This water is then treated and made available to the project's inhabitants. During major storm events when volume exceeds the capacity of cisterns, water overflows to the alleys where a sinuous and filtering journey to the Willamette begins, heading first to the east and away from the river before turning 180 degrees west and continuing its slow, inexorable trip. In the process the water is cooled, cleansed and oxygenated, in dramatic contrast to sheet flow characterizing the hydrology at present.

What the aforementioned schemes share is a notion that the 'sociability' diagram for a project, how its basic organization supports social life and quality of interaction, is one and the same as an 'environmental response' diagram for a project, that is to say, how it gathers energy and water, and how it uses passive means to distribute (or dissipate) heat, air and light. Where people collect, water collects. The medium of water, its flows and lingering presences assume primacy in architectural expression.

A final example suggests a full and yet not fully realized potential, one indicating a better path for water in the hyperdense city. With his 'urban courtyardism' projects, Hong Kong-based architect, Weijen Wang, subtracts volumes out of dense, high-story urban buildings; these spaces in the sky thus created are deployed every several stories so dwellers have equitable access to the light and air they provide.[24] Inhabitants of the city now typically move from subway and street to lobby and elevator to upper-story units slotted in like shoeboxes. Wang believes these multi-story green spaces, open to the elements, can serve as the 'middle ground' so needed for forming community in the contemporary high-rise city. These spaces can also incorporate water and serve as inflection points for the grey, green and blue. It would be possible, for example, to direct harvested rainwater to cisterns associated with these 'sky parks'. Vertical neighbourhood sub-water districts provide social and ecological critical mass within today's megacity.

Hydrological Down to the Detail

> *The trick will be whether Oregonians, living today on silicone and petroleum, can fashion our water future from the lessons of obsidian and cedar.*
>
> (Basatch 2006: 167)

Richard White helps us to visualize the journey of salmon from tributaries to the ocean and back as a transference of enormous sums of energy made possible by a conduit of a river. As we have seen, to the great detriment of aqueous beings, the flows, currents and overall metabolism of the system are dramatically altered by urbanization. A multi-scalar understanding must be brought to efforts to reconfigure these flows, from the larger water and urban sub-watershed down to the way architects and their collaborators configure the minutest of assemblages and choose materials at the level of the building detail.

If water is the medium connecting salmon, bears and forests, perhaps we can view building materials and details – the surfaces that we make cities out of – as the medium linking people, urban water bodies and a larger hydrologic reality. In viewing the architectural detail as an element in the urban pipeshed, designers are tasked to evolve construction systems beyond those dependent on rubberized membranes like EPDM and metals like copper and zinc and to embrace a back-to-the-future material reality of low-and no-toxicity, obsidian-like and cedar-like building skins.[25] Our assemblies can become mediums benefitting a diversity of species, with the design of details exhibiting watershed-scale thinking. Following this path, designers become disciples like Dreiseitl, Grau and Ludwig who initiated this chapter's journey: let the wisdom and instructional prowess of the learned, well-travelled waters inform our studies of the substances, structures and systems of the city (See Figure 16.8).

FIGURE 16.8: Owyhee River as it courses its way through southeastern Oregon.

NOTES

1. Adapted and republished from *Blue Architecture: Water, Design, and Environmental Futures* by Brook Muller, Copyright © 2022, by permission of the University of Texas Press.
2. Dreiseitl et al. (2001: 42). It is interesting to note that designers and scholars like Dreseitl, Grau and Ludwig, who have dedicated their careers to water related issues and design possibilities, would use the words 'teacher' and 'lessons' in describing this wondrous medium.
3. An anadromous species is one that migrates from ocean water to spawn in fresh water.
4. See *Ecological Urbanism* (Mostafavi and Doherty 2016) and note the term 'net positive watershed impact' was coined by the Portland, Oregon-based non-profit, Salmon-Safe. See https://salmonsafe.org. Accessed 14 February 2023.
5. Note in the 2014 Student Performance Criteria of the National Architectural Accrediting Board, the entity that reviews professional architectural degree programs, there is no mention of the word 'water'. 'Moisture' is mentioned in B.7 *Building Envelope Systems and Assemblies* and 'mechanical' and 'plumbing' are included in B.9 *Building Service Systems*. It exacerbates a view that architects are space makers; engineers come to the process later and add water (systems). Water, in other words, is no concern of architects.
6. Kenneth Vigil, correspondence with the author, 15 March 2017.
7. The quote is from Johnson's keynote talk during the Association for Pacific Rim Universities Sustainable Cities and Landscapes Hub Conference, Portland, OR, 15 September 2017. Note Ronald H. Pulliam and Johnson (2002) speak to the value of intermediate disturbance regimes; events that may cause damage and yet instantiate productive and regenerative ecological processes.
8. Sarah Hinners, correspondence with the author, 10 October 2017. Hinners is Director of the Ecological Planning Center and Research Assistant Professor in the Department of City & Metropolitan Planning, College of Architecture and Planning, University of Utah.
9. From a presentation at the Oregon Association of Clean Water Agencies Annual Summit, 9 May 2018.
10. This can be expensive in addition to being ecologically destructive. Postel (2017: 62) notes:

    > A recent survey of 37 U.S. treatment plants by the nonprofit American Water Works Association (AWWA) found that the conversion of 10 percent of a watershed from forest to housing or other devegetated uses increases water treatment costs by an average of 8.7 percent.

11. The landscape architect and emeritus professor Randy Hester would encourage architects to choose and save a species over the course of their career!
12. McIntyre is an aquatic toxicologist with Washington State University at Puyallup's Salmon Toxicology Lab. Scholz is a marine biologist, zoologist and manager of the multidisciplinary Ecotoxicology Program at NOAA's Northwest Fisheries Science Center. McIntyre, Scholz and other leading scientists studying the impacts of urban runoff on Coho salmon and other

forms of aquatic life have recently isolated the chemical 6 p-phenylenediamine-quinone (6PPD-quinone), a hardening agent used in the production of car tires, as a major cause of lethal eco-toxicological effects. Unfortunately, the reaction products of this chemical were not tested before the tires went to market.

13. See Salmon-Safe (2017: 1):

> Zinc and copper are highly toxic to salmon, other fish, and aquatic invertebrates serving as fish food sources (although still toxic, they pose less problems to humans and other mammals). Both are also quite soluble relative to other metals. It is not unusual to find 50 percent of more of the zinc and copper in a water sample to be dissolved. Zinc coming from galvanizing is particularly soluble, measured as much as nearly 100 percent. The soluble forms are much more toxic to aquatic life than the same metals in particulate forms, because they can enter and circulate through the body via respiration and digestion. And yet particulate zinc and copper are also physiologically damaging, since they produce harmful biochemical reactions when contacting sensitive gill tissues with the suspended particulates to which they are attached.
>
> Beyond these short-term effects, zinc and copper are also agents of longer-term aquatic ecological harm. Metal-bearing particulates eventually deposit in bottom sediments, and dissolved metals can be taken into the sediments through chemical processes. There, they threaten bottom-dwelling life. They can also be remobilized back into the water column to harm pelagic organisms, like salmon, through entrainment by high flows or dissolution into the soluble forms. Furthermore, these metals can 'biomagnify' through the food web: prey organisms accumulate some quantity without excretion; they are consumed by predators who accumulate peak of food webs, at least as adults, and thus are subject to the aggregation of metals throughout the web.

14. From Fleming's presentation at the Oregon Association of Clean Water Agencies Annual Summit, Lane Community College, Eugene, 9 May 2018.
15. Sarah Hinners, correspondence with author, 10 October 2017.
16. Quote is from Cahill's presentation in my 'Practices of the Futures of the City' architecture seminar, University of Oregon, 9 October 2018.
17. Miners actually broke the top of the hill to access the great quantities of silver to be found there. You have to appreciate the candor.
18. Iwai is with the Multnomah County (Oregon) Department of Community Services; correspondence with author, 16 May 2018.
19. From Moffett's presentation, 'Streets as Streams and Street Trees as BMPs', at the Oregon Association of Clean Water Agencies Annual Summit, 9 May 2018.
20. GHB Landscape Architects furnished us with another innovative urban example of flood control as civic placemaking with their Tasinge Plads project in Copenhagen, Denmark, in 2014.

21. It should be noted that both overabundance of groundwater in the floodplain and under-abundance near the foothills are highly problematic in a context of inevitable seismic activity.
22. Brent Bucknum, correspondence with the author, 19 July 2019.
23. Exploring the rivers and streams of Portland as a child, Duncan offers through his autobiography indelibly riparian reflections on the city.
24. 'Urban Courtyardism' was the title of a lecture Wang delivered at the University of Oregon at Portland, 2 October 2018.
25. EPDM stands for ethylene propylene diene terpolymer and is a durable, synthetic rubber roofing membrane.

REFERENCES

Basatch, Rick (2006), *The Oregon Water Handbook: A Guide to Water Management, Revised Edition*, Corvallis, OR: Oregon State University Press. [First published 1998.]

Childs, Craig (2000), *The Secret Knowledge of Water: Discovering the Essence of the American Desert*, Boston, MA: Little, Brown and Company.

Dreiseitl, Herbert, Grau, Dieter and Ludwig, Karl H. C. (eds) (2001), *Waterscapes: Planning, Building and Designing with Water*, New York: Princeton Architectural Press.

Duncan, David James (2001), *My Story as Told by Water: Confessions, Druidic Rants, Reflections, Bird-watchings, Fish-stalkings, Visions, Songs and Prayers Refracting Light, From Living Rivers, in the Age of the Industrial Dark*, Berkeley, CA: Counterpoint Press.

Hester, Randolph (2006), *Design for Ecological Democracy*, Cambridge, MA: The MIT Press.

Karvonen, Andrew (2001), *Politics of Urban Runoff: Nature, Technology and the Sustainable City*, Cambridge, MA: MIT Press.

Mostafavi, Mohsen and Doherty, Gareth (eds) (2016), *Ecological Urbanism*, Zurich: Lars Mueller Publishers.

Pulliam, H. Ronald and Johnson, Bart R. (2002), 'Ecology's new paradigm: What does it offer designers and planners?', in B. R. Johnson and K. Hill (eds), *Ecology and Design: Frameworks for Learning*, Washington, DC: Island Press, pp. 51–84.

Rahr, Guido (2018), 'Why protect salmon?', *Wild Salmon Center Web*, 26 December, https://www.wildsalmoncenter.org/work/why-protect-salmon/. Accessed 14 February 2023.

Salmon-Safe (2017), *Negative Impacts of Metallic Building Materials and Alternatives to Their Use (Information Sheet)*, Portland, OR: Salmon-Safe, pp. 1–2.

Walsh, Christopher, Roy, Allison, Feminella, Jack, Cottingham, Peter, Groffman, Peter and Morgan, Raymond (2005), 'The urban stream syndrome: Current knowledge and the search for a cure', *Journal of the North American Benthological Society*, 24:3, pp. 706–23.

White, Richard (1995), *The Organic Machine: The Remaking of the Columbia River*, New York: Hill and Wang.

17

Satoyama and the Art of Rural Regeneration

Diane Durston

The traditional rural landscape of Japan[1] is known as *satoyama* – where farmers maintain woodlands and streams, forage for bounties of wild edibles, gather firewood and carefully cultivate their rice terraced hillsides with abundant water from the mountains and continue to harvest modest crops. Satoyama (里山) refers to the borderland between mountains and arable land. *Sato* (里) means village, and *yama* (山) means hills or mountains (Takeuchi et al. 2003).

The word first appeared in records from 1759 when a forester named Hyoemon Terauchi reported the livelihoods of rural mountain and woodland communities using the word to describe the human management of landscapes surrounding those communities (Horiuchi et al. 2009; see Figure 17.1). A related term is *satoumi* (里海), which conjoins sato (里) noted above and umi (海) meaning sea or ocean, or in other words, 'Harmonious relationships between people and the sea.' (See also Uehara et al. 2019.)

As Catherine Knight (2010) observes:

> The term satoyama gained wide currency in Japan in postwar decades as a term that describes a sphere of 'encultured' nature that has traditionally existed on the periphery of rural settlements, but which is increasingly threatened by industrialisation, urban development, rural depopulation and changing lifestyles. [...] [A] primary reason for the appeal of this term is that it represents a sphere of nature in which nature and culture intersect, and in that it is reminiscent of a more idyllic rural lifestyle of the past, when the Japanese 'lived in harmony with nature'. In this sense, it overlaps with the term *furusato* (literally meaning 'old village', but more generally used to refer to a traditional agrarian lifestyle and landscape, and a place of spiritual belonging).
>
> (410–22; see also Ishizawa 2018)

FIGURE 17.1: Satoyama depicted in a facsimile of a Ukiyo-e style woodblock print during the Edo period by Katsushika Hokusai (葛飾北斎, *c*.1830–32), 下目黒(*Shimo Meguro* [*Lower Meguro*]), from the series, 冨嶽三十六景(*Fugaku sanjūrokkei* [*Thirty-six Views of Mount Fuji*])/ Freer Gallery of Art Study Collection, Smithsonian Institution, Washington, DC: Gift of Alan, Donald and David Winslow from the estate of William R. Castle, FSC-GR-112 (*c*.1930, no. 10).

While 80 per cent of the land on Japan's volcanic islands is mountainous, survival on the remaining 20 per cent of arable land had to be carefully managed over millenia. Satoyama was an environment in which the *biodiversity* found in the woodlands and fields was integral to the life of farmers and farmland. For instance, if trees were taken from the mountain, *not all* the trees were taken at once. If sap was removed from lacquer trees in the forest, not every tree was tapped each year. If water was diverted to feed the rice fields, it was returned to the river. Koi fish spawned in the manmade wetlands and returned to the rivers. Birdlife flourished. Salamanders and frogs lived there – and frogs kept the mosquito population down, rather than pesticides (Iwatsuki 2008; Kambu 2008; Katoh et al. 2009; Satsuka 2012; Jiao et al. 2019). Life in satoyama was never easy. This volcanic archipelago was fraught with frequently devastating earthquakes, typhoons, floods and the famines that followed. The life of a farmer in these mountain villages was hardscrabble. The forces of nature were to be both feared and respected. Today much of that way of life has changed leaving little land that reflects satoyama ideals.

In Japanese, *shizen* (自然) means 'nature' or 'spontaneity' (and in Chinese, 'natural world') and is regarded as a force that encompasses both people and wilderness and how they intertwine. Nature is seen as a universal phenomenon, a quality of all things, wholeness and harmony. Shizen is to be fostered by the people who inhabit the land – however, the number of people who inhabit satoyama landscapes are dwindling and young people continue to move to cities for more exciting and well-paying jobs. In many rural communities, only older people remain to protect the balance of nature. According to one satoyama resident, 'Forests have been preserved by the residents. So, we need to preserve a social system to enable such preservation, but it has been collapsing. If an old person dies, a forest dies' (Chakroun and Droz 2020: n.pag.).

Respect for nature has been cultivated in Japan through the three main spiritual traditions –Shintō, Buddhism and Daoism – that inform Japanese culture and teach the value of living in harmony with the unpredictable forces of nature.

> Shintō [...] in its original form considers all of nature to be divine. While the various gods, or *kami*, reside in all variety of natural phenomena – mountains, rivers, wind, and sea – forests are important elements of shrines as they are thought to contain the spirits of the tutelary deities of the sanctuary.
>
> (Isoya 2020: n.pag.)

In order to placate the often seemingly angry gods, farm communities honoured their deities throughout the year, bringing families together, thanking them for a successful harvest and appeasing them with offerings and ritual dance when times were bleak.

In the indigenous Shintō tradition, shrines both large and small were typically located in natural settings, surrounded by trees wherever possible. These 'guardian forests' maintained alongside sacred shrines are known as *chinju no mori* (鎮守の7 森). One remarkable example is Meiji Shrine, located in the heart of the booming metropolis of Tokyo, where 162 acres of forest surround the shrine dedicated to the spirits of the Emperor Meiji (明治天皇, 1852–1912) and Empress Shoken (昭憲皇后, 1849–1914). Built in the 1920s, the forest was completely manmade through the donation of 122,000 trees shipped from all over the country, and the planting efforts of over 110,000 volunteers. Today it serves as a model of biodiversity and an unlikely wildlife sanctuary in the centre of the busy Yoyogi district of Tokyo.

Satoyama Inspires the Arts and Informs Aesthetics

In the first decades of the twentieth century, artists and poets together with a handful of scholars in the burgeoning field of *forest ecology* first responded to the increasing loss of satoyama landscapes and ways of life. Poet and architect Michizō Tachihara (立原道造, 1914–39) captured this sense of loss in this poem:

> My dream always returns
> To a lonely village at the foot of a mountain
> Wind sighing over the knot-grass
> Skylark singing and singing,
> To a forest path in quiet noon,
> In the blue sky, sun shining clear.

Similarly, the aesthetics of the traditional art forms of *ikebana* (生け花), the art of flower arrangement (Durston 2020a), and *sado* (茶道), the tea ceremony, were also derived from an early awareness that nature is a source of inspiration. Concepts such as *mono no aware* (物の哀れ, the 'pathos' of things), *wabi* (侘, rustic poverty) and *sabi* (寂, solitariness), all originate in an acute awareness of the impermanent nature of life as it can be observed in the natural world. Wabi, a preference for austere, rustic simplicity and impermanence, was appropriated by Zen priests in the sixteenth century to define the guiding principles of tea ceremony as a vehicle for the practice of Zen (Durston 2020b). Those principles are:

> [T]o seek harmony (*wa*) with nature and other human beings; to show respect (*kei*) for all things and people; to reverse the purity (*sei*) of a clean and orderly environment; and to enter a state of tranquility (*jaku*) amid the chaotic world in which we live.
>
> (Durston 2006: 229)

The humble way of life of simple farmers that made the best of limited resources and treasured the muted colours and rough feel of handmade things is reflected in *wabi-cha* (侘茶), as the practice of this form of tea ceremony is now high art in Japan and around the world. Wabi-cha aesthetics reflect satoyama environments, where simple structures made of natural materials (bamboo, clay, paper, wood and reeds) constitute the teahouse – the same materials from which a satoyama farmer's house would have been built.

Satoyama and the Garden Arts

The design of Japanese gardens draws inspiration from the natural landscape of the countryside, particularly the gardens of practitioners of the tea ceremony who sought to recreate natural environments for teahouses that brought to mind the settings of humble farmers' huts on the edge of the forest (Durston 2018a, 2018b).

FIGURE 17.2: Working rice fields connect landscaped imperial gardens and villas at Shugakuin Imperial Villa. Photo used with permission from JapanGuide.com.

A garden that clearly illustrates the role that satoyama played in the development of the garden arts is Shugakuin Imperial Villa (修学院離宮) in Kyoto. Shugakuin is the kind of understated rural aesthetic sought by the great tea masters of the seventeenth century. One of three Imperial Villa gardens in Kyoto, the design of Shugakuin is unique because of the placement of upper, middle and lower teahouses, amid working rice fields that since the seventeenth century have provided a truly rural agricultural landscape as an essential part of the aesthetics of the estate (see Figure 17.2; also see Mari-Hiro Architects' Office 2020).

In 1964, these surrounding farmlands were bought by the Imperial Household Agency and are leased to local farmers who continue to work the fields (IHA n.d.). Shugakuin reflects the love for the sight of terraced paddy fields, even among members of the aristocracy.

The Struggle to Save Satoyama

In the 1960s, forest ecologist Tsunahide Shidei (1911–2009) wrote in favour of the revival of traditional satoyama farming methods to counter the increased use of chemical fertilizers. The Japanese government began to take note of some of the problems caused by rapid economic growth throughout Japan (Horiuchi et al. 2009).

Shidei recognized the importance of conserving quintessential rural landscapes of Japan, which consist of small villages of thatched- or tile-roofed farmhouses surrounded by rice fields, nestled in tranquil valleys at the foot of forested mountains with rivers running through them. Such places are still treasured by twenty-first-century conservationists hoping to preserve what is left of a once sustainable ecosystem that had existed in Japanese farming communities for generations.

International awareness of the ecological implications of Japanese ideas about natural foods and farming followed with the publication of *Zen Macrobiotics: The Art of Rejuvenation and Longevity* (Ohsawa ([1960] 2013) and *The One Straw Revolution: An Introduction to Natural Farming* (Fukuoka 1978). Elsewhere in the western world, the notion of *permaculture* was introduced in the 1970s by Bill Mollison and David Holmgren, two Australian academics. Their original intention was to create a more 'permanent culture' encompassing traditional knowledge, ecology, sustainability, organic food production, efficient energy use, natural home design, recycling, appropriate technology, ecological economics, and much more (Permaculture Design 2016; PEC 2023).

To further raise public awareness of the issues facing rural communities, Japan's national broadcaster, Nippon Hōsō Kyōkai (NHK), partnered with the BBC (British Broadcasting Corporation) to produce two popular documentaries: *Satoyama: Japan's Secret Garden* (Kikuchi and Murata 1999/2000) and *Satoyama: Japan's Secret Watergarden* (Mizunuma 2004), narrated by David Attenborough. According to NHK's publicity, the satoyama series 'poetically portrays the close relationship between people and nature in traditional rural environments in Japan' (NHK 2005). The documentaries use the term satoyama but not specifically relating to a category of nature. Rather the term is used in its most general sense of 'a traditional agrarian landscape', inferring that there is a harmonious human–nature relationship (Knight 2010: 428).

Totoro and Satoyama

The disappearance of satoyama landscapes came to the attention of the general public through an unexpected force of nature in the form of a giant forest spirit named Totoro (トトロ), who lives in a satoyama forest. The anime film, *My Neighbor Totoro*, was created in 1988 by renowned writer, director and animator Hayao Miyazaki (宮崎駿), who used the satoyama landscape of the Sayama Hills region 35 km outside of Tokyo as a backdrop for his internationally acclaimed animated film (Miyazaki 1988, 2002; also see 2006, 2018).

It tells the story of two sisters who seek solace in the forests that surround their rural farmhouse while their mother is hospitalized with a serious illness. Miyazaki drew inspiration from an actual native beech forest called Sayama Hills, which became the setting for the film.

The conservation of delicate broadleaf beech forests is key to ensuring continued biodiversity in Japan. While the largest pristine beech forest in the world has been preserved as a World Heritage Site in northern Japan, most of the remaining indigenous trees have been lost to development during the years of Japan's rapid modernization (Kanert 2017). While the conservation of indigenous forests is key to ensuring continued biodiversity (Morimoto 2011), what remains of natural forest land was clear-cut and reforested with cash-crop trees, such as the non-native *sugi* cedar trees popular for use in residential construction following the Second World War. When non-native plants invade forests, it upsets the ecological balance, destroying animal and insect habitat important to maintaining balance.

Facing increasing urban sprawl from the suburbs of Tokyo, the community around Sayama Hills district found inspiration from Miyazaki's film. In 1990, The Totoro no Furusato Foundation (2022: n.pag.) was established by Miyazaki and others, who had been donating funds for the preservation of the forest for more than a decade. An auction of Totoro-related art and donations from the community, together with donations from Miyazaki and others, funded the purchase and preservation of a total of 8,650 acres of land, acquired one parcel at a time to be kept as undeveloped satoyama landscape. A part of the area they preserved is now a public park which encourages families to bring their children for walks in the legendary satoyama landscape of Totoro Forest. Criticized for its Disney-like aspects, revenues from admissions nonetheless help to cover maintenance, and new generations are being introduced to the joys of this special satoyama environment through a robust educational program (JNTO n.d.).

The Satoyama Initiative

Though some progress is being made with public projects and funding, it is not surprising that a number of private and community-based projects – many of them in the field of the arts and crafts, and 'geotourism' – are contributing significantly to the promotion of the satoyama landscape and way of life.

Movements have begun among younger generations who have sought to escape the life of the urban 'salaryman', by rejecting long hours, as well as the concrete and steel environments they inherited from previous generations driven by the need to modernize and rebuild the economy following the devastation of the Second World War. Increased attention brought by the NHK-BBC documentaries and the public embrace of Totoro led to the establishment of the Satoyama Foundation, which has supported subsequent citizen-led and academic supported movements.

Public pressure led the Japanese government to reconsider the rapidly vanishing satoyama way of life as a possible international model of sustainable agriculture, and the Satoyama Initiative was born and brought to the attention of the United Nations (UN) by the time the headquarters campus of the UN University was built in Tokyo in 2010 (see Watanabe n.d.). The International Partnership for the Satoyama Initiative (IPSI) was established in October 2010, during the 10th Conference of the Parties to the Convention on Biological Diversity (CBD COP 10) held in Nagoya, Aichi, Japan, and annually thereafter. To date, more than 300 Satoyama Initiative projects have been undertaken worldwide, fuelling renewed interest in research of traditional farming models that have originated in such wide-ranging places as Africa (including Niger), Central America, France, Hawaii, India and Nepal (including Kathmandu) (UNU-IAS and IGES 2016; see also Satoyama Initiative 2023).

Revival of Traditional Crafts: Mashiko

One of the earliest revitalization projects aimed at regenerating the Japanese countryside began in the 1930s when Tokyo ceramic artist Shoji Hamada (濱田庄司, 1894–1978), later designated as a Jūyō Mukei Bunkazai Hojisha (重要無形文化財保持者, or Preserver of Important Intangible Cultural Properties), realized that regional craft villages were in serious decline and set out to change that by moving his studio and his kiln to the old pottery village of Mashiko in rural Toshigi Prefecture.

Early pottery production in Mashiko dates back to ancient times. Modern *Mashikoyaki* dates only to 1853, when a potter discovered that the local clay was ideal for ceramics. The style was further popularized by Hamada's student, Tatsuzō Shimaoka (島岡達三, 1919–2007), who was also designated as a Living National Treasure and worked in Mashiko from 1953 until his death.

Mashiko is a folkware kiln site that is unlike some of the other older kiln sites around Japan (see Figure 17.3). The town is open to newcomers – both foreign and domestic – whether or not they are potters or of other professional backgrounds in arts, sciences and education. In Japan, craftsmen are usually born into their profession, but in Mashiko, anyone can aspire to become a potter. Following Shoji Hamada, people looking to return to a more traditional Japanese lifestyle settled in the area. Twice a year, coinciding with the Golden Week Holidays in the first week of May, and again for the first week of November, there is a pottery and crafts festival where potters and craftsmen from Mashiko and surrounding areas come to the town and set up stalls (Holmes 1982: 12). The efforts of Hamada and Shimaoka have resulted in the revitalization and thriving of Mashiko through the efforts of a new generation of fine artist-potters, in spite of the disastrous earthquake that destroyed many of their kilns in 2011.

FIGURE 17.3: Mashiko ware (益子焼) kilns during Ceramics Fair. Noborigama (登り窯, climbing kiln) in Mashiko (益子町) of the Tochigi Prefecture (栃木県). Photo by NEXT – EXIT (2007) via Japanese Wikipedia.[2]

Gungendo

Gungendo, a World Heritage site, is another such successful revitalization effort started in the 1990s by a couple who restored many of the old farmhouses in historic Omori-cho to create a new village of artisans to revitalize crafts that were on the verge of being lost forever, while revitalizing old satoyama village life through the arts (see Figure 17.4). According to Gungendo founders, Tomi and Daikichi Matsuba:

> Omori-cho prospered as a silver mining town from the 17th to the early 20th century. With the decline of the silver mine, the townsfolk left to seek work elsewhere and the population began to fall. The current population [...] stands at just 400. Our mission is to create a way of life which incorporates food, clothing [including fibers and dyes from local resources] and tradition that can be passed down to future generations [...] and live a life that values its past traditions, but is also filled with joy for the present while looking towards the future.
>
> (Gungendo n.d.: n.pag.)

FIGURE 17.4: The philosophy of Gungendo is 'lifestyle with roots'. Photo used with permission of gungendo.com.

The efforts of the Matsuba's and their extended family of friends have made Gungendo a travel destination for a new generation of those who love fine handmade crafts and hope to spend time breathing the revitalized air of this satoyama village.

Revitalization through the Arts

The use of contemporary art as a catalyst for the revitalization of satoyama lands began in Japan in the mid 1980s, perhaps most notably with a project initiated by Tetsuhiko Fukutake, the founder of a major publishing company in Okayama, working with Chikatsugu Miyake, Mayor of Naoshima, an island in the Seto Inland Sea. Their shared dream was to revitalize the vanishing culture of the islands that had been declining rapidly with the industrialization of Japan following World War II.

In 1992, Soichiro Fukutake further realized his father's vision opening the Benesse House Museum designed by Pritzker-Prize winning architect Tadao Ando, based on 'coexistence of nature, art and architecture'. Today, dozens of new museums, outdoor installations and art venues are situated in various locations around the islands of Naoshima, Teshima, Inujima, Shodoshima and Megijima (Benesse Art Site Naoshima n.d.: 1, 2). These venues and the internationally acclaimed website continue to provide a wealth of innovative opportunities to experience contemporary art in the context of traditional village life, while engaging enthusiastic support of local farmers and fishermen. On reflection, the islands have become a canvas.

> If art with a message to modern society were placed [...] where the unspoilt landscape of Japan still remains, it could change a region that was losing its vitality. With this in mind, the world's first symbiosis between islands and art began on Naoshima.
> (BASN n.d.: 3)

Echigo-Tsumari Art Triennial

Following the success of Gungendo, satoyama revitalization programs have begun in various regions throughout Japan. According to the Japan Times Satoyama Consortium, part of the newspaper's Sustainable Japan project, an exciting arts-based project established a model for revitalizing satoyama – the Echigo-Tsumari Art Triennale – founded in 2000 by Fram Kitagawa, a Tokyo-based producer of exhibitions, concerts and events (Kitagawa 2015).

Originally, it was envisioned as a way of revitalizing the Echigo-Tsumari region of Niigata Prefecture, the fabled *Snow Country* ([1948] 1956) of Nobel Prize in Literature recipient, Yasunari Kawabata (川端康成, 1899–1972). It is a region of Japan suffering from pronounced population decline with the highest proportion of elderly citizens. Kitagawa served as the general director (or mediator) of the triennial. Every three years since 2000, the festival has brought together art, ecol ogy and communities on 300 square miles of land dotted with over 200 distinct villages. Remarkably, recent editions of the festival have come to attract half a million visitors annually to the region. Local people resisted in the beginning, but after many unsuccessful tries at stimulating interest in satoyama life, most have fully embraced the renewed vitality and financial resources that the festival has brought their villages (Muraoka 2021; Ha 2022).

Keihoku

As recently as 2021, young Japanese artisans and promoters of satoyama environments began a project to revitalize an agricultural location in the mountains north of Kyoto in Keihoku. This area once was a major source for the production of *urushi*, a natural lacquer used in the production and decoration of fine Kyoto lacquerware for generations. The decline of the younger population in Keihoku left no one to extract lacquer from urushi trees in the woodlands at the edge of satoyama villages.

FIGURE 17.5: 'Young urushi in a production area planted by general supporters who sympathize [with] their goal'. (Image and caption in Matsuyama 2019, with permission from Sachiko Matsuyama.)

Tokyo-born Sachiko Matsuyama[3] arrived in Kyoto with a mission to help rescue what might have been the last generation of fine artisans in the ancient capital once famous for its refined culture and craftsmanship. She began working with a group of young artisans to introduce their fine products to new audiences abroad. In the process, she met Takuya Tsutsumi, who had inherited a lacquer refinery business that was in need of help. With a dwindling supply of urushi at hand, they began to work on a plan to reforest part of the satoyama landscape by planting a grove of urushi trees to ensure that the primary material needed could still be available for future generations (see Figure 17.5). In a complicated collaborative process, they managed to revitalize the artform itself, eventually collaborating with the famous Australian surfboard maker Tom Wegener to produce strikingly new products with this 10,000-year-old craft (Forest of Craft 2021a, 2021b).

Matsuyama and Tsutsumi are convinced of lacquer's value, not just for its redolent surface but also as a gift and bond between people and nature. Their work won an award for creativity in the first Japan Craft 21 competition held in 2021. As Matsuyama explains:

> *Shokunin* (artisans) have pursued a completely sustainable living and working style, and they always take good care of their environment before taking anything from it with full gratitude. Our appreciation towards nature has been declining since the world has become so urbanized and digitized, and it was a great lesson that let us

be aware and thankful about our environment just like our ancestors had been for centuries.

(Matsuyama 2019)

Shunran -No-Sato

Another project in Ishikawa Prefecture is called Shunran-no-Sato and involves revitalization through *geotourism* or 'tourism that sustains or enhances the geographical character of a place, such as the environment, culture, aesthetics, heritage, and the well-being of the residents' (Beaver 2012).[4] It offers guestrooms in 40 villagers' homes and an opportunity to take part in traditional agricultural activities, such as rice harvesting and mushroom picking in autumn. The project was established by a group of elderly people, residents from two nearby communities, Miyachi and Sakeo, that belonged to the old administrative unit of Yamada District in Noto Town. The group's objective is to keep their villages alive and conserve their traditional agriculture.

A larger project to save the terraced rice fields in the Noto Peninsula was undertaken by urban dwellers who travelled 100 km from Kanazawa City to help repair the broken ridges of the terraces, preserving the historical landscape and keeping the culture of satoyama alive (Nagata and Chen 2013).

Satoyama revitalization projects such as these are being undertaken throughout Japan, including *sustainable ecotourism*,[5] encouraging visitors to journey outside cities and experience the satoyama landscapes. The Japan National Travel Organization (JNTO) has worked with organizers to host tours and bring income to small towns and villages.

Conclusion

For centuries, the satoyama landscape of rural Japan provided an integrated, sustainable system in which farmers could grow their crops, drawing upon the bounty of water from the mountains for the fields and wood for housing and fuel, while caring for the forests surrounding them. The increase in industrialization that fuelled Japan's economic success in the twentieth century consumed large tracts of satoyama land. Golf courses and residential development projects made satoyama little more than a dream-like landscape from the past. Its disappearance is challenging communities to join forces to save parts of rural landscapes to show future generations a more regenerative alternative to their present hectic urban lives (Takeuchi 2016).

The satoyama model and initiative have stimulated research into sustainable agricultural models in other countries and cultures around the world (Berglund et al. 2014). Critics of the program have suggested that the Satoyama Initiative is little more than a means of promoting Japan's own questionable environmental record, pointing at issues of industrial pollution, widespread use of chemical fertilizers and pesticides and concrete construction projects throughout the land (Ruud 2020). However, the achievements of the program have been significant, advancing efforts to create a more sustainable way of life and fuelling renewed interest in research in other traditional models of sustainable farming from Indigenous cultures worldwide (Satoyama Initiative 2023).

Perhaps what is making even more progress in restoring appreciation and support for a vision of thriving life is coming from younger generations tiring of urban life and seeking ways to take back their place in an organically renewed satoyama. These new paths are moving further towards revitalizing country life by integrating rural arts events and replanting satoyama woodlands. In this way, they are saving not only the aesthetic value of the Japanese landscape but the cooperative way of life that may be key to the survival of the planet itself.

Acknowledgements

I would like to acknowledge and thank Jeremy Swartz, University of Oregon, for his advice, research, and collaboration throughout the development of this project. My thanks also to Janet Wasko, University of Oregon; Makoto Suzuki, Principal, Tokyo NODAI Green Academy and Professor Emeritus, Tokyo University of Agriculture; Tsutomu Hattori, Assistant Dean, Department of Landscape Architecture Science/Landscape Planning Laboratory, Tokyo University of Agriculture; Sachiko Matsuyama Takamuro, Perspectives: Forest of Craft co-founder, Keihoku, Japan; Amy Katoh, author and promoter of rural Japanese crafts, Tokyo, Japan; Steve Bloom, CEO, Portland Japanese Garden, Portland, Oregon; and, Sadafumi Uchiyama, Chief Curator, Portland Japanese Garden, Portland, Oregon.

NOTES

1. The names for Japan are Nihon and Nippon, both written in logograms or kanji as (日本).
2. Link to *Wikimedia Commons* image and GFDL/CC license: https://commons.wikimedia.org/wiki/File:Climbing_kiln_Mashiko.jpg. Accessed 1 August 2023.
3. Sachiko Matsuyama was married in 2021 and now goes by both the name Matsuyama and her married name Takamuro.

4. Note that Japan also has a growing number of geoparks – 'A designated area of land principally focused on promoting awareness of geodiversity and sustainable geotourism' (Manley et al. 2019; also see Vafadari and Cooper 2020).
5. Sustainable ecotourism is differentiated from ecotourism, by 'giv[ing] equal weighting to ecological integrity, economic viability, and social justice' (Mayhew 2015).

REFERENCES

Beaver, Allan (2012), 'Geotourism', in A. Beaver (ed.), *A Dictionary of Travel and Tourism*, Oxford: Oxford University Press.

Benesse Art Site Naoshima (BASN) (n.d.a), 'Art & Architecture', Naoshima, Japan: Benesse Holdings, Inc. and Fukutake Foundation, https://benesse-artsite.jp/en/art/. Accessed 18 March 2022.

——— n.d.b, 'Benesse House Museum', Naoshima, Japan: Benesse Holdings, Inc. and Fukutake Foundation, https://benesse-artsite.jp/en/art/benessehouse-museum.html. Accessed 18 March 2022.

——— n.d.c, 'Philosophy', Naoshima, Japan: Benesse Holdings, Inc. and Fukutake Foundation, https://benesse-artsite.jp/en/about/. Accessed 18 March 2022.

Berglund, Björn E., Kitagawa, Junko, Lagerås, Per, Nakamura, Koji, Sasaki, Naoko and Yasuda, Yoshinori (2014), 'Traditional farming landscapes for sustainable living in Scandinavia and Japan: Global revival through the Satoyama Initiative', *Ambio,* 43, pp. 559–78.

Chakroun, Leila and Droz, Laÿna (2020), 'Sustainability through landscapes: Natural parks, satoyama, and permaculture in Japan', *Ecosystems and People*, 16:1, pp. 369–83.

Durston, Diane (2006), *Wabi Sabi: The Art of Everyday Life*, North Adams, MA: Storey Publishing.

——— (2018a), *Garden as Window*, Portland, OR: North American Japanese Garden Association.

——— (2018b), 'Portland Japanese Garden', in *Oregon Encyclopedia of History and Culture*, Portland, OR: Portland State University and the Oregon Historical Society, 17 March, https://www.oregonencyclopedia.org/articles/portland_japanese_garden/. Accessed 2 July 2021.

——— (2020a), *IKASU! Come Alive!*, Portland, OR: Portland Japanese Garden, 11 May, https://japanesegarden.org/2020/05/11/ikasu-come-alive/. Accessed 9 September 2021.

——— (2020b), *Ura no Tomoya: The Humble Fisherman's Hut*, Portland, OR: Portland Japanese Garden, 19 May, https://japanesegarden.org/2020/05/19/ura-no-tomoya-the-humble-fishermans-hut/. Accessed 9 September 2021.

Forest of Craft (2021a), 'About forest of craft [The cycle of craftsmanship]', https://www.forest-of-craft.jp/en/about-forest-of-craft. Accessed 9 August 2021.

——— (2021b), 'Ten core principles of craftsmanship', https://www.forest-of-craft.jp/en/ten-core-principles-of-craftsmanship. Accessed 9 August 2021.

Fukuoka, Masanobu (1978), *The One-straw Revolution: An Introduction to Natural Farming* (ed. L. Korn, trans. C. Pearce, T. Kurosawa and L. Korn), Emmaus, PA: Rodale Press. (First published Shizen Noho Wara Ippon No Kakumei, Tokyo: Hakujusha Co., Ltd.)

Gungendo (n.d.), 'Gungendo', homepage, Iwami Ginzan Seikatsu Bunka Kenkyusho Corporation, https://gungendo.com. Accessed 23 September 2021.

Ha, Thu-Huong (2022), 'Art came alive in the great outdoors in 2022', *Japan Times*, 17 December, https://www.japantimes.co.jp/culture/2022/12/17/arts/art-triennale-nature/. Accessed 29 January 2023.

Holmes, Ann Sommer (1982), *The Transition of the Artisan-potter to the Artist-potter in Mashiko: A Folkware Kiln Site in Japan*, New York: New York University Press (ProQuest Dissertations Publishing).

Horiuchi, Mio, Kai, Toshiya, Brand, Kaori and McDonald, Anne (2009), 'Harvest time in Satoyama', *Our World*, United Nations University, 13 January, https://ourworld.unu.edu/en/harvest-time-in-satoyama. Accessed 23 September 2021.

IHA (Imperial Household Agency) (n.d.), 'Shugakuin Imperial Villa: Sakyo-ku. Kyoto', *The Imperial Household Agency*, https://sankan.kunaicho.go.jp/multilingual/lang/en/shugakuin/index.html. Accessed 23 September 2021.

Ishizawa, Maya (2018), 'Cultural landscapes link to nature: Learning from Satoyama and Satoumi', *Built Heritage,* 2, pp. 7–19.

Isoya, Shinji (2020), 'Creating serenity: The construction of the Meiji Shrine Forest', Nippon.com, 8 July, https://www.nippon.com/en/japan-topics/g00866/. Accessed 12 September 2021.

Iwatsuki, Kunio (2008), 'Sustainable use of biodiversity, with reference to the Japanese spirit of worshipping nature', in N. Furuta (ed.), *Conserving Nature: A Japanese Perspective*, Tokyo: Biodiversity Network Japan in collaboration with Countdown 2010 and Nippon Keidanren Committee on Nature Conservation, pp. 4–11.

Jiao, Yuanmei, Ding, Yinping, Zha, Zhiqin and Okuro, Toshiya (2019), 'Crises of biodiversity and ecosystem services in Satoyama landscape of Japan: A review on the role of management', *Sustainability, 11*:2, p. 454.

JNTO (n.d.), 'Totoro Forest トトロの森', *Japan National Tourism Organization* (JNTO), https://www.japan.travel/en/spot/314/. Accessed 23 September 2021.

Kambu, Alphonse (2008), 'The need to conserve the vanishing satoyama heritage', in N. Furuta (ed.), *Conserving Nature: A Japanese Perspective*, Tokyo: Biodiversity Network Japan in collaboration with Countdown 2010 and Nippon Keidanren Committee on Nature Conservation, pp. 12–19.

Kanert, Michael (2017), 'Japan's 20 UNESCO World Heritage Sites', *AllAboutJapan*, 2 November, https://allabout-japan.com/en/article/4511/. Accessed 23 February 2022.

Katoh, Kazuhiro, Sakai, Sumire and Takahashi, Toshimori (2009), 'Factors maintaining species diversity in satoyama, a traditional agricultural landscape of Japan', *Biological Conservation,* 142:9, pp. 1930–36.

Kawabata, Yasunari ([1948] 1956), *Snow Country* (雪国), New York: Knopf.

Kikuchi, Tetsunori and Murata, Shinichi (dirs.) (1999/2000), *Japan's Secret Garden*, PBS NOVA,https://www.pbs.org/wgbh/nova/transcripts/2716satoyama.html. Accessed 23 September 2021.

Kitagawa, Fram (2015), *Art Place Japan: The Echigo-Tsumari Art Triennale and the Vision to Reconnect Art and Nature (trans. A. Matsuo and B. J. Monsma)*, New York: Princeton Architectural Press.

Knight, Catherine (2010), 'The discourse of "encultured nature" in Japan: The concept of satoyama and its role in 21st-century nature conservation', *Asian Studies Review,* 34:4, pp. 421–41.

Manley, Will, Foot, Katharine and Davis, Andrew (2019), 'Geopark', *A Dictionary of Agriculture and Land Management*, Oxford: Oxford University Press.

Mari-Hiro Architects' Office (2020), *Visit to Shugakuin Imperial Villa in Kyoto*, Bunkyo-ku, Tokyo: Mari-Hiro Architects' Office, https://mari-hiro.work/news-blog/21070802/. Accessed 9 December 2021.

Matsuyama, Sachiko (2019), 'Why a Japanese lacquer master sought a surfing legend', *Garland Magazine, 29 May*, https://garlandmag.com/article/urushi-surf/. Accessed 1 November 2021.

Mayhew, Susan (2015), 'Ecotourism', in *A Dictionary of Geography,* 5th ed., Oxford: Oxford University Press.

Miyazaki, Hayao (dir.) (1988), *Tonari no Totoro* (となりのトトロ), Tokyo: Studio Ghibli for Tokuma Shoten [English dub, *My Neighbor Totoro* (1990); re-dub by Walt Disney Pictures (2004)].

——— (dir.) (2002), *Mei to Koneko basu* (めいとこねこバス), Tokyo: Studio Ghibli for Ghibli Museum (English title: *Mei and the Baby Cat Bus*; mini-sequel to *My Neighbor Totoro*).

——— (dir.) (2006), *Mizugumo Monmon* (水グモもんもん), Tokyo, Japan: Ghibli Museum (English title: *Mon Mon the Water Spider*).

——— (dir.) (2018), *Kemushi no Boro* (毛虫のボロ), Tokyo: Ghibli Museum (English title: *Boro the Caterpillar*).

Mizunuma, Masumi (dir.) (2004), *Satoyama: Japan's Secret Water Garden*, London: BBC (British Broadcasting Corporation) Natural World, https://youtu.be/BIW88_a2csU. Accessed 23 September 2021.

Morimoto, Yukihiro (2011), 'What is satoyama? Points for discussion on its future direction', *Landscape and Ecological Engineering,* 7:2, pp. 163–71.

Muraoka, Maiko (2021), 'Art triennial revitalizes rural Echigo-Tsumari region', *Japan Times*, 9 June, https://sustainable.japantimes.com/satoyama/38. Accessed 29 October 2021.

Nagata, Akira and Bixia Chen (2013), 'Urbanites help sustain Japan's historic rice paddy terraces', *Our World* (United Nations University), 22 May, https://ourworld.unu.edu/en/the-people-who-sustain-japans-historic-terraced-rice-fields. Accessed 18 December 2021.

Ohsawa, George ([1960] 2013), *Zen Macrobiotics: The Art of Rejuvenation and Longevity,* 5th ed., (ed. C. Ferré), Chico, CA: George Ohsawa Macrobiotic Foundation.

Permaculture Design (2016), 'What is permaculture?', Permaculture Design Publishing, https://www.permaculturedesignmagazine.com/what-is-permaculture. Accessed 23 September 2021. [Adapted from documents by Steve Diver for Appropriate Technology Transfer to Rural Areas (ATTRA), Fayetteville, AR.]

PEC (2023), 'What is permaculture', Permaculture Education Center (PEC), https://permacultureeducation.org/what-is-permaculture/. Accessed 29 October 2021.

Ruud, Jon Magnus (2020), 'Problems with the concept *Satoyama*: A critical review of satoyama as a concept in conservationism', Master's thesis in Modern Japan, Oslo: University of Oslo.

Satoyama Initiative (2023), 'IPSI Collaborative Activities: Overview', Secretariat of the International Partnership for the Satoyama Initiative, Institute for the Advanced Study of Sustainability, United Nations University, https://satoyama-initiative.org/activities/ipsi-collaborative-activities/. Accessed 28 January 2023.

Satsuka, Shiho (2012), 'Biodiversity in satoyama conservation: Aesthetics, science and the politics of knowledge', in G. Martin, D. Mincyte and U. Münster (eds), *Why Do We Value Diversity? Biocultural Diversity in a Global Context, RCC Perspectives* 9, Munich: Rachel Carson Center for Environment and Society, pp. 79–82, https://www.environmentandsociety.org/sites/default/files/layout_issue_9_final_web.pdf. Accessed 23 September 2021.

Takeuchi, Kazuhiko, Brown, Robert D., Washitani, Izumi, Tsunekawa, Atsushi and Yokohari, Makoto (eds) (2003), *Satoyama: The Traditional Rural Landscape of Japan*, Tokyo: Springer-Verlag. (Original Japanese edition published in 2001 by University of Tokyo Press.)

Takeuchi, Kazuhiko, Ichikawa, Kaoru and Elmqvist, Thomas (2016), 'Satoyama landscape as social–ecological system: Historical changes and future perspective', *Current Opinion in Environmental Sustainability*, 19, pp. 30–39.

The Totoro no Furusato Foundation (2022), 'About Totoro Fund: The origins and development of the Foundation', Tokorozawa-shi, Saitama-ken: Totoro Fund, https://www.totoro.or.jp/totorofund/info.html. Accessed 14 September 2023.

Uehara, Takuro, Hidaka, Takeshi, Matsuda, Osamu, Sakurai, Ryo, Yanagi, Tetsuo and Yoshioka, Taisuke (2019), 'Satoumi: Re-connecting people to nature for sustainable use and conservation of coastal zones', *People and Nature*, 1:4, pp. 435–41.

UNU-IAS and IGES (Institute for Global Environmental Strategies) (eds) (2016), *Satoyama Initiative Thematic Review Vol. 2: Mainstreaming Concepts and Approaches of Socio-ecological Production Landscapes and Seascapes into Policy and Decision-making*, Tokyo: United Nations University, Institute for the Advanced Study of Sustainability (UNU-IAS).

Vafadari, Kazem and Cooper, Malcolm J. M. (2020), 'Community engagement in Japanese geoparks', in B. N. Sadry (ed.), *The Geotourism Industry in the 21st Century: The Origin, Principles, and Futuristic Approach*, Boca Raton, FL: Apple Academic Press, pp. 357–73.

Watanabe, Tsunao (n.d.), 'International Satoyama Initiative (ISI) works to promote the Satoyama Initiative', Tokyo: United Nations University, https://unu.edu/projects/international-satoyama-initiative.html. Accessed 29 January 2023.

18

Metaliving[1]

Jeremy Swartz

> *Worldmaking [...] starts from worlds already on hand; the making is a remaking.*
>
> — Nelson Goodman

The implications of the term 'plant' in Marshall McLuhan's ([1964] 2003) discussion of automation and 'learning a living' have multiple profound pathways. Consider a plant, first in its actuality and as metaphor, and next as a factory (an industrial facility) and as an organism (specifically a tree). McLuhan conflates the two when he notes how 'an automated plant is almost like a tree in respect to the continuous intake and output, it is a tree that can change [...] as required' (469).

My reading is that he acknowledged a transdisciplinary view including intra- and inter-dependencies amongst lifeforms, industries and economies. His perspective is echoed in multiple conceptual frameworks related to the natural, cognitive and social, including ecologies, enterprises, metabolisms, registers and islands (see p. 6 in this volume). This awareness of factory, organism or ecosystem as 'automated plants' helps us to understand 'learning a living' as adaptation to emerging environments. Furthermore, 'living as learning' is an active integration of the environment via experiential education (Garrison et al. 2014: 93). These logics are exemplified by research that has been diffused over at least a century. Human, nonhuman and more-than-human deliberations and kinships reinforce an increasing need for systems education. It recognizes life, mind, nature and enactments of societies' enduring and evolving values (Johnson 2023; Johnson and Schulkin 2023), and can foster multimodality (e.g. experientially and automatically).

After almost two centuries, we now need to honour the warnings noted by Edgar Allan Poe ([1841] 1920) and again by McLuhan ([1979] 1996) of the oncoming known and unknown risks posed by the vortices of human ingenuity and nature. We are caught in interconnected tsunamis of biological and representational information, debris and system communications and ecologies, arguably unlike other periods in human history. My conceptualization of metaliving not only recognizes patterns but may foster experiential wisdoms; time and space where opportunities for evading destruction and/or converging paths may emerge where nature, humans and more-than-humans come together (see Appendix).

The research presented in this chapter brings together a recognition of *life* and *metalife*. It recognizes that 'life is living, and living consists of [...] interrelated activities or functions in which *environing* energies are operatively involved' (Dewey 2012: 322, emphasis added). The chapter offers a brief overview of this volume, followed by discussions of integrative systems, pattern recognition and the study of *futures* at various scales from the local and regional to the national and global. I then engage possibilities for pluralities of futures, reflecting on artful living, or in other words, living in the midst of or in co-operation with emergence, and how scholars, researchers, communities and societies are embodying – bringing forth or keeping close – their heritages, stories and worlds.

For over a century, scholars have formulated ecological modes of inquiry. I employ these modes to foster more responsibility and accountability for our embodied engagements with environments, or in other words, myriad ways in which we inhabit and habituate, in this instance, postulating about the future.

I have previously highlighted definitions in Greek of μετά (*metá*) as *with, along with, in cooperation* and *together with* – not above or beyond (Swartz 2021: 265–66); it is also defined as *in the midst of* (Liddell and Scott 1940). In this way, I consider *life* as medium/environment/habitat ~ in co-operation with ~ *living* as mediating/environing/habituating. This points to *metaliving* as artful being and becoming in the midst of ongoing transformative shifts, feedbacks and feedforwards, happening all around us. This endeavour emerges from a synthesis of object-based *and* process-based ontologies, illustrated (metaphorically) through William James' notion of birds' perchings *and* flightings (or static and dynamic states of systems), in addition to methodological pluralism and metadisciplinarity as philosophical *co-inquiry*.

Volume Overview

While interdisciplinarity is the comparing and contrasting of disciplines, transdisciplinarity, an integration of disciplines/knowledge, is presented in the clustering of chapters and by networking them together in this volume. Ecological approaches and futures research are engaged via the lenses of the natural sciences, while the cognitive turn, pattern recognition and mediated systems bring us further into the material structures that make up environments/technologies/media that we inhabit and habituate within societies.

The field of communication and media is in the midst of attempting to resurface (via e.g. pattern recognition) from recent decades of algorithmization and more recently, rapid automation. Klaus Krippendorff (2018, see also 2021c) has identified this power of algorithms and their stakeholders and how they can create 'epistemological traps'. He offered transdisciplinary discourses and conversations as a way to facilitate ecological contexts which can co-create '*realities in anticipation of being held accountable*' (original emphasis).

By reengaging experientialist learning and education, ecological discourses are being regenerated and brought forth across and through bioinspiration, biophilia, human, nonhuman and more-than-human design. This includes views, practices, forms of judgment, institutions, experimental communicative practices, along with what constitutes significant phenomena in art–science, such as metaliving.

The conceptual frameworks of life and living are interrogated in this volume to explore how they lead to dynamic, radical and reconstructive ways of knowing, projects and inquiries that are working to move communication and media beyond habituated modes. The trilogy more broadly reintegrates fundamental concepts in a curational co-inquiry that can strengthen critical attention to, for example, the notion of imbricating *media* as both technologies and environments (also see Turner 2013).

The tapestry of this book continues this process, while embracing materiality, embodiment and experience/culture, which also can be expressed as *nature–bodymind–society* (see Chapter 7). It also creates the conditions for richer modes of deliberation and moral imagination that can serve for an account of *embodied aesthetics and ethics*. This work utilizes the resources of the past and present together, with a goal of elaborating pragmatism and ways to foster pattern recognition and deliberation in, for and about *living futures*. Clearly, art–science, transdisciplinarity and a philosophic orientation offer opportunities to link clusters in the collection.

The volume has also elaborated integrative cybernetic and systems histories with special attention to the fine and applied arts as well as language arts (e.g. second language writing, Matsuda et al. 2003; Matsuda 2021). Education, especially in science, technology, engineering, art and mathematics (STEAM), brings us full circle to what has constituted 'the ecological' and sustainability praxes. In terms of research, engineering and synthesis are converging or emerging via systems biology, biomimicry, ecological media and/or wetware. More specifically, we should be aware of *biodesign*[2] projects, especially three categories of 'metalife', including artificial life (ALife) or life 'as it could be'; biological art (BioArt) or 'life as it is'; and, synthetic biology (SynBio) or 'life as it might be or as it will be?' (Bureaud 2014; also see Johung 2019; Jones 2022; Ohlenschlager et al. 2023; Weibel 2023). After mapping the domains of the environment, economy and society together via the arts, are we in the midst of uncovering and recovering architectures and architectonics of making and knowledge creation?

As Chilean biologist, Humberto Maturana (1997: n. pag.) puts it, '[E]volutionary transformations [...] can occur through design, in the intentional use of prosthetic means that create new dimensions of interactions for an organism which thus become new sensory domains for them'. Do we need more ways to talk about design[3] and life to engage in more-than-merely self-reflective inquiry by taking responsibility for our (human) desires? Or, in other words, *meta-design*.[4] What constitutes artful living futures or metaliving plays an important role in recognizing not only humans in relationship with the natural world and built environments, but also in relationship with our nonhuman and more-than-human kin. Similarly, as human ingenuity continues to transform the world, embodied aesthetics and ethics are always involved in constructing realms (e.g. digital and virtual), as we are increasingly experiencing what emerges back into the actuality of the natural world through automatic, somatic and allopoietic processes.

More specifically, this expansion of the conceptual frameworks for information and communication has been presented across large clusters of categories. For example, envision information viewed and understood as physical/material as well as symbolic/representational, converged as a hybrid. These *virtual embodiments* (e.g. scanning and mapping elements and environments from the lifeworld) and *embodied virtualities* (e.g. printing and manufacturing from computing digital worlds) can be understood through tapestries (maps), temperaments (behaviors) and topologies (spaces of meaning) of information (Swartz 2021: 268).[5] Information can thus be understood *as* physical, *for* instruction and *about* epistemic systems. Therefore, information extends from documents to data; from facts and fictions to pattern recognition; and from differential equations to physical information and then back again to the analogue (Floridi 2010a; also see Hayles 1999; Myles 2021; Parihar et al. 2021).

Now imagine communication understood as models of exchange and modes of inquiry, as well as meanings of/for/by community. Traditional communication research has been conceptualized as models of transportation, transmission and ritual, as well as characterized by modes of sharing, imparting, connecting and participating. Therefore, communication is phenomenological and sensorial[6] and can involve different types of entities (humans, nonhumans and the more-than-humans, including plants, animals and machines). Most importantly, communication imbues meanings. Indeed, life or from its etymological roots, body, can be considered a 'communicative' ecosystem (as in the study of microbial ecology and biocommunication). But other forms of life also involve distinct *living and lifelike technologies* and are sometimes presented in concert with what constitutes patentable characteristics, tendencies and patterns (Bedau 2020; Bedau et al. 2013, 2018, 2019; Zwick 2023).

Integrative Systems

'Metamedia' (Swartz 2021) presented, in one sense, an account of integrative, immersive environments as instruments, or what could be colloquially described as plena, swimming pools, fish bowls or vessels for human–computer symbioses.[7] It also offers an account via the philosophy of language as container logics, blurring boundaries, for example, across *cyber* and *physical* by means of tangible and metaphorical schemata.[8] This can then also be conceptualized as elements of (or a score for) a gigantic orchestra with interdisciplinary musical instruments and transdisciplinary research collaboratories. In many ways, integrating conferences, experiences and exhibitions together can also make way for spaces, places and traces to engage in the bringing together of scholars, professionals and publics.

More attention is given in this chapter to these integrative systems, more specifically to *convergence* and *sustainability*. Indeed, the understanding and application of biotechnology, information science, nanotechnology and cognitive science (BINC, or *bio-info-nano-cogno*)[9] has been foregrounded for over a quarter century by institutions such as the National Science Foundation (NSF) and National Academies of Science, Engineering, and Medicine (NASEM) in the United States, and similar agencies in other parts of the world, as well as non-governmental organizations (NGOs).[10] Other areas related to *machine learning* and *sustainability* have extended the BINC model (see Roco et al. 2013; Rasmussen et al. 2018; Roco 2020).[11]

This emphasis on convergence and sustainability, and similarly on regeneration and thrivability, is also evident across other regions, countries and an increasing number of domains, for example, in the European Union (see Muench et al. 2022), United Kingdom (see RSA 2022), Japan (see AIST), Australia (Ellyard 2001, 2002), New Zealand (see RSTA 2022), China (see NSFC 2021–) and Costa Rica (SINAC n.d.), as well as The World Future Council (see WFC n.d.). It is also important to note that some of the largest *professional services networks* and global accounting firms actively work in this area, as well.[12]

More specifically, the focus of BINC convergence rests upon four points. The first is biological (bio), as applied to genomic and microbial research, biomedical, health and wellness, as well as *bioethical*, *bioeconomic* and *biopolitical* issues. The second is information (info) as in IT infrastructures, machine learning, data science, increasingly faster wireless connectivity, as well as *information ethics* and *infopolitics* (Floridi 2010b, 2013; Koopman 2018, 2019, 2020). Attention to the nanoscale (nano), the manipulation of matter at the atomic and molecular levels, has brought forth a wide range of industries and institutions, e.g. through the use of transmission electron microscopy (TEM) and increasing attention to nanoethics and nanoregulation. This area also overlaps with research at the Large Hadron Collider (LHC) at the Organization européenne pour la recherche nucléaire (CERN), as well as to the conceptualization and development of qubits and ethical quantum computing. The fourth point relates to the cognitive sciences (cogno), pertaining to natural, artificial and hybrid neurons and synapses, in areas such as brain-inspired, neuromorphic computing, the ethics of cognitive wetware or organic neurons, and more broadly neuroethics (Mead and Mahowald 1988; Mead 1990; Nature Electronics 2022; Schuman et al. 2022).

Domains of convergence are seen as well through the lenses of governance and education, for example, in sustainability via policy and curriculum choices in schools, and as applied in workforce development in STEM (science, technology, engineering and mathematics) fields and programs. The inclusion of transdisciplinary practice, experimental publishing and knowledge production are also related developments (Garcia Topete et al. 2021: 3; Sellberg et al. 2021). This is also understood as 'transdisciplinary intelligence' (4), Art, Science, and Technology Studies (ASTS), or *art–science*, also known as 'third culture'.[13] Interestingly, John Dewey (1922) foresaw this more than a century earlier with his nondualistic account of science and art as intelligence, and by the time he published *Art as Experience* in 1934, he was advancing the need for an account of embodied aesthetics and ethics as related to art–science.

Patterns

The integrative sectors and domains above have been explored through what I call *speculative transformations* (interdisciplinary instruments and transdisciplinary collaboratories). As historian, Arunabh Ghosh observes, 'Arriving at some degree of knowledge about future events has motivated human thinking and action since time immemorial' (2019: 127).

Dewey ([1922] 2008) considered forecasting processes in terms of *moral deliberation* as an environment for contemplation or *dramatic rehearsal* as a form of 'reflective intelligence to revise one's judgments', or in the imaginative playing through of a situation important to opening possibilities (Fesmire 2003; Hamington 2010). Dewey invoked a lawyer, a business person, a military general and others in explaining that

> In all cases of deliberation, judgment of *value* enters; the one who engages in it is concerned to weigh values with a view to discovering the better and rejecting the worse. In some cases, the value of ends is thought of and in other cases the value of means. Moral deliberation differs from other forms not as a process of forming a judgment and arriving at knowledge but in the kind of value which is thought about.
> (Dewey [1932] 2008: 275, original emphasis)

This process involves dialogical environments necessary to create spaces for *pattern recognition* – a significant and broadly practiced process that deserves special attention especially as applied to the study of history, the present day and multiplicities of futures.

A well-known observation by McLuhan notes that the message of a medium is the change of scale, pace or pattern (McLuhan [1964] 2003: 20). *Patterns* are reviewed through the lens of formation, recognition and completion, which are all considered to be aspects of cognition and intelligence. First, *pattern formation* involves shapes, flows and branching structures (see Ball 2009; also see Thompson [1961] 1992). This is obvious in plants and trees but also applicable to human bodies and environments symbolically, mirrored in binary (digital) systems. 'When we are faced with information overload, [McLuhan] taught that the mind must resort to pattern recognition to achieve understanding' (Gordon 2003: xiii–xiv).

Pattern recognition more broadly is 'grasping interrelationships' (McLuhan 2003: 53), as well as perceiving the actions, more specifically 'of the ström' (212; McLuhan et al. [1979] 1996). It is interesting that *ström* (or malestrom) can be understood in many ways, as flow or current, as in a fluid and an electric charge, or as a stream as in air or any steady flow, e.g. radio signals or words, moving waters or a sequential read-write repository of data (Svenska Akademien 2022). Overall, pattern recognition is the process of, for example, auditory and/or visual detection of characteristics or other information (Schwab and Nusbaum 2013a, 2013b), as well as to 'make sound and reasonable decisions about the categories of the patterns' (Jain et al 2000: 4). Some applications include bioinformatics, data mining, document classification, industrial automation, biometric recognition and remote sensing (5). Additionally, for a discussion of issues of control related to pattern recognition, see David Savat (2009).

While there is no shortage of applications, pattern formation and recognition are relevant in ecological monitoring and assessment, network medicine (e.g., Barabási et al. 2011) and fractal fluency (Taylor et al. 2018). Fractals are patterns that repeat at various scales so as to create shapes of rich visual complexity (Mandelbrot 2004). Mid-range fractal complexity has been shown to contribute to health, wellbeing and performance, especially in natural and built environments. It may also foster *biophilia* (see Robles et al. 2021; Taylor 2021; Brielman et al. 2022; Taylor and York 2022) – an 'urge to affiliate with other forms of life' (Wilson 1984: 85).

These patterns are also relevant to an integrative framework of art–science, exemplified in works by Leonardo da Vinci. In Figure 18.1, observe the similarities among the fractal patterns or branchings at various scales, from lilies (plant) to veins (human) to rivers (environment) across natural systems. (See Appendix.)

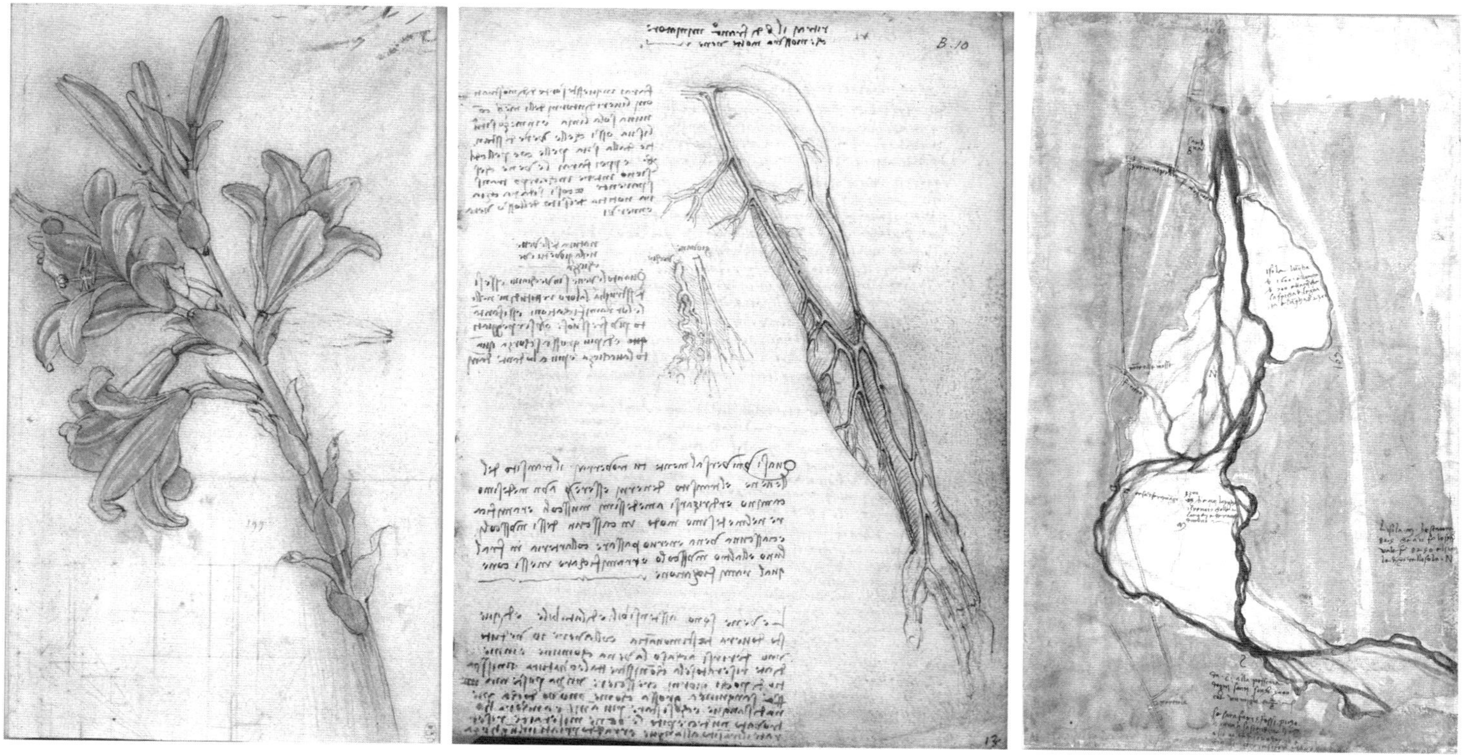

FIGURE 18.1: Three drawings by Leonardo da Vinci. (left) *A lily (Lilium candidum)*, *c.*1475, Stylus, pinpointing, leadpoint, pen and ink, brown wash, ochre wash and rubbing, white heightening (partly discoloured), pricked through, 31.4 × 17.7 cm (sheet of paper). (center) *The veins of the arm; and notes on the death of a centenarian*, *c.*1508, Pen and ink over traces of black chalk, 19.2 × 14.1 cm (sheet of paper). (right) *A map of the Arno west of Florence*, 1504, Pen and ink, blue and green wash, 42.2 × 24.2 cm (sheet of paper). Royal Collection Trust / © Her Majesty Queen Elizabeth II 2022.

As noted in the Introduction and various chapters in this volume, recognizing patterns is also present in the study of biomediation, biosemiotics and biocommunication, for instance, in comparing, contrasting and bringing together various forms of information (e.g. actual, biotic/instructional information; representational, Shannon information; and hybrid instantiations).

Meanwhile, N. Katherine Hayles (2021) identifies pattern completion as an 'ability to make inferences from *partial information*', especially 'in the context of forming episodic memories' (545), and refers to 'the brain as an "uncertain archive"' (544; see also Kreiman 2016; Liu et al. 2016). She explores how *unlearning* 'dissolves existing patterns of rigid and stereotyped behavior to make way for new forms [...] constructed communally [...] and makes them receptive to novel kinds of experience' (543). Alvin Toffler (1970) also presented learning, unlearning and relearning in relation to educational habits (414; see also Ratcliffe 2011: 381). This is echoed in attending to and acknowledging not merely stakeholders, but the environment, public health and wellbeing as 'a disciplined undoing of received assumptions' (Hayles 2021: 541).

According to Hayles, unthought or nonconscious cognition is important to understanding distributive, *cognitive assemblages*, opening possibilities — for *critical* analyses of complex human-technical systems (546) and remediation of ecosystems (Hegde et al. 2022). Interestingly, Francisco J. Varela (1999) noted the concept of *wu-wei* as a 'process of acquiring a disposition where immediacy precedes deliberation' (33).

Similarly, Dewey considered *nondoing* as 'a form of moral doing' and *active patience* as 'an especially thoughtful instance of pragmatic action' (Garrison et al 2014: 195). Qualities like these can become habituated, but according to Osamu Mizutani, they cannot be merely described; they must also be demonstrated (107–08).

Finally, Gregory Bateson commented that a '*pattern which connects is a metapattern*' (1979: 11, original emphasis). I offer the blend of metá and patterns, meaning *in the midst of patterns*, not above and beyond them. This was alluded to earlier in the discussion of system dynamics and decisionmaking pertaining to the 'limits to growth' (Schoenberg and Swartz 2023), which is a 'common system structure that produce[s] patterns of behavior' or a *systems archetype* (Meadows 2008: 187). Other examples include: drifting or eroding goals; escalation, policy resistance or fixes that fail; shifting the burden (to the intervenor) or addiction; success to the successful; tragedy of the commons; etc. (Meadows 2008; Herasymowych and Senko 2018; Hu et al. 2021). As a diagnostic, systems archetypes 'provide insight into the underlying structures from which behavior over time and discreet events emerge' (Braun 2002: 1; see also Stollberg-Rilinger 2018).

One ongoing 'design problem' revolves around the 'tragedy of the commons' systems archetype and *critical sustainability*, including merits and drawbacks of solutions-focused or responsibility-driven triple bottom line approaches (people, planet, profit). Instead, what could really constitute thrivability?

The mapping of these various modes and steps of pattern formation, recognition and completion, as well as anticipation and/or prediction involve the analysis of the *pace* of the pattern. They have benefited specifically from nineteenth and twentieth century developments, which McLuhan observed in the difference between 'the instant speed of electric information' and 'causal connections' (McLuhan [1964] 2003: 465). More recently, the blurring of time through processes of slow-motion recordings or time-lapse photography offer even more in-depth understanding and appreciation of pattern recognition.[14]

Futures

Scholars and professionals from a variety of disciplines have developed research programs dedicated to the study of futures, whether commodities, outcomes, prospects, propensities or accountability mechanisms. Each uses various theory-practices to interpret, forecast, predict and/or criticize various possible, probable, plausible and/or preferred futures (see Amara 1981). In addition, an entire discipline has evolved around the study of the future, an idea promoted by writer H. G. Wells at the turn of the twentieth century. This has been articulated interchangably as futurology,[15] future studies (Inayatullah 2019) and alternative futures,[16] or applied in art as *futurism*, for example, afrofuturism (Maurits 2019; Bruce 2020; Parikka 2022). Each in their own ways refer to 'the study of forces creating the future – both their possible outcomes and their implications for present decisions' (Schwartz and Becker [2001] 2005)[17] – including environmental trends. There are a wide range of approaches and techniques involved with the study of the future (see Chambers et al. 1971), for instance:

- *forecasting*: This process has generally been understood most simply as pattern recognizing the weather and reporting on it. It also extrapolates past and present *trend analysis* (e.g. from polls, opinions, etc.), where 'data-bases are used to map social patterns and forecast social trends' (Pasquinelli 2018: 255; see also Ghosh 2019).
- *backcasting*: 'Working backwards from a possible future state to determine how it might unfold' (Smart 2019: n.pag.; also see Bibri 2018).
- *scenario planning*: An interactive process for exploring alternative, plausible futures and what those might mean for strategies, policies and decisions, or in essence, 'what-if' questions (Scearce et al. 2004; Turner 2006: 186–87; Vermeulen 2018: 30; Smart 2019; Inayatullah 2019: 140).
- *horizon scanning* or *environmental scanning*: 'A systematic method of looking for drivers that influence the future. [...] "Environmental" here is not

restricted to the natural environment, but covers all types of environment[s]' (Smart 2019: n.pag.).

- *insights* or *social network analytics* (SNA): SNA 'is the ability to gather and find meaning in data gathered from social channels to support business decisions – and measure the performance of actions based on those decisions through social media' (IBM n.d.; see also Scott 2017; Simon 2018: 402).
- *strategic* and *integrative design*: This deals with possibilities for problem-solving creatively and ways to facilitate 'innovative' systems including objects and processes (see Shanks 2016, linked to Carleton et al. 2013; Inayatullah and Sweeney 2021; Roumani 2022).[18]
- *retroinformation*: 'Information that can be communicated between different eras and scales' (Munnecke 2021).[19]
- *science fiction prototyping*: Futuristic stories that explore a fusion of fact, fiction and emerging theories related to technology and society (see Johnson 2011; Oppermann 2018; Robinson 2020).
- *science fiction storytelling and multimedia productions*: Future worlds imagined and presented by writers and artists via various multimedia formats.[20]
- *critical worldbuilding* is 'a form of collaborative knowledge construction [that] bridges critical thinking with creative production with an emphasis on viewing the world through multiple perspectives [...] through simulations and role-play scenarios' (Hergenrader and Slota 2023; see also Ludwig and Wiskopf 2019).

Other approaches include: (1) understanding and design, similar to research and development (R&D) and can be especially applied to systems biology as knowledge-driven and synthetic biology as application-driven (Green 2022); (2) design frameworks of observation, prototyping and visualization; and (3) validating models of concentration and differentiation, especially when attempting to study probabilities, anticipations and/or predictions of various past–present–future states.

In recognizing patterns in past histories and bringing them together with the present experience, those in positions of power might conceivably have necessary comprehension of a given situation and potential for designing ways out of given problematics. These types of approaches may afford possibilities as well as implications for what can be referred to as *glimpses* and *glimpsing* (recognizing object/noun and action/verb), which may enjoin philosophical co-inquiry as *metadisciplinarity*, as well as capacities and tendencies for critical futures research (see also Ingwersen 2022; Paul 2019). Whether a swift decision by an expert in a given situation or sensemaking acquired via experience, glimpsing takes a comprehensive view. In French, *coup d'œil* or a 'stroke of [the] eye' and in German, *fingerspitzengefühl*, or 'finger tips feeling', exemplify this embodied metaphor of *sensemaking* (see also Urquhart et al. 2020; Stokols et al. 2020).

More importantly, it implies an experientialist or enactivist 'situational awareness', one that can include, for example, multiple ways of knowing: as *tactical*, in the sense of strategy and tact that is both moral and instrumental; as a form of feedforward neural networks[21] validating *empirical* research and a form of differentiation; and as a form of swordsmanship that is active selection and *responsible* glimpsing of experience. I would also bring attention to 'tactfulness', especially related to artful or mindful modes of living and inquiry. In what ways does morally imaginative glimpsing exist at various scales?

Large-Scale Institutions

Forecasting techniques and programs have also been employed by governments, corporations, universities, institutes and hybrid institutes formed by universities and industry, including various conferences, workshops and programs that prepare and present future scenarios.[22] These and other institutions, projects or concepts can be thought of as *lookouts* or *lighthouses* or *distant early warning* systems (e.g. McLuhan's 1969 deck of cards). Only a few examples will be discussed here.

The first exemplar is the NSF, an important player in issues related to BINC (as noted previously), as well as some of the strategic objectives of diversity, inclusion and environment/climate. These include ensuring 'accessibility and inclusivity' especially for underrepresented communities in STEM (NSF 2022: 30–31) and ways to grow and unleash 'a diverse STEM workforce to advance the progress of science and technology...' (32–33). In regard to climate impact, NSF documents note 'the importance of reducing direct and indirect emissions of greenhouse gases' and 'ways in which it can reduce the impacts of its own operations' (47).

Also noted previously, NSF has been especially engaged in promoting convergence of science, technology and society. In a 2013 report, the metaphor of an orchestra conductor was employed:

> the conductor represents the coordination required to orchestrate the complex vision, with each converging field of science and technology functioning like one of the instruments in the orchestra, playing together in different combinations for various passages in the musical score. Like the conductor, societal governance must actively and harmoniously engage in coordination of the total composition of complex scientific, technical, and societal convergences that are taking place and gaining momentum.
>
> (Roco et al. 2013: xxx–xxxi; 549; also see Turner 2009)

The conductor metaphor has been previously used by various writers. For instance, Denis Noble employs the idea in illustrating *downward causation* in his book, *The Music of Life: Biology Beyond Genes* (2006: 50–51). However, questions arise regarding this metaphor, for instance, is the conductor writing the score? Or is the composer conducting the orchestra? Interestingly, this orchestration notion was updated by Roco to a 'holistic view' in 2020,

> exploiting the interdependence and unity in nature and society. [...] The behavior of a system is a function of its components and interactions between those components. [...] Each *converge platform* (foundational S&T fields, Earth-scale, human-scale and societal-scale) is characterized by a set of concepts, group of participants, and specific investigative tools.
>
> (Roco 2020: 321–22, emphasis added)

Another area of interest for NSF has been wellness, '[a] conceptualization of human health that emphasizes prevention of disease more than cure, and proactive enhancement of general physical and cognitive well-being more than merely dealing with problems reactively when they arise' (Roco et al. 2013: 551). An example is regenerative medicine or '[r]estoring the cells and organs of the human body to support normal functioning, whether by stimulating the body's own repair mechanisms or by culturing person-specific replacement tissues initially outside the body' (550).

An overall indicator of NSF's priorities is represented by the topics of the winners of the NSF 2026 Idea Machine prize competition, which involved submissions that presented 'grand challenges' requiring fundamental research in science, engineering or STEM education [...]' (NSF n.d.). Prize winner topics included engineered living materials, from thinking to inventing, public carbon capture and sequestration as well as emergence.

The second set of exemplars have built on the NSF's emphasis on social progress. Attempts to assess 'social progress' have been common throughout history and seen as 'the capacity of a society to meet the basic human needs of its citizens, establish building blocks that allow citizens and communities to enhance and sustain the quality of their lives, and create the conditions for all individuals to reach their full potential' (SPI 2022: n.pag.).

In 1893, Thomas Henry Huxley (aka 'Darwin's Bulldog'), an English biologist and anthropologist specializing in comparative anatomy, added the notion of ethics to the process of assessing social progress, which he explained ...

> [...] means a checking of the cosmic process at every step and the substitution for it of another, which may be called the ethical process; the end of which is not the survival

> of those who may happen to be the fittest, in respect of the whole of the conditions which obtain, but of those who are ethically the best.
>
> (Huxley 1893: 33)

What constitutes 'those who are ethically the best' is highly-context specific and would require significant further discussion.

Over a half a century later, various projects emerged that have explored the concept of social progress through scenario planning and other methods. One example was the Global Business Network (GBN), created in the United States in 1987 by a group of entrepreneurs as a 'consulting firm that specialized in helping organizations to adapt and grow in an uncertain and volatile world' (Wikipedia 2022: n.pag.; GBN 2010), using tools such as scenario planning and experiential learning. The organization represented a 'particular blending of countercultural and techno-cultural organizational styles' (Turner 2006: 184) and mostly 'reversed the countercultural antipathy toward industry and the military' (194). After the Monitor Group took over GBN in 2000, they declared bankruptcy in 2012. Deloitte then acquired Monitor in 2013, forming a non-profit organization called The Social Progress Imperative, which publishes annual assessments of how nations provide social and environmental needs of citizens. Created in 2014, *The Social Progress Index®* (SPI) 'measures the well-being of a society by observing social and environmental outcomes directly rather than the economic factors'. This organization and others have also produced reports projecting the future, e.g., *Social Progress in 2030, Global Trends in 2040* and *The World in 2050.*

Small-Scale Community Organizations

Overall, knowing that these seemingly metasystem level organizations actually exist is different than not knowing. But how can history and especially philosophy, help ameliorate some of the emerging problems of the 'superfluous products' of our own creation (Dewey [1925] 2008: 212) – meaning some of the material and environmental outcomes of these entities.

It might be helpful to look for possible community-based solutions or alternatives in the open, local, cooperative universe, where many emerging issues are being confronted on a daily basis. For instance, *food security* has been addressed through public domain plant breeding (Food Farmer Earth [2010] 2020; Roach 2022; also see Hope 2008) and searchable patent databases (i.e. lens.org). Sheryl D. Breen (2015) notes that there are different scales and methods of seed saving projects that exist from the Svalbard Global Seed Vault (or genebanks) in Norway to Indigenous peoples of North America 'traditional ways of local seed conservation [with] anthropological evidence of seed caches that maintain viability for long periods' (46). Other community projects involve farmers and growers markets, co-ops and community-sponsored agriculture (or CSA).

Health and wellbeing initiatives are also important concerns dealt with by community organizations that strive for 'a collective environment organized to enable people to gain control of their social, emotional, and physical well-being through direct service, education, and community', as well as 'compassionate, humanistic healthcare, and supportive services', so everyone receives the care they need. A specific example is WhiteBird Clinic and Lane County's Crisis Intervention Service in Eugene, Oregon.[23]

Emergency intervention is another area where local non-profit organizations are sometimes active. For example, Crisis Assistance Helping Out On The Streets (CAHOOTS) is a mobile crisis and medic response team for Eugene's Public Safety System, which has received national attention and has become a model for other local communities.[24] Somewhat related issues are involved with *disaster mitigation*, which involves Community Organizations Active in Disaster (COAD) and involves bridge building between the local, state and national levels.

Meanwhile, *intersection repair* has emerged as community-based projects around the world (see Luca 2020; Bertoini 2020). A specific example is the City Repair Project in Portland, Oregon. From a field action report, Jan C. Semenza (2003) notes:

> In an attempt to invigorate neighborhood stewardship, the community organized and created a public gathering place; together, they painted a gigantic sunflower in the middle of an intersection and installed several interactive art features [in other words,] collective actions of 'place-making' [and] a sense of well-being.
>
> (1439)

This idea is being brought together with the larger conceptualization of *lateral public health* and responses to crisis with systemic resilience.

> Community members worked alongside city officials, civil engineers and city commissioners to design, permit and implement these features [...]. The process entailed community outreach, informal meetings, design workshops, project development, (building) permitting process, and construction workshops [(Semenza 2007)]. The interventions were intended to attenuate the negative consequences of climatic events [...] by physically improving the built environment.
>
> (Semenza 2021: 3)

Lateral public health is intended to develop communities' capacities to reduce risk by *desiloing* and creating connections between governmental and nongovernmental actors and agencies across civil society, the academy and other organizations and individuals. This mode has proven highly successful in dealing with 'the *convergence* of two compounding risks: [pandemics and environmental disasters]. Risk is defined as a function of hazard, exposure and vulnerability' (Semenza 2021: 2, emphasis added). Mitigation efforts have included: community health education, collaboration and forecasting, as well as other capacity building initiatives including collaboration with 'scientists to design, develop and implement an early warning system for climatic events' and strengthening community resilience by, for example, community-based waste water surveillance – upgrading water treatment and distribution systems to avoid waterborne outbreaks or improving urban drainage for vector abatement (3; see also Kirby et al. 2021: 1242).

These organizations demonstrate adaptability and agility tethered to their bioregions more than many large organizations, and thus help to ameliorate some of the ongoing and emerging problems in near real time response.

In the Midst

This chapter has reflected on integrated systems, recognizing patterns, futures research and socially responsible vs. accountable organizations, especially through the lenses of the natural, cognitive and social sciences, as well as their convergence and ongoing emergence. This brief discussion has contrasted scales, from large international and national organizations to smaller local/community institutions, as well as pointing to various ways to potentially adjust paces or speeds of our observations and interactions with (living) systems.

Although this section does not represent a conclusion, I offer a few brief ruminations on metadisciplinarity, pragmatism, multi-stakeholder arrangements and critical futures — to inspire re-magnifying and re-widening kinship via pluralistic epistemologies and ontologies, and critical worldmaking.

As noted in this volume, life can be understood as mediums, environments and multimodal habitats. It is always in ongoing relationship with living – for example, as mediating, environing and habituating. I acknowledge Humberto Maturana and Bernhard Poerksen's ([2004] 2011) ethical awareness of a biology of cognition and love of life. This reimagines living as artful being and becoming in the midst of, in co-operation with, life. Thus, metalife and meta-living are open-ended, calling for curational co-inquiry, enacting Dewey's criticism of criticism and acknowledging James's 'much-at-onceness' ([1910] 1987: 1008) and McLuhan's 'all-at-onceness' ([1967] 2001: 63). For example,

> The immediacy of [...] computers [...] shift the very nature of time: instead of the one-thing-after-another, linear time of print, electronic media fostered an 'all-at-onceness' that would characterize [...] the Internet era [....] McLuhan's work anticipated [...] reduced attention spans. 'He paid such attention to the present', his son Eric [McLuhan] says, 'and that enabled him to understand what would necessarily happen in the future'.
>
> (Scott 2011)

Similarly, 'the internet' is typically conceptualized as a thing, but should more importantly be recognized as a process, as *internetworking* (Koopman 2010). So when considering the internet as a web of relationships, then metaliving can be understood as 'pattern recognition' or 'simultaneous [planetary] feed and counter-feed' (e.g. looping-back and looping-forward in machine learning). New questions arise surrounding the paradoxes of attaching properties of certain scales to assumptions that are then applied to other scales; this also goes for how we perceive time and the paces of systems we experience.

Living in a time of metalife and metaliving requires metadisciplinary, curational modes of co-inquiry. In this sense, it is not enough to take into account mere *means* or mere *ends* in the advancement of society but also a deep listening, an ongoing 'letting appear' as an *ends-in-view*. In other words, an openness to contemplating in a multiplicity of ways, how sometimes doing nothing, instead of something, can lead to what has been described as *effortless action*. This would be a recognition of the pattern of *rest* – for example, like a musical break from sonorizing, performing, gabbing, thinking – for our own health and wellbeing and that of society. In this way, we could cultivate and embrace artful living, the ways in which nature supports us, our families, our immune systems, our everyday life and living. Consequently, we might be able to contemplate and practice an artful embodied ethics and aesthetics of metalife.

This suggests larger questions emerging around what constitutes *pragmatism*. It is not merely 'pragmatic', 'practical' and/or 'aesthetic', but a wider recognition of a naturalistic/materialist epistemology, object (static) and process (dynamic) oriented ontologies, as well as living art–science. From pattern formation, recognition and completion to various futures, a pragmatist curational turn integrates maintenance and prevention as care, as well as the capability of rendering emergency crisis mitigation/management or interventions as repair becomes ever more poignant in these challenging times. For example, Chris Cook, in conversation with John Wood (2022c), argues that there should be a four-fold relationship to multi-stakeholder arrangements: agents (including a *custodian* and a *steward*), rights and obligations (responsibilities) and being held accountable regarding shared risks and rewards across communities. In a presentation on critical cybernetics, Klaus Krippendorff (2021a, 2021b) emphasized this relationship across responsibilities and accountability that we need to take to heart.

Metaliving or artful living futures are environmental and ecological – a much-at-onceness. In the sense of a plenum or swimming pool, today's immersivities and speculative transformations attenuate responsibility for our desires, especially as applied to the study of futures including imaginative scenarios (see Dunne and Raby 2013: 169).

One way to do so is through *critical future studies*, which includes materialist media theory, critical posthumanisms, critical disability studies, energy humanities and petrocultures, environmental humanities and science fiction studies.[25] For example, Technische Universität Dresden (TU Dresden) offers a program in the 'interdisciplinary and transcultural study of literatures from the U.S., Canada, and Indigenous Nations', focusing on North American future studies. It 'highlights entanglements among humans, technology, and environments in critical imaginations of the future' (Ingwersen 2022).[26]

Similarly, there are research centers and journals dedicated to these areas, including the environmental humanities, described as:

> An engaged, scholarly response to madness—an attempt to address the systemic pathology of a species disconnected from the conditions of its world. Becoming-with offers a metaphysics grounded in connection, challenging delusions of separation—the erroneous belief that it is somehow possible to exempt ourselves from Earth's ecological community.
>
> (Wright 2014: 278)

An example is the Center for Environmental Futures (CEF) at the University of Oregon (UO), in cooperation with the larger Eugene area, which reflects a longstanding investment in environmentalism and social justice. CEF 'contextualizes and complements environmental science and policy with a focus on narrative, critical thinking, history, cultural analysis, aesthetics, and ethics'.[27] Another coordinated effort at UO is the Just Futures Institute's (JFI) approach, which consists of principles such as relationship building (not extractive research models), community and capacity building and respect for expertise, accountability, protection of intellectual property, and awareness of social location, empathy and compassion (Reyes-Santos 2022).[28]

Critical future studies also resonates with decolonizing speculation (Bahng 2017) and the ways it 'captures [...] finance and surveillance, increasingly entangled with algorithmic media' (Parikka 2022: 6; see also Vishmidt 2023).

It also includes organizations at the forefront of these and other areas, such as the Initiative for Indigenous Futures (IIF).[29] These are just some initial steps towards integrating the modes of a block universe or plural universes or spectrum together. Interestingly, rephrasing Edmund Leach ([1967] 1968), Stewart Brand (1968) stated that '*We are as gods and might as well get used to it*' (original emphasis). However he re-evaluated his statement acknowledging that 'we are as nature' (see Alvarado and Sussberg 2021), which Indigenous peoples have recognized for countless generations. This is also embodied in the statement, 'Water is the source of life, not a resource' (Chief Arvol Looking Horse [2018] 2020: 171).

These connections recognize co-curation as a kinship and values of friendship and loyalty among all living systems, identified as human, nonhuman and more-than-human kinship, as well as kinship with nature (respect for wilderness and acknowledging ancient kinships between plants, forest and ecosystems). These kinships are all represented in Aldo Leopold's notion of *biotic community* and *land ethic* that 'enlarges the boundaries of the community to include soils, waters, plants and animals, or collectively: the land'. Leopold notes that 'the role of Homo sapiens [changes] from conqueror of the land-community to plain member and citizen of it [... and a] respect for his fellow-members and also respect for the community as such' (Leopold [1949] 1989: 204). In other words, where nature, nonhumans, humans and more-than-humans together continue to re-imagine our moral compass, and where we humans are not merely stewards but are neighbours – metaliving *in the midst of* multiplicities of worlds.

NOTES

1. This is a glimpse into a research project in progress.
2. *Bio Design*, from The Museum of Modern Art exhibition, is an evolving form of organic design that 'harnesses living materials, whether they are cultured tissues or plants, and embodies the dream of organic design' (Antonelli [2012] 2018: 7). It

> goes further than the many biology-inspired approaches to design and fabrication. Unlike biomimicry, cradle-to-cradle, and the popular but frustratingly vague 'green design,' *biodesign* refers specifically to the incorporation of living organisms or ecosystems as essential components, enhancing the function of the finished work. It goes beyond mimicry to *integration*, dissolving boundaries between the natural and built environments and synthesizing new *hybrid typologies*.
>
> (Myers [2012] 2018b: 8, emphases added)

Biodesign is meant to 'incite discussion and careful consideration of the potential unintended consequences [...] in which biological and biomimetic processes replace those that are mechanized and digitized today' (17).

3. For the integration of intelligence, design and decisions, in addition to considering 'how things are' from the view of normative science and the possibility for 'how things ought to be' via the natural sciences, see Herbert A. Simon ([1969] 2019). Krippendorff (2007) offers another shift in design through the 'dimensions of the sematic turn from technology-centered design to human-centered design' (39) and postulates that '*Design strategies that go against the ecological wisdom of a culture are likely to fail*' (205, original emphasis). Instead, he intones, '*Interfaces follow recognizable meanings*' and that 'creates possibilities and proposes collaboratively realizable futures' (322, original emphasis). In addition, see Fred Turner (2016), who offers a view of the landscape of attachments to contemporary foci on *prototypes*.
4. Elisa Giaccardi in the journal, *Leonardo*, historicizes metadesign as follows:

> [Gene] Youngblood [(1986)] defined metadesign as a strategy for instigating a revolution in communication world and overcoming the broadcast style of mass culture. As such, metadesign deals with the creation of context [meaning] rather than content [symbolic systems]; it is a mode of integrating systems and setting actions in order to create environments in which people may cultivate 'creative conversations' and take control of the context of their cultural and aesthetic production.
>
> (Giaccardi 2005: 343)

   She also points to metadesigners as *systems integrators* (Giaccardi 2005: 347) and their relationships to '"open-ended systems" (OES), in which interaction takes place within networked and evolving systems that put the user or the environment in control of the interaction itself' (348; also see Maturana 1997; Varela 1999; Wood 2022a, 2022b).
5. Information can also be seen through the lenses of the social and technical, mathematical and semantic, physical and biological, economic and political, cultural and environmental.
6. Senses include, but are not limited to, the auditory, visual, kinaesthetic, tactile, olfactory, gustatory and interoceptive.
7. See examples of terrariums and aquariums (Ohlenschlager et al. 2023).
8. Also see discussions of *cyber-physical systems* (Höftberger 2016).
9. For an explanation of the use of BINC, instead of NBIC, see Leif Rasmussen et al. 2018. Other acronyms have been utilized including Bits, Atoms, Neurons and Genes (BANG), Genetics, Robotics, Artificial Intelligence and Nanotechnology (GRAIN) and Genetics, Nanotechnology and Robotics (GNR).
10. Also see Roco and Bainbridge (2002), Bainbridge and Roco (2006, 2016), NSF (n.d., 2017–2022) and NASEM (2014, 2019).
11. The term sustainability has been applied at an ecological level of industry scale, and typically referred to as *industrial ecology*, which refers to '[a] science for assessing and minimizing the impact of industry on the environment, using such methods as [1] *material flow*

analysis, [2] *life-cycle analysis*, and [3] *design for environment*' (Roco et al. 2013: 547, emphasis added; see also endnote 19 on pp. 15–16 in this volume).

12. These include KPMG International Limited (Klynveld Peat Marwick Goerdeler), founded 1987, London, UK, oldest component 1818, headquartered in Amstelveen, Netherlands; Deloitte Touche Tohmatsu Limited (Deloitte), founded 1845, London, UK; PricewaterhouseCoopers International Limited (PwC), founded 1998, oldest component 1854, London, UK; Ernst & Young Global Limited (E&Y), founded 1989, London, UK.
13. There are many examples of the idea of *third culture* (e.g. see Snow and Snow 1959; Vesna 2001; Malina et al. 2015; Garcia Topete 2019, 2021; Cizek and Uricchio 2022; Rogers 2022).
14. For instance, hummingbird flights or fungal growth (see Schwartzberg 2021), the construction of various social and cultural structures and patterns over time (see Reggio 2012; Reggio and Kane 2023) and remixed audio-video visualizations of future states of the universe (e.g. melodysheep 2019).
15. Futurology is defined as 'The activity of predicting the state of the world at some future time, by extrapolating from present trends. Mainly a pseudo-science, given the complexities of social, political, economic, technological, and natural factors' (Blackburn 2016).
16. See Jim Dator's ([2009] 2019) work at the University of Hawaii Manoa as an example of 'alternative futures'. Also see Paul Bracken (2021) interview, 'Exploring alternative futures'.
17. Commonly used methods in future studies (Schwartz and Becker [2001] 2005) include 'conventional forecasting, rigorous mathematical models predict future conditions [as well as] gathering and synthesizing expert opinion about the future. Finally, scenario planning methods utilize speculative, but carefully constructed, descriptions of future world'. More methods are described in this section.
18. 'Components of a design mindset' would include synthesis, modelling, speculation, facilitation and implementation with special attention to transdisciplinary collaboration and recognition of an evolving mindset (Mejía et al. 2022).
19. This is reminiscent of a passage from *The Medium is The Massage*: 'When faced with a totally new situation, we tend always attach ourselves to the objects, to the flavor of the most recent past. We look at the present through a rear-view mirror. We march backwards into the future' (Fiore and McLuhan [1967] 2001: 74–75).
20. A very wide variety of imagined future worlds have built by early writers such as Mary Shelley, Jules Verne, H. G. Wells and on to Ursula K. Le Guin, David Brin, Octavia Butler, Kim Stanley Robinson, Cherie Dimaline, Sue Burke, etc. These authors as well as other creators and directors, highlight ecological themes of interconnected tapestries, symbiotic relationships of physical, biological, cultural, empathetic, ethical, moral, spiritual and social systems and dynamics (Lucas 2018; cf. Scott 2012; Hershman Leeson 2022). Also, immersive storytelling design environments can include LED (light-emitting diode) volume walls.
21. See Schoenberg and Swartz (2023). For additional interpretations of feedforward, see I. A. Richards (1968), Terrence W. Gordon (2010: 23) and Mark B. N. Hansen (2015).

22. This is only a partial list of examples: Art | Sci Center (UCLA), Cascade Institute, Centre for Sustainable Transitions (Stellenbosch Univ.), Center for Systems Science and Engineering (Johns Hopkins Univ.), Cognitive Neuroscience of Creativity Laboratory (Penn State Univ.), Complex System Society, Engineering Systems, Design and Innovation (Univ. of Bristol), Foresight (Univ. of Houston), Global Ecovillage Network, Good Systems (Univ. of Texas at Austin), International Council on Systems Engineering, International Federation for Systems Research, International Society for System Sciences, Institute for the Study of Complex Systems, Institute for Emerging Ethics and Technology, Leonardo/International Society for the Arts, Sciences and Technology, Max Planck Institute for Biological Cybernetics, MIT Media Lab, New England Complex Systems Institute, Osher Rainforest (California Academy of Sciences), Planetary Collegium (Plymouth Univ.), Rodale Institute, Santa Fe Institute, SpeculativeEdu (funded by ERASMUS+), STARTS (Science, Technology and Arts; European Commission), Systems Science and Industrial Engineering (Binghamton Univ., SUNY), TreePeople, United Nations Development Programme Data Futures Platform, Univ. of the Underground, Vermont Complex Systems Center, Waterloo Institute for Complexity and Innovation, and World Future Council; as well as channels and conferences: Anticipation (Arizona State Univ.), Ars Electronica, Care Ethics Otherwise (Rock Ethics Institute), Complexity Digest (Univ. Nacional Autonoma de Mexico), Complexity Explorer, Right Livelihood Award; and a list of journals too long to include here.
23. White Bird Clinic (n.d.), 'About Us', https://whitebirdclinic.org/about/. Accessed 14 February 2023.
24. White Bird Clinic (n.d.), 'CAHOOTS In the News', https://whitebirdclinic.org/services/cahoots/cahoots-in-the-news/. Accessed 14 February 2023. See also Mark Obbie (2021), Tatiana Parafiniuk-Talesnick (2021), Andrew Selsky (2022) and Nayo Macauley (2021).
25. 'Critical Future Studies', Technische Universität Dresden (TU Dresden), https://tu-dresden.de/gsw/slk/anglistik_amerikanistik/na-literatur/die-professur/profile-critical-future-studies. Accessed 14 February 2023.
26. 'North American Literature with a Focus on Future Studies', TU Dresden, https://tu-dresden.de/gsw/slk/anglistik_amerikanistik/na-literatur#intro. Accessed 14 February 2023.
27. 'About CEF', Center for Environmental Futures, University of Oregon, https://cef.uoregon.edu/about-cef/. Accessed 14 February 2023. See also Dipesh Chakrabarty, Ursula K. Heise and Rob Nixon (2022).
28. 'Vision', Just Futures Initiative, University of Oregon, https://jfi.uoregon.edu/about/vision/. Accessed 14 February 2023.
29. The Aboriginal Territories in Cyberspace (AbTeC), a research network based at Concordia University conducts the Initiative for Indigenous Futures (IIF), 'developing multiple visions'. 'About', AbTeC, https://indigenousfutures.net/about/. Accessed 22 May 2023. See also a position paper by Jason Edward Lewis (2020).

REFERENCES

AIST (The National Institute of Advanced Industrial Science and Technology), Global Zero Emission Research Center, Koto-ku, Tokyo, Japan: AIST Department of Energy and Environment, https://www.gzr.aist.go.jp/en/. Accessed 14 February 2023.

Alvarado, David and Sussberg, Jason (dir.) (2021), *We Are As Gods*, Structure Films.

Amara, Roy (1981), 'The futures field', *The Futurist* 15:1, pp. 25–29.

Antonelli, Paola ([2012] 2018), 'Vital design', in *Bio Design: Nature + Science + Creativity, Revised and Expanded Edition*, New York: Museum of Modern Art, pp. 6–7.

Bahng, Aimee (2017), *Migrant Futures: Decolonizing Speculation in Financial Times*, Durham, NC: Duke University Press.

Bainbridge, William Sims and Roco, Mihail C. (2006), *Managing Nano-Bio-Info-Cogno Innovations: Converging Technologies in Society*, Dordrecht: Springer, https://www.researchgate.net/profile/James-Baker-24/publication/226441804_Nanotechnology_for_Biology_and_Medicine/links/56eadcab08ae2a58dc49bb42/Nanotechnology-for-Biology-and-Medicine.pdf. Accessed 14 February 2023.

——— (eds) (2016), *Handbook of Science and Technology Convergence*, Switzerland: Springer International Publishing.

Ball, Philip (2009), *Shapes: Nature's Patterns: A Tapestry in Three Parts*, Oxford: Oxford University Press.

Barabási, Albert-László, Gulbahce, Natali and Loscalzo, Joseph (2011), 'Network medicine: A network-based approach to human disease', *Nature Reviews Genetics, 12*:1, pp. 56–68.

Bateson, Gregory (1979), *Mind and Nature: A Necessary Unity*, New York: Bantam Books.

Bedau, Mark A. (2020), 'Dynamic natural kinds: Open-ended evolution in the joints of nature,' in ALIFE 2020: The 2020 Conference on Artificial Life, Cambridge, MA: MIT Press, pp. 199–201.

Bedau, Mark A., Gigliotti, Nicholas, Janssen, Tobias, Kosik, Alec, Nambiar, Ananthan and Packard, Norman (2018), 'Detecting the on-going emergence of technological innovations [extended abstract]', in Third Workshop on Open-Ended Evolution (OEE3), ALIFE 2018 Conference, Tokyo, Japan, 25 July.

——— (2019), 'Open-ended technological innovation', *Artificial Life, 25*:1, pp. 33–49.

Bedau, Mark A., McCaskill, John S., Packard, Norman H., Parke, Emily C. and Rasmussen, Steen R. (2013), 'Introduction to recent developments in living technology', *Artificial Life,* 19:3&4, pp. 291–98.

Bertolini, Luca (2020), 'From "streets for traffic" to "streets for people": Can street experiments transform urban mobility?', *Transport Reviews, 40*:6, pp. 734–53.

Bibri, Simon Elias (2018), 'Backcasting in futures studies: A synthesized scholarly and planning approach to strategic smart sustainable city development', *European Journal of Futures Research,* 6:1, pp. 1–27.

Blackburn, Simon (2016), 'Overview: Futurology', in *The Oxford Dictionary of Philosophy*, 2nd ed., Oxford: Oxford University Press, https://www.oxfordreference.com/view/10.1093/oi/authority.20110803095839389. Accessed 14 February 2023.

Bracken, Paul (2021), 'Faculty viewpoints: Exploring alternative futures', *Yale Insights*, Yale School of Management, 15 September, https://insights.som.yale.edu/insights/exploring-alternative-futures. Accessed 14 February 2023.

Brand, Stewart (1968), 'Purpose', *Whole Earth Catalog,* Fall, p. 3, https://archive.org/details/1stWEC-complete/page/n1/mode/2up. Accessed 14 February 2023.

Braun, William (2002), 'The system archetypes', *MIT CTL (Center for Transportation and Logistics), 27 February*, pp. 1–26, https://ctl.mit.edu/sites/ctl.mit.edu/files/attachments/tab%204c%20Pacheco%20-%20system%20archetypes.pdf. Accessed 14 February 2023.

Breen, Sheryl D. (2015), 'Saving seeds: The Svalbard Global Seed Vault, Native American seed-savers, and problems of property', *Journal of Agriculture, Food Systems, and Community Development* 5:2, pp. 39–52.

Brielmann, Aenne A., Buras, Nir H., Salingaros, Nikos A. and Taylor, Richard P. (2022), 'What happens in your brain when you walk down the street? Implications of architectural proportions, biophilia and fractal geometry for urban science', *Urban Science,* 6:1&3, pp. 1–35.

Bruce, Delan (2020), 'Afrofuturism: From the past to the living present', *UCLA Newsroom*, 3 September, https://newsroom.ucla.edu/magazine/afrofuturism. Accessed 14 February 2023.

Bureaud, Annick (2014), 'Introduction: Art at the future tense', in A. Bureaud, R. F. Malina and L. Whiteley (eds), *Meta-Life: Biotechnologies, Synthetic Biology, A-Life and the Arts*, Oakland, CA and Cambridge, MA: Leonardo/ISAST and MIT Press.

Carleton, Tamara, Cockayne, William and Tahvanainen, Antti-Jussi (2013), *Playbook for Strategic Foresight and Innovation*, Stanford, CA: Stanford MediaX Lab – Foresight.

Chakrabarty, Dipesh, Heise, Ursula K. and Nixon, Rob (2022), 'Reimagining Environmental Futures', *Global Dialogues Series, Stanford Global Studies*, 11 February, https://sgs.stanford.edu/events/global-dialogues-series-reimagining-environmental-futures. Accessed 10 January 2023.

Chambers, John C., Mullick, Satinder K. and Smith, Donald D. (1971), 'How to choose the right forecasting technique', *Harvard Business Review*, July, https://hbr.org/1971/07/how-to-choose-the-right-forecasting-technique. Accessed 14 February 2023.

Chief Arvol Looking Horse ([2018] 2020), 'Standing rock is everywhere: One year later, a call for continued efforts to protect our water and our Earth', in K. García-Antón, H. Gaski and G. Guttorm (eds), *Let the River Flow: An Indigenous Uprising and its Legacy in Art, Ecology and Politics*, Norway and Amsterdam: Office for Contemporary Art and Valiz. [First published in 2018 in *The Guardian*, Kings Place, London, 22 February.]

Cizek, Katerina and Uricchio, William (2022), *Collective Wisdom: Co-Creating Media for Equity and Justice*, Cambridge, MA: MIT Press.

Dator, Jim ([2009] 2019), 'Alternative futures at the Manoa School', in (2019), *Jim Dator: A Noticer in Time, Selected Work, 1967–2018*, Cham: Springer Nature, pp. 37–54. [First published in the *Journal of Future Studies*, November 2009, 14:2, pp. 1–18.]

Dewey, John ([1922] 2008), *Human Nature and Conduct*, in J. A. Boydston (ed.), *The Middle Works, 1899–1924,* vol. 14, Carbondale, IL: Southern Illinois University Press and InteLex Electronic Edition.

——— ([1925] 2008), *Experience and Nature*, in J. A. Boydston (ed.), *The Later Works, 1925–1953*, vol. 1, Carbondale, IL: Southern Illinois University Press and InteLex Electronic Edition.

——— ([1932] 2008), *Ethics*, rev. edition, in J. A. Boydston (ed.), *The Later Works, 1925–1953*, vol. 7, Carbondale, IL: Southern Illinois University Press and InteLex Electronic Edition.

——— ([1934] 2008), *Art as Experience*, in J. A. Boydston (ed.), *The Later Works, 1925–1953*, vol. 10, Carbondale, IL: Southern Illinois University Press and InteLex Electronic Edition.

——— ([1939] 2008), *Freedom and Culture*, in J. A. Boydston (ed.), *The Later Works, 1925–1953*, vol. 13, Charlottesville, VA: InteLex Electronic Edition.

——— ([1944] 2008), 'The democratic faith and education', *1942–1948, Essays, Reviews, and Miscellany*, in J. A. Boydston (ed.), *The Later Works, 1925–1953*, vol. 15, Charlottesville: InteLex Electronic Edition. [First published in Antioch Review, 4: June 1944, pp. 274–83, from an address read by Jerome Nathanson before the conference on The Scientific Spirit and Democratic Faith, at the Ethical Culture School, New York, 27 May 1944.]

——— (2012), *Unmodern Philosophy and Modern Philosophy*, P. Deen (ed.), Carbondale, IL: Southern Illinois University Press.

Dewey, John and Childs, John L. ([1933] 2008), 'Philosophy', *Essays and How We Think*, rev. ed., in J. A. Boydston (ed.), *The Later Works, 1925–1953*, vol. 8, Carbondale, IL: Southern Illinois University Press and InteLex Electronic Edition. [First published as Chapter 2 in *The Educational Frontier*, W. H. Kilpatrick (ed.), New York, London: Century Co., 1933, pp. 32–72.]

Dunne, Anthony and Raby, Fiona (2013), *Speculative Everything: Design, Fiction, and Social Dreaming*, Cambridge, MA: MIT Press.

Ellyard, Peter (2001), *Planning for Thrival and Thrivability in a Planetist 21st Century*, Perth, WA: Keynote Address, WA Technology and Industry Council.

——— (2002), *Leadership for 21st Century Thrival and Thrivability*, Australia: Saxton Business Speakers Bureau.

Fesmire, Steven (2003), *John Dewey and Moral Imagination: Pragmatism in Ethics*, Bloomington, IN: Indiana University Press.

Fiore, Quentin and McLuhan, Marshall ([1967] 2001), *The Medium is The Massage*, Corte Madera, CA: Gingko Press.

Floridi, Luciano (2010a), *Information: A Very Short Introduction*, Oxford: Oxford University Press.

——— (ed.) (2010b), *The Cambridge Handbook of Information and Computer Ethics*, Cambridge: Cambridge University Press.

——— (2013), *The Ethics of Information*, Oxford: Oxford University Press.

Food Farmer Earth ([2010] 2020), 'Alan Kapuler – Open Source Public Domain Plant Breeding', YouTube, https://youtu.be/lOtL-nG-vd4. Accessed 14 February 2023.

Garcia Topete, Alex (2019), *Transdisciplinary Intelligence: Training Hybrids and Amphibians*, Dallas: University of Texas at Dallas, 10 July, 10.13140/RG.2.2.30915.96805/1. [Findings are part of the HERMES (Highly Effective Research Merging Epistemic Systems) Project.]

Garcia Topete, Alex Josue (2021), 'Transdisciplinary intelligence: Linking factors of knowledge production and the cultural dimensions of industries', Ph.D. dissertation, Dallas, TX: Arts, Technology and Emerging Communication, The University of Texas at Dallas.

Garcia Topete, Alex, Lilly, Chaz, Nazir, Cassini and Malina, Roger F. (2021), 'ArtSciLab: Experimental publishing and knowledge production in collaborative transdisciplinary practices', in H. S. Rogers, M. K. Hapern, D. Hannah and K. deRidder-Vignone (eds), *Routledge Handbook of Art, Science, and Technology Studies*, London: Routledge, pp. 359–68.

Garrison, James W., Hickman, Larry A. and Ikeda, Daisaku (2014), *Living as Learning: John Dewey in the 21st Century*, Cambridge, MA: Dialogue Path Press.

GBN (Global Business Network) (2010), 'Where We Started', *GBN.com*, https://web.archive.org/web/20100105183523/http://gbn.com/about/started.php. Accessed 13 February 2012.

Ghosh, Arunabh (2019), 'Forecasting', in H. Paul (ed.), *Critical Terms in Futures Studies*, Cham: Springer Nature, pp.127–30.

Goodman, Nelson (1978), *Ways of Worldmaking*, Hassocks, UK: Harvester Press, p. 6.

Gordon, W. Terrence (2003), 'Introduction to the critical edition', in McLuhan, Marshall ([1964] 2003), *Understanding Media: The Extensions of Man, Critical Edition* (W. T. Gordon, ed.), Corte Madera, CA: Gingko Press, pp. xi–xxi.

——— (2010), 'Background, context, definitions, and ... stumbling blocks', in *McLuhan: A Guide for the Perplexed*, New York: Continuum, pp. 1–28.

Green, Sara (2022), 'Philosophy of systems and synthetic biology', in E. N. Zalta (ed.), *The Stanford Encyclopedia of Philosophy*, Summer, Stanford, CA: Metaphysics Lab, Stanford University, https://plato.stanford.edu/archives/sum2022/entries/systems-synthetic-biology/. Accessed 14 February 2023.

Hamington, Maurice (2010), 'Care ethics, John Dewey's "dramatic rehearsal," and moral education', *Philosophy of Education Archive*, pp.121–28.

Hansen, Mark B. N. (2015), *Feed-forward: On the Future of Twenty-First-Century Media*, Chicago: University of Chicago Press.

Hayles, N. Katherine (1999), *How We Became Posthuman: Virtual Bodies in Cybernetics, Literature and Informatics*, Chicago: University of Chicago Press.

——— (2021), 'Unthought', in N. B. Thylstrup, et al. (eds), *Uncertain Archives: Critical Keywords for Big Data*, Cambridge, MA: MIT Press, pp. 541–50.

Hegde, Gurudatt M., Aditya, Sanghmitra, Wangdi, Dechen and Chetri, Bimal Kumar (2022), 'Mycoremediation: A natural solution for unnatural problems', in V. R. Rajpal, I. Singh and S. S. Navi (eds), *Fungal Diversity, Ecology and Control Management*, Singapore: Springer Nature Singapore, pp. 363–86.

Herasymowych, Marilyn and Senko, Henry (2018), '"Positive" systems archetypes', *Systems Thinker*, https://thesystemsthinker.com/positive-systems-archetypes/. Accessed 14 February 2023.

Hergenrader, Trent and Slota, Stephen (2023), 'The Worldbuilding Workshop: Teaching Critical Thinking Through World Modeling, Simulations, and Play', *CriticalWorldbuilding.com,* Cambridge, MA: MIT

Hershman Leeson, Lynn (dir.) (2022), *Logic Paralyzes the Heart*, USA: Hotwire Productions.

Höftberger, Oliver (2016), *Report on the AMADEOS Workshop on Emergence in Cyber-Physical Systems-of-Systems (CPSoS)*, Forschungsbereich Cyber-Physical Systems, http://hdl.handle.net/20.500.12708/38829. Accessed 14 February 2023.

Hope, Janet (2008), *Biobazaar: The Open Source Revolution and Biotechnology*, Cambridge, MA: Harvard University Press.

Horn, Eva (2018), *The Future as Catastrophe: Imagining Disaster in the Modern Age* (trans. V. Pakis), New York: Columbia University Press.

Hu, Min, Liem, William and Ballard, Ellis (2021), *System Archetypes. Methods Brief Series 1.07: Systems Thinking Foundations*, St. Louis, MO: Social System Design Lab.

Huxley, Thomas H. (1893), *Evolution and Ethics*, New York: Macmillan and Co.

IBM (International Business Machines) (n.d.), 'What is social media analytics?', *IBM*, https://www.ibm.com/topics/social-media-analytics. Accessed 14 February 2023.

Inayatullah, Sohail (2019), 'Futurology', in H. Paul (ed.), *Critical Terms in Futures Studies*, Cham: Springer Nature, pp. 139–43.

Inayatullah, Sohail and Sweeney, John A. (2021), 'From strategic to transformative foresight: Using space to transform time', *World Futures Review,* 13:1, pp. 27–33.

Ingwersen, Moritz Andree (2022), 'Critical future studies', North American Literature with a Focus on Future Studies, Technische Universität Dresden (TU Dresden), 29 June, https://tu-dresden.de/gsw/slk/anglistik_amerikanistik/na-literatur/die-professur/profile-critical-future-studies. Accessed 14 February 2023.

Jain, Anil K., Duin, Robert P. W. and Mao, Jianchang (2000), 'Statistical pattern recognition: A review', *IEEE (Institute of Electrical and Electronics Engineers) Transactions on Pattern Analysis and Machine Intelligence,* 22:1, pp. 4–37.

James, William ([1910] 1987), 'Percept and concept—The import of concepts', in *Some Problems of Philosophy,* in *William James: Writings 1902–1910*, Canada: Penguin Books Canada Ltd., pp. 1007–1020.

Johnson, Brian David (2011), *Science Fiction Prototyping: Designing the Future with Science Fiction*, Williston, VT: Morgan & Claypool Publishers.

Johnson, Mark L. (2023), 'From life to mind', in J. Swartz and J. Wasko (eds), *LIFE: A Transdisciplinary Inquiry*, Bristol: Intellect Ltd., pp. 185–208.

Johnson, Mark L. and Schulkin, Jay (2023), *Mind in Nature: John Dewey, Cognitive Science, and a Naturalistic Philosophy for Living*, Cambridge, MA: MIT Press.

Johung, Jennifer (2019), *Vital Forms: Biological Art, Architecture, and the Dependencies of Life*, Minneapolis, MN: University of Minnesota Press.

Jones, Caroline A., Bell, Natalie and Nimrod, Selby (eds) (2022), *Symbionts: Contemporary Artists and the Biosphere*, Cambridge, MA: MIT Press.

Kirby, Amy E., Walters, Maroya Spalding, Jennings, Wiley C., Fugitt, Rebecca, LaCross, Nathan, Mattioli, Mia, Marsh, Zachary A., Roberts, Virginia A., Mercante, Jeffrey W., Yoder, Jonathan and Hill, Vincent R. (2021), 'Using wastewater surveillance data to support the COVID-19 response—United States, 2020–2021', *Morbidity and Mortality Weekly Report,* 70:36, p. 1242.

Koopman, Colin K. (2010), 'How does internetworking work? (Talk at Metaphi.)', *Requiem for Certainty*, 5 October, https://cwkoopman.wordpress.com/2010/05/10/how-does-internetworking-work-talk-at-metaphi/. Accessed 14 February 2023. [First presented at 'Digital Humanities: Real-Virtual Dialogue', University of Oregon Lawrence Hall 177, 10 May 2010.]

——— (2018), 'Infopolitics, biopolitics, anatomopolitics: Toward a genealogy of the power of data', *Graduate Faculty Philosophy Journal,* 39:1, pp. 103–28.

——— (2019), *How We Became Our Data: A Genealogy of the Informational Person*, Chicago: University of Chicago Press.

——— (2020), 'Coding the self: The infopolitics and biopolitics of genetic sciences', *Hastings Center Report,* 50, pp. S6–S14.

Kreiman, Gabriel (2016), 'On biologically plausible mechanisms for pattern completion', MIT CBMM (Center for Minds, Brains and Machines), YouTube, 8 November, https://youtu.be/BLx_iHVYLiA. Accessed14 February 2023.

Krippendorff, Klaus (2007), 'The cybernetics of design and the design of cybernetics', *Kybernetics,* 36:9&10, pp. 1381–92.

——— (2018), 'Escaping entrapments in uni-verses', *What is Universe*?, University of Oregon Portland, Oregon, 21 April.

——— (2021a), 'ASC Speaker Series #3 - Towards a Critical Cybernetics by Klaus Krippendorff', *American Society for Cybernetics* (ASC), 28 February, https://youtu.be/JGkeoIAFTgY. Accessed 28 February 2023.

——— (2021b), ' ISSS 2021 Paper session: In Search of a Critical Cybernetics (Part 1)', *International Society for the Systems Sciences* (ISSS), 10 July, https://vimeo.com/585868761/ee13ba771f. Accessed 5 February 2023.

——— (2021c), 'From uncritical design to critical examinations of its systemic consequences', in *Proceedings of Relating Systems Thinking and Design* (RSD10) *2021 Symposium,* Delft, The Netherlands, 2–6 November.

Leach, Edmund ([1967] 1968), 'A Runaway World', The Reith Lectures, British Broadcasting Company (BBC), London, https://monoskop.org/images/e/e9/Leach_Edmund_A_Runaway_World_1968.pdf. Accessed 14 February 2023.

Leopold, Aldo ([1949] 1989), *A Sand County Almanac, and Sketches Here and There*, Oxford: Oxford University Press.

Lewis, Jason Edward (ed.) (2020), *Indigenous Protocol and Artificial Intelligence Position Paper*, Honolulu, HI: The Initiative for Indigenous Futures and the Canadian Institute for Advanced Research (CIFAR), https://www.indigenous-ai.net/position-paper/. Accessed 14 February 2023.

Liddell, Henry George and Scott, Robert (1940), 'μετά', in *A Greek-English Lexicon* (rev. and augmented by H. S. Jones and R. McKenzie), Oxford: Clarendon Press, http://www.perseus.tufts.edu/hopper/text?doc=Perseus:text:1999.04.0057:entry=meta/. Accessed 1 April 2023.

Liu, Kathy Y., Gould, Rebecca L., Coulson, Mark C., Ward, Emma V. and Howard, Robert J. (2016), 'Tests of pattern separation and pattern completion in humans—A systematic review', *Hippocampus,* 26:6, pp. 705–17.

Lucas, George (2018), interviewed by James Cameron, in Randall Frakes, Brooks Peck, Sidney Perkowitz, Matt Singer, Gary K. Wolfe and Lisa Yaszek (eds), *James Cameron's Story of Science Fiction*, San Rafael, CA: Insight Editions, pp. 91–119.

Ludwig, David and Weiskopf, Daniel A. (2019), 'Ethnoontology: Ways of world-building across cultures', *Philosophy Compass* 14:9, e12621, pp. 1–11.

Macauley, Nayo (2021), 'Alternative response to mental health-related 911 calls', PCOM Capstone Projects, Master of Science in Biomedical Sciences, Public and Population Health Leadership Concentration. Philadelphia, PA: Philadelphia College of Osteopathic Medicine.

Malina, Roger, Strohecker, Carol, LaFayette, Carol and Ione, Amy (On behalf of SEAD Network Contributors) (2015), *Steps to an Ecology of Networked Knowledge and Innovation: Enabling New Forms of Collaboration among Sciences, Engineering, Arts, and Design*, Cambridge, MA: MIT Press.

Mandelbrot, Benoit B. (2004), *Fractals and Chaos: The Mandelbrot Set and Beyond*, New York: Springer.

Maurits, Peter J. (2019), 'Futurism', in H. Paul (ed.), *Critical Terms in Futures Studies*, Cham: *Springer Nature*, pp. 131–37.

Maturana, Humberto R. (1997), '*Metadesign*', Santiago: Instituto de Terapia Cognitiva (INTECO).

Maturana, Humberto R. and Poerksen, Bernhard ([2004] 2011), *From Being to Doing: The Origins of the Biology of Cognition*, Heidelberg: Carl-Auer-Systeme.

Matsuda, Paul Kei (2021), 'Forum: On the disciplinary identity of second language writing', *Journal of Second Language Writing* 52:100796.

Matsuda, Paul Kei, Canagarajah, A. Suresh, Harklau, Linda, Hyland, Ken and Warschauer, Mark (2003), 'Changing currents in second language writing research: A colloquium', *Journal of Second Language Writing*, 12:2, pp. 151–79.

McLuhan, Marshall ([1964] 2003), 'Automation: Learning a living,' in *Understanding Media: The Extensions of Man, Critical Edition* (W. T. Gordon, ed.), Corte Madera, CA: Gingko Press, pp. 459–73.

——— ([1979] 1996) 'York University Lecture, 1979', in M. McLuhan, T. Wolfe, S. McLuhan and M. Productions (eds), *The Video McLuhan*, Toronto: McLuhan Productions.

——— (2003), *Understanding Me: Lectures and Interviews*, S. McLuhan and D. Staines (eds), Cambridge, MA: MIT Press.

Mead, Carver (1990), 'Neuromorphic electronic systems,' *Proceedings of the IEEE* 78:10, pp. 1629–163, https://doi.org/10.1109/5.58356. Accessed 14 February 2023.

Mead, Carver A. and Mahowald, Misha A. (1988), 'A silicon model of early visual processing', *Neural Networks* 1:1, pp. 91–97.

Meadows, Donella H. (2008), 'System traps ... and opportunities', in *Thinking in Systems: A Primer*, White River Junction, VT: Chelsea Green Publishing, pp. 111–41.

Mejía, G. Mauricio, Henriksen, Danah, Xie, Yumeng, García-Topete, Alex, Malina, Roger F. and Jung, Kendon (2022), 'From researching to making futures: A design mindset for transdisciplinary collaboration,' *Interdisciplinary Science Reviews*, pp. 1–32.

melodysheep (2019), 'Timelapse of the Future: A Journey to the End of Time', YouTube, 20 March, https://youtu.be/uD4izuDMUQA. Accessed 14 February 2023.

Mendell, Madeleine, Hogan, Mél and Verhoeven, Deb (2022), 'Matters (and metaphors) of life and death: How DNA storage doubles back on its promise to the world', *The Canadian Geographer/Le Géographe canadien,* 66:1, pp. 37–47.

MIT (2016), *Convergence: The Future of Health*, Washington, DC: MIT Washington Office, June, http://www.convergencerevolution.net/s/Convergence-The-Future-of-Health-2016-Report-55pf.pdf. Accessed 14 February 2023.

Muench, Stefan, Stoermer, Eckhard, Jensen, Kathrine, Asikainen, Tommie, Salvi, Maurizio and Scapolo, Fabiana (2022), *Towards a Green and Digital Future, EUR 31075 EN*, Luxembourg: Publications Office of the European Union, https://publications.jrc.ec.europa.eu/repository/bitstream/JRC129319/kjna31075enn_1.pdf. Accessed 14 February 2023.

Munnecke, Tom (2021), 'Retroinformation ideas', *ECLECTICA*, 21 October, http://www.munnecke.com/2021/10/retroinformation-ideas.html. Accessed 14 February 2023.

Myers, William ([2012] 2018), *Bio Design: Nature + Science + Creativity, Revised and Expanded Edition*, New York: Museum of Modern Art.

Myles, Eileen (2021), 'Why Lynn Hershman Leeson is always ahead of her time', *Aperture*, 23 June, https://aperture.org/editorial/why-lynn-hershman-leeson-is-always-ahead-of-her-time/. Accessed 14 February 2023.

——— (2019), *Fostering the Culture of Convergence in Research: Proceedings of a Workshop*, Washington, DC: The National Academies Press, https://nap.nationalacademies.org/catalog/25271/fostering-the-culture-of-convergence-in-research-proceedings-of-a. Accessed 14 February 2023.

Nature Electronics (2022), 'Artificial neurons emulate biological counterparts to enable synergetic operation', *Nature Electronics* 5, pp. 721–22, https://doi.org/10.1038/s41928-022-00862-3. Accessed 1 February 2023.

Noble, Denis (2006), *The Music of Life: Biology Beyond Genes*, Oxford: Oxford University Press.

NRC (National Research Council) (2009), *A New Biology for the 21st Century*, Washington, DC: The National Academies Press, https://nap.nationalacademies.org/catalog/12764/a-new-biology-for-the-21st-century. Accessed 14 February 2023.

NSF (National Science Foundation) (n.d.), *Futures of the Scientific Imagination*, Alexandria, VA: National Science Foundation, https://www.nsf.gov/news/special_reports/futures/. Accessed 14 February 2023.

——— (n.d.), *Learn About Convergence Research*, Alexandria, VA: National Science Foundation, https://beta.nsf.gov/funding/learn/research-types/learn-about-convergence-research. Accessed 14 February 2023.

——— (2017–2022), *NSF's 10 Big Ideas: Understanding the Rules of Life*, Alexandria, VA: National Science Foundation, https://www.nsf.gov/news/special_reports/big_ideas/life.jsp. Accessed 14 February 2023.

——— (2022), *Leading the World in Discovery and Innovation, STEM Talent Development and the Delivery of Benefits from Research – NSF Strategic Plan for Fiscal Years (FY) 2022–2026*, Alexandria, VA: National Science Foundation.

——— (n.d.), *The NSF 2026: Idea Machine*, Alexandria, VA: National Science Foundation, https://www.nsf.gov/news/special_reports/nsf2026ideamachine/index.jsp. Accessed 14 February 2023.

NSFC (National Natural Science Foundation of China) (n.d.), *Fundamental Research* [open access journal], China: Center for Science Communication and Achievement Transformations, https://www.nsfc.gov.cn/csc/20345/FR/index.html. Accessed 15 August 2023.

Obbie, Mark (2021), 'The power of asking, "Is there a better way?"', *The Whole Story: Ideas and Dispatches from the Solutions Journalism Network*, 21 June, https://thewholestory.solutionsjournalism.org/the-power-of-asking-is-there-a-better-way-c8483ab4c9be. Accessed 14 February 2023.

Ohlenschlager, Karin, Weibel, Peter and Weidinger, Alfred (eds) (2023), *Christa Sommerer & Laurent Mignonneau: The Artwork as a Living System 1992–2022*, Cambridge, MA: MIT Press.

Oppermann, Serpil (2018), 'Storied Matter', in R. Braidotti and M. Hlavajova (eds), *Posthuman Glossary*, New York: Bloomsbury Publishing, pp. 411–14.

Parafiniuk-Talesnick, Tatiana (2021), 'CAHOOTS launches mobile crisis response 101 course as cities across the country look to start similar programs', *KGW*, 15 December, https://www.kgw.com/article/news/local/cahoots-mobile-crisis-response-101-course/283-5b6564be-2037-4969-94e0-fd8dd882918b. Accessed 14 February 2023.

Parihar, Arpana, Pandita, Vasundhara, Kumar, Avinash, Parihar, Dipesh Singh, Puranik, Nidhi, Bajpai, Tapas and Khan, Raju (2021), '3D printing: Advancement in biogenerative engineering to combat shortage of organs and bioapplicable materials', *Regenerative Engineering and Translational Medicine*, 8:19, July, pp. 1–27.

Parikka, Jussi (2022), 'Counter-futuring', *COUNTER-N*, 16 January, pp. 1–10, https://edoc.hu-berlin.de/bitstream/handle/18452/25126/counter-futuring-FINAL.pdf. Accessed 14 February 2023.

Pasquinelli, Matteo (2018), 'Metadata society', in R. Braidotti and M. Hlavajova (eds), *Posthuman Glossary*, New York: Bloomsbury Publishing, pp. 253–56.

Paul, Heike (ed.) (2019), *Critical Terms in Futures Studies*, Cham: Springer Nature.

Peters, Michael A. (2020), 'Afterword: Critical philosophy of technological convergence: Education and the nano-bio-info-cogno paradigm', in M. Stocchetti (ed.), *The Digital Age and Its Discontents: Critical Reflections in Education*, Helsinki: Helsinki University Press, https://library.oapen.org/handle/20.500.12657/41638. Accessed 14 February 2023.

Peters, Michael A., Jandrić, Petar and Hayes, Sarah (2021), 'Biodigital philosophy, technological convergence and postdigital knowledge ecologies', *Postdigital Science and Education,* 3, pp. 370–88, https://doi.org/10.1007/s42438-020-00211-7. Accessed 14 February 2023.

Poe, Edgar Allan ([1841] 1920), *A Descent into the Maelström, A Tale* (illus. M. Roux), Paris: Devambez Publishers, https://archive.org/details/descentintomaels00poee/page/n11/mode/2up. Accessed 14 February 2023.

Rasmussen, Leif, Stover, Piper, Bedau, Mark, Beinhocker, Eric, Crokett, Molley, Farmer, Doyne, Tainter, Joseph and Rasmussen, Steen (2018), 'Santa Fe Institute Working Group "Envisioning new modes of cultural and technological change" 31 July to 2 August 2017', 30 November, http://sfi-edu.s3.amazonaws.com/sfi-edu/production/uploads/resource_link_files/SFI_Working_Grp_Summary_V_November_30_2018_db5975.pdf. Accessed 14 February 2023.

Ratcliffe, Susan (ed.) (2011), *Oxford Treasury of Sayings and Quotations*, Oxford: Oxford University Press.

Reggio, Godfrey (dir.) (2012), *The Quatsi Trilogy* [Blu-ray], New York: Criterion Collection. [The special edition includes: *Koyaanisqatsi: Life Out of Balance* (1983), *Powaqqatsi: Life in Transformation* (1988), and *Naqoyqatsi: Life as War* (2002), as well as *Anima Mundi* (1992).]

Reggio, Godfrey and Kane, Jon (dirs.) (2023), *Once Within A Time*, USA: opticnerve™.

Reyes-Santos, Alaí (2022), 'PNW Just Futures Institute for Racial and Climate Justice: Vision, Mission, and Justice Principles', *Pacific Northwest Just Futures Institute, University of Oregon*, January, https://jfi.uoregon.edu/files/2022/01/JFI-Research-and-Data-Justice-Principles-1.pdf. Accessed 5 February 2023.

Richards, Ivor Armstrong (1968), 'The secret of "feedforward"', *Saturday Review*, February, pp. 14–17.

Roach, Margaret (2022), 'Where adventurous gardeners buy their seeds', *New York Times*, 13 January.

Robinson, Kim Stanley (2020), *The Ministry for the Future*, London: Orbit Books.

Robles, Kelly E., Roberts, Michele, Viengkham, Catherine, Smith, Julian H., Rowland, Conor, Moslehi, Saba, Stadlober, Sabrina, Lesjak, Anastasija, Lesjak, Martin, Taylor, Richard P., Spehar, Branka and Sereno, Margaret E. (2021), 'Aesthetics and psychological effects of fractal based design', *Frontiers in Psychology*, 12, Article 699962, April, pp. 1–21.

Roco, Mihail C. (2020), 'Principles of convergence in nature and society and their application: From nanoscale, digits, and logic steps to global progress', *Journal of Nanoparticle Research,* 22:11, pp. 1–27, https://link.springer.com/content/pdf/10.1007/s11051-020-05032-0.pdf. Accessed 14 February 2023.

Roco, Mihail C. and Bainbridge, William Sims (2002), *Converging Technologies for Improving Human Performance: Nanotechnology, Biotechnology, Information Technology and Cognitive*, Dordrecht: Springer Science+Business Media.

Roco, Mihail C., Bainbridge, William Sims, Tonn, Bruce and Whitesides, George (eds) (2013), *Converging Knowledge, Technology and Society: Beyond Convergence of Nano-Bio-Info-Cognitive*

Technologies, WTEC (World Technology Evaluation Center, Inc.), Dordrecht, Heidelberg: Springer, https://ndc-ipr.org/media/ndc_old/documents/NBIC2-FinalReport-WEB.pdf. Accessed 14 February 2023.

Rogers, Hannah Star (2022), 'Preface: The rise of art-science', in H. S. Rogers, M. K. Hapern, D. Hannah and K. deRidder-Vignone (eds), *Routledge Handbook of Art, Science, and Technology Studies*, London: Routledge.

Roumani, Nadia (2022), 'Integrative design: A practice to tackle complex challenges', *The Stanford d.school: Shared Ideas from The Hasso Plattner Institute of Design at Stanford University, Medium.com*, 29 March, https://medium.com/stanford-d-school/integrative-design-a-practice-to-tackle-complex-challenges-a07fc3e636c4. Accessed 14 February 2023.

RSA (Royal Society for Arts, Manufactures and Commerce) (2022), *Design for Life: To Regenerate People, Place and Planet: A New Mission for the RSA*, London: RSA, https://www.thersa.org/globalassets/_foundation/new-site-blocks-and-images/approach/rsa_design-for-life-paper.pdf. Accessed 14 February 2023.

RSTA (Royal Society of Te Apārangi) (2022), *Highlights Te Tau 2021*, Wellington: Royal Society Te Apārangi, https://www.royalsociety.org.nz/who-we-are/our-reports/our-annual-reviews/highlights-te-tau-2021/. Accessed 14 February 2023.

Savat, David (2009), 'Deleuze's objectile: From discipline to modulation', in M. Poster and D. Savat (eds), *Deleuze and New Technology*, Edinburgh: Edinburgh University Press, pp. 45–62.

Scearce, Diana, Fulton, Katherine, and the Global Business Network Community (2004), *What If?: The Art of Scenario Thinking for Nonprofits*, Emeryville, CA: Global Business Network, a member of the Monitor Group.

Schoenberg, William and Swartz, Jeremy (2023), 'System dynamics, machine learning and structural validation', in J. Swartz and J. Wasko (eds), *LIFE: A Transdisciplinary Inquiry*, Bristol: Intellect, pp. 289–308.

Schuman, Catherine D., Kulkarni, Shruti R., Parsa, Maryam, Mitchell, J. Parker, Date, Prasanna and Kay, Bill (2022), 'Opportunities for neuromorphic computing algorithms and applications', *Nature Computational Science* 2:1, pp. 10–19.

Schwab, Eileen C. and Nusbaum, Howard C. (eds) (2013a), *Pattern Recognition by Humans and Machines: Volume 1: Speech Perception*, Cambridge, MA: Academic Press.

—— (2013b) *Pattern Recognition by Humans and Machines: Volume 2: Visual Perception*. Cambridge, MA: Academic Press.

Schwartz, Peter and Becker, Miles B. ([2001] 2005), 'Future studies', in D. Cuff and A. S. Goudie (eds), *Encyclopedia of Global Change*, Oxford: Oxford University Press, https://www.oxfordreference.com/view/10.1093/acref/9780195108255.001.0001/acref-9780195108255-e-0172. Accessed 14 February 2023.

Schwartzberg, Louie (2021), 'How mushroom time-lapses are filmed', *WIRED* [magazine] on YouTube, https://youtu.be/5yq0_mqN97s. Accessed 14 February 2023.

Scott, John (2017), *Social Network Analysis*, 4th ed., London: SAGE Publications.

Scott, Luke (dir.) (2012), 'The Peter Weyland Files: TED Conference, 2023', in R. Scott (dir.), *Prometheus* [Blu-ray], USA: 20th Century Fox Home Entertainment, https://youtu.be/E4SSU29Arj0. Accessed 14 February 2023.

Sellberg, My M., Cockburn, Jessica, Holden, Petra B. and Lam, David P. M. (2021), 'Towards a caring transdisciplinary research practice: Navigating science, society and self', *Ecosystems and People* 17:1, pp. 292–05.

Selsky, Andrew (2022), 'Oregon will be the 1st state to receive federal support for mental health crisis teams', *KGW*, https://www.kgw.com/article/news/health/us-support-mental-health-crisis-teams/283-1ebd97f2-340e-4473-8b6e-e2357c57b0a7. Accessed 14 February 2023.

Semenza, Jan C. (2003), 'The intersection of urban planning, art and public health: The sunny-side piazza', *American Journal of Public Health,* 93:9, pp. 1439–41.

——— (2021), 'Lateral public health: Advancing systemic resilience to climate change', *The Lancet Regional Health–Europe*, 9, p. 100231.

Semenza, Jan C., March, Tanya L. and Bontempo, Brian D. (2007), 'Community-initiated urban development: An ecological intervention', *Journal of Urban Health, 84*:1, pp. 8–20.

Shanks, Michael (2016), 'Design foresight', *Michael Shanks ~ Archaeologist*, 7 July, https://web.archive.org/web/20220611045431/https://mshanks.com/foresight/. Accessed 14 February 2023.

Simon, Herbert A. (1969 [2019]), *The Sciences of the Artificial,* reissue of the 3rd ed., Cambridge, MA: MIT Press.

SINAC (Sistema Nacional de Áreas de Conservación) (n.d.), *National System of Conservation Areas*, San José, Costa Rica: MINAET – Ministerio de Ambiente, Energía y Telecomunicaciones, https://www.sinac.go.cr/EN-US/conozca/Pages/default.aspx. Accessed 14 February 2023.

Smart, John (2019), 'Appendix 3. Resources – Media and Tool for Better Futures: Foresight Glossary', Foresight University website, https://www.foresightguide.com/foresight-glossary/. Accessed 14 February 2023.

Snow, Charles Percy and Snow, Baron (1959), *The Two Cultures and the Scientific Revolution*, vol. 960, Cambridge: Cambridge University Press.

SPI (Social Progress Imperative) (2022), '2022 Social Progress Index®: Executive Summary', SocialProgress.org, 26 September, https://www.socialprogress.org/static/8a62f3f612c8d40b-09b3103a70bdacab/2022%20Social%20Progress%20Index%20Executive%20Summary_4.pdf. Accessed 13 February 2023.

Stokols, Daniel, Misra, Shalini and Lotrecchiano, Gaetano R. (2020), 'Emerging Directions in the "Science of Team Science": An Interview with Professor Daniel Stokols', in G. R. Lotrecchiano and S. Misra (eds), *Communication in Transdisciplinary Teams*, Santa Rosa, CA: Informing Science Press, pp. 309–28.

Stollberg-Rilinger, Barbara (2018), 'Decision', in H. Paul (ed.), *Critical Terms in Futures Studies*, Cham: Springer Nature, pp. 73–78.

Svenska Akademien (2022), *Ordböcker*, Stockholm, Sverige: Svenska Akademien, https://svenska.se/tre/?sok=str%C3%B6m. Accessed 5 February 2023.

Swartz, Jeremy (2021), 'Metamedia', in J. Swartz and J. Wasko (eds), *MEDIA: A Transdisciplinary Inquiry*, Bristol: Intellect.

Taylor, Richard P., Juliani, Arthur W., Bies, Alexander J., Boydston, Cooper R., Spehar, Branka and Sereno, Margaret E. (2018), 'The implications of fractal fluency for biophilic architecture', *Journal of Biourbanism*, 6, pp. 23–40.

Taylor, Richard P. and York, Richard (2022), 'Fractal shifts and esthetic rifts: Climate change and emotional well-being', *Climatic Change*, 173:1, pp. 1–9.

Thompson, D'arcy Wentworth ([1961] 1992), *On Growth and Form*, Cambridge: Cambridge University Press.

Toffler, Alvin (1970), *Future Shock*, New York: Bantam Books.

Turner, Fred (2006), *From Counterculture to Cyberculture: Stewart Brand, the Whole Earth Network, and the Rise of Digital Utopianism*, Chicago: University of Chicago Press.

——— (2009), 'Burning Man at Google: A cultural infrastructure for new media production', *New Media & Society*, 11:1&2, pp. 73–94.

——— (2013), *The Democratic Surround: Multimedia and American Liberalism from World War II to the Psychedelic Sixties*, Chicago: University of Chicago Press.

——— (2016), 'Prototype', in B. Peters (ed.), *Digital Keywords*, Princeton, NJ: Princeton University Press, pp. 256–68.

Urquhart, Christine, Lam, Louisa Mei Chun, Cheuk, Bonnie and Dervin, Brenda L. (2020), 'Sense-making/sensemaking', in P. Moy (ed.), *Oxford Bibliographies Online (Communication)*, Oxford: Oxford University Press, https://www.oxfordbibliographies.com/display/document/obo-9780199756841/obo-9780199756841-0112.xml. Accessed 9 January 2023.

Varela, Francisco J. (1999), *Ethical Know–How: Action, Wisdom, and Cognition*, Stanford, CA: Stanford University Press.

Vermeulen, Timotheus (2018), 'Altergorithm', in R. Braidotti and M. Hlavajova (eds), *Posthuman Glossary*, New York: Bloomsbury Publishing, pp. 29–31.

Vesna, Victoria (2001), 'Toward a third culture: Being in between', *Leonardo*, 34:2, pp. 121–25.

Vishmidt, Marina (ed.) (2023), *Speculation, Documents of Contemporary Art*, London and Cambridge, MA: Whitechapel Gallery and MIT Press.

Weibel, Peter (ed.) (2023), *BioMedia: The Age of Media with Life-like Behavior*, Leipzig: ZKM | Center for Art and Media Karlsruhe and Spector Books.

WFC (World Future Council) (n.d.), *Our Publications*, Hamburg: World Future Council, https://www.worldfuturecouncil.org/media-room/publications/. Accessed 14 February 2023.

Wikipedia Contributors (2022), 'Global Business Network', *Wikipedia, The Free Encyclopedia*, https://en.wikipedia.org/w/index.php?title=Global_Business_Network&oldid=1111758747. Accessed 25 September 2022.

Wilson, Edward O. (1984), *Biophila*, Cambridge, MA: Harvard University Press.

Wood, John (2022a), 'From conservation to regeneration' [unpublished draft], *metadesigners*, https://metadesigners.org/Conservation-to-Regeneration. Accessed 14 February 2023.

——— (2022b), 'Writing the paradigm' [a brief overview of *Metadesigning Designing in the Anthropocene*'], *metadesigners*, https://metadesigners.org/Writing-The-Paradigm. Accessed 14 February 2023.

——— (2022c), 'Chris Cook in conversation', *Journal of Writing in Creative Practice*, 15:1, pp. 41–47, https://metadesigners.org/Chris-Cook-Interview-Nondominium. Accessed 14 February 2023.

Wright, Kate (2014), 'Becoming-with', *Environmental Humanities*, 5:1, pp. 277–81.

Youngblood, Gene (1986), 'Metadesign: Toward a postmodernism of reconstruction', in *Ars Electronica Catalog*, Linz: Linzer Veranstaltungsgesellschaft.

Zwick, Martin (2023), *Elements and Relations: Aspects of a Scientific Metaphysics*, Cham: Springer Nature.

Appendix
Exhibition • Experience • Nature

Exhibition

FIGURE A.1: *LIFEWORLDS* exhibition poster art by Wade Larsen.

As noted in the Appendix in *MEDIA*, the first volume of the trilogy: 'By recognizing patterns of life in an ecology of inquiry, media are environments and communication is art'. The *LIFEWORLDS* exhibition continued this focus on integrated views of communication, where lives are ecosystems and culture is experience — planting seeds for further investigations into systems, as well as environmental and ecological approaches. Growing from that rich soil, different spaces, places and traces were imagined – those that we inhabit during our lives and the nature of the worlds we understand and work, envision and create.

There are many worlds of life, or *life-worlds* – for example, analogue, living, breathing entities; digital codes, programs and interactive visualizations; as well as transactions across metabolisms. In this sense, worlds are constantly emerging, novelizing, nourishing and necessitating questions, such as what constitutes life-worlds?

The concept of *life-world* or *Lebenswelt* by Edmund Gustav Albrecht Husserl refers to the sum of immediate or everyday experiences, interactions and more – the world of lived experience (*OED Online* 2023). Husserl's view 'has been used as a way of integrating science with wider forms of human activity' (Sokolowski 1999: 407; also see Husserl 1970) and also involves our *surroundings*, or in German *Umgebung*. The life-world resembles Jakob Johann Freiherr von Uexküll's *Umwelt* (Tønnessen 2009: 58; cf. Husserl 1954), and according to Maurice Merleau-Ponty, is 'an intermediary reality between the world such as it exists for an absolute observer and a purely subjective domain' (2003: 167). Therefore, *Umwelt* or environment can be understood as 'a model of an apparently existing *phenomenal* world' (Tønnessen 2009: 60, emphasis added). For Uexküll, this represents a mode of knowledge where 'we no longer see where behavior begins and where mind ends' (Merleau-Ponty 2003: 178; also see von Uexküll [1937] 2001).

This volume and exhibition engaged syntheses of living, biological media; semiotic or symbolic, representational media; and synthetic approaches, hybrid media, in other words a biosemiotic account relevant to life-worlds and information theory. '[B]iosemiotics depicts a world of the living in which all living beings are to be regarded as true subjects actively engaged and involved in (and through) their life-worlds' (Tønnessen 2009: 58).

LIFEWORLDS featured art–science installations, continuing engagements of artist–artifact–audience as a process of meaningful discourse. The notion of *embodiment* was also echoed in syntheses and synergies of the senses, or *sensemaking*. This implies modulating our sense system by paying attention, being aware, offering a glimpse of, and/or imagining specific circumstances where seemingly disparate frameworks/concepts can be understood as interwoven.

These and other forms of communication via our senses were explored through the 1) anatomical (body), botanical (plant) and hydrological (water) and the

FIGURE A.2: Overview of the Light Court Commons, University of Oregon Portland.[1]

utilization of various; 2) technologies, platforms (drawings, photography and miniature trees and containers) and, 3) juxtaposition via multisensory installation art (audio-visual and olfactory). (See Figure A.2.)

Overall, the exhibition explored a spectrum of world-views across biocentric, anthropocentric and ecocentric perspectives.[2] Moreover, it continued an inquiry into a wide range of themes with material, historical and environmental ramifications.[3]

The artists and selected works are presented with brief descriptions.

Leonardo da Vinci has been discussed in several chapters, especially as an early example of a systems philosopher. His work was featured through a selection of prints of his drawings of anatomical and botanical subjects and water vortices. 'Nature as a whole was alive for Leonardo, and he saw similar patterns and processes in both the macrocosm of the living earth and the microcosm of an individual organism' (Capra 2013: 281). 'He portrays nature's forms in ceaseless movement and transformation, recognizing that living forms are continually shaped and transformed by underlying processes' (68). 'Leonardo's ultimate goal – in his science as well as in his art – was always to understand the nature of life' (xii), 'analyzed today in the separate sciences of fluid dynamics, geology, botany, and ecology' (129; also see Capra 2008, 2017).

FIGURE A.3: Leonardo da Vinci, *The Head of Leda*, *c.*1505–08. Black chalk, pen and ink. 20.0 × 16.2 cm (sheet of paper). Royal Collection Trust/© Her Majesty Queen Elizabeth II.

Leonardo's work was discussed in Chapter 18 as an example of pattern recognition, which was fundamental to his intricate and dynamic representations of humans, nonhumans and the environment. As he wrote: 'La Natura è piena d'infinite ragioni che non furono mai in isperienza' (Ms. I, folio 18r). (Nature is full of infinite causes that never were found in experience.)

One example of the curated drawings was *The Head of Leda* (Figure A.3),[4] which was for his lost painting of *Leda and the Swan* (Clayton 2018). The Leonardo images were provided courtesy of The Royal Collection Trust, with the collaboration of Picture Library Manager Karen Lawson and Digital Imager Daniel Partridge.

Representations of nature through legacy artistic technologies, such as photography, have been a common theme, as exemplified by the Ilfochrome photographs of Gary Tepfer. He has been a professional photographer since 1970, and his subjects have covered local, regional and international locations, including the Pacific Northwest and Southwest, as well as Mongolia in East Asia (see Figure A.4). Many of Gary's images are the result of braving harsh weather and

FIGURE A.4: (above) Gary Tepfer, *Harsalaa, Tsagaan Gol, Mongolia*, circa 2000; and (below) *Cliffrose, Four Corners*, 2000. Photos by Gary Tepfer.

perilous travel conditions. Deep time and geological thinking are always at play in Tepfer's work,[5] as well as in the work of his wife, Esther Jacobson-Tepfer, author of *The Hunter, the Stage, and the Mother of Animals: Image, Monument, and Landscape in Ancient North Asia* and *The Anatomy of Deep Time*. Their work – collectively and individually – presents inspiring views of the inextricability of culture and nature.

Carla Bengtson's multisensual art piece, *Euglossa* sp., was created with pollination biologist Peter Wetherwax and orchid bee researcher Santiago Ramirez and features a 'hand blown bottle containing a reproduction of the perfume male orchid bees display to attract a mate'[6] (see Figure A.5). The title references orchid bees' genus. Bengtson (2018: n.pag.) explains:

> Male euglossine, or orchid bees, found in central and South America, fly long distances, up to 30 miles, and visit multiple odor sources, to create an attractive scent that is a blend of 30-40 different scent compounds. Once the bees have accumulated sufficient quantities of their precious scent compounds, they perform an elaborate courtship display in which they expose and disperse their scents for females, who mate with males displaying the most attractive perfume blends. [...] As each orchid species has a single, species-specific pollinator, these displaced compounds suggest

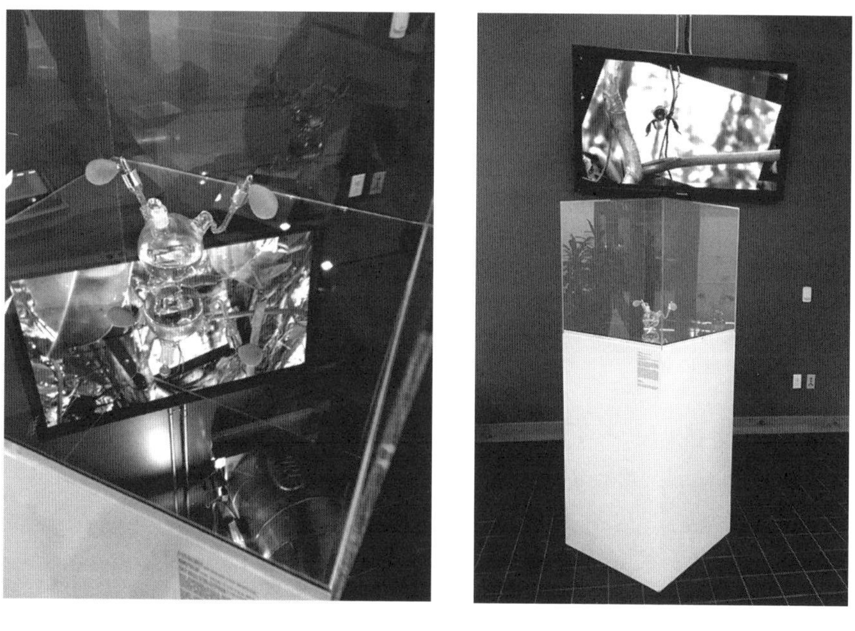

FIGURE A.5: (left) Carla Bengtson, Peter Wetherwax and Santiago Ramirez, *Euglossa* sp., 2015; (right) *Euglossa* sp. with Carla Bengtson's *Mimetic Excess* (video), 2014.

> profound consequences for over 200 described species of orchids and bees, as well as yet to be described species that are in danger of going extinct even as they are being discovered.

Euglossa sp. brings together glass, metal, cloth and scents through interdisciplinary collaboration. The perfume bottle was created in collaboration with Sky Glass. The piece was juxtaposed with Bengtson's *Mimetic Excess* (2014), '[a] looped video of an orchid bee performing an aggressive display towards its own image reflected in the lens of a camera was played on an iPad hung in a fig tree'.[7] Thus, the installation can be considered ecomedia that works across auditory, olfactory, tactile and visual experiences and boundaries meshing materially, digitally, culturally, ecologically and politically.

Various bonsai or tray plants (Figures A.6 and A.7) were presented by Ryan Neil, an artist and bonsai professional, who received his training under several notable bonsai masters (Masahiko Kimura,[8] Ben Oki and Harold Sasaki). Neil is the owner of Bonsai Mirai in St. Helens, OR, and founder of The Artisans Cup, one of the premier showcases for American bonsai.[9] His miniature tree artistry has been compared to the sculptural: 'Instead of using a chisel to shape stone, he holds scissors and wires in his hands to sculpt small, wild pines, junipers and other familiar trees into artistic, natural forms that are bold and nontraditional' (Eastman 2016; PJG 2017).

FIGURE A.6: Ryan Neil, *Hollywood Juniper*. Species: *Juniperus chinensis*. Age: 85 years. Region: California Central Coast.

FIGURE A.7: Ryan Neil, *Bald Cypress*. Species: *Taxodium distichum*. Age: 125 years. Region: Florida Everglades.

Neil not only studied with bonsai masters but has also diligently worked to understand and appreciate unique characteristics of regional American forests – not merely the trees, but the ecosystems and their cultural contexts. Indeed, the bonsai tradition connects with some of the themes that have emerged in this volume, as bonsai mastery and appreciation embody both material and symbolic forms of ecomedia. As noted by bonsai master John Naka, 'It must have philosophy, botany, artistry, human quality behind it to be a bonsai'. Naka further explains the integrative nature of the art form: 'The bonsai is not you working on the tree; you have to have the tree work on you'.[10]

The exhibition was presented in the Light Court Commons '[a]t the intersection of two historic buildings [....] an exhibition space. A place to gather. A place to rest'. *LIFEWORLDS* was curated by Jeremy Swartz in collaboration with the White Box Gallery[11] from 6 April to 6 May 2017 at the University of Oregon (UO) Portland.

Experience

The Experience: *LIFE • LIFEWORLDS • LIFEWORKS* at the Portland Japanese Garden (PJG) brought together culture, art and nature, where biophilia and ecodesign were poignantly enacted in the fusion of natural and built environments. Participants were guided through the space, learning about the grounds, facility and garden maintenance and repair in co-inquiry with Garden curator, Sadafumi Uchiyama and Arlene Schnitzer Curator of Culture, Art & Education, Diane Durston.[12]

Beginning with introductions at the Jordan Schnitzer Japanese Arts Learning center, the group slowly walked through the Nezu Gate, down across the Moon Bridge to learn about the Kashintei Tea House, then back around to contemplate the Strolling Pond Garden, and up to the Flat Garden and Pavilion Gallery. (See Figure A.8.) Participants then had more time to visit and reflect on these locations again, as well as the Natural Garden and the Sand and Stone Garden.

The history and mission of the garden echo themes of life-worlds and the ways in which our lifeworks are simultaneously material, social, ecological and contemplative. 'Designed in 1963, it encompasses 12 acres with eight separate garden styles, and [...] is a place to discard worldly thoughts and concerns and see oneself as a small but integral part of the universe'.[13] During that time period,

> Japanese gardens were founded across the country as a way to build cultural understanding. Needing no translation, an American could experience firsthand Japanese ideals and values, *communicated* simply through nature. [...] Professor Takuma Tono of Tokyo Agricultural University was retained to design the Garden. [...] His intention was to represent different historical developments in Japanese garden architecture and through that *communicate* Japanese culture to create a cultural exchange.
> (PJG 2022, emphases added; see also 2020)

Thus, natural environment and cultural exchange are media of communication that integrate nature and culture non-dichotomously as *natureculture*. It is in this way that the significance of healing gardens again comes to mind as a therapeutic process offered by such settings. As PJG Chief Executive Officer, Steve Bloom has noted, 'We stand as a place where every single person in our community is welcome and our role is to be a place where anyone can come to center, reset, and heal'.[14]

Ecomedia and Ecofair

The themes from all of these experiences continued to be explored through the transdisciplinary environments of the *Reimagining Sustainability* conference and

FIGURE A.8: (center) The group gathers in front of Zig Zag Bridge next to Heavenly Falls. (left) Garden curator, Sadafumi Uchiyama, offers insights into the gardens and their cultures. (right) A stone pebble path leading towards cherry blossoms. Photo by Connie Chandler (BeKnownTrusted).

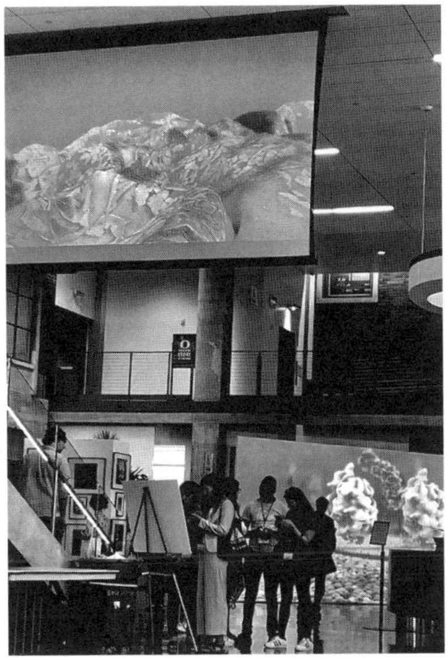

FIGURE A.9: An overview of the Oregon Plaza in the Erb Memorial Union (EMU), UO in Eugene. Photo by John Higgins (ProfHigginsJ).

Ecomedia arts festival. The theme was inspired in part by the United Nations Sustainable Development Agenda and ways of 'revising and remixing theories, practices, and institutions', as well as their relationships 'to openness, transparency, accessibility, and the recomposition of power [...] in an interconnected world'.[15] It was hosted by the International Association for Media and Communication Research (IAMCR), in cooperation with the UO Eugene and the Jordan Schnitzer Museum of Art from 20 June to 24 June 2018.

Ecomedia included artists who reconceptualized media/environments from ecological, political and systems approaches. Utilizing legacy and emerging technologies, techniques and platforms, each in their own way elaborated developing contemporary conditions – through communication/arts-based research – investigating materiality, virtuality and hybridity. (See Figure A.9.) Ecological media, in other words, are relational. Some examples include the distinctions between actual/symbolic, physical/virtual, analogue/digital, nature/culture and territory/map, including parallel discourses related to a reappraised, integrative notion of information noted in this volume.

In a physical sense, energy–matter constitutes part of the materiality of media, as noted in the introduction to this volume. Therefore, if media are environments, then what actually comprises ecological technologies? Are they gardens, forests, both or other? How do they relate to appropriate, responsible, resilient and/or other sustainability-related rhetorics? Are there ways towards dehybridization and disintegration?

Contexts and relationships of, for and about organisms and environments can always be taken under an ecological lens, especially when considering lost, forgotten and/or stolen life-worlds. One example was media artist Violet Ray's looped projection of a galloping, *Buffalo Reborn* (2018).[16] The site-specific video installation was projected onto a selection of cedars in the Old campus Quad South within the UO arboretum, which consists of approximately 4000 trees of more than 500 species on 295 acres.[17] Are ecomedia ecological (meaning environmentally friendly) or simply synthetic forms of media?

Another example was the multimodal performance art of *Tesla: Light, Sound, Color* by Harmonic Laboratory (2018: n.pag.), in Silva Concert Hall, Hult Center for the Performing Arts, which acted as an opening ceremony

> bring[ing] lightning to the stage with live physics demonstrations, digital animation, an original string and electronic musical score, and contemporary choreography. [...] Serbian immigrant Nikola Tesla continues to captivate the public with his electrical inventions that includes alternating current, contributions to radio transmission, the Tesla coil, and 280 other patents. The production [brought] Tesla's enigmatic story to the stage by combining the science of his inventions with mixed media representations of his complex and tumultuous life.

Finally, *Ecomedia* featured several shorts, including animations, virtual worlds and environmental communication and media,[18] as well as the long-form art documentary, *WASTE LAND* (2010). The film examines 'socioartistic sustainability' in the Global South through the work of artist Vik Muniz. Directed by Lucy Walker, with co-directors João Jardim and Karen Harley, it was introduced and screened by Denize Araujo. The work featured:

> the world's largest garbage dump, Jardim Gramacho, located on the outskirts of Rio de Janeiro. There [Muniz] photographs an eclectic band of 'catadores' – self-designated pickers of recyclable materials. [His] initial objective was to 'paint' the catadores with garbage. However, his collaboration with these inspiring characters as they recreate photographic images of themselves out of garbage reveals both the dignity and despair of the catadores as they begin to re-imagine their lives.
>
> (Walker 2010: n.pag.)

The arts festival was curated by Jeremy Swartz including selected works from co-curator, Melody Owen's *Translating the Animal* exhibition.[19] Additionally, the *Ecofair* was organized to highlight local Eugene and UO campus organizations, including the student-produced *FLUX* and *Envision* magazines, Student Sustainability Center, Multicultural Center and others.[20]

In a multiplicity of ways, *LIFEWORLDS*, the Portland Japanese Garden experience, the *Ecomedia* arts festival and *Ecofair* imbricated ecological, economic, environmental, political and social issues.

Nature

This volume has offered merely a glimpse of how nature and culture, the natural and societal, can be understood as interwoven, meshed or entangled at various scales. By recognizing patterns of life, we continue to explore, understand and (de)hybridize ways of knowing in an on-going inquiry into how nature connects life and universe. Even if conceptualized as a fabric or weaver, composition or composer, artwork or creator, ethical and aesthetic implications for various entities' roles and relationships in nature are still open-ended questions, as has been the case since time immemorial.[21]

Creative endeavors that illustrate and integrate interpretations of life and nature were presented as material media, as environmental, informational, organizational and communicative. For millennia, the study of nature – as an object and as socially produced – has invigorated imaginations across geography, philosophy, political ecology and more. Art, language, music and science also offer embodied metaphors of and transdisciplinary practices within nature. However, what constitutes nature continues to be elusive, contested and in many ways paradoxical, requiring profound humility – perhaps these are some of nature's greatest lessons.

In this way, these brief remarks are humbly offered for further research in the midst of futures yet to come. A fuller, richer, more diverse and inclusive, co-curational conceptualization is undoubtedly necessary – not only for nature, but also for what constitutes universe.[22] In the meantime, by recognizing networks of relationships in cosmologies of inquiries, we may further appreciate that *lives are ecosystems and culture is experience*.

NOTES

1. All photos in the Appendix by Jeremy Swartz, except where otherwise noted.
2. These can be defined as *biocentricism*: 'the rights and needs of humans are not more important than those of other living things' (*OED Online* 2023), where the scale is attributed to 'all manifestations of life' (ten Have and Neves 2021a: 159), including non-humans, microorganisms, plants, mountains and rivers; *anthropocentrism*: humanity is the 'primary or exclusive focus' (*OED Online* 2023), where the scale of the system is human life; and

ecocentrism: 'environmental concerns should take precedence over the needs and rights of human beings' (*OED Online* 2023; also see Park and Allaby 2017a), which 'considers planet Earth as the house in which all creatures dwell' (ten Have and Neves 2021b: 449), and includes ecosystems, habitats, species and populations.

3. This would include data, food, health, economics, rights/privacy, audiences, platforms and living technologies.
4. Leonardo da Vinci, 'The Head of Leda', Royal Collection Trust, rct.uk, https://www.rct.uk/collection/912516/the-head-of-leda/. Accessed 7 September 2022.
5. Tepfer, Gary, 'Harsalaa, Tsagaan Gol, Mongolia', and 'Cliffrose, Four Corners', gtepfer.com, http://gtepfer.com/Asia/gt129.html and http://gtepfer.com/SouthWest/gt076.html. Accessed 11 October 2020.
6. Bengtson, Carla, 'Mimetic Excess Archive', carlabengtson.com, https://www.carlabengtson.com/category/with-species-slides/mimetic-excess/. Accessed 7 February 2023.
7. Bengtson, Carla, 'Mimetic Excess', carlabengtson.com, https://www.carlabengtson.com/mimetic-excess/. Accessed 7 February 2023.
8. Bonsai Empire, 'Masahiko Kimura', bonsaiempire.com, https://www.bonsaiempire.com/locations/bonsai-artists/masahiko-kimura. Accessed 7 February 2023.
9. Neil, Ryan, 'Mirai: Learn Bonsai, Tools, Trees', bonsaimirai.com, https://bonsaimirai.com/; 'The Artisans Cup', http://www.theartisanscup.com/. Accessed 5 June 2018. Also see Moor, Robert (2022), 'The beautiful, brutal world of bonsai', *New Yorker*, 14 November, https://www.newyorker.com/magazine/2022/11/21/the-beautiful-brutal-world-of-bonsai. Accessed 5 February 2023.
10. Bonsai Empire, 'John Naka', bonsaiempire.com, https://www.bonsaiempire.com/locations/bonsai-artists/john-naka. Accessed 5 February 2023.
11. Cris Moss facilitated the White Box Gallery's showing of *MADE IN OREGON*, an exhibition of the National Council on Education for the Ceramic Arts, 6–9 April 2017.
12. 'In 2022, the government of Japan recognized the transformational impact that two leaders of Portland Japanese Garden have made by bestowing the highest of honours. Diane Durston, Curator Emerita, was awarded the Order of the Rising Sun, Gold and Silver Rays. Sadafumi Uchiyama, Chief Curator [and Director of the International Japanese Garden Training Center], was awarded the Foreign Minister's Commendation' (PJG 2023).
13. PJG, 'About', Portland Japanese Garden, https://japanesegarden.org/about-portland-japanese-garden/. Accessed 5 February 2023.
14. PJG, 'History', Portland Japanese Garden, https://japanesegarden.org/about-portland-japanese-garden/history/. Accessed 5 February 2023.
15. IAMCR2018, 'Theme: Reimagining Sustainability', 2018 International Association for Media and Communication Research Conference, https://oregon2018.iamcr.org/theme.html. Accessed 20 August 2018.
16. Violet Ray, 'Buffalo Reborn', *Violet Ray Graphics: Advertising the Contradictions*, http://www.violetraygraphics.com/search/label/BUFFALO%20REBORN. Accessed 20 August 2018.

17. CPFM, 'Campus Trees of Interest', Campus Planning & Facilities Management (CPFM), University of Oregon, https://cpfm.uoregon.edu/sites/cpfm2.uoregon.edu/files/tree_tour_dec_2020.pdf. Accessed 5 February 2023.
18. Short videos included Christina Demenshina, *The Last Gardener* (*Der Letzte Gärtnerl*), 2018; Melody Owen, *Brackish Reports*, 2018; and a series of videos curated by Andy Opel, *Apalachicoal River Project*, 2018, featuring co-directed pieces by Hayle Wilson and Kassi Bell, *Apalachicola River—Introduction*, Rillian DeSalla, Hayle Wilson and Kassi Bell, *The Politics of the Water Wars*, Aleks Iricanin and Sammi Simons, *Tommy Ward and the Decline of the Oysters*, and Kyle Jones and Jake Liotta, *Apalachicola River and the Gulf Fisheries*.
19. Selected works from Melody Owen's *Translating the Animal* (see https://ir.library.oregonstate.edu/downloads/br86b8856) included director/animator Andrew Benson and composer Kaitlyn Aurelia Smith, *Riparian*, 2016 (see https://youtu.be/goegShPV3i0); Catherine Clover, *Calling the Birds II*, 2010 (see https://youtu.be/r7k1d_Ib6cc); Rui Costa and Maile Colbert, *Birds of Maio*, 2008 and 2014 (see https://constellationssounds.org/sounds/birds-of-maio) and *Binaural*, 2004; Christina Gruber, *16.660 base pairs of O. mykiss*, 2018; Julia Oldham, *Timber*, 2009 (see https://vimeo.com/171691978); Eva Papamargariti, *Precarious Inhabitants*, 2017 (see https://youtu.be/kK-R7l0jpyg&t=139s); Lisa Schonberg, *Atta: Ant Acoustics in Amazonas*, 2017 (see http://www.lisaschonberg.com/ant-acoustics-in-amazonas); Rick Silva, *Silva Field Guide to Bird of a Parallel Future*, 2015 (see http://silvafieldguide.com); and Andy Thomas, *Nightingale & Canary*, 2014 (see https://youtu.be/h-x0Wou8U4o).
20. See IAMCR2018, 'Ecofair', 2018 International Association for Media and Communication Research Conference, https://oregon2018.iamcr.org/ecofair.html. Accessed 20 August 2018.
21. For instance, see various authors and entries on nature, including spatiality, temporality and ecophenomenology (cf. Whitehead 1938; Williams 1992; Merleau-Ponty 2003; Hadot 2004; Grosz 2008; Rogers et al. 2013; Blackburn 2016; Jenkins 2016; Ducarme and Couvet 2020; Toadvine 2022; *OED Online* 2023).
22. See *UNIVERSE: A Transdisciplinary Inquiry*, Bristol: Intellect Ltd. (forthcoming).

REFERENCES

'anthropocentrism, n.' (2023), *OED Online*, Oxford: Oxford University Press, https://www.oed.com/dictionary/anthropocentrism_n. Accessed 24 July 2023.

Bengston, Carla (2018), 'Perfumes - BEYOND HUMAN', *Beyond Human Perfume*, https://www.beyondhumanperfume.com/perfumes.html. Accessed 7 February 2023.

'biocentrism, n.' (2023), *OED Online*, Oxford: Oxford University Press, https://www.oed.com/dictionary/biocentrism_n. Accessed 24 July 2023.

Blackburn, Simon (ed.) (2016), 'nature', in *The Oxford Dictionary of Philosophy*, 3rd ed., Oxford: Oxford University Press, https://www.oxfordreference.com/display/10.1093/acref/9780198735304.001.0001/acref-9780198735304-e-2135. Accessed 7 August 2023.

Capra, Fritjof (2008), *La botanica di Leonardo: Un discorso sulla scienza delle qualità* (trans. E. Biagini), Città di Castello: Aboca Edizioni.

——— (2013), *Learning from Leonardo: Decoding the Notebooks of a Genius*, San Francisco, CA: Berrett-Koehler Publishers.

——— (2017), 'Leonardo's Science of Living Forms', comments at opening of *LIFEWORLDS* exhibition, Light Court Commons, University of Oregon Portland, 6 April.

Clayton, Martin (2018), *Leonardo da Vinci: A Life in Drawing*, New York: Rizzoli Publications.

Ducarme, Frédéric and Couvet, Denis (2020), 'What does "nature" mean?', *Palgrave Communications* 6:1, pp. 1–8.

Eastman, Janet (2016), '"Unbridled" American bonsai: Portland Japanese Garden's living art exhibit', *Oregon Live | The Oregonian*, 11 May, https://www.oregonlive.com/hg/2016/05/portland_japanese_garden_ryan.html. Accessed 28 April 2017.

'ecocentrism, n.' (2023), *OED Online*, Oxford: Oxford University Press, https://www.oed.com/dictionary/ecocentrism_n. Accessed 24 July 2023.

Grosz, Elizabeth (2008), *Chaos, Territory, Art*, New York: Columbia University Press.

Hadot, Pierre (2004), *Le voile d'Isis: essai sur l'histoire de l'idée de nature*, Paris: Gallimard.

Harmonic Laboratory (2018), 'Tesla: Light, Sound, Color', *Harmonic Laboratory*, https://harmoniclab.org/portfolio/tesla-light-sound-color/. Accessed 20 August 2018.

Husserl, Edmund (1954), 'Die Krisis der europaischen Wissenschaften und die Tranzendentale Phänomenologie' (Husserliana: Gesammelte Werke 6), Haag: Martin Nijhoff.

——— (1970), *The Crisis of European Sciences and Transcendental Phenomenology: An Introduction to Phenomenological Philosophy*, Evanston, IL: Northwestern University Press.

Jenkins, Willis (2016), 'Nature', in R. C. Anderson (ed.), *Berkshire Encyclopedia of Sustainability*, Great Barrington, MA: Berkshire Publishing Group, pp. 283–84.

'Lebenswelt, n.' (2023), *OED Online*, Oxford: Oxford University Press, https://www.oed.com/dictionary/lebenswelt_n. Accessed 24 July 2023.

Leonardo da Vinci (1986–90), *I manoscritti dell'Institut de France: Edizione in facsimile sotto gli auspici della Commissione Nazionale Vinciana e dell'Insitut de France*, diplomatic transcript and criticism by Augusto Marinoni, Florence: Giunti Barbèra.

'Life-world, n.' (2023), *OED Online*, Oxford: Oxford University Press, https://www.oed.com/dictionary/life-world_n. Accessed 24 July 2023.

Merleau-Ponty, Maurice (2003), *Nature: Course Notes from the Collège de France,* compiled D. Séglard, trans. R. Vallier, Evanston, IL: Northwestern University Press. [First published in French in 1995, *La Nature: Notes Cours Du Collège De France*, Paris: Éd. du Seuil.]

'nature, n.' (2023), *OED Online*, Oxford: Oxford University Press, https://www.oed.com/dictionary/nature_n. Accessed 24 July 2023.

Park, Chris and Allaby, Michael (2017a), 'Ecocentrism (biocentrism)', in C. Park and M. Allaby (eds), *A Dictionary of Environment and Conservation,* 3rd ed., Oxford: Oxford University Press.

——— (2017b), 'Human exceptionalism paradigm (HEP)', in C. Park and M. Allaby (eds), *A Dictionary of Environment and Conservation,* 3rd ed., Oxford: Oxford University Press.

PJG (Portland Japanese Garden) (2017), 'American bonsai: The unbridled art of Ryan Neil', Portland Japanese Garden, 3 July, https://japanesegarden.org/2017/07/03/american-bonsai-unbridled-art-ryan-neil/. Accessed 5 February 2023.

——— (2020), 'A garden's power to heal', Portland Japanese Garden, 7 February, https://japanesegarden.org/2020/02/07/a-gardens-power-to-heal/. Accessed 5 February 2023.

——— (2022), 'Portland Japanese Garden reopens providing peace, healing', Portland Japanese Garden, 5 April, https://japanesegarden.org/2022/04/05/2022-cherry-blossom-watch/. Accessed 5 February 2023.

——— (2023), 'A transformational impact | Celebrating the legacy of Diane Durston and Sadafumi Uchiyama', YouTube, 26 January, https://youtu.be/UQRJR9nBuSk. Accessed 6 February 2023.

Rogers, Alisdair, Castree, Noel and Kitchen, Rob (eds) (2013), 'nature', in *A Dictionary of Human Geography*, Oxford: Oxford University Press, https://www.oxfordreference.com/display/10.1093/acref/9780199599868.001.0001/acref-9780199599868-e-1255. Accessed 7 August 2023.

Sokolowski, Robert (1999), 'Husserl, Edmund', in R. Audi (ed.), *The Cambridge Dictionary of Philosophy*, 2nd ed., Cambridge: Cambridge University Press.

ten Have, Henk and Neves, Maria do Céu Patrão (2021a), 'Biocentrism', in *Dictionary of Global Bioethics,* Cham: Springer.

——— (2021b), 'Ecocentrism', in *Dictionary of Global Bioethics,* Cham: Springer.

Toadvine, Ted (2022), 'Anthropocene time and the memory of the world', *Chiasmi International* 24, pp. 171–90.

Tønnessen, Morten (2009), 'Umwelt transitions: Uexküll and environmental change', *Biosemiotics,* 2:1, pp. 47–64.

von Uexküll, Jakob ([1937] 2001), 'The new concept of Umwelt: A link between science and the humanities' (trans. G. Brunow), *Semiotica*, 134:1&4, pp. 111–23. [First published in 1937, 'Die neue Umweltlehre: Ein Bindeglied zwischen Natur- und Kulturwissenschaften', *Die Erziehung* 13:5, pp. 185–99.]

Walker, Lucy (dir.) (2010), 'WASTE LAND: Synopsis', Waste Land Movie, https://www.wastelandmovie.com/synopsis.html. Accessed 29 May 2018.

Whitehead, Alfred North (1938), *Modes of Thought*, New York: The Macmillan Company.

Williams, George C. (1992), *Natural Selection: Domains, Levels, and Challenge*, New York: Oxford University Press.

Notes on Contributors

Mark A. Bedau is a professor of Philosophy and Humanities at Reed College and an adjunct professor of Systems Science at Portland State University in Oregon. Bedau is co-editor of *The Nature of Life: Classical and Contemporary Perspectives from Philosophy and Science* with Carol E. Cleland (Cambridge University Press, 2010), *The Ethics of Protocells: Moral and Social Implications of Creating Life in the Laboratory* with Emily C. Parke (MIT Press, 2009) and *Emergence: Contemporary Readings in Philosophy and Science* with Paul A. Humphreys (MIT Press, 2008). Bedau was an editor-in-chief of the MIT Press journal, *Artificial Life*, and co-founder of the European Center for Living Technology in Venice, Italy. He was also a partner in the European Union-funded Programmable Artificial Cell Evolution program and co-organizer of the Eleventh International Conference on the Simulation and Synthesis of Living Systems. Bedau has been a visiting professor in the Life Sciences Ph.D. Program at the European School of Molecular Medicine in Milan, Italy.

Tom Bennett is an associate professor in Plant Signalling and Development at the University of Leeds. He is a developmental biologist and geneticist, interested in understanding how long distance signalling within the plant body allows their growth and development to be precisely regulated in both space and time. Bennett received his B.Sc. from the University of Leeds, and his Ph.D. from the University of York, working on the regulation of shoot branching under the supervision of Professor Ottoline Leyser. He subsequently worked at the University of Utrecht in the Netherlands and the University of Cambridge, before returning to Leeds to establish his research group, which included a research theme on how plants sense and respond to their neighbours.

Roza D. Bilas is a Ph.D. candidate in the Center for Plant Science at the University of Leeds. She is interested in a variety of plant interactions with other organisms, especially with other plants. In the past Bilas worked on plant-plant parasitism at the University of Virginia and on plant-mycorrhiza symbiosis at the University of Cambridge. The main focus of her work is the identification of the genetic and molecular factors underpinning plant-plant signalling that allows neighbour detection and recognition. Bilas uses the model wild plant Arabidopsis thaliana to bridge the gap between recent findings in other species and enrich available phenotypic studies with missing genetic characterization.

Brendan J. M. Bohannan is James F. and Shirley K. Rippey Chair in Liberal Arts and Sciences, professor of Environmental Studies and Biology and director of the Center for Ecology and Evolutionary Biology at the University of Oregon (UO). He joined the UO faculty after nine years at Stanford University. Bohannan's numerous publications have focused mostly on microbial biodiversity and have been in the 'most highly cited' category among ecologists and evolutionary biologists. His work (lab/research group) has received extensive grants from national research funding agencies and has attracted widespread professional and public attention.

Amanda Bretman is Dean of Research Quality and professor of Behavioural Ecology at the University of Leeds, and previously worked at the University of East Anglia and University of Exeter. Her work focuses on the evolution of sexual and social behaviours. She is particularly interested in the effects of social environments, especially in species not classically considered social. Bretman mainly uses *Drosophila* fruits flies to assess the consequences of variation in the social environment for traits such as reproductive strategies, cognition and ageing, and to explore mechanisms including genes and epigenetics, and the microbiome. A major theme of her work is investigating high temperature limits of reproduction.

Tessa J. Brown is a Ph.D. candidate in the Department of Communication, Media and Film at the University of Calgary, Canada. She is a researcher in the Environmental Media Lab and manager and lead co-editor for *Heliotrope: Writing toward the light*, a journal that includes work at the intersection of culture, media, technology and the environment.

Fritjof Capra is a physicist, educator and co-founding director of the Center for Ecoliteracy in Berkeley, California. He serves as a Fellow of Schumacher College in the United Kingdom, a member of the Earth Charter International Council and faculty of the Amana-Key executive education program in São Paulo, Brazil. Capra is the author of *Patterns of Connection: Essential Essays from Five Decades* (University of New Mexico Press, 2021), and three books on Leonardo da Vinci, including *The Science of Leonardo* (Anchor Books, 2008), *La Botanica di Leonardo* (Aboca Museum Editions,

2009) and *Learning from Leonardo* (Berrett-Koehler, 2013). He is also co-author with Ugo Mattei of *The Ecology of Law* (Berrett-Koehler, 2015) and a contributor to the edited volume, *Systems* (Whitechapel and MIT Press, 2015). Capra's online course, CapraCourse.net, is based on the multidisciplinary textbook co-authored with Pier Luigi Luisi, *The Systems View of Life* (Cambridge University Press, 2014).

Carol E. Cleland is a professor of Philosophy, director of the Center for the Study of Origins and a member of the Center for Astrobiology at the University of Colorado Boulder. She is a SETI (Search for Extraterrestrial Intelligence) Institute Affiliate, and the author of *The Quest for a Universal Theory of Life: Searching for Life As We Don't Know It* (Cambridge University Press, 2019) and co-editor (with Mark A. Bedau) of *The Nature of Life: Classical and Contemporary Perspectives from Philosophy and Science* (Cambridge University Press, 2010). Cleland was also involved as co-investigator and key collaborator on several science teams of the now-disbanded National Aeronautics and Space Administration (NASA) Astrobiology Institute (NAI).

Diane Durston is curator emerita of the Portland Japanese Garden, where she served as Arlene Schnitzer Curator of Culture, Art, and Education (2007–2018). In 2022, she received the Order of the Rising Sun, Gold, and Silver Rays from the Government of Japan, and in 2018, Certificate of Commendation from the Consul General of Japan in Portland. Durston is a writer, lecturer, cultural consultant and educator, who lived in Japan (1977–1996), including eight years in a satoyama village in northern Kyoto. Durston is the author of three books and numerous essays on the art, culture and traditions of Kyoto. Her book, *Old Kyoto: A Guide to Traditional Shops, Restaurants, and Inns, Revised and Updated* (Kodansha International, [1986] 2013), is now in its 15th printing. Other books include *Kyoto: Seven Paths to the Heart of the City* (Kodansha International, 1987) and *Wabi Sabi: The Art of Everyday Life* (Storey Publishing, 2006). She has also contributed essays to *Kyoto: The Forest Within the Gate* (White Pine Press, 2017); the *Encyclopedia of Japan*; *Japan, The Cycle of Life*; the *Japan Crafts Sourcebook* and the *Oregon Encyclopedia*.

Robert Este received his Ph.D. from the University of Calgary and is the author of *Raising the Alarm: An Examination of Innovation and Philosophical Denial* (Sense Publishers, 2013). He also is a co-author of several articles in journals such as *Biology and Philosophy*, *Cybernetics & Human Knowing*, *Journal of Physics A: Mathematical and Theoretical* and *Physics Letters A*.

Divina Frau-Meigs is a professor of Media Studies at Sorbonne Nouvelle University. She is a world-renowned specialist in media and information literacy and an expert with United Nations Educational, Scientific and Cultural Organization (UNESCO) and the European Union. She holds the UNESCO chair Savoir Devenir in Durable Digital Development. Frau-Meigs is the author of *Media Matters in the Cultural Contradictions of the Information Society: Toward a Human Rights-based Governance* (Council of Europe Publishing, 2011) and co-author of *Digital Citizenship Education: Volume I: Overview and News Perspectives* (Council of Europe Publishing, 2017). She also is a lead editor of *Public Policies in Media and Information Literacy in Europe: Cross-Country Comparisons* (Routledge, 2017), *The Handbook of Media Education Research* (Blackwell, 2020) and *MOOCs and the Participatory Challenge: From Revolution to Reality* (Springer, 2021).

Randy Goebel is a professor in the Department of Computing Science, Associate Vice President Research & Innovation and Associate Vice President, and fellow and co-founder of the Alberta Machine Intelligence Institute at the University of Alberta in Edmonton, Canada. He is also the principal investigator for xAI (explainable artificial intelligence) Lab and a scientific advisor for the German Centre for AI Research, the National Institute of Informatics (Tokyo), the Japan Science and Technology Agency and numerous AI companies. Goebel's recent research is focused on the formalization of visualization, algorithm complexity, systems biology and natural language processing, including applications in legal reasoning and medical informatics.

David Hobill is a professor emeritus of Physics and Astronomy at the University of Calgary. He received his Ph.D. in Physics at the University of Victoria (1980) and held postdoctoral positions at the Institut Henri Poincaré (Paris) and Steven Institute of Technology (Hoboken, NJ). Before moving to Calgary, he was a Research Scientist at the University of Illinois' National Center for Supercomputing Applications, a hub of transdisciplinary research and digital scholarship. His research interests are in general relativity (black holes, gravitational lensing, cosmology, gravitational waves), high-performance computing and nonlinear dynamical systems (plasma physics, chemistry, biology).

Mél Hogan is the director of the Environmental Media Lab and an associate professor of Communication, Media and Film at the University of Calgary. She is also an associate editor of the *Canadian Journal of Communication*. Hogan's research has been published in a wide variety of journals. She is also the host and editor of The Data Fix. Since 2012, her research has been on the political–social implications and environmental impacts of server farms and data centres, culminating in the Data Center Industrial Complex (2021) and Critical Studies of the Cloud (2022). As an extension of research that looks at alternatives to water- and energy-intensive data storage, she is a Principal Investigator about DNA-based data storage following a previous project about genomics in the cloud (genomics.media). Hogan presented the 2020 McLuhan lecture at the Canadian Embassy in Berlin and a plenary at Transmediale 2020. She is currently exploring the use of massive data centers to power humanoid robots, embodied AI, holograms and more.

MARK L. JOHNSON is a professor emeritus of Philosophy and Philip H. Knight Professor of Liberal Arts and Sciences, Emeritus, at the University of Oregon. He is co-author of *Mind in Nature: John Dewey, Cognitive Science, and a Naturalistic Philosophy for Living* (MIT Press, 2023) with Jay Schulkin, *Out of the Cave: A Natural Philosophy of Mind and Knowing* (MIT Press, 2021) with Don M. Tucker, as well as *Metaphors We Live By* (University of Chicago Press, 2003) and *Philosophy in the Flesh: The Embodied Mind and Its Challenge to Western Thought* (Basic, 1999) with George Lakoff. He is the author of *Embodied Mind, Meaning, and Reason: How Our Bodies Give Rise to Understanding* (UCP, 2017) and *The Aesthetics of Meaning and Thought: The Bodily Roots of Philosophy, Science, Morality, and Art* (UCP, 2018). In addition, Johnson is the author of numerous articles and book chapters on a broad range of topics.

STUART KAUFFMAN is a professor emeritus of Biochemistry at the University of Pennsylvania and an Affiliate Faculty at the Institute for Systems Biology, a non-profit scientific research organization located in Seattle, Washington. Originally a medical doctor, Kauffman is a seminal member and an external professor of the Santa Fe Institute, a MacArthur Fellow and a Trotter Prize winner, as well as the director of the Institute for Biocomplexity and Informatics. His publications include *Investigations* (Oxford University Press, 2000), *Reinventing the Sacred: A New View of Science, Reason, and Religion* (Basic Books, 2008), *Humanity in a Creative Universe* (OUP, 2016) and *A World Beyond Physics: The Emergence and Evolution of Life* (OUP, 2019). His areas of expertise include developmental biology, combinatorial chemistry and molecular evolution.

ROBERT K. LOGAN is a professor emeritus of Physics at the University of Toronto and has conducted research in elementary particle physics, media ecology, complexity theory, information theory, systems biology, environmental science, linguistics and industrial design. He published and collaborated with Marshall McLuhan and continues his McLuhan studies research. Logan is the author or editor of sixteen books, 23 book chapters and over 135 articles. He is the chief scientist of the Strategic Innovation Lab at OCAD (Ontario College of Art & Design) University, a fellow at the University of St Michael's College and on the faculty of the School of the Environment, University of Toronto. In June 2011, he received the Walter J. Ong Award for Career Achievement in Scholarship from the Media Ecology Association.

Pier Luigi Luisi is a professor emeritus of Chemistry in the Department of Material Sciences at the Swiss Federal Institute of Technology (ETH) Zürich, Switzerland. From 2003 through 2012, he was a professor of Biochemistry at the University of Rome, Italy. His research focuses on the origin of life and the synthetic biology of minimal living cells. Luisi has authored over 500 reviewed papers and dozens of books, including *The Systems View of Life* (Cambridge University Press, 2014) with Fritjof Capra and *The Emergence of Life: From Chemical Origins to Synthetic Biology* (CUP, 2nd ed., 2016). In 1985, he created the Cortona Week, a transdisciplinary residential week at the ETH Zürich for graduate students who are working with artists, psychologists and religious leaders to foster a new generation of world leaders with a more integrated and balanced view of life. Since 1987, Luisi has been associated with the Mind and Life Institute, created by Francisco Varela and the Dalai Lama, and devoted to dialogue between modern science and Buddhism.

Ed Madison is an associate professor at the University of Oregon (UO) School of Journalism and Communication and has affiliated faculty status with the College of Education, Agora Journalism Center and Center for Science Communication Research. He serves as a member of the Board of Trustees of UO, and is a media professional with a 30-year track record as an executive producer/director and entrepreneur.

Salma Monani is the chairperson and professor of the Department of Environmental Studies at Gettysburg College and a faculty affiliate in Cinema and Media Studies and Peace and Justice Studies. She is a co-editor, with Steve Rust and Seán Cubitt, of *Ecocinema Theory and Practice* and its sequel (Routledge and AFI, 2013 and 2023) and *Ecomedia: Key Concepts* (Routledge/Earthscan 2015), as well as lead editor (with Joni Adamson) on *Ecocriticism and Indigenous Studies: Conversations from Earth to Cosmos* (Routledge/Interdisciplinary Perspectives in Literature 2017). She has also published extensively on explorations of film and environmental justice, Film Festival Studies and Indigenous eco-activism, and is currently working on a monograph: *Indigenous Ecocinema: Decolonizing*

Media Landscapes (contract with the University of West Virginia Press' Salvaging the Anthropocene series).

NICOLAE MORAR is an associate professor of Environmental Studies and Philosophy and an associate member of the Institute of Ecology and Evolution at the University of Oregon. His publications include numerous peer-reviewed journal articles and book chapters, in addition to the following edited volumes: *Pierre Klossowski, Living Currency* (Bloomsbury Press, 2017) with Vernon W. Cisney and Daniel W. Smith, *Between Foucault and Derrida* (Edinburgh University Press, 2016) with Vernon W. Cisney, Yubraj Aryal and Christopher Penfield, *Between Deleuze and Foucault* (EUP, 2016) with Thomas Nail and Daniel W. Smith, *Biopower: Foucault and Beyond* (University of Chicago Press, 2015) with Vernon Cisney, and *Perspectives in Bioethics Science and Public Policy* (Purdue University Press, 2013) with Jonathan Beever. Books in progress include *Biology, BioEthics and BioPolitics: How to Think Differently About Human Nature* and *Ethical Dimensions of Science and Engineering: A Broader Conception*, co-authored with Jonathan Beever, Bryan Cwik and Nancy Tuana.

BROOK MULLER is Charles Eliot chair in Ecological Planning, Policy and Design at the College of the Atlantic, and former dean of the University of North Carolina at Charlotte College of Arts + Architecture. His research and practice focus on the design implications of resilient, ecologically responsive urban water systems. Muller is the author of *Blue Architecture: Water, Design and Environmental Futures* (University of Texas Press, 2022) and *Ecology and the Architectural Imagination* (Routledge, 2014). With Behnisch & Partner Architects of Stuttgart, Germany, he served as a co-project leader for the design of the National Institute for Forestry and Nature Research in Wageningen, The Netherlands, a European Union pilot project for an environmentally friendly building (1993–96). Muller has worked on water and sustainable development-focused projects in Egypt, the Madeira Islands and Tanzania.

THOMAS NAIL is a distinguished scholar in the College of Arts, Humanities and Social Sciences and professor of Philosophy at the University of Denver. He is the author of numerous books, including *The Figure of the Migrant* (Stanford University Press, 2015), *Returning to Revolution: Deleuze, Guattari and Zapatismo* (Edinburgh University Press, 2015), *Theory of the Border* (Oxford University Press, 2016), *Being and Motion* (OUP, 2018), *Theory of the Image* (OUP, 2019), *Theory of the Object* (EUP, 2021), *Theory of the Earth* (SUP, 2021), *Marx in Motion: A New Materialist Marxism* (OUP, 2020), *Lucretius I: An Ontology of*

Motion (EUP, 2018), *Lucretius II: An Ethics of Motion* (EUP, 2020) and *Lucretius III: A History of Motion* (EUP, 2022).

Jussi Parikka is a professor in Digital Aesthetics and Culture at Aarhus University, where he teaches in the Department of Digital Design and Information Studies and leads the AUFF-funded project 'Design and Aesthetics for Environmental Data' (2022–24). Parikka was elected as a member of Academia Europaea in 2021. His books include the media ecology-trilogy *Digital Contagions* (Peter Lang Inc. 2nd ed., 2016), the award-winning *Insect Media* (University of Minnesota, 2010) and *A Geology of Media* (UMP, 2015), which addresses the environmental contexts of technical media culture. Parikka is also author of *What is Media Archaeology* (2012) and *Operational Images: From the Visual to the Invisual* (UMP, 2023), as well as co-author of *The Lab Book: Situated Practices in Media Studies* (UMP, 2022) with Darren Wershler and Lori Emerson and editor of several collections. In addition, he has worked in many curatorial contexts and is currently on the curatorial team of the Transmediale art and digital culture festival (2023).

Stephen Rust is a senior instructor in the Department of English and affiliated faculty in Digital Humanities at the University of Oregon. He is co-editor with Salma Monani and Seán Cubitt of *Ecocinema Theory and Practice* (Routledge and AFI, 2013), *Ecocinema Theory and Practice 2* (Routledge, 2023), and *Ecomedia: Key Concepts* (Routledge and Earthscan, 2015) and the *Routledge Handbook of Ecomedia Studies* (Routledge, 2023). Rust is a founding advisory board member of *Media+Environment* and the *Journal of Environmental Media*.

William Schoenberg is in the Faculty of Social Sciences and System Dynamics Group at the University of Bergen, Norway, and holds a Ph.D. in system dynamics from the same university. He is a key member of the team that designs and develops Stella Architect and the isee Exchange. He is the scientist behind the Loops that Matter model analysis method and the lead developer of software in isee systems that enable modelers to take advantage of the method in their analysis of complex, dynamic models.

Ilya Shmulevich is a professor at the Institute for Systems Biology, a non-profit scientific research organization in Seattle, Washington. Shmulevich directed a Genome Data Analysis Center within The Cancer Genome Atlas (TCGA) consortium. He also directed one of three National Cancer Institute (NCI) Cancer Genomics Cloud Pilots, which is now operating as an NCI Cancer Genomics Cloud Resource (isb-cgc.org). Shmulevich is a co-author or co-editor of six books in the area of computational biology. He holds affiliate professor appointments in the Departments of Bioengineering and Electrical Engineering at the University of Washington and has held affiliate appointments at Tampere University of Technology in Finland and Strathclyde University, Glasgow, UK.

Jeremy Swartz is founder of Metamedia @ UofO: A Metadisciplinary Collaboratory, curator and co-director of the *What is…?* series, courtesy research associate in Media Studies at the University of Oregon, and adjunct assistant professor in Communication at Southern Oregon University. He is co-editor of *MEDIA: A Transdisciplinary Inquiry* (Intellect, 2021) with Janet Wasko. Swartz is a vice-chair of the Environmental Impact Committee for the International Association for Media and Communication Research (IAMCR) and has served as an NSF/HHMI Scientific Teaching Fellow. He has been associated faculty at the UO Institute of Cognitive and Decision Sciences, and the Department of Communication at Portland State University. His research explores pattern recognition in complex systems via philosophy, communication, technology, information, biology, ecology, music, art and law.

Janet Wasko is a professor in Media Studies and Knight Chair in Communication Research Emeritus at the University of Oregon. She is the author or editor of 23 books including *MEDIA: A Transdisciplinary Inquiry* (Intellect, 2021) with Jeremy Swartz, *Understanding Disney: The Manufacture of Fantasy*, 2nd ed. (Polity, 2020); *A Companion to Television*, 2nd ed. (Wiley-Blackwell, 2020) with Eileen R. Meehan; *Global Media Giants* (Routledge, 2017) and the *Global Media Giants* book series (Routledge and CRC Press, 2019–) with Benjamin Birkinbine and Rodrigo Gómez; and *The Handbook of Political Economy of Communications* (Wiley-Blackwell, 2011) with Graham Murdock and Helena Sousa. Wasko also co-edits the *Global Handbooks in Media and Communication Research* series (IAMCR and Wiley Blackwell) with Karin Gwinn Wilkins. She is an International Communication Association fellow and founder and co-director of the *What is…?* series, as well as serving as president of IAMCR.

Index

Figures are indicated by *italic* page numbers; table by **bold**.
Endnotes are indicated only where the information would not be readily found from the page and are given as page and note number, e.g. 427n2.
Authors named in discussion are indexed.